# AFRICA: GEOGRAPHIES OF CHANGE

*Richard Grant*

PROFESSOR OF GEOGRAPHY,
UNIVERSITY OF MIAMI, CORAL GABLES, FL, USA

ADJUNCT SENIOR RESEARCH SCIENTIST TO THE EARTH
INSTITUTE'S MILLENNIUM CITIES INITIATIVE,
COLUMBIA UNIVERSITY IN THE CITY OF NEW YORK

VISITING PROFESSOR, DEPARTMENT OF GEOGRAPHY,
ENVIRONMENT MANAGEMENT AND ENERGY STUDIES,
UNIVERSITY OF JOHANNESBURG,
SOUTH AFRICA

NEW YORK OXFORD
OXFORD UNIVERSITY PRESS

Oxford University Press is a department of the University of Oxford.
It furthers the University's objective of excellence in research, scholarship,
and education by publishing worldwide.

Oxford New York
Auckland Cape Town Dar es Salaam Hong Kong Karachi
Kuala Lumpur Madrid Melbourne Mexico City Nairobi
New Delhi Shanghai Taipei Toronto

With offices in
Argentina Austria Brazil Chile Czech Republic France Greece
Guatemala Hungary Italy Japan Poland Portugal Singapore
South Korea Switzerland Thailand Turkey Ukraine Vietnam



For titles covered by Section 112 of the US Higher Education
Opportunity Act, please visit www.oup.com/us/he for the
latest information about pricing and alternate formats.

Published in the United States of America by
Oxford University Press
198 Madison Avenue, New York, NY 10016
http://www.oup.com



**Library of Congress Cataloging-in-Publication Data**
Grant, Richard, 1964- author.
Africa : geographies of change / Richard Grant.
pages cm
Includes bibliographical references.
ISBN 978-0-19-992056-3
1. Human geography--Africa, Sub-Saharan. 2. Africa, Sub-Saharan--Economic
conditions--21st century. 3. Africa, Sub-Saharan--Social conditions--21st century.
4. Urbanization--Africa, Sub-Saharan. 5. Rural development--Africa, Sub-Saharan.
6. Sustainable development--Africa, Sub-Saharan. 7. Social change--Africa,
Sub-Saharan. I. Title.
GF701.G73 2014
304.20967--dc23

2014003254

Printing number: 9 8 7 6 5 4 3 2 1

Printed in the United States of America
on acid-free paper

# DEDICATION

*To Adriana Cala, Sofia and Natalia Grant who let me go to numerous places in Africa and always welcomed me home, and to my mother, Alva Grant, who passed in 2014 but remains a major inspirational figure. I am deeply grateful to these generous individuals for their encouragement, support and feedback.*

# CONTENTS

# PREFACE

Africa has become more important. After several centuries of Western dominance and development failures, Africa is no longer what has been imagined as "Africa." Negative stereotypes and pessimism in combination with a myriad of real crises (economic, health, political, conflict, and humanitarian) undermined the potential of the region and focused the world's positive attention on other world regions. But times have changed.

In the 21st century, there has been a dramatic turnaround in interest in and engagement with Africa. Scholars, analysts, and the media now speak of "Africa's turn," "Africa rising," "emerging Africa," "resurgent Africa," "renaissance Africa," and "Africa's moment." This increased global attention on the part of those who want to secure access to Africa's resources is leading to a "new scramble" for Africa. This time around, though, the players are not only Western countries but increasingly China, Brazil, India, the Gulf states, and other nontraditional players (e.g., Malaysia and Turkey) who are jostling for African concessions. Oil, natural gas, metallic mineral deposits, and abundant farmland are transforming the world's images of Africa from a "lost continent" to a continent of "strategic opportunity."

The specter of Africa as a theater for terrorism is also gaining traction and in many ways is a return to the older "coming anarchy" argument. Various jihadist groups are active in the region (e.g., Al-Qaeda, Al-Qaeda in the Islamic Maghreb and Al-Shabaab) and the frequency of their attacks is raising concern. "Ungoverned" and "wild spaces" in Africa are now under closer scrutiny and surveillance from international counterterrorism organizations—although, to be fair, the Africa terrorist frontline scenario is restricted geographically and is less applicable to most of the region.

Making sense of these new and competing Africa narratives requires thoughtful and informed analysis. Meta-narratives can help us to reframe a region but can highlight only a particular aspect of African events. This book emphasizes that using such meta-narratives to explain and predict African events is inadequate and that "thick" moments in African history should not be combined with "thin" geographies. Unraveling these complex geographies, this book provides a unique balance of knowledge with some forward-thinking analyses to reveal the continuity and change that are under way in different parts of Africa. This text takes a penetrating look at Africa through the use of major themes that are most acute in the urban and rural transformation (e.g., water, climate change, health, food, migration, informality, China, sustainable development) and provides rich and detailed case studies representative of different parts of Africa.

The central argument of the book is that an urban revolution is under way in Africa and a rural revolution is imminent. This is a profound transformational story. The region in the past 50 years has recorded the fastest population growth in the entire history of our planet. Africa is no longer a continent of villages and towns; it comprises a wide diversity of urban settlements, from mega-cities through progressively smaller urban settlements. Most of Africa's population lives in cities of fewer than 500,000 people, yet the urban primacy and the dominance of the colonial capitals remains a key feature of African economies and societies. Urbanites require food, fuel, water, and ample supplies of other resources, and, as such, cities extend their reach over widening tributary areas and territories. Shortly after 2030, Africa will surpass the milestone threshold where more than 50% of its population will reside in urban settlements. This demographic metamorphosis is already affecting rural Africa and it will influence it ever more profoundly in coming years.

Rural and urban Africa are becoming increasingly interconnected, not just in terms of migration flows but also in other ways, including family finances and the pervasive effects of urbanism, cosmopolitan beliefs, and consumer culture. Indeed, urban centers outside the region are also increasingly linked to rural Africa by diaspora remittances, by institutional investors buying African farmland and water rights, and by mobile phone networks that bind rural and urban actors in ways that do not require travel. There are spaces where rural and urban processes blend, such as in the peri-urban environments of African cities, the sponge-like periphery of the city that attracts migrants who live an in-between existence. Global development organizations and agrocorporations have bold plans to further transform rural Africa by launching a "green revolution" to accelerate food production, improve productivity, and feed African urban populations (and possibly others). Rural infrastructure is slowly improving and is attracting investors; there is considerable room for change, but major obstacles to mobility persist. International development initiatives that are under way to deploy mobile technologies and platforms to deliver health and educational services to rural Africa may be especially transformative.

Of course, interconnecting urban and rural Africa and thinking through the interactions between African physical environments and human change are ambitious intellectual projects and represent a formidable challenge. Effort is made in the chapters to link urban and rural phenomena, but for conceptual clarity separate chapters on rural Africa and Africa's environments are included.

Why write this book? First, Africa should no longer be regarded as a mysterious and unknown region. The fabled, deeply troubled continent of the past no longer exists; in fact, it existed more in representation than in reality. This book draws on the large and expanding knowledge base on Africa and provides a coherent, timely synthesis that takes stock of this diverse, complex, and changing region. Second, this is a particularly opportune time to study Africa. Rapidly growing economies, expanding national populations, increasing business interest, growing global connections, and expanding civil society participation are transforming the region. By 2030, only Africa and Asia will be growing in terms of population rates (other regions will have reached a plateau). Third, Africans are becoming more self-assured and insistent about being partners in development. Fourth, knowledge about Africa is evolving and becoming more sophisticated. The 21st century ushered in more positive assessments about the region and changed the conversation about Africa. This conversation continues to evolve in 2014 as the United Nations moves toward the implementation of its Sustainable Development Agenda, replacing the Millennium Development Goals (MDGs), whose implementation period was from 2000 to 2015. From 2015 to 2030, sustainable development initiatives will encourage new thinking and innovation in Africa and about Africa. And, fifth, there is a pressing need for a book that is structured around geographical perspectives in assessing the scale and scope of Africa's contemporary transformations.

*Africa: Geographies of Change* provides a topical focus but is keenly sensitive to regional, national, and local actors and contexts. This book purposely does not engage in a regional survey of the continent; rather, it follows the contemporary geographical tradition of moving beyond regional description toward more interdisciplinary interpretation. At the same time, considerable effort went into the selection of diverse case studies that highlight important local, national, and regional topics (e.g., coltan mining in eastern Democratic Republic of Congo, e-waste in Accra, Ghana, circular migration in the Burkina Faso–Côte d'Ivoire corridor, mobile phone technological innovations in Kenya, money transfers to Somalia, sustainable development in Nigeria, South Africa, and the Great Lakes area, HIV/AIDS in Uganda, and China's relations with Zambia).

As the discipline of geography advances, it is incorporating new and more complex perspectives from urban studies, public health, political economy, environmental studies, political geography, political ecology, and sustainable development. These, in turn, provide ever more nuanced understandings of African issues, thereby deepening our knowledge of Africa and Africans in the global economy. Africa can no longer be understood by looking through a single prism; accordingly, chapter topics have been carefully selected to develop a comprehensive approach to explaining contemporary Africa as well as shedding light on the processes that will shape its future. Therefore, this book should be of considerable interest in fields beyond Geography, especially African Studies, Environmental Studies, Development Studies, Urban Studies, Globalization Studies, Political Science, and International Studies.

The book's recurrent leading themes are the historical legacies of colonialism; the profound demographic transformation unleashed by the growing shift toward urban centers; widening urban–rural spheres of interactions; the growing complexity and diversity in the range of options for African states to secure external support for national development projects; and rising self-confidence among Africans, who want to be partners in development plans and projects. At the same time, persistent poverty (sometimes extreme), a burgeoning informal economy, inferior infrastructure, resource overdependence, and the challenge of climate change are also taken into consideration.

There are numerous good books about Africa, but most of them focus on a different era: yesterday's Africa. Africa in the 21st century is an increasingly different place than it was in the 20th century. The overriding objective of this book is to sharpen awareness that Africa is no longer the place that has been imagined but a complex region taking its place on the world stage; it is place where many changes are being wrought and where alternative futures are being contested.

Hallmarks of *Africa: Geographies of Change*

- Emphasis on contemporary Africa to excite students and invigorate a new generation of educated global citizens to think differently about alternative and more positive possibilities for African development
- Relevant historical knowledge to underpin the assessment of contemporary issues and advance student thinking to contemplate future development and policy trajectories and outcomes
- Current material on China in Africa, the mobile phone revolution, the Millennium Development Goals, sustainable development, "land and water grabs," food security, informal livelihoods, the "Green Revolution," new satellite cities, and African terrorist groups
- Incorporation of uniquely African perspectives into each chapter to ensure that the book does not draw exclusively from literatures produced in the Global North
- Carefully selected case studies to vividly illustrate topics being discussed, which are critical to informing a broad-based geographical perspective
- Case studies of fair trade products and coltan mining to show how resources in Africa are connected to everyday consumer products in the Global North in order to make connections to the region that can be less obvious
- Accessible writing, synthesis of scholarly and policy debates, links to videos that help bring material and contexts to life, and a comprehensive bibliography
- Multiphrenic information about Africa to stimulate active student engagement with the subject matter, including a listing of social media and the leading Internet sites on reporting about Africa

Ultimately, the major goals of this book are to provide a broad and comprehensive knowledge base regarding 21st-century Africa so that students can critically reflect on the material, stimulating in-depth learning.

# ACKNOWLEDGMENTS

Writing a textbook on contemporary Africa is a major undertaking, and during my work I received advice, guidance, and input from various experts. First and foremost, I was provided with excellent feedback from the reviewers who pushed me to develop and extend my analysis to rural Africa. This book is a lot better because of their encouragement to link rural and urban Africa. They are:

Fenda Akiwumi, University of South Florida
Seth Appiah-Opoku, University of Alabama
Raymond Asomani-Boateng, Minnesota State University – Mankato
Donovan C. Chau, California State University Bernadino
Amy Cooke, University of North Carolina – Chapel Hill
James Craine, California State University Northridge
Gregory DeFreitas, Hofstra University
Sheila J. Gibbons, Salem State University
Joel Hartter, University of New Hampshire – Durham
Joshua Inwood, University of Tennessee – Knoxville
Ezekiel Kalipeni, University of Illinois – Urbana Champaign
Margaret Kidd, Texas Southern University
Calvin Masilela, Indiana University of Pennsylvania
Garth A. Myers, Trinity College
Benjamin Neimark, Old Dominion University
Kefa M. Otiso, Bowling Green State University
Jennifer Rogalsky, State University of New York Geneseo
Elizabeth Edna Wangui, Ohio University – Athens
John Warford, Florida A&M University

In addition to the external reviewers, several colleagues engaged me in conversations about Africa that broadened and deepened my understanding of the region. I had the pleasure of engaging many of these colleagues in the field in Accra, Cape Town, Gaborone, Johannesburg, and other African cities as well as to learn from their experiences that went beyond areas I had the pleasure of traveling to. Many colleagues commented on chapters and helped make the presentation of material better. Other colleagues provided a sounding board to develop my thinking on important issues. This book is much better because of these conversations. For their scholarly engagement, I thank:

Japhet Aryiku, Adakum Educational Foundation
Susan Blaustein, Columbia University
Kevin Chaplin, Amy Biehl Foundation Trust

Jonathan Crush, Balsillie School of International Affairs,
Heather Dalton, Flower Valley
Pádraig Carmody, Trinity College, Dublin
Mike Gaines, University of Miami
Bill Green, University of Miami
Chris Hanson, University of Miami
Maria Huchzermeyer, University of the Witwatersrand
Francis Koti, University of North Alabama
Mazen Labban, Rutgers University
Nico Kotzé, University of Johannesburg
Adam Levy, University of Colorado at Boulder
Isaac Luginaah, The University of Western Ontario
Peter Muller, University of Miami
Joe Melara, Columbia University
Bill Moseley, Malacaster College
Jim Murphy, Clark University
Martin Murray, University of Michigan
Jan Nijman, University of Amsterdam
Franklin Obeng-Odoom, University of Technology, Sydney
Martin Oteng-Ababio, University of Ghana
Sophie Oldfield, University of Cape Town
Francis Owusu, Iowa State University
George Owusu, University of Ghana
Gordon Pirie, University of Cape Town
Jyotika Ramaprasad, University of Miami
Chris Rogerson, University of Johannesburg
Jayne Rogerson, University of Johannesburg
Jeffrey Sachs, Columbia University
Elliot Sclar, Columbia University
Shouraseni Sen Roy, University of Miami
Sarah Smiley, Kent State University
Justin Stoler, University of Miami
Daniel Thompson, University of Miami
Ivor Turok, Human Sciences Council, South Africa
Ian Yeaboah, Miami University
Leo Zulu, Michigan State University

The book has also benefited tremendously from the team of professionals at Oxford University Press, and I am grateful to be associated with them. Here I thank my editorial team: Dan Kaveney, Executive Editor, for shepherding the project from the idea stage through to publication, and Assistant Editor Christine Mahon and Editorial Assistant Nathaniel Rosenthalis, for helping to attend to the many details involved in preparing a manuscript for production and for ably assembling the diverse and helpful review panels mentioned above. The book you are holding wouldn't have been possible without Production Editor Christian Holdener's steady hand on the tiller coordinating the myriad of tasks required to turn a manuscript into composited, edited pages supported by a terrific art program. Finally, I wish to thank Marketing Manager David Jurman for his tireless work and travel in support of the book.

# CHAPTER 1

# INTRODUCTION

## SETTING THE AFRICAN STAGE

Africa is the cradle of humanity. Its historic human significance is immense: it is the origin of the earliest human beings and the land from which peoples migrated to populate Europe, Southeast Asia, and much later the Americas. Archaeological, genetic, and linguistic research indicates that peoples migrated from Africa to nearby regions around 130,000 years ago. Despite archaeological records chronicling 7 million years of human traditions, we know comparatively little about Africa from 5,000 years ago until the onset of European colonialism around 1500 CE. Much more is known about the European, Asian, and American descendants of the original African peoples than about their African descendants.

Much of Western understanding and contemporary framing of the region has roots in the European colonial experience. The colonial era had many enduring legacies, and it is notorious as the time when misconceptions about Africans became entrenched. Colonialism put forward a myth that Africans were peoples without history. Ancient civilizations (e.g., the Nubia) and cities dating back to the 11th century (e.g., Great Zimbabwe) are just two of many possible examples that demonstrate unequivocally Africa's rich and diverse historical past. It is important to consider that when universities were getting off the ground in Oxford and Cambridge (UK) in the 12th century, Timbuktu (Mali) already had three established universities. Colonialism obliterated numerous artifacts and destroyed many African traditions, and the European written word and European laws and plans were emphasized at the expense of African oral traditions and means of administration.

Two major misconceptions of Africa are that it is a single country or entity and a marginal area. Geographically, Africa is an enormous region, roughly three and half times larger than the United States. The African continent constitutes the second largest landmass on Earth, after Eurasia, with one fifth of the planet's land surface. It is larger than the United States, China, India, Japan, and the entire European region combined (Fig. 1.1).

Stretching from the tip of Bizerate, Tunisia, to the base at Cape Agulhus, South Africa, Africa is 5,000 miles (8,000 km) from north to south; from the furthest point west on Cape Verde to the furthest point east in Somalia (Cape Gwardafuy) it is 4,600 miles (7,400 km) east to west. The expansive region contains a range of climates and biomes that have created conditions conducive to the formation of diverse peoples and cultures.

A fascinating map produced by anthropologist George Murdock in 1959 illustrates the complex ethnic mosaic of ethno-linguistic groups in Africa, with 835 distinct groupings (some linguistic experts argue for even more groupings) (Fig. 1.2). Whatever the true number of ethnic groupings in Africa, it is clear that

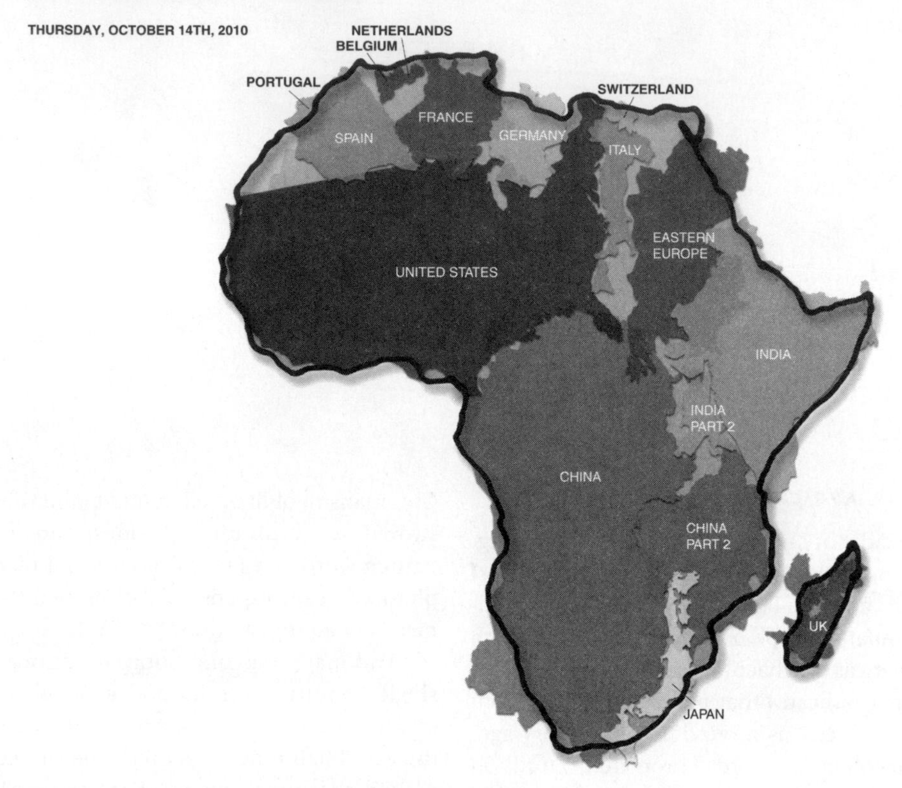


**FIGURE 1.1** True Graphic of Africa's Size.

colonial partition was enforced with little regard for the spatial boundaries of ethnic groups (their histories, languages, and economic linkages), and this gave rise to multi-ethnic states and partitioned ethnic groups (e.g., the Maasai of Kenya and Tanzania) in an arbitrarily determined quilt-work of national borders (compare Fig. 1.2 and Fig. 1.3). There was considerable diversity among ethnic groups with regard to size, areal extents, and regional power and regional ties and extra-regional ties. Some societies had large population concentrations and some areas maintained ties within Africa and with distant places in the Middle East, China, and India, all before the European colonists arrived, but many groups and areas were sparsely populated and isolated. Nations are far more than multicolored patches on a map: they have internal coherence and historical significance and deep ties to places and environments.

Contemporary Africa is an amalgamation of many regions and subregions. Dividing it into subregions is a useful starting point, but justifying a contemporary regional breakdown of Africa south of the Sahara is fraught with difficulties. Any attempt to delineate subregions is going to be heavily critiqued. No major topographical features partition Africa into regions, and historic migrations of peoples within Africa led to shifting regional distributions of the population over time. Regional geographers (e.g., DeBlij, Muller, and Nijman 2014) interpret maps of environmental distribution, ethnic patterns, and cultural landscapes to yield a four-region structure—West Africa, East Africa, Central Africa, and Southern Africa—and various African countries are subsumed into these broad regions (Fig. 1.4). These regional classifications provide the basis for the management of major international organizations' reporting (e.g., the United Nations [UN]) and variously

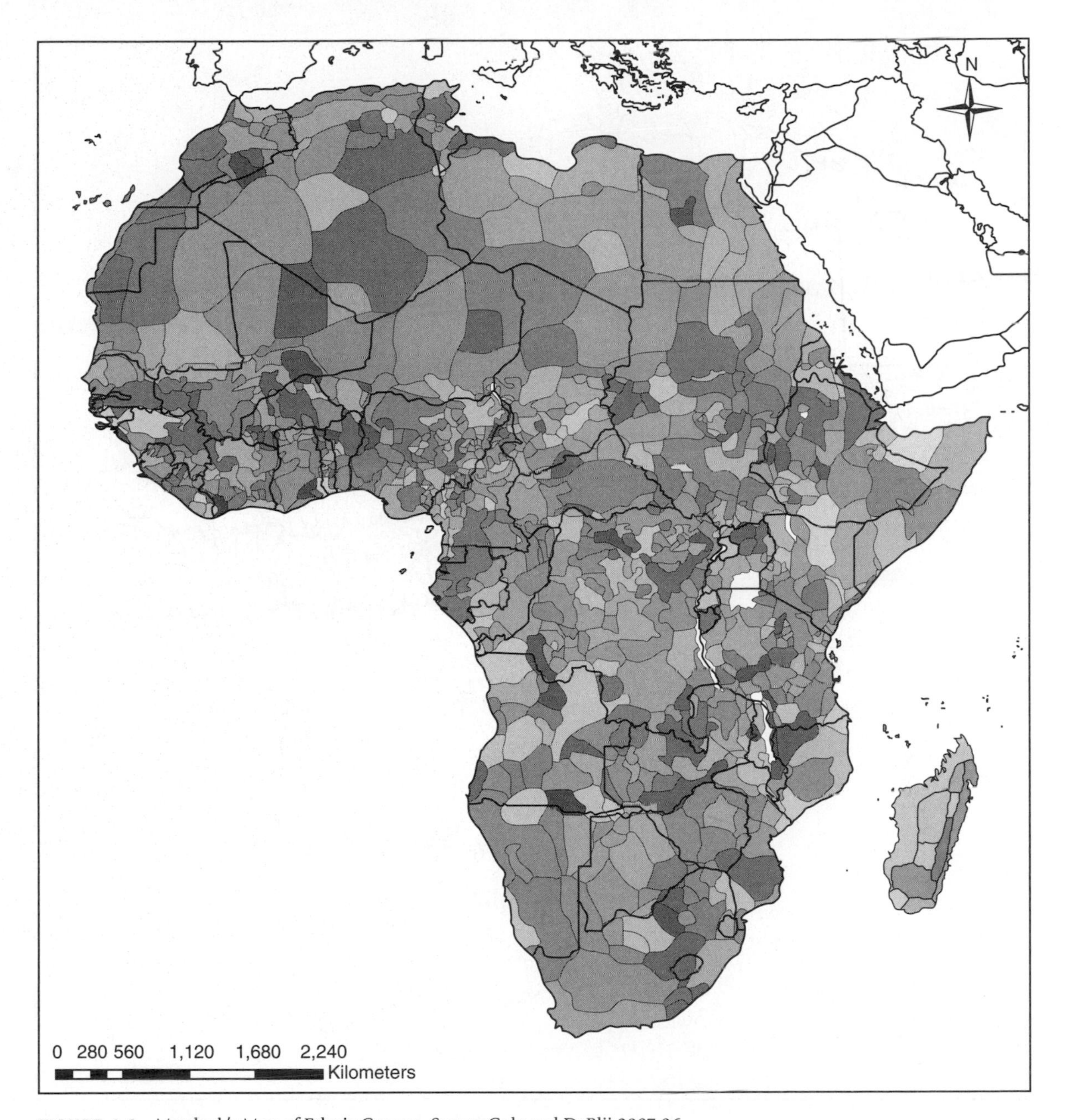


**FIGURE 1.2** Murdock's Map of Ethnic Groups. *Source:* Cole and DeBlij 2007:96.

constitute membership in several prominent regional economic and political organizations. The African Union (AU) holds ambitious plans to complete an African Economic Union by 2027, uniting these four regions of Africa and the North African region to facilitate an economic union with a common currency, full mobility of the factors of production (labor, capital, and technology), and free trade among the 54 countries.

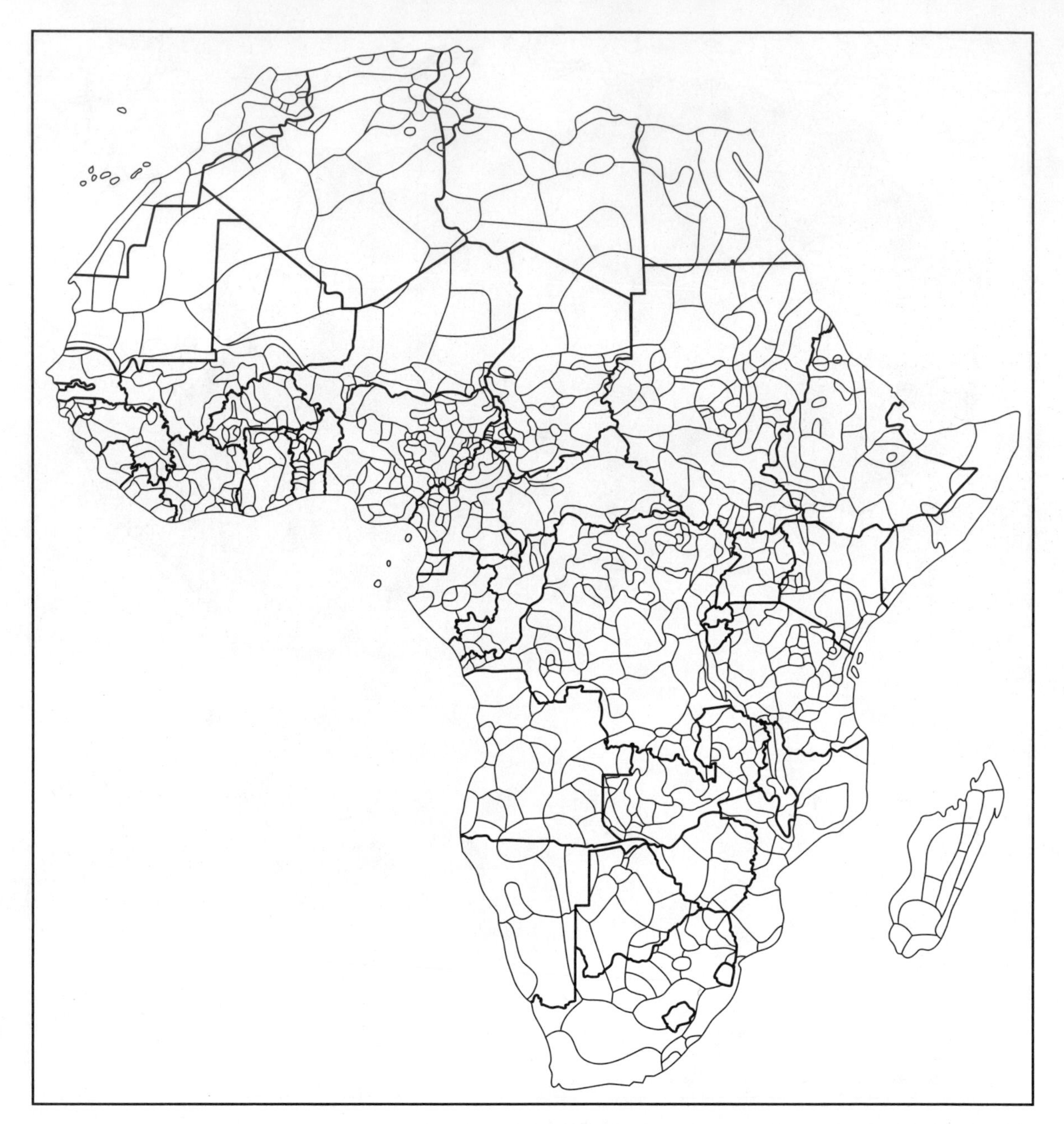

**FIGURE 1.3** Colonial Boundaries: Heavy Lines Superimposed on Murdock's 1959 Ethnolinguistic Map of Africa.
*Source:* Larson 2011.

Paradoxically, none of the existing regional trade groupings demonstrates the unity assumed by broad regional classifications, and most regions are composed of several subregional groupings. There are also countries and territories whose locational affiliation and appropriateness is debated. For example, Chad lies in a transition zone between West, East, and Central Africa, and its northern section is more aligned

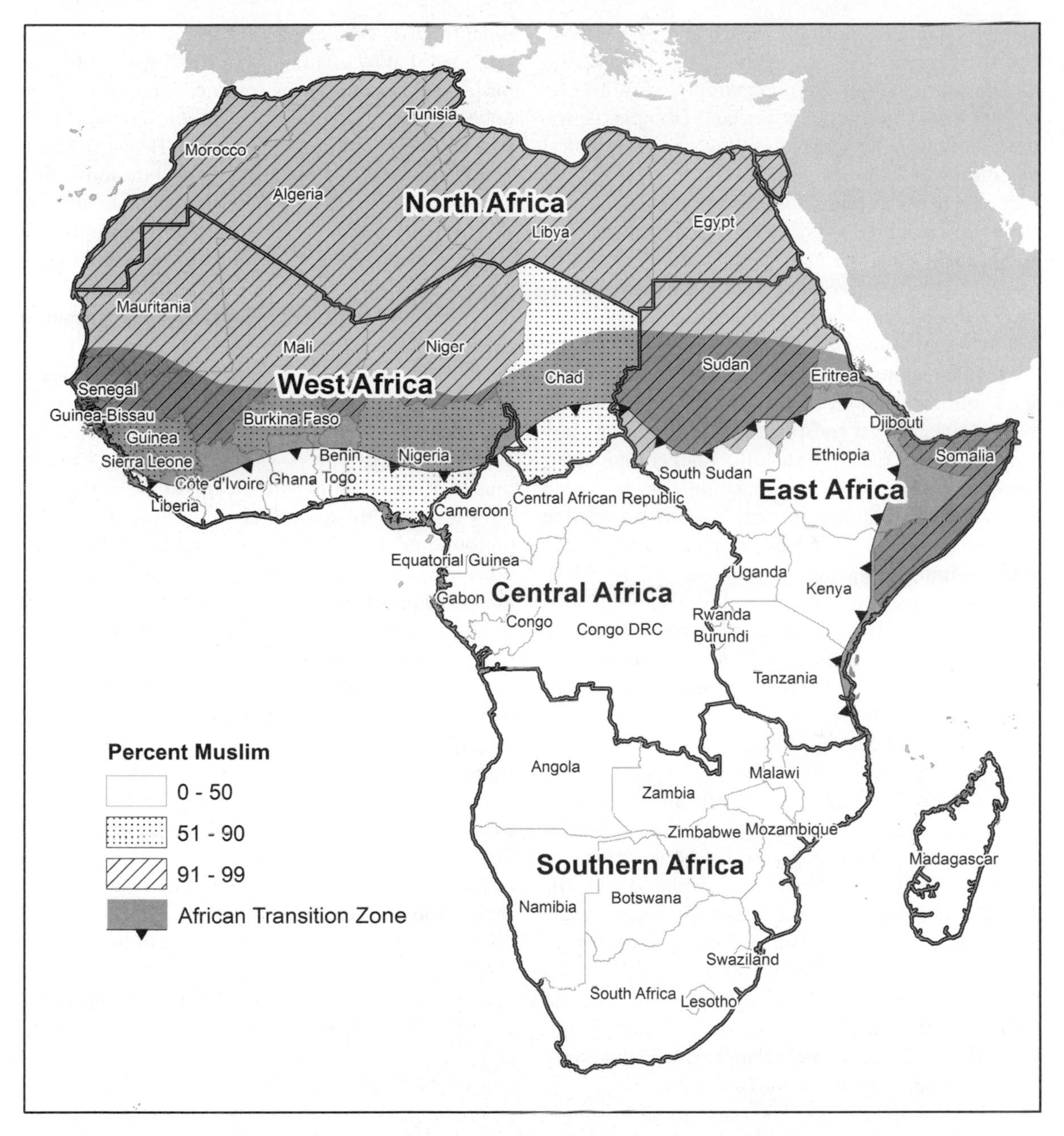


**FIGURE 1.4** Africa's Regions. *Source:* Adapted from "Maps of African Transition Zone" (DeBlij 2012:200) and "Map of Sub-Saharan Regions" (DeBlij, Muller, and Nijman 2014:306).

with North Africa than the rest of the country is. Based on cultural-religious criteria, DeBlij (2012) proposes an addendum to the four-region classification, adding a middle belt or transition zone that straddles the region south of the Sahara from Guinea in West Africa to coastal Tanzania in the East. This is a zone of long-standing Islamic influence that completely dominates some countries (e.g., Somalia), dissects others, and creates Islamized northern areas and non-Islamic southern zones in some countries (e.g., Nigeria and Chad, where 51% of the population is Muslim) and coastal strips in others (e.g., lowland Ethiopia, Kenya, and Tanzania). Its social, political, and geopolitical significance is immense. It has been the site of numerous wars between Arabs and Africans and is currently the terrain in which several terror groups operate (e.g., Al-Shabaab, Boko Harman, and Al-Qaeda in the Islamic Maghreb). In recent decades this transition zone has expanded southward as the Islamic African population has increased.

While regions are a useful starting point in learning about the complexities of Africa, regions and subregions are artificial constructions, and contemporary dynamism in Africa is at the urban rather than at the regional scale, and urban transformation is having profound implications on rural areas (e.g., drawing rural resources such as food, minerals, and people into the functioning and expansion of African cities). This text works from the standpoint that an extended urban approach is better equipped to combine global, African, and local perspectives in efforts to understand the profound change that is occurring throughout urban and rural Africa.

In this text, I discuss Africa south of the Sahara (as opposed to the more racist "black Africa" or the Eurocentric "Sub-Saharan Africa"). North Africa is not included in the discussion as it merits a study in its own right. Throughout the pages of this book, "Africa" is used as shorthand for Africa south of the Sahara. The global positioning of Africa situates the region at the minimum aggregate distance to other major world regions as well as antipodal to the Pacific Ocean. In a 21st-century context, Africa is centrally located to Asia as well as to Europe, North America, and the Gulf states—a centrality important to the operation of markets and the reach of financial capital in the contemporary era.

## THE URBAN REVOLUTION AND EMERGING RURAL REVOLUTION IN AFRICA

The 21st century is the dawn of a new era in Africa. An urban revolution is unfolding. For much of its history, Africa was the least urbanized region of the world, and its societies and populations were predominately rural. In 2011, 40% of Africans live in cities, and it is predicted that by 2050 this figure will rise to 60%. In the last 27 years, the entire continent's population has doubled from 500 million to 1 billion, a number projected to reach 2 billion by 2050 (UN-HABITAT 2010). The simultaneous population explosion and rural–urban transformation within the region is nothing short of spectacular. In East Africa, for example, the urban population has doubled in the last seven years—the shortest doubling of population of any region in world history. Since current urbanization trends indicate that cities are soon to be the habitat for the majority of Africans, now is the time for a thoughtful examination of what is unfolding. The region's population explosion will change the political, social and economic dynamics within African countries and thus their relationship with the rest of the world.

The term "urban revolution" is used to portray profound urbanization and its wider societal implications (Parnell and Pieterse 2014; Myers 2011). An urban revolution occurs in environments with large gains in population (natural increase of population in urban areas coupled with high urban migration) within a short time period. At some tipping point (impossible to pin down, but when the level of urbanization reaches a significant threshold, perhaps around 40%), society shifts from a rural agrarian-based to an urban orientation. Of course, urban settlements vary in size and complexity, but urban influence extends well beyond the confines of the capital city. Urban settlements thrive on commodity production, exchange, diversification, and differentiation that transcend the ethos of household self-sufficiency that prevails in the countryside. Over time, the economic division of labor changes, and the links between residents and economic agents from the world beyond evolve. Eventually all of society, to varying degrees, becomes enmeshed in urban affairs. This is Africa's urban revolution.

Urbanization in Africa is not yet complete, but its trajectory seems inevitable. Urbanization is accompanied by sustenance and sustainability activities; stable food supply, water, shelter, security, transportation, and communications are all requisites for the densification of urban settlements. While urban centers are not the only players in statewide phenomena, national capitals dominate national decision-making processes and debates. These cities host the majority of national elites and institutions (ministries, courts, universities, banks, etc.) and the emerging middle class in Africa, as well as a range of stakeholders whose activities often involve them significantly in nationally important issues. Voluntary civic and social organizations, such as self-help groups, business associations, advocacy groups, nongovernmental organizations (NGOs), charities, professional associations, trade unions, student groups, community groups, and social networking groups, heavily populate the urban policy and advocacy terrain. They typically (but not necessarily) pursue more transformative agendas and can be rapid responders to unfolding urbanization challenges, sometimes at odds with the entrenched interests of the political, economic, and social elites. Civic and social organizations can provide checks and balances and encourage more transparent and accountable responsive governance. Serving as change catalysts is not always a smooth process: in many instances, certain organizations can be the targets of state repression. Urban actors extend their influence through information and communication technologies, which connect urban centers with the entire globe and link rural areas into an urban network. Perhaps the best example is the uptake of mobile phones, particularly in the past 10 years; they have connected urban and rural societies like never before in instantaneous communication. Africans now joke that the region has more mobile phones than working toilets.

The scale and scope of urban transformation calls for a rethinking of the rural–urban dichotomy that has underpinned African studies until very recently. A binary division of African societies into urban and rural tends to mask more varied dynamics, such as mobilities and links between rural and urban areas, most saliently expressed in migration flows, circulatory movements, and remittance flows through which urban migrants move monies and imported commodities from cities to rural hometowns. An extended urban reach is also under way: for example, the entry of urban actors into the agrarian scene, such as urban-based elites buying up rural land, signals a new phenomenon. Facebook's consideration of delivering Internet connectivity to rural Africa via the launch of solar-powered drones (if the project gets implemented and works) could profoundly change rural Africa (Wired 2014). Thus, rural environments are linked in diverse ways to local and transnational urban sites, and transformative change in rural Africa seems inevitable.

An urban vista allows us to confront the rural bias that affects many studies of Africa, particularly those in the English language. In the past, a major concentration within African studies (a multi-disciplinary association of scholars and specialists of Africa) has been rural affairs: experts in rural studies, peasant studies, and other related fields easily represented the remote village as the "true" Africa, and development models tended to reinforce this kind of thinking. It is hardly surprising then that most stereotypes of Africa evoke rural imaginaries of sorts. It is important to underscore again that Africa has a long urban tradition, a history that has until recently been neglected, and that the future of Africa is likely to be increasingly urban.

At the same time, we need to be mindful that not only people moving to cities change but also urbanites visiting rural Africa and maintaining family homes and/or second homes or indeed families are change agents. There are sparks of a parallel socio-spatial transformation in rural areas (Krause 2013): an African rural revolution in the making. Key elements include the spread of mobile phones to even remote African villages, inflows of remittances, foreign direct investment (FDI) in agricultural land, introduction of hybrid seeds and other agricultural technologies, growing diversification of rural livelihoods, the development of rural infrastructure, and the growth of rural towns and markets that are altering hitherto rural isolation and connecting virtually all areas to rural nodes and urban centers and even linking them to global circuits. The demands of global urban industrial society for various African resources (gold, diamonds, oil, uranium, timber, etc.) also means that rural Africa is integrated into global circuits, and newer global demands for African-sourced food, biofuels, and natural gas will deepen this global integration of rural spaces.

Mobile phones reduce the cost of communication over long distances and enable individuals to interact more easily and frequently with members of their social networks, making it easier and faster to ask for help as needed. Various combinations such as mobile phone and remittances are very transformative and powerful, providing a channel to tap funds for launching small businesses, deliver financial security to migrant workers' families, and provide emergency funds that help mitigate the effects of natural or political shocks. The latter provides the first-ever insurance mechanism in rural Africa.

Reductions in communication costs improve access to all kinds of information. For example, poor, rural, semiliterate farmers can now obtain almost instant information on the likely impact of a drought, food shortage, or cholera epidemic by accessing information via a mobile phone (Rothberg and Aker 2013). Farmers can now obtain real-time information about grain prices in proximate and distant markets, and traders can use this same market information to move grains from low-price to high-price areas so markets operate more efficiently and more equitably. A pioneering program called iCow (2013) in Kenya enables small farmers to access agricultural services and information (e.g., information about gestation, access to veterinarians and artificial inseminators) over the mobile phone. Dairy farmers using iCow services report significant increases in milk production and a gain of US$78 per month in rural Kenya (iCow 2013). Mobile phone technology therefore provides new benefits that are transforming the everyday lives of rural Africans.

Rural Africa was never a static place, and development policy interventions have always sought to improve societies but the rural arena is now in play more so than at any time since the introduction of colonial plantation agriculture. Global agro-corporations, international financial capitalists, and their domestic partners have turned their attention and investment funds toward rural Africa. The livelihoods and culture of millions of Africans are at stake. Rural Africa has become more unsettled, and tensions are escalating among agricultural interest groups (small family farms vs. large industrial farms, pastoralists vs. nomadic farmers, forest farmers vs. the timber industry, food producers vs. biofuel producers), raising the important question of what is the fastest way to lift most rural people out of poverty. Traditional agriculture is under threat: many of the new ideas on conservation, tourism, and commercial farming (e.g., producing food to be sent to food markets abroad) entail complete breaks with traditional activities and practices. Some new projects are being hailed as successes, such as the development of horticulture exports in Kenya for urban markets in Europe. Other projects, such as wildlife conservation efforts in Tanzania, have made the Maasai "strangers in their own land" (Goldman 2011:65). However, the socio-spatial transformation of rural Africa is still at a preliminary stage and therefore not easy to assess, but there are passionate debates about the changes.

Krause (2013) contends that there is a ruralization of the world that is subsumed within the urbanization of the world. This is spatially represented in the sponge of the urban fringe or peri-urban edge, a porous in-between zone that is neither urban nor rural in its character or governance (Pieterse and Parnell 2014). Examples abound of rural migrants practicing urban agriculture and adopting rural lifestyles and practices in Africa's cities, and urban phenomena in rural areas such as a growing diversity within the informal rural economy and international migrants moving to engage in commercial farm labor. One the most contentious issues in rural Africa is offshoring food farming, whereby investors from the Gulf states, India, and to a lesser extent China aim to develop commercial farms to produce food for urban markets outside of Africa.

## THE CONCEPT OF "AFRICA"

The origin of the word "Africa" is unknown, but experts trace it to Latin (*Africa* = sunny) and Greek (*Aphrike* = not cold). Cartographic evidence from the sixth century shows that the Romans identified a region as Africa on early maps of the world; however, they had knowledge only of the northern parts of the continent. Much later, around the 11th century, Europeans began to use "Africa" to refer to the land south of the Sahara. The evolution of Africa as a cartographic reality through delineating and refining the contours of the continent on maps contributed to the conception of "Africa" as a unified cultural reality that continues to permeate popular thought. Although similarities among peoples and groups are likely, we cannot deduce from their mere coexistence on the same continent that one cultural reality existed. Africa is not one country or place and Africans are not one people: the region comprises a rich mosaic of peoples, places, cultures, economies, languages, and political systems.

Indeed, Africa as a region is very much a geographical accident. Even African philosophers struggle with the concept of "Africa" and point out that interregional encounters have led to constructions of a wider spatial stage for Africa but not a unitary Afrocentric narrative. As evidenced by historical records, extensive dispersal of many peoples of Africa throughout the world meant that Africans were internationalizing well before the contemporary period of what academics and policy makers have termed "transnationalism." Migration, however, does not confirm peoples' consciousness of living in or originating from a region termed Africa. It takes time for complex networking to develop among cultures, economies, and societies and for territories and landscapes to be inscribed with common meanings. At most, groups created their own "little Africas," each laying the foundations of African identify across a much larger area. Clearly, this is no different than peoples in Europe and in the Americas struggling with their own regional spatial consciousness. In sum, for Africans, "Africa" was never a single place.

Unfortunately, Africa as a concept is as much a product of modern race thinking as it is an obvious cultural and historical region. Historically, Western societies have represented Africa in their desperate desire to assert Western difference from the rest of the world (the "West" category could also be unpacked, but that is beyond the scope of this book). In the Western imagination, Africa is often an empty category juxtaposed against the full, developed, modern West. Western writers have consistently depicted Africans as "others" (Edgar Rice Burroughs' *Tarzan*, set in fictionalized Africa, is a classic example; never having set foot in Africa did not hinder Burroughs' ability to capture the Western imagination of Africa or contribute to white outsiders' perspectives). Common fantasies and myths depicted Africa in the most derogatory ways: "a dark continent," "a mysterious place," with endless tropes of tribalism, savagery, and chaos. Wole Soyinka (2012:3), Nigerian Nobel Laureate in Literature, emphasizes that the darkness of Africa was due to "the cataract in the eye of the beholder." Writings that asserted the superiority of the West were an instrumental part of the domination of colonial peoples. Despite the obvious commercial imperatives of European colonialism, Europeans portrayed their conquests as noble altruism (e.g., referred to by English poet Rudyard Kipling as "the white man's burden").

In the past, deeply embedded assumptions and stereotypes about Africa too easily informed popular writings, academic research, and development policies and now hinder knowledge creation and real learning. Many contemporary narratives recycle old assumptions and clichés about Africa: failed states, development failures, lost decades, enduring poverty, hopelessness, horrific diseases, etc.

The most prominent anticolonial thinker of the 20th century, Frantz Fanon (1966), asserts the psychological effects of colonialism on subjects' engendered feelings of dependency and inferiority: subjects were taught to aspire to be Europeans but could never do enough to be accepted. Sartre (1966:7), in his preface to Fanon's *The Wretched of the Earth*, conveys the indecency of Europe's manufacturing of an indigenous colonial elite: "They picked out promising young adolescents, they branded them, as with a red-hot iron, with the principles of Western culture; they stuffed their mouths full with high-sounding phrases, grand glutinous words that stuck to the teeth. After a short stay in the mother country they were sent home, whitewashed. These walking lies had nothing left to say to their brothers." Extreme social distance was preserved by the continuation of Western narratives telling us what Africa societies are *not* and relating little about what they *are*.

Western media concentrate on negative crisis reporting about Africa (hunger, famine, starvation, endemic violence, conflict and war, and HIV/AIDS). Such one-sided, incomplete information about Africa is a self-perpetuating growth industry. The Western imaginary is further impeded by the preoccupation with the "animalization of Africa." Television channels (e.g., Animal Planet and Discovery) and numerous documentaries focus on wild Africa and safari Africa. Economic fundamentals of the Western media business explain some of this bias. Media conglomerates from the United States and Europe traditionally controlled much of the news that flows into and out of Africa. Media transnationals control what is reported about the entire world and how it is reported and exported to the African media. The commercialization and corporate control of the news mean that the Western media are profit-driven; they select coverage on what is good for the bottom line. Crisis-driven journalism that churns out fast, short-byte news stories and attention-grabbing headlines with little in-depth reporting or proper contextualizing is all too common in the West. A possible exception is the UK *Guardian* newspaper, which has the most in-depth and consistent coverage of the region. These days, however, the media landscape is becoming more populated in Africa, and CCTV (China) and Al-Jazeera (Qatar) have made great headway; more Africans and other people from around the world are learning about the region from non-Western media houses.

Western photojournalists have had a powerful effect in depicting conflict, disease, rape, and poverty as staples of their coverage. African photojournalists have been documenting other sides to life in Africa for years (e.g., African fashion, art, and buildings and more humanistic everyday lives), but their more hopeful and positive images do not fit the prevailing narrative. The power imbalances among those in the Global North seeing of a one-sided view of Africa, those being watched (the sufferers in Africa), and other Africans such as those who are thriving have never been so acute. A very low knowledge base on Africa (most U.S. students know only a handful of African countries) allows the media to continue their role in educating the masses about what kind of a place it is.

Lack of cultural understanding of Africa and African cultures has enabled the global fashion industry to usurp African designs and cultures. Cultural exploitation accompanied by cultural insensitivity is common. For example, Western companies have employed iconic cultural brands like the Maasai to make money, and little has flowed back to the community (80% of the Maasai live below the poverty line). Companies like Louis Vuitton (France) appropriated Maasai designs for their 2012 fashion collection (Fig. 1.5). All of this external commercial exploitation has promoted the Maasai to fight back and begin a process of trademarking their brand, illustrating the collision of regional culture, capitalism, and identify in 21st-century Africa. The Maasai Intellectual Property Initiative (MIPI 2013) in partnership with the U.S. NGO Light Years IP was launched in 2012 to reclaim the Maasai's economic rights to their iconic image (estimated to be worth several million dollars per year).

Negative and incomplete imagery allows groups in the West to claim they are "saving" Africa by their humanitarian interventions, a problematic contention. Some humanitarian groups do great work and change lives (e.g., the Millennium Villages Project has changed the lives of thousands of rural villagers), but there will always be well-meaning but frivolous groups that get in the way with misguided initiatives (e.g., Knickers 4 Africa, a small British women's group that collects used underwear for export to Africa). Negative, outdated, and misplaced representations of Africa mean that many

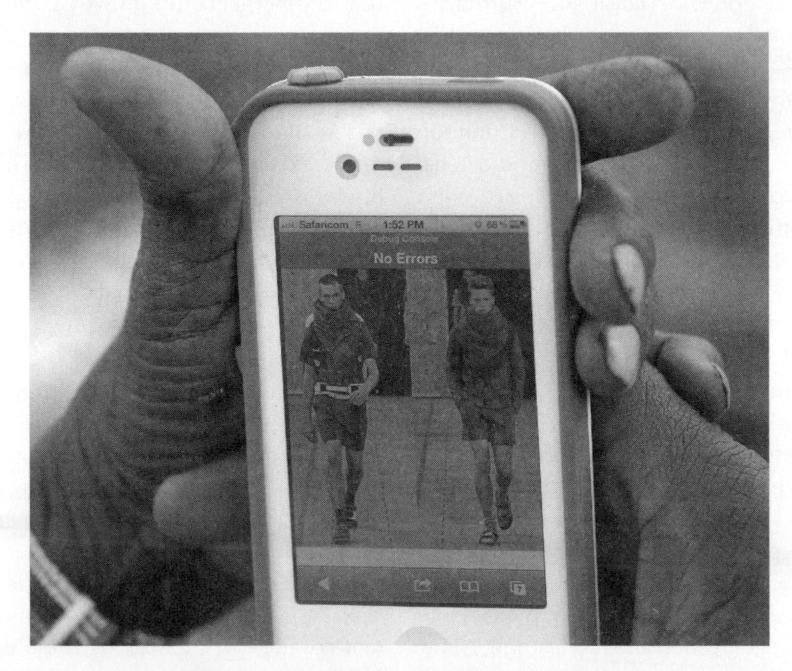


**FIGURE 1.5** Louis Vuitton's Maasai Collection. *Source:* © Thomas Mukoya/Reuters/Corbis.

governments (e.g., the United States) do not have clear and relevant Africa policies or a geographically informed citizenry that can debate and demand Africa policies.

Africa is a vast and complex region and defies easy definition. The closest to an accurate definition may be found perhaps in Sierra Leone poet Abisoeh Nicol's (1950) poem that speaks to the essence of an African definition of Africa as well as to a Euro-Afrocentric conceptualization:

> You are not a country, Africa,
> You are a concept,
> Fashioned in our minds, each to each,
> To hide our separate fears
> To dream our separate dreams.
>
> *—Abisoeh Nicol, "The Meaning of Africa" (ca. 1950)*

Despite the lack of a precise definition, Africa lives on as a real category. African states like all states are inventions, but the primary difference is that African states were not constructed by their inhabitants but rather from the outside by Europeans, many of whom never set foot in the region. Imposed and acted upon by international organizations and their official reports, it is a geographical entity. Wide arrays of actors on the continent understand their situation and act based on a concept of the region and Africa's place in the world. (Listen to the Institute of Development Studies' [IDS 2012] podcast on how the British media provide imbalanced accounts of Africa news.) Defying an easy and simple explanation, Africa has become an inconvenient continent to understand; it does not fit many of the major development paradigms, but it is a spatial stage that has increasing relevance in the 21st century.

Old representations of Africa need to be reconsidered. Several of them, overly reproduced in the Global North, grossly simplify and distort understandings of the complexity of Africa. First, there is the romanticized image of the "noble savage" living in primitive unspoiled rural settings in harmony with nature. This representation conjures up images of innocence and a lost paradise. Second, there is the representation of difference, which leads to a generalized "other." Third, an urban underdevelopment representation highlights poverty, suffering, and lack of employment, with images of helplessness and hopelessness. When these representations dominate, they can inadvertently imbue Global Northerners with feelings of superiority. There is power in who gets to tell the story and how the Western story has become the definitive story of Africa. We need more complex, nuanced, and balanced assessments of Africans and their continent as subjects in their own right rather than as objects. Africa is not a single country or a uniform place. Africans have diverse histories, multiple trajectories, and various and complex interactions among themselves and with outsiders; there are great differences in resources among the regions and states of Africa.

Even now, simplified narratives of Africa are retrofitted for today's youth. For instance, the documentary *Kony 2012* became a viral sensation and Facebook phenomenon (86 million people watched the 30-minute movie in March 2012). The documentary was made by a U.S.-based campaign organization, Invisible Children, and it focused on Ugandan warlord Joseph Kony and his role in child abductions. The video motivated young people across the world to care about an issue in a "distant land" and to act as a pressure group to bring Kony to justice in a strong example of pop humanitarianism (join the movement, sign petitions, donate, purchase "action kit" merchandise, put up fliers, and encourage governments to capture the fugitive). The video came under heavy criticism for its simple good-and-evil message and its promotion of blind action, overlooking the agency of locals and failing to understand the local context. Invisible Children's spending practices were critiqued (less than one third of its budget went to on-the-ground programs). The documentary did not resonate well with many groups, especially Northern Ugandans, who saw it as a simplified portrayal of complex events—not to mention that Kony had left the region for the Central African Republic five years earlier. Since the campaign, the United States has deployed 100 U.S. Special Forces personnel to the region to assist governments in tracking Kony, and a 2013 Crownfunding campaign ("Expedition Kony") fell way short of its campaign goal to raise US$450,000 to launch an expedition to the capture the fugitive. The Kony phenomenon may be supported by a well-intentioned youth movement with a social conscience, but it falls into the old trap of Westerners knowing what is best for Africa and a Western message that simplifies on-the-ground complexities.

A single story is always problematic because one storyteller has to begin somewhere and sometime before proceeding to a middle and an end. For instance, if we start the story of Africa with the arrival of the Europeans, we get one particular account. If we begin the story with the arrival of the Chinese, we get a very different account. If we start the story with the failure of African political leadership and with the arrival of Europeans, we arrive at yet another story. (Listen to the TED talk by Nigerian author Chimamanda Adichie [2009].) Any single story may well feed into stereotypes that emphasize difference rather than similarity, making it virtually impossible to engage with people or a region fairly. A fundamental problem with stereotypes is their incompleteness: only more in-depth examinations can bring forward a balance of stories.

A revisionist story has recently emerged about China engaging with Africa, particularly with East Africa, long before the Europeans. It has come to light, with the circulation of a rare map, "Great Ming Amalgamated Map of 1389," that Chinese cartographers produced the earliest, most complete Africa exterior map (Europeans had no cartographic knowledge of the southern part of Africa at that time). This map, drawn on a large tapestry, had been stored carefully within historical archives in Beijing, China. Less than a decade ago, the Chinese government decided to publicize the map's existence, authorizing the South African government to reproduce the tapestry for a 2004 map exhibition at the South African Parliament in Cape Town. The existence of the map showed that direct communications between China and Africa took place over 100 years earlier than the "discovery of the African continent" by Europeans. Of course, it is important to ask why the Chinese waited so long to divulge the map's existence and to counter Eurocentric narratives of Africa's "discovery."

The scramble for Africa and China's deepening engagement with many African states are important reasons. Also, this is a time when African states have increased self-confidence and there is a resurgence of interest in and attention to Africa's place in the world and to the world's place in Africa. Educated Africans are now more acutely aware of the power of the international media in creating a global imagination about Africa and how the media are prone to stereotyping Africans. (Listen to the BBC's [2011] podcast and award-wining play *Silhouettes* to hear an African account of Hollywood's depiction of African accents, vocabulary, and modes of reasoning.) Africans are now providing a fuller picture of Africa that is more easily accessible because of the Internet. African news magazines (e.g., *African Business* and the French-language *Jeune Afrique*) and various African news websites and blogs now provide direct information from Africa.

Unfortunately, the dominance of the West's negative representations of Africa is exemplified in how the popular Western media write about Africa (See Box 1.1) and it is also reflects the marginalization of Africa's social scientific scholarship. Africa's share of social science knowledge production (measured by papers on Webs of Knowledge [citation indexes]) has been declining for several decades. Many scholars in Africa publish via local outlets; by default, this typically results in making their scholarship invisible to mainstream North-centric

### BOX 1.1 HOW TO WRITE ABOUT AFRICA

Always use the word 'Africa' or 'Darkness' or 'Safari' in your title. Subtitles may include the words 'Zanzibar', 'Masai', 'Zulu', 'Zambezi', 'Congo', 'Nile', 'Big', 'Sky', 'Shadow', 'Drum', 'Sun' or 'Bygone'. Also useful are words such as 'Guerrillas', 'Timeless', 'Primordial' and 'Tribal'. Note that 'People' means Africans who are not black, while 'The People' means black Africans.

Never have a picture of a well-adjusted African on the cover of your book, or in it, unless that African has won the Nobel Prize. An AK-47, prominent ribs, naked breasts: use these. If you must include an African, make sure you get one in Masai or Zulu or Dogon dress.

In your text, treat Africa as if it were one country. It is hot and dusty with rolling grasslands and huge herds of animals and tall, thin people who are starving. Or it is hot and steamy with very short people who eat primates. Don't get bogged down with precise descriptions. Africa is big: fifty-four countries, 900 million people who are too busy starving and dying and warring and emigrating to read your book. The continent is full of deserts, jungles, highlands, savannahs and many other things, but your reader doesn't care about all that, so keep your descriptions romantic and evocative and unparticular.

Make sure you show how Africans have music and rhythm deep in their souls, and eat things no other humans eat. Do not mention rice and beef and wheat; monkey-brain is an African's cuisine of choice, along with goat, snake, worms and grubs and all manner of game meat. Make sure you show that you are able to eat such food without flinching, and describe how you learn to enjoy it—because you care.

Taboo subjects: ordinary domestic scenes, love between Africans (unless a death is involved), references to African writers or intellectuals, mention of school-going children who are not suffering from yaws or Ebola fever or female genital mutilation.

Throughout the book, adopt a sotto voice, in conspiracy with the reader, and a sad *I-expected-so-much* tone. Establish early on that your liberalism is impeccable, and mention near the beginning how much you love Africa, how you fell in love with the place and can't live without her. Africa is the only continent you can love—take advantage of this. If you are a man, thrust yourself into her warm virgin forests. If you are a woman, treat Africa as a man who wears a bush jacket and disappears off into the sunset. Africa is to be pitied, worshipped or dominated. Whichever angle you take, be sure to leave the strong impression that without your intervention and your important book, Africa is doomed.

Your African characters may include naked warriors, loyal servants, diviners and seers, ancient wise men living in hermitic splendour. Or corrupt politicians, inept polygamous travel-guides, and prostitutes you have slept with. The Loyal Servant always behaves like a seven-year-old and needs a firm hand; he is scared of snakes, good with children, and always involving you in his complex domestic dramas. The Ancient Wise Man always comes from a noble tribe (not the money-grubbing tribes like the Gikuyu, the Igbo or the Shona). He has rheumy eyes and is close to the Earth. The Modern African is a fat man who steals and works in the visa office, refusing to give work permits to qualified Westerners who really care about Africa. He is an enemy of development, always using his government job to make it difficult for pragmatic and good-hearted expats to set up NGOs or Legal Conservation Areas. Or he is an Oxford-educated intellectual turned serial-killing politician in a Savile Row suit. He is a cannibal who likes Cristal champagne, and his mother is a rich witch-doctor who really runs the country.

Among your characters you must always include The Starving African, who wanders the refugee camp nearly naked, and waits for the benevolence of the West. Her children have flies on their eyelids and pot bellies, and her breasts are flat and empty. She must look utterly helpless. She can have no past, no history; such diversions ruin the dramatic moment. Moans are good. She must never say anything about herself in the dialogue except to speak of her (unspeakable) suffering. Also be sure to include a warm and motherly woman who has a rolling laugh and who is concerned for your well-being. Just call her Mama. Her children are all delinquent. These characters should buzz around your main hero, making him look good. Your hero can teach them, bathe them, feed them; he carries lots of babies and has seen Death. Your hero is you (if reportage), or a beautiful, tragic international celebrity/aristocrat who now cares for animals (if fiction).

Bad Western characters may include children of Tory cabinet ministers, Afrikaners, employees of the World Bank. When talking about exploitation by foreigners mention the Chinese and Indian traders. Blame the West for Africa's situation. But do not be too specific.

Broad brushstrokes throughout are good. Avoid having the African characters laugh, or struggle to educate their kids, or just make do in mundane circumstances. Have them illuminate something about Europe or America in Africa. African characters should be colourful, exotic, larger than life—but empty inside, with no dialogue, no conflicts or resolutions in their stories, no depth or quirks to confuse the cause.

Describe, in detail, naked breasts (young, old, conservative, recently raped, big, small) or mutilated genitals, or enhanced genitals. Or any kind of genitals. And dead bodies. Or, better, naked dead bodies. And especially rotting naked dead bodies. Remember, any work you submit in which people look filthy and miserable will be referred to as the 'real Africa', and you want that on your dust jacket. Do not feel queasy about this: you are trying to help them to get aid from the West. The biggest taboo in writing about Africa is to describe or show dead or suffering white people.

Animals, on the other hand, must be treated as well rounded, complex characters. They speak (or grunt while tossing their manes proudly) and have names, ambitions and desires. They also have family values: see how lions teach their children? Elephants are caring, and are good feminists or dignified patriarchs. So are gorillas. Never, ever say anything negative about an elephant or a gorilla. Elephants may attack people's property, destroy their crops, and even kill them. Always take the side of the elephant. Big cats have public-school accents. Hyenas are fair game and have vaguely Middle Eastern accents. Any short Africans who live in the jungle or desert may be portrayed with good humour (unless they are in conflict with an elephant or chimpanzee or gorilla, in which case they are pure evil).

After celebrity activists and aid workers, conservationists are Africa's most important people. Do not offend them. You need them to invite you to their 30,000-acre game ranch or 'conservation area', and this is the only way you will get to interview the celebrity activist. Often a book cover with a heroic-looking conservationist on it works magic for sales. Anybody white, tanned and wearing khaki who once had a pet antelope or a farm is a conservationist, one who is preserving Africa's rich heritage. When interviewing him or her, do not ask how much funding they have; do not ask how much money they make off their game. Never ask how much they pay their employees.

Readers will be put off if you don't mention the light in Africa. And sunsets, the African sunset is a must. It is always big and red. There is always a big sky. Wide empty spaces and game are critical—Africa is the Land of Wide Empty Spaces. When writing about the plight of flora and fauna, make sure you mention that Africa is overpopulated. When your main character is in a desert or jungle living with indigenous peoples (anybody short) it is okay to mention that Africa has been severely depopulated by Aids and War (use caps).

You'll also need a nightclub called Tropicana, where mercenaries, evil nouveau riche Africans and prostitutes and guerrillas and expats hang out.

Always end your book with Nelson Mandela saying something about rainbows or renaissances. Because you care.

---

*Source:* Binyavanga Wainaina (2005).

social scientists. Moreover, South Africa, Nigeria, and Kenya account for 75% (mostly produced by leading universities within these countries) of Africa's academic publications in the social sciences. It is very worrisome that many African countries continue to publish a negligible proportion of research and writing on Africa.

## THE "NEW AFRICA" GROWTH STORY

In recent years, a number of decidedly upbeat economic examinations of leading African countries' performance and increased economic momentum have appeared, garnering very positive pro-business assessments. *The Economist* magazine reports that six African economies registered among the top ten fastest-growing economies in the world for the last decade (2001–11) (Angola 11.1%, Nigeria 8.9%, Ethiopia 8.4%, Chad 7.9%, Mozambique 7.9%, and Rwanda 7.6%), and the average African economy is forecasted to outpace its Asian counterparts over the next five years (Fig. 1.6).

In 2000, *The Economist* regrettably labeled Africa "the hopeless continent." However, the World Bank

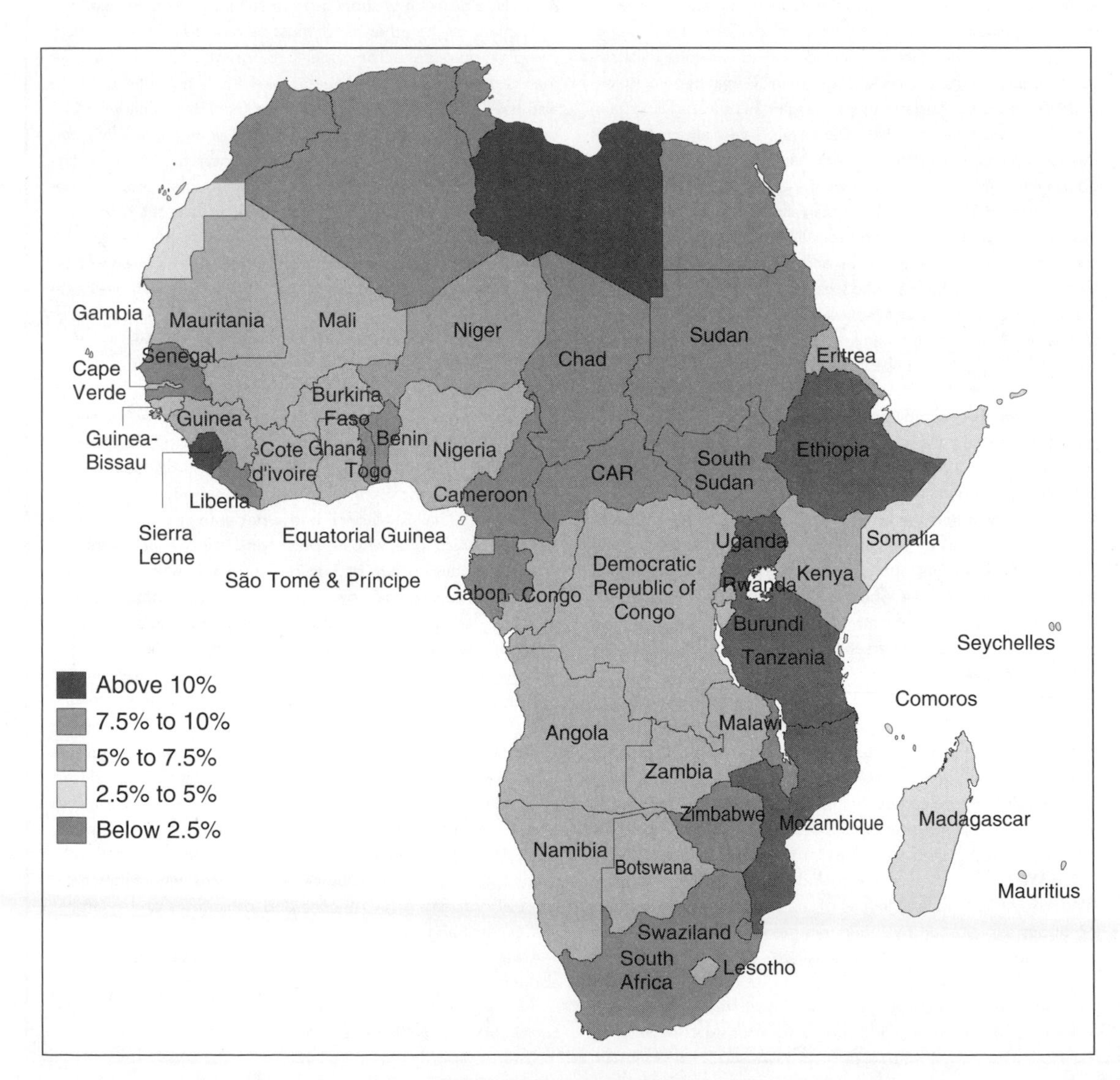


**FIGURE 1.6** Africa's GDP Growth (2012–2016 forecast). *Source:* EIU 2012:2.

believes that Africa could be on the brink of an economic takeoff, much like the one that China witnessed 30 years ago and the one that India has undergone since 1995. Some business stories have hyped Africa as the consumer market frontier. Some of largest global players (e.g., Coca-Cola) are steering the way with established footholds. Coca-Cola's market presence in Africa already surpasses its market penetration in China or India. Africa is "in play" again because international capital has access to the region (and this is not just limited to European capital, as was the case in the colonial period). *Barron's*, a financial news publication lauds Africa as "the final frontier" for investment and marketing opportunities. Wall Street is taking notice: the first fund exclusively focused on Africa was introduced in 2010 (Nile Pan Africa Fund trades on the stock market as NAFAX). In truth, it is not all of Africa but a cohort of 17 African countries (with 300 million people) that is leading the way, with another six countries showing promising but less steady improvement (Radelet 2010). In 2010, private equity firms raised US$1.5 billion for projects in Africa.

A segment of the business press focuses on bottom-of-the-pyramid opportunities (developing business models that target the poorest consumers, typically those that earn less than US$2.50 per day): the sum of Africa's poor population is very lucrative when businesses can tap into large volumes of small increments in purchasing power. The consumer product invasion of the world's poor, not surprisingly, is heavily critiqued. For example, major ethical as well as health objections have been raised about targeting soft drinks at people eking out a living (where discretionary spending is scarce and monies may be diverted away from more productive expenditures into the consumption of empty calories). Nevertheless, when money can be made, corporations will focus on the bottom line. Thinking that the poor of today could be the middle class of the future has led to a fundamental tilt in consumer marketing strategies and adds Africa to the marketing focus.

There is far more to the bottom of the pyramid than hype. Changing perceptions of Africans are altering corporate behaviors. For instance, Siemens, operating for over 150 years in Africa, for the first time in its history has established a separate African organization within the corporate structure; Coca-Cola switched its African regional office from Windsor, England, to Johannesburg, South Africa; and global corporations are promoting Africans to manage African markets. To avoid any claims of being outside invaders, global corporations are emphasizing corporate social responsibilities on the ground. For instance, Coca-Cola has committed US$30 million (2009–15) for a variety of community-based initiatives throughout Africa (e.g., hygiene education, HIV/AIDS and malaria prevention). Coca-Cola's sophisticated network of 900,000 retail outlets throughout the continent presents significant distribution opportunities (condoms have been distributed via some of Coca-Cola's channels) (Mahajan 2009).

Africa experts have put forward various strands of a more positive assessment on "Africa's turn." Four fundamental changes are igniting this turn. First, there has been a catching up and a closing of the gap between Africa's economic performance and that of more developed regions. From 2000 to 2008, continental Africa's annual Gross Domestic Product (GDP ) growth was 4.9%, making it the world's third fastest-growing region (after Asia and the Middle East). This continental growth rate is more than twice the average growth rate in each of the two preceding decades. The growth trajectory could be even more impressive, but infrastructure deficiencies trim an estimated 2% from growth per year.

Second, improved internal macroeconomic management in many African states, combined with many successes in ending political conflicts, has led to an improved business climate in the formal sphere (measured by reductions in the costs of doing business and in the time it takes to start up a new business), whereby African leaders have sought to harness global capital and expertise to promote free market economies in their respective states. The "peace dividend," after years and decades of conflict and military rule in some places, and the tilt toward democracy are producing a management dividend and making the region more attractive to international investors. According to the Economist Intelligence Unit (2012), two thirds of global companies (based on a survey of 217 global corporations) have prioritized expansion in Africa within the next decade.

Third, a new generation of policymakers, activists, and business leaders are emerging across Africa. They are Africans to the core but with a global perspective derived from the Internet age, easier international travel and communication, and greater exposure to current trends and international thinking. Fed up with unaccountable governments of the past, they are bringing new ideas, energy, and vision to the table. They have been called the "cheetah generation" for not relating to the old colonist paradigm, the slave trade, or the internal-oriented nationalism of the early independence days. (Listen to Ayittey's 2007 TED talk.)

Fourth, the African debt crisis is winding down, and improving interactions with donors are now evident. Relationships with the World Bank and the International Monetary Fund (IMF) are much less adversarial than they had been in the mid-1980s (when countries were forced to comply with IMF mandates) and are now healthier, moving from crisis management to more forward-thinking development.

Several components of Africa's economic turnaround are important. There is, without doubt, an element of learning from past mistakes: African governments are evolving from their roles as merely recipients of Western governmental aid and development industries and are becoming more assertive in their desire to act as "partners" in the development of their states. Fortunately, a partnership mantra also coincides with the changing rhetoric of the global development community. In recent resource deals that have taken place in the context of intensified global competition for commodities, African governments appear to possess a newfound bargaining power. Buyers are motivated to offer upfront payments, build infrastructure in return, and even share management skills and technology. African governments' enlarged role allows them to coordinate multiple suppliers across industries, secure financing, and strike multiple deals simultaneously, installing them in the driver's seat; this contrasts to their backseat position in economic development from 1980 to 2000. Democratic trends have resulted in investment dividends by promoting the region as less risky for international investment and have encouraged more domestic investment. Historically, domestic business had not been systematically involved in Africa's development. Following independence, governments tended to view the private sector with suspicion, and local entrepreneurs with the requisite capital and skills were scarce. In these contexts, state-owned enterprises emerged as the main institutions for promoting the development of Africa's industry and infrastructure. This nationalist model ran out of steam by the mid-1980s.

Of colossal importance, China has exploded onto the African scene, influencing trade, investment, aid, and diplomatic relationships. It is now Africa's largest trading partner, but its engagement is uneven: the bulk of China's trade and investment is concentrated in resource-rich countries (e.g., South Africa, Zambia, Angola, and Sudan in particular). China's infrastructure commitments alone now surpass those of the World Bank (the most prominent development organization in Africa until recently). The "China effect" has coincided with a surge of investment flows into the region, which peaked at US$87 billion in 2007 (before the global financial downturn that began in 2008). China's effect is everywhere: Figure 1.7 shows the African Union headquarters built by the Chinese.

Large numbers of new investors provided an important buffer that has protected much of Africa from the steep decline experienced by other world regions during the global recession. Despite the global economic downturn, engagement between China and Africa appears to be strengthening: Beijing has concluded various memoranda of understanding with African governments (e.g., US$23 billion of investment is committed to Nigerian oil refineries). China leads the way among the emerging economic powerhouses known as the "BRIC" countries (Brazil, Russia, India, and China). Along with other countries such as Turkey and Malaysia, it can engage Africa free of the North's

**FIGURE 1.7** African Union Headquarters, Addis Ababa. *Source:* © Wolfgang Kumm/dpa/Corbis.

historical baggage but with enough clout to offer an alternative to the Global North and the traditional international financial apparatus. A discourse of both difference and similarity is evoked to frame relationships between the BRIC countries and Africa. The bottom line is that Western development models are no longer "the only game in town": many African leaders have witnessed the rise of Asia in their own lifetime. A tilt toward Asia is also happening within the inner circle of national policymaking. For instance, all four international advisers on Kenya's key national planning body come from East Asia. Pamphlets used to train cadres in Ethiopia's ruling party, the Ethiopian People's Revolutionary Democratic Front, use the examples of Taiwan, South Korea, and China as best practices of state intervention.

The traditional Western aid cartel in Africa is being challenged on many fronts. India, Brazil, and the Gulf states play by different rules, with different assistance models closely linked to commerce. In 2009, a group of 31 businesses from Singapore sent a delegation to explore business opportunities in Africa; this would have been unthinkable 15 year earlier. International investments in Africa now surpass foreign aid flows, possibly breaking Africa's decades-old addiction to Western aid.

Initially investments flowed into extractive industries (mining, oil, and gas), underlining the importance of Africa's rich resource base. The continent's abundance of natural resources, including 10% of the world's oil reserves, 40% of its gold, and 80% to 90% of chromium and platinum, has boosted recent growth rates. However, investors are now diversifying into a range of sectors, including wholesale and retail, tourism, financial services, telecommunications, transportation, construction, and manufacturing. The technology sector is attracting international as well as African investors. Africa is breaking into light manufacturing and information technology, from call centers to mobile phone platforms to movies to video gaming. Nigerian companies are leading the way in developing Africa-themed movies and video-gaming content; for example, popular video games in 2013 include "My Village," "Lagos Traffic," and "The Tribes." Sizable technology clusters are evident, such as Bollywood (Nigeria), iHub (Kenya), and CyberCity (Mauritius). Newer projects are even more ambitious. New technological cities were launched in 2013: the Konza Technological City (Kenya) and Hope City (Ghana). The Hope City project, just outside the capital city of Accra, will construct Africa's tallest building in a 75-story mixed-use development serving several hundred technology companies, 50,000 workers, and 25,000 residents.

Africa's long-term growth will increasingly reflect interrelated social and demographic trends that are creating new engines of domestic growth. Of critical importance here are projections that forecast continued urbanization and the rise of middle-class African consumers and the transformation of all urban settlements (See Box 1.2 on urbanization and small town in East Africa). The middle class (arbitrarily defined as those earning US$5,000 and above) hovers around 85 million households but is projected to increase to 128 million households by 2020. Incomes of US$5,000 and above are of vital importance in consumer marketing because at this threshold people start spending roughly half their income on items other than food. The African middle class, defined in terms of wealth, and presumably in aspirations and attitudes as well, is expected to enlarge. Africa's emerging middle class is attracting the attention of multinational retailers and service providers. Shiny new shopping malls, international hotel chains, and global fast-food outlets are becoming more of a common sight in many African cities (Turok 2013). Middle-class members of the global village are mobilizing via social media and the Internet and deploying mobile technologies in innovative ways (Rotberg 2013). Middle-class Africans want urban jobs, not farm employment, and they strive to hold their leaders accountable to standards of responsible governance and transparency.

Africa's population is increasing significantly, and this is both an opportunity and a challenge. Half of all the persons born in the world from now until 2050 will be Africans (Rotberg 2013:12). An African region comprising 2 billion people will be accompanied by major shifts in the sizes of African countries and cities. For example, Nigeria, now home to 174 million people, will grow to 730 million by the end of the century, becoming the third most populous country in the world. In the decades ahead, there will be many more people aged between 15 and 35 years than in any other

## BOX 1.2 URBANIZATION IN SMALL TOWNS IN EAST AFRICA

East Africa is the least urbanized region in Africa, but population projections indicate that it will grow at the fastest rate in Africa in the coming decades.

East Africa provides an important illustration that most of Africa's population lives not in capital cities but in secondary cities and smaller towns, some of which are growing at stealth rates and will eventually become large settlements. The fact that migrants are showing signs of shunning capital and apex cities and instead relocating to small, coalescing urban settlements is noteworthy. Some of these smaller settlements are already experiencing rates of growth exceeding those of Africa's largest urban centers. They provide evidence of the emergence and consolidation of an urban hierarchy that is part and parcel of the urban transformation and its extensive reach.

Mining towns during their early stages of formation they typically experience exceptionally rapid rates of growth. Eyewitness accounts detail astounding numbers amassing at gold rush sites in a short period of time; later, town consolidation produces outmigration and triggers different in-migration streams (Bryceson 2011). Typically, this rapid frontier expansion occurs before accurate numbers can be captured by an official population census or the urban dynamics can be assessed. Mobile phone communications and artisanal miners' freedom of movement within mining frontiers can result in thousands arriving at a mineral strike site in a matter of months. (Johannesburg's famous gold rush in 1886 triggered a migration explosion that within three years made it the largest settlement in South Africa. By the start of the 20th century, Johannesburg's population climbed to 100,000, a stealth-mining period that drove migration into the city and its expansion.)

Sharp upward growth rates can also happen in non-mining towns. Market towns, for example, can move along spectacular trajectories of demographic expansion and economic growth. One such town is Katoro, a market town deep within Africa's populous and ethnically diverse Great Lakes region in northwest Tanzania. Its emergence has been stunning. Katoro's urban birth was facilitated greatly by its strategic location within a mining frontier on the road leading from the Geita gold mines in the vicinity of international borders with Uganda, Rwanda, Burundi, and the Democratic Republic of Congo. Its comparatively tranquil location near a regional hotspot of civil strife and instability has served as a major advantage, especially in the cross-border trade of foodstuffs. Severe droughts across the region compelled several governments, including that of Tanzania, to ban food exports in 2011 to increase national stocks, although indirectly this has resulted in an expansion of lucrative informal cross-border trade. Furthermore, the town's location within the Geita regional district but bordering the Biharamulo regional district spurred interregional investments that concentrated in Katoro. Biharamulo's restrictions on building in the local mining town of Matabe because of its location within a forest preserve resulted in a boom in Katoro as cash-flush miners sought easy, local investment outlets that would circumvent Biharamulo's taxes.

Within the past 25 years, Katoro evolved in this context from a village into a regional trading node, exhibiting higher-order urban functions at present. It is part of the rapidly expanding contiguous twin settlement of Katoro-Buseresere (by 2002 this settlement already contained 30,477 people). In Tanzania, a population of 15,000 generally constitutes a sufficient threshold to catalyze daily market exchange and cosmopolitan social interaction beyond familial ties. This urban site witnessed the proliferation of a division of labor among residents and also between the residents and economic agents from the world beyond. Katoro became the site of a bustling regional and international market (peaking on Saturdays), and the market was integrated into the regional capital of Mwanza, where traders converge on Thursday to procure stock for Katoro's Saturday market. On market day the urban settlement bustles as itinerant traders converge, along with visiting miners flush with cash. The buccaneering spirit of some of the initial investment activities (alcohol, prostitution, tax evasion, minerals, and cross-border exchanges) was subsequently accompanied by government expenditures on schools, clinics, etc., private investment in permanent residential and commercial buildings, and congregational investments in places of worship.

Katoro's urban evolution reflects the conscious choices to migrate, to reside in a specific place, to engage in economic activities, and to consume goods and services. The overall trend is movement away from rural livelihoods based on direct production toward diversification and fostering of commercial linkages with areas beyond.

world region, and this youth bulge will endure for many years.

Africa may benefit from its growing population in what demographers and economists term the "demographic dividend." With fertility rates dropping and family size shrinking, the share of working-age adults in a country rises, and this has a positive impact on economic growth. With the median age rising, today's mass of young people is expected to move into their productive work years (contingent on their securing jobs). This optimistic view contends that with more people working, more money circulates and economies prosper. This demographic dividend, crucial to the growth of Asian economies a generation ago, is expected to offer large opportunities to African economies in the next decades. It remains to be determined whether Africa can realize a happy demographic dividend: the downside of the population explosion is that there will be more people to feed, house, educate, care for medically, and socialize. Some demographers are neutral or even skeptical about the effects of population change on Africa's economic growth potential (drops in fertility rates are slow and uneven across Africa, and there is considerable cultural and political resistance to family planning). Larger populations are simply more costly and require substantial infrastructure improvements: more and better-equipped schools, roads, runways and railway lines, deeper ports, larger airports, enlarged sanitation and water systems and broadband connections. However, sustaining FDI and economic growth rates, along with continual improvements in health, education, and food productivity will be essential to ensure that development "works" in Africa, particularly with regard to efficiency, poverty reduction and gender equity.

The "big five" management consultancy firms are now jostling for African business with upbeat regional assessments. McKinsey's 2010 report, "Lions on the Move: The Progress and Potential of African Economies," has garnered much attention with its rigor. Ernst & Young's (2011) "It's Time for Africa" report adds to the chorus that Africa is now the second most attractive region (after Asia) for international investors. Citing an African proverb that "the best time to plant a tree is twenty years ago. The second best time is now," Ernst & Young is striving to drum up business for its management consultants. Different management consultancy firms highlight different opportunities: Accenture emphasizes the edge that South African businesses have due to their familiarity with the "Africa terrain" (institutional and cultural contexts) and so are best positioned to serve as a bridge to the rest of Africa; Boston Consulting (2011) lauds present-day "African capitalism," highlighting 40 companies that have emerged with the potential to become "players" beyond the region; and the Monitor Group (2011) profiles 439 ventures that have been able to grow while simultaneously alleviating poverty. The important point is that the notion that African firms cannot compete is wholly outdated—not to mention condescending. There are plenty of examples of African firms that have become very successful in international markets in the last decade (e.g., cut flower exporters in Kenya and mobile phone providers that operate throughout the region) (Bain & Company 2011).

Africans themselves are leading the growth in investments across the region. This optimism is underlined by a 21% compound growth rate in Africans investing in African countries since 2003, and their investments are flowing into a range of sectors. Also, governments are striking deals in which buyers make upfront payments, commit to make infrastructure improvements, and share management skills and technology, demonstrating that the investment model has evolved from the colonial extraction model. Some oil and gas producers in Africa (Nigeria, Ghana, Angola, and Gabon) are creating new "local content laws;" specifics vary by country, but they typically require quotas for local employment and managerial staff and specify that goods and services such as equipment, information technology (IT), and insurance be procured locally.

McKinsey's "Lions on the Move" report compares Africa's business environment today with that of the future (Table 1.1).

Pieterse and Parnell (2014:14) underscore that "despite the crude development policy thinking that shapes many of these glossy reports (with some notable exceptions), it seems that the new private sector actors have a real impact on the policy landscape, outstripping the influence of scholars, civil society pressure groups and the old-style development industry."

**TABLE 1.1 AFRICA'S BUSINESS ENVIRONMENT**

| *Africa Today* |
|---|
| US$1.6 trillion (collective GDP) |
| US$860 billion (consumer spending 2008) |
| 60% (share of uncultivated arable land) |
| 52 cities with populations over 1 million |
| *Africa Tomorrow* |
| US$2.6 trillion (collective GDP 2020) |
| US$1.4 trillion (consumer spending 2020) |
| 1.1 billion workers (2040) |
| 50% living in cities (2030) |

*Source:* McKinsey (2010).

The "Lions on the Move" scenario is based on a great deal of conjecture, but the key issue is this: Where are "the lions" and "the cheetahs" moving? Will it be to more unequal societies where only the top echelons of society reap the benefits, or to societies with a sizeable middle class? Will development benefit the poor? On macroeconomic evidence alone, Africa is growing, but there are rifts in the "Africa rising" narrative. A narrow elite is benefiting disproportionately and the poor and unemployed are being left behind. Current growth is neither inclusive nor democratic: unregulated, informal economic activities are very common in urban Africa, both in terms of the numbers of informally employed and in terms of the goods and services provided by the informal economy. Regional growth figures focus on the most advanced economies, such as South Africa, and on oil exporters that are diversifying (Angola and Nigeria), as well as on a handful of economies that are consolidating recent gains (Ghana, Kenya, Senegal). However, discussions of the region mask individual economies that are lagging (South Africa, Sierra Leone, Mali, Madagascar, Sudan, South Sudan, etc.).

Afrobarometer opinion polls (based on surveys of 51,605 respondents in 34 countries) reveal that one in five Africans lack food, health care, and clean water. Three quarters of Africans believe their governments are not doing enough in reducing income inequalities (Afrobarometer 2013). Africans are taking to the streets to air their social and political grievances. Women traders in Nigeria have protested tax increases; farmers and miners in South Africa have demonstrated against low wages; Ugandans have demonstrated against poor road conditions. Some urban demonstrations have been large and widespread, taking on multiple demands (e.g., the July 2013 protests in Burkina Faso against high prices, low wages, poor health and education services, and corruption). Harsch (2013) tallies 3,000 protests across Africa in the first six months of 2013 and protests about lack of job creation are becoming more frequent. Social protests are symptomatic of both the new democratic environment and an undercurrent of widespread social grievances. There is even some speculation that "an African spring is in the making" (Harsch 2012).

Africa is a net importer of food, and there will be more mouths to be feed in the future. In terms of food production, Africa produces less food per person now than it did in the 1960s as agriculture concentrated on export cash crops. Increasing agricultural production is a long-term worry for Africa. Conflict and climate change could quickly derail economic growth. Africa has a bulging youth population that will pose a serious economic challenge. An abundance of youth can be like gearing on a balance sheet: it makes good situations better and bad situations worse. By 2040, Africa's working population will be 1.1 billion (more than China or India), placing a huge burden on societies to prepare the workforce. Workers will require extensive skilling and up-skilling, and educational systems are currently unprepared. Education and skill development will have to be much improved: African businesspeople routinely complain about the shortage of skilled employees, and without better skills African workers cannot emulate their Asian counterparts. It is far from certain that Africa's growth trajectory can be realized.

## AFRICA AS A MILLENNIUM DEVELOPMENT FRONTIER

Africa has always been a key region to the development complex (the array of academics, practitioners, policymakers, civil society, and other humanitarian organizations and to a lesser extent businesses), aiming to make a difference and promote various paths toward development. Legacies of colonialism and persistent poverty, poor health, economic and infrastructure deficits, unemployment, and inequality coupled with population growth make a compelling case that Africa is in need of development.

In the 20th century, "big push" Western ideas dominated African development policy. In the 1950s and 1960s the modernization paradigm promoted urbanization and industrialization and Western technology and values to transform society so that economic growth could "take off" and lead to development and eventually achieve "a level of high mass-consumption" (Rostow 1960). However, the model, based on the historical trajectory of the United States after its independence, could not be replicated in a single African country. Its grand failure to deliver development lead to major pullbacks in the 1970s (e.g., Tanzania, Ghana, Zambia and

Ethiopia). Following the writings of dependency theorists (e.g., Wallerstein 1974) and a tilt toward socialism, many African leaders questioned the relevance of the Western development model in Africa and saw it as perpetuating a dependent relationship where Africa was overly reliant on the West, or neo-colonialism.

In the 1980s and 1990s the external big push returned with structural adjustment policies and liberalization policies mandated by the World Bank and the IMF following economic crises, balance-of-payment deficits, insolvency, and widespread conditions of underdevelopment. Structural adjustment policies opened up African markets to global competition and global investors took center stage. These policies were very transformative in ushering in market forces and accelerated urbanization, but they also inflicted a lot of suffering on populations and were very unpopular.

In the early 21st century development policies started to focus on alleviating poverty. The UN was a driving force in developing the Millennium Development Goals (MDGs) in 2000. The MDGs comprise eight international development goals that have been approved by all 189 members of the UN and most international organizations and NGOs. The internationally agreed-on frameworks represent the broadest consensus on development priorities ever achieved, and they are inspiring efforts to improve the lives of poor people around the world (Fig. 1.8).

The MDGs will expire in 2015, and the UN and the international community are launching Sustainable Development Goals to follow on from the MDGs. Although the MDGs have not been fully achieved, many African states have registered successes on many of the targets (e.g., primary school enrollments and maternal care). A renewed commitment is necessary to continue along this development track. However, a central difference this time around is that when the Sustainable Development Goals are implemented, all countries will be pioneers, not just Africa and the developing world (as was the case in the MDGs). Sustainable development will be the next development frontier, and low-carbon technologies, greener living, eco-friendly buildings, designs, and communities, protection of biodiversity, and adaptation to climate change will be crucial elements. These new targets will require an overhaul of existing energy, industrial, urban, transport, food, and natural resource systems. Local solutions to local problems may be designed that will be able to draw on global technology and expertise. The increasingly digital world enables African entrepreneurs and development practitioners to connect in new ways. This may open up possibilities for contributions to stem from ordinary Africans, who no longer have to wait for help to trickle down from the center and to be diffused in their direction.

The resilience and creativity of many Africans to devise affordable solutions to problems (doing more with less) may bode well for the global sustainable-development tilt. Olopade (2014:1) emphasizes that "Africa is a bright continent" where solution-oriented ordinary people are taking the reins of development and making empowered contributions to solve immediate development challenges. The creativity of Africans is expressed in a flurry of recent innovations and adaptations, which will be discussed in detail in subsequent chapters—for example, the Ushahidi platform developed in Kenya to monitor elections; the formal Dahabshill and the informal Hawala systems of money transfers for remitting money between the members of the Somali diaspora and relatives back home; the uptake in mobile phone technologies for delivering various content (e.g., education, health, market prices); "green shacks" (incorporating solar power and other green technologies) and the incorporation of agriculture into urban living. The sustainable-development frontier may allow solutions to flow from Africa to other world regions.

## URBANIZATION IN AFRICA IN CONSIDERATION

The scale and velocity of Africa's urbanization dwarf that which occurred in the Global North in an earlier time. For example, London in 1910 was seven times larger than it had been in 1800; Kinshasa and Lagos are now 40 times bigger than they were in 1950. While cities of the Global North are all too often seen as paradigms for understanding urbanization everywhere in the world, urbanization in the Global South is playing out very differently: populations are larger, timeframes are shorter, and modes of integration into the global economy differ. A major underlying difference is that

## The Millennium Development Goals

In September of the year 2000, leaders of 189 countries met at the United Nations in New York and endorsed the Millennium Declaration, a commitment to work together to build a safer, more prosperous and equitable world. The Declaration was translated into a roadmap setting out eight time-bound and measurable goals to be reached **by 2015**, known as the Millennium Development Goals, namely:

**1. Eradicate extreme poverty and hunger**
- Reduce by half the proportion of people whose income is less than $1 a day
- Achieve full and productive employment and decent work for all, including women and young people
- Reduce by half the proportion of people who suffer from hunger

**2. Achieve universal primary education**
- Ensure that all boys and girls complete a full course of primary schooling

**3. Promote gender equality and empower women**
- Eliminate gender disparity in primary and secondary education preferably by 2005, and in all levels of education no later than 2015

**4. Reduce child mortality**
- Reduce by two thirds the mortality of children under five

**5. Improve maternal health**
- Reduce maternal mortality by three quarters
- Achieve universal access to reproductive health

**6. Combat HIV/AIDS, malaria and other diseases**
- Halt and reverse the spread of HIV/AIDS
- Achieve, by 2010, universal access to treatment for HIV/AIDS for all those who need it
- Halt and reverse the incidence of malaria and other major diseases

**7. Ensure environmental sustainability**
- Integrate principles of sustainable development into country policies and programmes; reverse the loss of environmental resources
- Reduce biodiversity loss, achieving, by 2010, a significant reduction in the rate of loss
- Halve the proportion of people without access to safe drinking water and basic sanitation
- Improve the lives of at least 100 million slum dwellers by 2020

**8. Develop a global partnership for development**
- Develop further an open, rule-based, predictable, non-discriminatory trading and financial system
- Address special needs of the least developed countries, landlocked countries and small island developing States
- Deal comprehensively with developing countries' debt
- In cooperation with pharmaceutical companies, provide access to affordable essential drugs in developing countries
- In cooperation with the private sector, make available the benefits of new technologies, especially information and communications technologies

**For more information**, please visit: www.un.org/millenniumgoals

Issued by the UN Department of Public Information

**FIGURE 1.8** The Millennium Development Goals (MDGs).

urbanization in the Global North was triggered by increases in agricultural productivity; this has not happened to date in Africa. It is no longer possible to defend the position that cities in Africa will evolve according to Eurocentric "norms" or that the urban form in Africa is a consequence of the all-encompassing influences of the dominant global powers (the colonial urban anchor and the ex-colonial city as central engines in the domestic economy). Macro-thinkers who conceptualize late 20th- and early 21st-century urbanization as a diffusion of urbanism from a European center outward and those who sketch urban histories in terms of an all-determining system of global capital exploitation miss too many of the details on the ground in urban Africa.

Until recently, urbanization in Africa was conceptualized as "abnormal" (Obeng-Odoom 2010). Three arguments provided the intellectual scaffolding for this widely held viewpoint. First, the dependency claim is that Africa's urbanization is atypical because it is based on colonialism, which structured a particular type of incorporation into the international system that was particularly advantageous to external powers. Second, the over-urbanization argument posits that capitalism's deep penetration in the countryside pushes people off the land and into cities because conditions are worse in rural environments. Third, the urban bias claim is that urban development is the result of targeting spending on inefficient pro-urban public policy that detracts from resource investment in the countryside. The hegemony of the "abnormal" thesis had profound implications. Hostile perspectives toward African cities were common. Development organizations and many African governments developed policies with an eye to reducing the rate of urbanization and slowing and/or reversing rural migration. Many assumed that urbanization and economic development could only be negatively correlated in African contexts.

The "abnormal" thesis is increasingly being challenged. First, recent growth figures provide evidence that in some African countries urbanization is not decoupled from economic growth. Second, GDP values do not account for the large size of African informal economies: informal economic activities can account for one half added to GDP and growth rates in specific informal economic activities (e.g., processing of e-waste and retailing cellphone time) and may even surpass the top economic performances in the world (rapidly expanding informal sectors may grow at 12% to 20% per annum). Third, most official data-collection efforts in Africa (as elsewhere) focus on national-level rather than city-level data. Therefore, there is much conjecture in relating national data to urban trends and in apportioning (with any accuracy) an individual city's role within the national experience. In terms of the availability of city-level data to assess urbanization processes accurately, we are in "the twilight zone" of statistical accuracy. Turok (2013) emphasizes the dearth of econometric studies that could elucidate the complex relationship between urbanization and economic development in Africa. Fourth, if we operationalize development with an alternative methodological lens, such as the Human Development Index (HDI), a different picture emerges. HDI measures the overall progress of a country along three dimensions: health, knowledge, and a decent standard of living. Urbanization and HDI are positively linked in Africa.

Based on this evidence, the ground is shifting under the "abnormal" thesis. For instance, the World Bank (2000, 2009) switched its prognosis of African urbanization without industrialization to urbanization and growth in services bypassing the development of manufacturing. Nevertheless, on balance, there is a great diversity of urban experiences throughout Africa. A recognition and empirical examination of these experiences unsettle the notion of African urban exceptionalism.

Africa's urbanization is not accidental. If urbanization in Africa is not comparable to that in the West, are there similarities and differences between African cities and other cities in the Global South? There is evidence of similarities, but important differences remain. First, Africa's infrastructure lags behind its peers in the developing world. There are critical differences in terms of missing regional links in infrastructure within and between among African countries. Africa has more landlocked countries than any other of the world regions: 15 countries are without sea access and are only weakly linked in regional air hubs and trans-African road networks. The fragmentary regional African infrastructure further isolates many small countries, hindering urban centers from harnessing efficient large-scale technologies and participating more successfully in the global economy. To put this in perspective, Africa's road density network per 386.1 miles$^2$ (1,000 km$^2$) is five times less than that of BRIC countries. Overall, Africa has a huge infrastructure deficit, and 30% of its existing infrastructure is in dire need of rehabilitation. Its annual per capita expenditure on urban infrastructure is low compared to the BRIC countries (e.g., US$34 compared to US$116 in China). Power is problematic, and 30 African countries experience intermittent power cuts; rolling blackouts to avoid widespread power disruptions occurs even in South Africa, the country with the best infrastructure. Poor infrastructure restrains Africa's GDP growth by 2% per year. The African Development Bank emphasizes that it would take US$93 billion of spending until 2020 to rehabilitate Africa's infrastructure. Current average logistics costs in Africa are twice that in the BRIC countries.

Second, Africa lacks a manufacturing tradition comparable to its peers. Africa's industrialization, fostered by nationalist government policies in the aftermath of independence, never took off but rather went into reverse in the context of the post-1985 economic liberalization environment. Export-processing zones (free trade areas that offer special concessions for international firms to locate there) based on assembly jobs and call centers have by and large been unsuccessful (although Mauritius and Madagascar have used these zones successfully). Low FDI until recently meant that urban authorities could not finance urban infrastructure improvements, so the infrastructure catch-up predicament further deterred investors.

There is evidence of a broad convergence of general urban characteristics as well as drivers of urbanization. First, there is the phenomenal growth of slums in Africa, both in terms of absolute numbers (166 + million) and as a proportion of the total urban population (62%), now surpassing the total slum population in Latin American and the Caribbean. Africa's cities have in common with India and Brazil that they house extreme poverty. Some commentators even speculate about the emergence of mega-slums (slums that house 1 million people) in Lagos and Kinshasa. Second, urban corridors or city-regions are emerging in Africa (e.g., the Gauteng city-region centered on Johannesburg and the Greater Ibadan-Lagos-Accra [GILA] corridor in West Africa) that are similar to the city-regions that have been identified in China's Beijing–Tianjin corridor and São Paulo in Brazil. Third, there is a blurring of the distinction between city and countryside around many of the urban centers in Africa (akin to many Asian cities). Heterogeneous mixes characterize this hermaphroditic landscape: rural and urban features coexist in environmental, socioeconomic, and institutional terms. Fourth, Africa's concentration on resource exploitation is driving much of the urban expansion, and urban growth is not based on diversified economies. Clearly, there is a need to add to the value of natural resources before they are exported (Botswana is leading the way in adding value to its diamonds) and to diversify into manufacturing and its more technologically advanced components as well as a range of financial and producer services (e.g., commercial banking, engineering, marketing and product design) (Turok 2013).

Drivers in addition to regular employment migration streams are evident. For one, climate change and environmental degradation are contributing to the movement of people to cities. There are no reliable figures on these movements, but urban growth has paralleled a threefold increase in the number of storms, droughts, and floods over the past 30 years. Gradual climate changes appear to have a greater impact on the movement of people than extreme events. The International Organization for Migration estimates that the number of environmental migrants will climb to 200 million by 2050. Also, conflict and war accelerate urbanization as people flee the violence in their home area to seek refuge in the city. In receiving cities outside of migrants' home states, displaced people keep a low profile, avoiding registration, enumeration, and profiling exercises. Those who are displaced but remain in their home state can add to the urban footprint when displacement camps on the outskirts of cities become integrated socially and economically into cities (e.g., one third of Darfur's population is located in camps on the perimeter of cities). Overall, relationships between displacement and urbanization as well as climate and environmental change are topics that urban planners, demographers, and development specialists have not addressed to a significant degree.

## CONCLUSIONS

Africa is still perceived far too negatively in most of the world. The region remains in an intellectual limbo compared to other world regions. Africa is no longer the fabled, deeply troubled, conflict-ridden, "dark continent," but the amazing distance between Westerners and Africans remains a chronological gulf. In most representations in the West, Africans remain confined to a past era and are mislaid in the contemporary era. Westerners' knowledge about Africa is laden with stereotypes, decontextualized images, and sound-bites about particular places and particular people: it is as shallow as it is incomplete. Only recently has the West's knowledge of Africa begun to improve.

Africa is rising in terms of many economic criteria, both absolutely and relatively. Many African economies are at an inflection point, and this is the basis for tempered optimism but not premature exuberance. There is no denying that many risks loom (the global financial situation, climate change, national political change, geopolitical stability, food security, etc.). The current slowdown of economic growth in the BRIC countries may hurt African economies, especially China's apparent transition from a heavy emphasis on resource investment toward domestic consumption. Perhaps only the story is changing: the fundamental structure of relations remains intact. Simply put, improved economic performance may be due more to a new scramble for Africa than anything else. Even Africa opportunity boosters acknowledge the lack of a spectacular regional role model of economic success comparable to the Singapore model, which helped power the Asian surge. South Africa, the most sophisticated economy in Africa and the engine that powers the southern African region, still faces considerable challenges in the years ahead both politically and economically (specifically with regard to its high unemployment level).

Putting Africa into perspective is still fraught with problems. Models used to represent, analyze, and plan in the Global North have little relevance to urban and rural Africa. African economies are characterized by very different articulations with the global economic system and substantial informality. The countries of the region are engaging with a wider range of powers today than at any point in the past. China, Brazil, India, the Gulf states, Malaysia, and a range of other corporate financial players worldwide, as well as traditional partners from Europe and the United States. are jostling for ties with the region, providing considerable room for African agency and maneuverability.

In a radical departure from earlier thinking, urbanist Michael Cohen (1996) has proposed the urban convergence hypothesis, postulating that cities in the Global North are converging toward cities of the Global South in the contemporary age of globalization (growing unemployment, declining infrastructure, deteriorating environment, collapsing social compact, and increasing institutional weakness; we can also add that growing informal economic activities in all urban environments are features of 21st-century urbanism). Obviously it is difficult to predict what will happen in the next half-century, but this kind of thinking goes hand in hand with urban Africa theorists' arguments that urban theory needs to grow from looking through the prism of urban Africa. Understanding African cities as part of the Global South experience is a framing that is more relevant than ever to understanding 21st-century urbanism as opposed to previous prisms that better explained Europe's urbanization in the 19th century and that of the United States in the 20th century.

The urban change unfolding in Africa is drawing the region closer to the forces that shape many cities outside of the region. Urbanization in Africa is normalizing in the sense that it is diversifying away from the colonial spatial imprint that was firmly centered on national cities. Now, more than ever, the factors that draw cities in Africa into the global capitalist economy are also shaped by local agents, forces, and circumstances. At the same time sparks of a rural revolution are discernible. Still, many in the Global North continue to understand Africa in an outdated and partial way—as an inexpensive source for raw materials and/or as a rescue mission for their pop humanitarian impulses. Being caught in a loop between these positions has prevented the West from expanding its thinking about Africa, and it prevents a balance of stories about Africa and Africans.

**REFERENCES**


Afrobarometer. 2013. "Public Attitudes on Democracy and Good Governance in Africa." Afrobarometer Round 5, 2011/2012. Available at http://www.afrobarometer-online-analysis.com/aj/AJBrowserAB.jsp (accessed November 2, 2013).

Bain & Company. 2011. "Investing in Africa and Succeeding." Available at http://multimedia.avusa.co.za//view_video.php?viewkey=180214c23b9e55da4f7.

Boston Consulting. 2011. "The African Challengers. Global Competitors Emerge from the Overlooked Continent." Available at http://www.bcg.com/documents/file44610.pdf

Bryceson, D. 2011. "Birth of a market town in Tanzania: Towards narrative studies of urban Africa." *Journal of Eastern African Studies* 5(2):274–293.

Cohen, M. 1996. "The hypothesis of urban convergence: are cities in the North and South becoming more alike in the globalization era?" In *Preparing for the Urban Future: Global Pressures and Local Forces*, eds. M. Cohen, B. Ruble, J. Tulchin, and A. Garland, pp. 23–38. Baltimore: The John Hopkins University Press.

Cole, R., and H. DeBlij 2007. *Survey of Sub-Saharan Africa: A Regional Geography*. New York: Oxford University Press.

De Blij, H. 2012. *Why Geography Matters*, 2nd ed. New York: Oxford University Press.

De Blij, H., P. Muller, and J. Nijman. 2014. *Geography. Realms, Regions and Concepts*, 16th ed. New York: John Wiley.

Economist Intelligence Unit (EIU). 2012. *Africa Cities Rising. Forecasting Data and Analysis from the EIU*. London: EIU.

Ernst & Young. 2011. "It's Time for Africa. Ernst and Young's 2011 Africa Attractiveness Survey." Available at http://www.ey.com/Publication/vwLUAssets/2011_Africa_Attractiveness_Survey/$FILE/11EDA187_attractiveness_africa_low_resolution_final.pd

Fanon, F. 1966. *Wretched of the Earth*. English trans. New York: Grove Press.

Goldman, M. 2011. "Strangers in their own land: Maasai and wildlife conservation in Northern Tanzania." *Conservation & Society* 9(1):65–79.

Harsch, E. 2012. "An African spring in the making. Protest and Voice across a continent". *Whitehead Journal of Diplomacy & International Relations* 13(1): 45–61.

Harsch, E. 2013. "Social protest: an African perennial." *African Futures*. Available at http://forums.ssrc.org/african-futures/2013/10/07/social-protest-an-african-perennial (accessed November 2, 2013).

iCow. 2013. "Mobile technology spreads seeds of information to farmers." Available at http://www.icow.co.ke/blog/item/22-mobile-tech-spreads-seeds-of-information-to-farmers.html (accessed August 30, 2013).

Krause, M. 2013. "The ruralization of the world." *Public Culture* 25(2):233–248.

Larson, P. 2011. "African conflict and the Murdock Map of Ethnic Boundaries." Available at http://peterslarson.com/?s=MURDOCK (accessed December 4, 2013).

Mahajan, V. 2009. *African Rising. How 900 Million African Consumers Offer More Than You Think*. New Jersey: Wharton School Publishing.

McKinsey. 2010. "Lions on the Move: The Progress and Potential of African Economies." Available at http://www.mckinsey.com/mgi/publications/progress_and_potential_of_african_economies/index.asp.

Monitor Group. 2011. "Report on Market-based solutions to Poverty in Africa from the Monitor Group. "Available at http://www.monitor.com/Portals/0/MonitorContent/imported/MonitorUnitedStates/Articles/PDFs/Monitor_Promise_and_Progress_Exec_Summary_May_24_2011.pdf

Murdock, G. 1959. *Africa: Its Peoples and Their Culture Histories*. New York: McGraw-Hill.

Myers, G. 2011. *African Cities. Alternative Visions of Urban Theory and Practice*. New York: Zed Press.

Nicol, A. 1950. "The Meaning of Africa." Available at http://afrilingual.wordpress.com/2011/08/18/the-meaning-of-africa-–-abioseh-nicol/

Obeng-Odoom, F. 2010. "'Abnormal' urbanization in Africa: A dissenting view." *African Geographical Review* 29(2):13–40.

Olopade, O. 2014. *The Bright Continent: Breaking Rules & Making Change in Modern Africa*. New York: Houghton Mifflin Harcourt.

Parnell, S., and E. Pieterse, eds. 2014. *Africa's Urban Revolution.* New York: Zed Books.

Pieterse, E., and S. Parnell. 2014. "Africa's urban revolution in context." In *Africa's Urban Revolution,* ed. S. Parnell and E. Pieterse, pp. 1–17. New York: Zed Books.

Radelet, S. 2010. *Emerging Africa: How 17 Countries Are Leading the Way.* Baltimore: Brookings Institution Press.

Rostow, W. 1960. *The Stages of Economic Growth: A Non-Communist Manifesto.* Cambridge: Cambridge University Press.

Rotberg, R. 2013. *Africa Emerges: Consummate Challenges, Abundant Opportunities.* Cambridge: Polity Press.

Rothberg, R., and J. Aker 2013. "Mobile phones: Uplifting weak and failed states." *Washington Quarterly* 36(1):111–125.

Sartre, J.-P. 1966. "Preface." In *Wretched of the Earth.* English trans, F. Fanon, pp. 7–31. New York: Grove Press.

Soyinka, W. 2012. *Of Africa.* New Haven: Yale University Press.

Turok, I. 2013. "Securing the resurgence of African cities." *Local Economy* 28(2):142–157.

UN-HABITAT/ 2010. *The State of African Cities 2010. Governance, Inequality and Urban Land Markets.* Nairobi: UN-HABITAT.

Wainaina, B. 2005. How to write about Africa. *Granta* 92 (Winter):93–95. Available at http://www.granta.com/Magazine/92/How-to-Write-about-Africa/Page-1 (accessed December 4, 2013).

Wallerstein, I. 1974. "Dependence in an Interdependent World: The Limited Possibilities of Transformation within a Capitalist World Economy." *African Studies Review* 17(1):1–26.

Wired (2014). Facebook wants to connect rural Africa using drones. Available at http://www.wired.co.uk/news/archive/2014-03/04/facebook-drones (accessed March 6, 2014).

World Bank. 2000. *World Development Report.* Washington, D.C.: World Bank.

World Bank. 2009. *World Development Report.* Washington, D.C.: World Bank.

## WEBSITES

BBC. 2011. African Performances 2011: Silhouettes by Africa Ukoh. http://www.bbc.co.uk/worldservice/africa/2011/09/110922_african_performance_2011_second_prize_silhouettes.shtml) (accessed December 4, 2013). Play about a Nigerian actor who moves to Hollywood in the hope of making it big in the movie business.

Chimamanda Adichie. 2009. The danger of s single story. http://www.ted.com/talks/chimamanda_adichie_the_danger_of_a_single_story.html (accessed December 4, 2013). Nigerian author's commentary on Western representations of Africa and Africans.

George Ayittey. 2007. "Africa's cheetahs versus hippos. "Available at http://www.ted.com/talks/george_ayittey_on_cheetahs_vs_hippos.html. (Accessed December 8, 2013). Ghanaian economist on Africa's cheetahs versus hippos.

Institute of Development Studies. 2012. http://www.ids.ac.uk/news/ids-film-examines-how-british-media-portray-global-south?em=NE) (accessed December 4, 2013). Podcast provides a commentary on how the British media provide imbalanced accounts of Africa news.

Invisible Children. 2013. http://www.invisiblechildren.com (accessed December 4, 2013). Nonprofit organization that achieved notoriety with its media campaign to capture the African warlord Joseph Kony.

iCow. 2013. http://www.icow.co.ke (accessed December 4, 2013). This application provides farmers with agricultural information to help increase their productivity.

Maasai Intellectual Property Initiative (MIPI). 2013. http://maasaiip.org/ (accessed December 4, 2013). Nonprofit organization that works to protect the intellectual property of the Maasai people.

CHAPTER 2

# REFRAMING AND REREPRESENTING AFRICAN AFFAIRS

## INTRODUCTION

The myth of Africa as a "dark continent," a mysterious place populated by deprived and depraved African natives, is deeply embedded in Western consciousness. European writers in the 19th century projected this fictional Africa representation. The people of Africa were characterized by Westerns as deprived, lacking Western civilization, education, culture, religions, industry and progress. This was accompanied by advancing various myths of savagery and chaos in the region. Joseph Conrad's *Heart of Darkness* (an 1899 serialized magazine article that was later published in book format and regularly makes lists of the top 100 books of the 20th century) is a good example of a fictional text in this genre (Fig. 2.1). *Heart of Darkness* is a controversial work of fiction (heavily critiqued as a racist), and it is regularly reprinted; the latest edition is 2014. Thoughtful scholars have helped correct some myths about Africa and Africans, but misrepresentation and unfair negative representation have endured. Indeed, the darkness metaphor may teach us a lot more about the political development of Western modernity than it does about Africa (Popke 2001).

Africa is an inconvenient region for generalization, comprising a multiplicity of states, more than 1,000 spoken languages, and thousands of cultural groups. Lumping together such diversity and homogenizing African contexts and/or selectively picking cases is problematic. Too often Africa is portrayed in general, with a lack of distinction between different countries and peoples, their paths and options, their different histories, and without understanding the whys and wherefores of current situations.

This chapter outlines the legacy of the darkness representation and stereotyped images of Africa that are continually peddled by the media and some self-interested communities. The crisis and chaotic Africa narrative is challenged by Africa evidence indicating a changing region and a movement out of the shadows, in different and often unanticipated directions, that diverge greatly from what it is expected in negative depictions of the region. The Africa growth narrative (reviewed in Chapter 1) coincided with the explosion of Africa interest and the multiplicity of resources on the Internet. As Africa's economies boom so does Africa knowledge, and the latter is becoming more diverse and democratic. Africa's current and future transformation is the subject of deep reflection and debate. In this chapter seven perspectives of Africans' revisioning Africa are reviewed to illustrate how Global Northern frameworks can be rethought. The metaphor and organization of this chapter is Africa moving from darkness into shadows and then into splintering light.

## REPRESENTING AFRICA: THE DARKNESS LEGACY

In the 21st century it needs to be underscored that Europe is not represented by the humanitarian tragedy that took place in Bosnia; Myanmar/Burma is not

**FIGURE 2.1** Front Cover of Joseph's Conrad's *Heart of Darkness*, 2014 Edition.

representative of the Asian investment environment; and Venezuela is not considered illustrative of Latin America. Yet, when it comes to Africa, extreme cases (e.g., evidence of kleptocrats in Nigeria and Guinea-Bissau, humanitarian tragedies in Somalia and Ethiopia, and pirates and terrorists in the Horn of Africa and Nigeria) are taken to represent all of Africa in every context. Such overgeneralization is barely tolerated in other world regions. It is, however, totally mistaken to argue that extreme suffering, bad governments and failing states do not exist. Unfortunately, there is abundant evidence supporting many negative characterizations in Africa, but normative standards ("proper states," "good economies," "transparent regimes," and "responsible leadership") may also be unfair benchmarks: all societies of the world are imperfect under the microscope.

A failure to capture Africa's heterogeneity is a central problem in understanding Africa. Because Africa's diversity is rarely captured in media reporting, major information gaps hinder public knowledge. The international media fail to provide balanced and comprehensive Africa coverage; many places are even ignored. For instance, Angola—as well as its capital and largest city, Luanda—were hardly reported on in 2011. Angola is overlooked by two of the largest media corporations that report on Africa (BBC and Al-Jazeera). Both the London and Qatari corporations maintain Johannesburg offices, but no one in Al-Jazeera's Johannesburg bureau speaks Portuguese, let alone an indigenous Angolan language. The BBC is also short-staffed in Portuguese language speakers, a situation only compounded by the BBC budget cutbacks that ended its Portuguese for Africa service in 2011. As a consequence, the media are unsure of the changing Angolan narrative, which further undermines responsible coverage. Media and programming space requires financing, and most media outlets are not willing to prioritize funds to make Africa coverage more complete. The general public can understand a country and a city more fully only when provided with more complex and subtle reporting by well-informed journalists and scholars who spend considerable time in the field.

Accessibility to African knowledge is restricted to an elite (scholars, policymakers, practitioners). The knowledge gap between experts and the general public is massive, and the media are culpable. Reporting on only those aspects of African affairs deemed most important to Western audiences, the media are prone to select stories according to Western values. Mainstream TV and Internet coverage is restricted to special programming on PBS, BBC, France 24, and Al-Jazeera, and to occasional news slots—for example, *60 Minutes,* with its formulaic news safaris by helicoptered reporters engaging in one-off exposés composed of the hunt, the chase, and the triumph of good over evil. *AC-360,* with acclaimed journalist Anderson Cooper, is among the worst offenders: CNN reporters relate the news abroad rather than provide actual reports on the locals and their contexts. Scholarly studies of U.S. network coverage of Africa show that the large networks focus on the major story (South Africa in the 1980s and southern Sudan in 2011 and 2014) to the detriment of broader and balanced coverage. In the 1980s, three major U.S. networks (ABC, NBC, and CBS) reported on South Africa but underreported on the region, limiting Africa beyond South Africa to a few minutes of total coverage. In 2013 the Westgate Mall tragedy in Nairobi and the rise of Al Shabaab dominated U.S. coverage of the region. Time and again the foremost international news story is taken to represent a subregion and even the entire region.

As a consequence, African successes measured according to African values never see the light of day. For example, a water pump in an urban slum or a rural area may transform community livelihoods and health, but it hardly makes international news. International press coverage marginalizes events deemed "ordinary" or uniquely African. In their defense, journalists, when allotted 800 words or less to explain a complex Africa event, find it difficult to do much more than peddle a few facts within a thin historical and geographical framework. Mall massacres, coups, wars, and other human tragedies make better copy when succinctly reported.

The public receives little news reflecting grassroots Africa. Most Africa news flows from North to North and not from Africa to North. What is often marketed as news from Africa is, for the most part, Northern reporters and/or African reporters on Northern payrolls adding to Africa's poor image. The manner in which international correspondents pitch and ultimately frame their story and secure its backing may indeed tell us more about the global news business or the inertia of ideas about Africa in the Global North than about contemporary Africa. Extending this line of reasoning, the information that is recorded about Africa is actually the return of Northern ideas to a global marketplace, dominated by the Global North.

There are groups in Africa (e.g., some reporters and photojournalists, aid officials and informal entrepreneurs) who are co-conspirators in perpetuating negative narratives depicting political turmoil, corruption, violence, poverty, and humanitarian and personal tragedy. For instance, West African Internet scammers promote these dramatic and stereotyped representations of Africans and the region. Internet fraud scammers, particularly advance fee scams, also known as "419" scams ("419" is the number of the Nigerian criminal code that deals with fraud), deploy these representations. Numerous Internet scams purport to share inheritance windfalls and/or include the lucky recipient in a lucrative business or export venture as a "business partner." African scammers have received personal gain but in process fed the unruly and chaotic narrative of Africa integrated but not abiding by any rules in their participation in the information superhighway. No doubt scammers would neither have been able to get Westerners' attention nor dupe victims with more authentic, balanced, and place-sensitive narratives.

Academics have failed to contribute more nuanced accounts to public understandings of Africa. Their glaring disconnect in the Africa conversation is indirectly abetted by their ineffectiveness and by the persuasiveness of simplistic apocalyptic narratives in the popular press such as Robert Kaplan's "The Coming Anarchy" (Box 2.1). Astonishingly, individuals working

## BOX 2.1 "THE COMING ANARCHY"

Robert D. Kaplan's essay "The Coming Anarchy," published in the *Atlantic Monthly* (1994), achieved notoriety in the Global North. The thesis was grounded on the journalist's travel from West Africa to Turkey to Asia. Written in 1994, it took up the timely topics of the end of the Cold War and the search for an alternative macro scenario for the future of the world and, to a lesser extent, the place of Africa in this future world. Kaplan is a renowned journalist who has consulted for the U.S. military, and presents himself as "a master global strategist". He captures attention because of stellar book sales, his profile in public discourses, and his ability to influence the upper echelons of policymaking. "The Coming Anarchy" was faxed to every U.S. embassy in Africa, and then-president Clinton found this article "stunning," remarking, "It makes you really imagine a future like one of those Mel Gibson Road Warrior movies" (Kagen 2000:1). However, Kaplan's numerous critics exposed him as one of America's top pundits who got it totally wrong (Bestman and Gusterson 2005).

"The Coming Anarchy" presents a terrifying and unambiguous portrayal of the contours of the post–Cold War world. The essay's primary themes are "Third World" anarchy, with West Africa as an epicenter, and the threat to (Western) readers' safety, health, and culture by an impending collapse of "Third World" countries. West Africa's implosion is offered as an over-the-edge event of the impending global future. Kaplan is fixated on a fear-provoking Africa with numerous impressionistic anecdotes of his experience in West African cities (with no details about his length of stay in days or hours [!] or about whom he spoke with and in what language), detailing corruption, slums, crime, disease, and pollution. His article recounts the children "as numerous as ants" and the "nightmarish Dickensian spectacle" of Guinea's capital, complete with garbage floating in puddles, dead rats, and "scabrous" homes "coated with black slime." He envisions infrastructure as continuing to fall apart and dangerous disease-ridden coastal trading ports becoming infectious diffusion gateways to the rest of the world. He believes that there is a "re-primitivized man" in Africa's urban slums and because of weak social bonds and unstable social systems the region is on "the verge of igniting—producing hordes of young men who turn to violent crime." He speculates that the West African countryside is draining into dense, coastal slums, and he contends that the region's rulers will be forced to reflect the values of shanty dwellers. Kaplan's future world is nothing short of apocalyptic.

However, his doom-saying is unfounded and the deformities of his thinking are sadly not unique. He gets away with it because of a deficient base of knowledge about Africa and because scholars are not fulfilling the roles of public intellectuals. Kaplan has no Africa credentials. His thesis is empirically selective in assembling facts and embellishing loose collections of myths that do not appear under the microscope. Punditry is a business where the object is to hold audiences' attention through hurling ideas, large claims, and colorful statements. "The Coming Anarchy" falls into the media trap of building on audiences' existing prejudices rather than on unsettling assumptions about the world and/or trying to present accurate knowledge.

Kaplan is a modern-day mythmaker. At best, his thesis is misleading; at worst, it promotes dangerous ideas about Africa and shapes an ill-conceived future course for U.S. foreign policy on Africa. It is evident that not much has changed since colonial narratives and the construction of an image of Africa as the repository of the West's greatest fears. The colonial image has become the media image. The coming anarchy and similar scenarios (e.g., Mike Davis' *Planet of Slums* also portrays West Africa as the largest footprint of poverty on the planet and disillusioned slum dwellers as the next big threat to global peace and stability that could be recruited by Al-Qaeda and its various African affiliates, such as Al-Qaeda in the Islamic Maghreb [AQUIM] and Al Shabaab) could become self-fulfilling unless these overgeneralized theses are scrutinized and people talk back and act to deconstruct the negative prophecies. In every instance, Africa is better off with more complex narratives and multiple scenarios and alternatives.

on the ground in Africa allowed themselves to be bystanders in broader conversations about Africa until recently. Their reluctance to speak broadly and loudly meant that broad-stroked journalistic accounts and one-size-fits-all policy papers occupied center stage. However, social media are changing the conversation, and Africans and Africa experts are using this format to contribute knowledge and to inform members of the engaged public as well as connect to a new generation of an Africa-interested public. Academics and practitioners are becoming more active in an increasingly multiphrenic environment of knowledge production, and blogs, tweets, shorter more accessible articles, and op-ed pieces are taking on entrenched myths and misunderstandings of Africa realities (see Moseley 2011; Samatar 2013).

Presenting a picture of Africa's darkness as the only reality is as inappropriate today as it has always been (and even more so, given past mischief). More thoughtful, balanced assessments are needed so Africa can be rendered less dark in the eyes of the world. James Ferguson, a prominent Africanist (a term used to describe scholars engaging in primary research on Africa) and anthropologist, put forward a "global shadows" (2007) metaphor as an alternative framing of contemporary Africa. Although the shadows metaphor is abstract, it sheds nuanced light on Africa in four different ways. (1) African aspirations of development and modernity have always been shadowed by questions about the authenticity of the copy and whether the duplicate is too different from the original (a faux copy) or not different enough (merely an empty derivative). (2) The bond and relationship producing the shadow are inseparable from that to which it is bound: the Global North (this bond is much more than an empty space). (3) Shadow relations prevail within Africa states (a shadow/informal economy, a shadow state occupied by clandestine as well as civil society networks, and shadow soldiers existing alongside official armies). (4) The shadows of Africans in the world involve extensive African mobilities of peoples, monies, and commodities (some legal, some illegal). African governments' desire and aspirations to move out of the shadows toward full membership and full inclusion in the world is a long-term goal. Recognizing the ambiguity of the generalized construct that is "Africa," the idea of Africa is still very consequential and it informs global policy, geopolitical events, interregional interactions, and visions of influential leaders (e.g., Nkrumah, Kenyatta, and Mbeki). The concept of Africa must be accurate and accepted so that Africans can rightly claim social and economic rights that are informed by an implicitly moral demand for full membership of the global community.

## MOVING FROM A WORLD OF SHADOWS INTO SPLINTERING LIGHT

Much that is happening in many parts of Africa is positive. Leading economic historians underscore that Africa has a better chance of economic success now than any time in the last five centuries. Given the weight of recent historical experiences and representation, this evidence needs a much higher profile. Two African women (Ellen Johnson Sirleaf and Leymah Gbowee, both Liberians) were recipients of the Nobel Peace Prize in 2011 for their nonviolent struggle for the safety of women and for women's rights to full participation in peace-building work. They add to the growing list of Africans awarded Nobel prizes since 1960 (John Lutuli, 1960; Desmond Tutu, 1989; Nelson Mandela, 1993; Frederick William de Klerk, 1993; Kofi Annan, 2001; Wangari Maathai, 2004, and literature prizes have been awarded to Wole Soyinka, 1989; Nadine Gordimer, 1991; and John Coetzee, 2005). More prizes have been awarded to Africans than to Latin Americans.

Numerous recent events challenge and overturn many widely held assumptions. Mo Ibrahim, the Sudanese tycoon, is an African philanthropist who is endowing scholarships at London's Business School as well as London's School of Oriental and African Studies (the latter was founded as a place where British officials learned how to run their empires). As many as 500 African companies have been growing at more than 8% per year since 1998, and some are well placed to break out beyond the region onto the global stage. African academics are writing in peer-reviewed journals more than ever before. Africa Journals OnLine (AJOL), a South Africa–based portal, compiles over 400 scholarly peer-reviewed African journals, allowing a diverse body of scholarship to flourish within Africa and Africa-originating research to percolate to the rest of the world, in theory at least. Angola's president Dos Santos' pledge of assistance for Portugal, its former colonizer, to help Lisbon deal with its financial crises is aid in reverse. Increasing Portuguese migration for new opportunities in former Portuguese colonies in Africa (e.g., Angola, Guinea-Bissau, and Mozambique) is new movement and brain drain in reverse.

Africans are gaining greater global recognition outside of their home region, not just as scholars, writers, business and political leaders, and sports stars but increasingly in diverse professions. African-born architects are building prominent international buildings in Europe and North America (e.g., David Adjaye designed the Nobel Peace Center in Oslo and is the lead designer for the National Museum of African American History and Culture in Washington, D.C., scheduled to open in 2015). African Catholic priests and parishioners are contributing to a remapping of Europe's religious landscape, and African churches are also flourishing. Catholic priests from Africa attend to the faithful in Italy, the United Kingdom, Netherlands, and Belgium, and evangelical African preachers minister to congregations across Europe. The spectacle of Christian missions in reverse—Africans converting the descendants of those who ventured to the "dark continent" now based in a secular Europe—entails a deep irony. One can only wonder how inverting dominant historical relations will affect the future imagining of Africa.

Regional demographic trends of fewer people living in Europe compared to Africa are altering the historical balance of the population. With the European and U.S. populations reaching a plateau and projections of worker shortfalls, there is good reason to believe that Africans will continue to have an impact beyond their home region. Indices of income distribution are improving in many African states: only 12 African states register as having more unequal family income distributions than the United States (in other words, in most African countries wealth is more evenly distributed among society as opposed to concentrating at the top, whereas in the United States, the top 1% holds 90% of the wealth). According the CIA's (2012) own rankings, the United States is now more unequal than Nigeria and Ethiopia.

One third of Africa's university graduates now make their professional living outside of Africa. Africans in the diaspora are active in the development of their home regions. Remittances sent to Africa in 2012 reached US$60.4 billion, and perhaps an extra 50% of this figure gets sent back via informal channels (United Nations Economic Commission for Africa 2013). Most of this money keeps extended families afloat, while some funds are channeled into productive investment in houses and businesses and microdevelopment projects. Most of the latter projects (building toilets and community development centers, stocking libraries, equipping Internet cafés) target very local areas and are not always delivered effectively nor completed. Nevertheless, diaspora hometown initiatives demonstrate a continued commitment to a locality as well as a modest effort to ameliorate infrastructure deficits. An important impact of this new dynamic is that communities acquire goods and services with extraordinary value through connections beyond their locality.

These are contemporary developments that the ancestors of Africans and European colonialists could not have imagined. Still, the representation of Africa in the Global North has remained fixed in the colonial period. Dethroning misguided groups in the Western academy, many of them highly influential in Africa "chaos and crisis speak," is still a work in progress, and a balance of narratives about Africa's place in the world is only now emerging.

One highly positive sign is that academic output is expanding in Africa. For a long time, Africa-based scholars found outlets in local journals, nongovernmental organization (NGO) reports, and government and private consultancy reports. These outlets are still important in Africa-based research, but there is a movement toward wider and more public dissemination. This reflects the revitalization of academic life and improved academic freedom across Africa. AJOL is now adding journals at a rapid rate, and journals from 30 countries in Africa are included in its online portal. Even AJOL's listing is not exhaustive; additional Africa-based journals are published by international publishing houses (e.g., *Urban Forum* is published by Springer). High-profile Africa scholars in the diaspora are reasonably well represented in editorial membership and research output in top-ranked journals (according to impact factor rankings, which is only one [and some argue biased] yardstick of quality based on average number of recent article citations and cannot assess relevance), based in the United States and Europe. Journals published in Africa are now entering international impact rankings; for example, *South African Medical Journal* is ranked in the upper third of English journals in medicine worldwide, and *Southern African Development*, a leading Southern Africa development journal, joined the impact-ranking system for the first time in 2010. The economics and politics of research funding and university rankings mean that it may be impossible to dislodge the preeminent position of U.S. and European universities as gatekeepers of knowledge production about Africa. And, of course, it would be wrong to dichotomize African studies scholarship by Africans and the rest. For instance, 10,000 Nigerian academics are employed in the United States, indicating the extent of the circulation of African human capital. Africa research output and knowledge production now include an Africa-based stream, but the study of Africa is still a long way from cross-fertilization and coproduction of knowledge.

### SOCIAL MEDIA: AFRICA REPORTING, BLOGGING AFRICA, AND AFRICA PODCASTING

Electronic space is a medium that is becoming richer and more diverse and has the potential to greatly expand the Africa conversation. A growing number of sites concentrate on Africa and report and analyze issues, host African expert reflections, and offer blogs and podcasts about Africa topics. Table 2.1 provides a summary of the leading sites on Africa (multi-blogging and individual bloggers) and links to media at the forefront of Africa reporting and podcasting. Emphasis is on academic bloggers and practitioners who have a research and policy focus, but not all bloggers are traditional academics with university posts.

Several geographers are active bloggers (Ed Carr, Christian Kull, Bill Moseley, Abdi Samatar, and Daniel Thompson) but also included in Table 2.1 are blogs by other commentators from research organizations, NGOs, consultancies, business, and so forth. There is no mechanism to rate blog quality, and thinking in this way defeats the process of democratizing knowledge. An example of a highly successful blog with many followers and contributors on an Africa theme is Deborah Brautigam's "China in Africa: the Real Story." The Africa-focused social media landscape is ever-changing: new sites appear frequently, some efforts run out of steam, and blog initiatives may end after a hiatus of intensive activity and focus on a particular topic. For instance, William Easterly's "Aid Watchers" ended in May 2011 and Shanta Devarajan's "Africa Can End Poverty" ended in June 2013 but has since been continued by other World Bank staffers; both blogs are important as historical markers in documenting how the Africa conversation changes in a particular time period. In some emerging areas such as the technology–Africa interface, technology insiders are at the knowledge frontier (see "White African" and "Africa Musings" blogs).

Blogging offers a new genre of authoritative and accessible academic textual production, and in this way it is changing the nature of what it is to be a 21st-century academic practitioner and student learning about Africa. The advantages of the medium are its currency and ability to disseminate on-the-ground reporting and coverage of material that is sparsely covered or not covered at all in conventional texts. In this format ideas can be developed and expressed, often in a concise and accessible form quite different from the traditional lengthier academic books, written for expert audiences. As such, social media are on the way to becoming a forum for academic generation of knowledge outside of the classroom and traditional libraries and professional forums (e.g., public lectures, workshops, and conferences). These media are changing the way Africa and African topics are written about, and they represent a new contribution to an increasingly multiphrenic environment of academic production of knowledge on Africa.

The downside of social media is its huge size and its varying quality and coherence (it is devoid of quality filters, and topic postings are often random and reflect the interests of the blogger as much as anything else), making it difficult to navigate and to decipher

**TABLE 2.1 AFRICA IN SOCIAL MEDIA**

***Multi-blogging Sites***

| Blog Name | URL | Themes |
|---|---|---|
| African Arguments | http://africanarguments.org | international affairs, business, politics, book reviews |
| Africa at LSE | http://blogs.lse.ac.uk/africaatlse/ | development, economics, health, cities, politics, international affairs, gender |
| Africa News Blog | http://blogs.reuters.com/africanews | news, politics, business, lifestyle |
| Africa in Words | http://africainwords.com | literature and writing |
| Africa on the Blog | http://www.africaontheblog.com | diaspora, development, politics |
| Al-Jazeera Blog Africa | http:/blogs.aljazeera/blog/Africa | news, analysis |
| Footprints Blog | http://www.sustainabilityinstitute.net/newsdocs/footprints | sustainability, sustainable development, sustainable urbanism |
| Millennium Villages Project | http://blogs.ei.columbia.edu/category/millennium-villages | rural development, MDGs |
| Millennium Cities Initiative | http://mci.ei.columbia.edu/blog/ | urban development, MDGs |

***Individual Bloggers***

| Blog Name | Author | URL | Themes |
|---|---|---|---|
| Al-Jazeera Opinion | Abdi Ismail Samatar | http://www.aljazeera.com/indepth/opinion/profile/abdi-ismail-samatar.html | development, Somalia |
| Africa Can End Poverty | Shanta Devarajan | http://blogs.worldbank.org/africacan (June 2013) | development, combatting poverty |
| Africa Political Ephemera | Sara Rich Dorman | http://africanpoliticalephemera.blogspot.co.uk | post-colonial political objects, fashion, posters |
| Africa is a Country | Sean Jacobs | http://africaisacountry.com/latest/ | reinventing the Africa narrative, visual commentary on Africa |
| Aidnography | Tobias Denkus | http://aidnography.blogspot.de | development, anthropology, communications |
| Africa Musings | Juliana Rotich | http://afromusing.com | Ushahidi, renewable energy, tech Africa |
| Aid Watchers | William Easterly | http://aidwatchers.com (May 2011) | foreign aid policy and approaches, transparency |
| China-Africa Real Story | Deborah Brautigam | http://www.chinaafricarealstory.com | China-Africa myths and realities, Chinese workers, research ideas |
| Christian Kull | Christian Kull | http://christiankull.net | geography, environment, development |
| Informal City Dialogues | Sharon Benzoni | http://nextcity.org/informalcity/city/accra | urban development |
| Isthmus & Strait | Kelsey Jones Casey | http://isthmusandstrait.com/blog | community development, land, gender, racial and economic justice |
| Map East Africa | Daniel Thompson | http://mapeastafrica.com/blog/ | geography, mapping, urban and political trends, East Africa |
| Mats Utas | Mats Utas | http://matsutas.wordpress.com/about/ | informality, Liberia, Sierra Leone, Somalia |
| Open the Echo Chamber | Ed Carr | http://www.edwardrcarr.com/opentheechochamber | rural development, foreign aid, climate change, food security |

| | | | |
|---|---|---|---|
| Paul Collier | Paul Collier | http://www.theguardain.com/profile/paulcollier | foreign aid, development, population |
| Researching African States | Iván Cuesta | http://africanstates.wordpress.com | African politics, research |
| Texas in Africa | Laura Seay | http://texasinafrica.blogspot.com | politics, development, advocacy |
| uthinkafrica | Richard Grant | uthinkafrica.com | geography, development, urban Africa |
| White African | Eric Hersman | http://whiteafrican.com | Africa and technological innovation |
| Zimbabweland | Ian Scoones | http://www.ianscoones.net/Blog.html | rural development, agriculture, Zimbabwe |
| ZSpace | Patrick Bond | http://www.zcommunications.org/zspace/patrickbond | social change, activism |

***Podcasts***

| Name | URL | Thematic Focus |
|---|---|---|
| China Africa | http://china.buzzsprout.com | China-Africa current topics weekly |
| BBC Africa Today | http://www.bbc.co.uk/podcasts/series/africa | news roundup 5 days per week |
| The Guardian | http://www.theguardian.com/global-development/series/global-development-podcast | weekly development topics |

***Media Reviews***

| Outlet | URL | Perspective |
|---|---|---|
| The Guardian | http://www.theguardian.com/world/africa/roundup | progressive |
| Think Africa Press | http://thinkafricapress.com | progressive |
| Pambazuka | http://www.pambazuka.org/en/ | radical |
| African Strategic Center | http://africacenter.org | U.S. military security |
| China Daily | http://www.chinadaily.com.cn/world/africa.html | Chinese |
| France 24 | http://www.france24.com/en/africa | French |
| Good News About Africa | http://www.goodnewsaboutafrica.com | South African |
| One | http://one.org/africa/blog | grassroots advocacy and campaign to end poverty |
| Knowledge, Technology and Society | http://knotsids.blogspot.co.uk | development, agriculture, rural development |
| This is Africa, our Africa | http://www.ourafricablog.com | cultural Africa, photography, film, style |

whom and what is important and how it fits into or deviates from other knowledge. Without a good background knowledge and general framework, social media postings about Africa can be as confusing as they are provocative and informative.

The growing number of daily compilations of Africa news dispatches (from different perspectives) makes it possible to stay current on various events unfolding in Africa. *The Guardian* consistently has the most comprehensive general Africa coverage. Topic coverage varies greatly among many organizations. For example, the African Center for Strategic Studies provides daily security briefs from a U.S. and pro-U.S. African perspective. At the other end of the political spectrum, Pambazuka offers a platform for critical perspectives (mostly African) by disseminating news about freedom and social justice issues in Africa. It provides a broad base of analyses and has earned a reputation as being among the top 10 websites that are changing the world of the Internet and politics. Pambazuka operates from three offices in Africa as well as from London and provides a platform for 2,600 organizations and individuals to produce Africa news in multiple languages, making it the most pan-African web news source. Another useful source is "Think Africa Press," a London-based magazine staffed by a large number of Africa-based writers; it has become one of the leading Web magazines about Africa. The publication focuses on gender, health, environment, development, and culture and represents an alternative to the coverage offered by some of the mainstream media houses.

Inspiring Africans and Africa experts are making a contribution to the uplift of the region by focusing on advancing the humanity of Africans. African Arguments, the multi-blogging site funded by the Royal Africa Society and the Social Sciences Research Council in the United Kingdom, has a wide expert base (journalists, academics, and businesspeople) and aims to present the most vigorous Africa debate on the Internet. Several good news sites are available. For example, Africa Good News is a South Africa-based portal, and Good News About Africa is a London-based site; both assemble only positive news about Africa without understating the challenges.

Information about Africa is more diverse, splintered, and voluminous than ever before. Social media platforms allow opportunities for different and sometimes alternative voices. The Web offers a different platform of hope and possibility to change popular perceptions about Africa: decades of academic research have not effectively communicated different ideas about Africa to nonacademics. NGOs using the Internet can offer competing constructions of reality and can even allow their affiliates and supporters to participate in the knowledge process itself. Mapping using the Ushahidi (discussed in Chapter 6) is a good example of how NGOs and communities create their own geographical information in real time. The Internet is better equipped than traditional static media because it extends beyond immediate media space by using extensive networks of hyperlinks and tags for readers to follow, stimulating reader interest and debate, crystallizing issues, and keeping them alive. This arena has the potential to diffuse the gatekeeping function of traditional media conglomerates.

Obvious drawbacks to Africa on the Web and social media exist: information is fragmentary, and it takes time and effort to provide coherence to the multiple sources of information. Without some basic knowledge about Africa and consideration of the various objectives and agendas, many students will struggle to make sense of the various social constructions of African reality. It is hard to support the case that the Internet nullifies the old representation of Africa: even the most prominent sites attract niche audiences. Overall, Internet space has the potential to enlarge the positive representation of Africa by providing increasing amounts of new and alternative knowledge by an increasing numbers of sources.

## AFRICANS' REVISIONING OF AFRICA

There is no off-the-shelf model to solve complex Africa challenges. Virtually everyone acknowledges that Africa has challenges—economic, leadership, environmental, and social. Despite Africa's vast natural resources, many of its peoples remain mired in the deadly grip of poverty, squalor, and destitution, and buffeted by environmental degradation. Although macroeconomic indicators show improvement in some national contexts (growth rates, international investment, human development index [HDI], Gini coefficients, etc.), hard-hitting critiques claim Africans are worse off today than they were at independence in 1960. A good portion of this blame has to be shouldered by Africans themselves: leaders, politicians, and intellectuals have failed Africa.

Ake (1991:14) claims that "most African regimes have been so alienated and so violently repressive that their citizens see the state as enemies to be evaded, cheated and defeated, if possible, but never as a partner in development." He believes that African leaders are so totally engrossed in coping with hostilities that they misrule and unleash repression. Their dependence and policy reliance on external powers, coupled with their domestic priority of holding on to power at all costs, afford them little energy for anything else. Ultimately, African leaders' compulsion to gear policies to win international approval rather than to meet the needs of their citizens is a failure of the highest magnitude.

The ire of many African commentators is understandable. Monga (1996:38–39, vii–ix) points out, "When it comes to Africa, one can afford to indulge in approximations, generalizations, even illiteracy. Africa's overall image is so negative that only the most pessimistic types of discourse conform to the logic that governs understanding of the continent. Publications as 'prestigious' as the *Financial Times*, *Der Spiegel*, or *Time*, can publish cover stories and surveys built upon falsehoods and factual errors without stirring up a storm of protest, no doubt because 'experts' on Africa know that rebuttals will not damage their professional reputations." Monga goes on, "the more I read, the more frustrated I become because neither the academics nor the journalists were able (or willing) to capture the determination of people at the grassroots level to engage in the political arenas, at any cost, in order to bring about some positive changes in the way they had been ruled for several centuries." The weight of history and the psychological damage, not to mention the economic and political realities that stem from history and its various continuities, are hard to overstate.

Difficult to dismantle is the international supposition that Africa is broken, off the global map, and unwholesome and that the international development industry must lead Africans on what to do. Multilateral agencies, international governments, and most NGOs participate in large efforts that result in Africa being dependent on external assistance. Tens of thousands of Westerners, spanning a range from volunteers to highly paid consultants based out of the top hotels, would need to find a new line of work if Africa stopped needing development advice. Their collective weight makes it very difficult for Africans to maneuver and to recover their humanity. Owomoyela (2010:48) puts it, "Let us imagine, as an analogy, a man/woman is on trial for certain offences. The tribunal is quite disposed to be considerate because its members believe that the accused had a rough and deprived upbringing. They also hold the belief that the charges are quite valid, consistent with what is known about the accused and established patterns of behavior. Tribunal members are of the opinion that the best course of action is to admit guilt, plead for leniency and the opportunity to reform. But suppose the accused pleads innocence and insists on explaining why those who assume guilt are wrong. The more the accused strives to persuade the tribunal the more the tribunal may be alienated and consequently hold the opinion that the accused is irredeemable and deserving of the most draconian lesson." This is the kind of framing that led the International Monetary Fund to impose structural adjustment policies on African governments and for market-based policies to be rolled out across Africa from the mid-1980s onward.

Marxist ideas have held sway among African scholars for decades. In the 1970s, there was a tilt toward the left after the first generational wave of academics returned from the Global North (some of whom were messengers of the new design) and witnessed development failures firsthand. University library shelves contain many books written by African scholars that frame Africa's situation from a Marxian perspective, a theoretical stance oddly out of style with fashionable topics in the highest-ranked research journals. This Marxist tilt was reinforced by significant material cuts at African universities, where scholars lacked access to the latest theoretical debates, effectively silencing African voices in the production of global knowledge on Africa. It is a deep paradox that the greatest opening of African societies has occurred during the last 30 years, a span when African academics were cut off from global scholarship and vice versa. Store shelves filling up with imported global commodities as library shelves were cut off from global materials became an African predicament of enormous magnitude. The left's critique of market-led policies was largely silenced by press censorship, which has only recently begun to lessen.

Africans are seeking an answer to a complex question. African academics, policymakers, and commentators have been arguing for some time that Africa's future must be founded upon indigenous roots and true partnerships. A wide spectrum of ideas on alternative development and on alternatives to development has been put forward. It is not possible to present all of these ideas here, but seven prominent visions put forward by Africans are presented in the next sections of this chapter (various other visions are presented throughout this book). Africans are battling to promote different ideas (many of them incompatible with each other) that prioritize (1) restructuring global economic arrangements (e.g., Samir Amin); (2) increasing financial transparency and ending the looting of Africa (e.g., Patrick Bond and others); (3) promoting better political leadership (Mo Ibrahim); (4) supporting sustainable environmental development (Wangari Maathai); (5) ending aid (Dambisa Moyo); (6) promoting civil society participation in development (e.g., Slum/Shack Dwellers International); and (7) promoting alternative ways to understand Africans and African urban society by a postcolonial lens (e.g., Jennifer Robinson and AdouMaliq Simone).

Common elements among these diverse perspectives include the recognition of the complexity of African societies; the need to move beyond project-oriented development; the necessity for Africans to develop fresh ideas, not replicate models from the United States, Europe, China, or elsewhere; and the aim for Africans to become active partners in sustainable development (rather than passive development subjects) and for development to be more inclusive of the poor.

## UNCHAINING AFRICA

Samir Amin (1990) is a highly acclaimed Egyptian political economist who emphasizes delinking from the global economy as the only viable choice for Africa. His radical perspective is based on an the assumption that the international financial system is unfair and disadvantages Africans who, as it happened, were never even consulted about the financial regime and its functioning. Neocolonialism, exploitative relations, and lack of African peoples' autonomy means that only maldevelopment (in medical terms this is brain maldevelopment of the fetus) can take root. For Amin, economic development is only a tool, but all too often it is misconstrued as the end goal of development, and concentrating on economic growth rather than fairness is misguided.

Amin highlights that Africa is adversely incorporated into the global economy, shackled by five external monopolies: high technologies, control over the financial systems, control over environmental discourse, the mass media, and weapons of mass destruction. Economic projects in Africa have not brought about human and social development: instead, Africans are generally poorer than the rest of the world. Amin's position over the years, as well as that of many of his followers, has shifted from self-reliance and "autarky"—a complete delinking from the international system—toward a revamping and restructuring of the relationship.

The Amin camp believes that the international system is rigged and results in a systematic looting of Africa's resources because most African governments have assumed ownership of all valuable resources. Claiming to hold them in trust for the people, they actually utilize them for the needs of multinational corporations (MNCs), various local cronies, and their own purposes. Alternative development and a new departure for Africa can occur only under conditions of a better balance in the political and economic organization of the global economy, a real multiplicity, and not just a system that benefits the elite in the Global North. Radical African political economists now argue that alternative thinking about Africa's resources is needed, even a consideration of leaving them in the ground.

## ENDING THE LOOTING OF AFRICA

Patrick Bond is a Marxist who contends that the most horrid chapter in the global economy since slavery is the gargantuan illicit financial flow out of Africa. His book, *Looting Africa: The Economics of Exploitation* (2006), is a tour de force Marxian analysis that explains why Africans are poor and becoming poorer. Africa is being drained of resources by a pinstripe financial brigade of bankers, accountants, lawyers, and corporate officers who build the instruments that make an offshore world. "Offshore" is defined as a legal space that decouples the legal and real location of a transaction with the aim of avoiding some or all regulations. This financial brigade operates from luxury offices in the heart of global financial centers (e.g., London, Paris) as well as from an offshore financial network of tax havens (involving secrecy jurisdictions, disguised corporations, anonymous trust accounts, fake foundations, etc.), integrated into an offshore labyrinth.

Offshoring reaches deep into African societies via corrupt officials, suave insiders, rogues, and various companies that do all within their means to avoid paying taxes. Individual interests at the expense of societal interests rule the day: secrecy, greed, and excessive profiteering knit local and international players together. Their bonds cement in-group trust as well as an external wall of opacity. The outcome is that money leaves and flows elsewhere, circulating and hiding in a shadow economy without being identified from anywhere. Money escaping from Africa is not only a Marxist concern: Christian Aid, Oxfam, and civil society organizations such the Basel Institute of Governance, as well as journalists and whistle-blowers, are engaged in tracing the hidden billions from development (Shaxson 2011). A task force on financial integrity and economic development was launched in 2009 to coordinate activities among civil society organizations and national governments (some African governments are members). International organizations (e.g., World Bank and Basel Institute of Governance) are now starting to put more emphasis on mechanisms for asset recovery and transparency. The largest identified official culprits in Africa are presented in Table 2.2.

**TABLE 2.2 LARGEST AFRICAN KLEPTOCRATS**

| Country | Head of State | Amount (US$) |
|---|---|---|
| Nigeria | Sani Abacha | 4.3 billion |
| Côte d'Ivoire | Felix Houphouét-Biogny | 3.5 billion |
| Nigeria | Ibrahim Babangida | 3.0 billion |
| Zaire (Congo) | Mobutu Sese Seko | 2.2 billion |
| Mali | Moussa Traoré | 1.8 billion |
| Côte d'Ivoire | Henri Konan Bédié | 200 million |
| Congo | Denis Sassou Nguesso | 120 million |
| Gabon | Omar Bongo | 50 million |
| Cameroon | Paul Biya | 45 million |
| Ethiopia | Haile Mariam | 20 million |
| Chad | Hissène Habré | 2 million |

*Source:* Based on information reported from Basel Institute of Governance, Tax Justice Network-Africa, and Pambazuka.org.

Many asset recovery efforts are ongoing. Asset recovery efforts from Abacha's reign have led to the retrieval of US$1.1 billion from banks in the United Kingdom, Switzerland, Lichtenstein, and Luxembourg. Asset portfolios attributed to Mobutu Sese Seko, one of the worst resource plunderers, are staggering and still under examination. His known property constellation included a vineyard in Portugal, a 32-room mansion in Switzerland, a castle in Spain, and a magnificent first-floor apartment in Paris close to the Arc de Triomphe and to the furrier who made his leopard-skin hats. The *pièce de résistance* was his marble palace in his home village of Gbadolite.

Major banks (e.g., Citibank, Barclays, ABN Amro) participate as pass-through conduits connected to occluded financial intermediaries in a complex labyrinth. Money does not flow in obvious and transparent geographies. Illicit transfer evidence is mounting, an elephant in the room, bizarrely ignored in mainstream analyses and mysteriously under the radar partly because it is veiled in secrecy that makes it impossible to see more than a minor part of the system at any one time and is endlessly shifting as competition among financial centers produces competition to enlarge the loopholes. The offshore financial terrain is now even locating to African tax havens. For example, Mauritius is becoming a conduit haven for investment into India and into Africa extractive industries from China and from London. Gaborone, Monrovia, and Seychelles are emerging as additional on/offshore hubs, but the government of Ghana terminated its offshore financial arrangement with Barclays Bank of London in 2011 after Organization for Economic Cooperation and Development (OECD) criticism.

According to Global Financial Integrity (GFI) data, illicit flows hemorrhaged from US$854 billion to US$1.8 trillion between 1970 and 2008, growing at rates of about 12% per year (Kar and Cartwright-Smith 2010). Other estimates of the draining of Africa calculate that between US$0.50 and US$10 leaves for every aid dollar sent. Draining may be facilitating a revolving door of odious debts: in-debt African governments have to accept aid as well as allow unrestrained resource extraction to meet the demands of their creditors. An exact value of illicit flows may never be determined, but the size of the racket is mind-boggling. Mounting evidence reveals that Africa is a net-creditor region to the rest of the world: the region's net external assets now vastly exceed what is owed in debts.

We are still in an early phase of education in attempts to see, understand, and measure the offshore system. The Tax Justice Network-Africa (TJN-A) is a pan-African organization that provides data, reports, and commentary for academics, students, and those interested in the harmful tax and financial activities that are depleting African societies. As with much of the gaze on Africa, the Western media have focused on capital flight from Africa, pointing an accusatory finger at the political-economic establishments in Gabon, Nigeria, South Africa, Congo, and Côte d'Ivoire in particular. However, each capital flight location in Africa needs to have a corresponding inflow destination somewhere else.

It is worth reversing the gaze to ask why the emphasis is not on the inflows, especially as richer countries have more resources and technical expertise in forensic accounting to go after this illicit money. It is not just clandestine operations that are occluded, but routine MNC behavior utilizes transfer price mechanisms to avoid paying African taxes. Palen, Murphy, and Chavagneux (2010:175) cite a Deloitte report that underscores the lack of a single successful African challenge to the transfer pricing behavior of any MNCs operating in Africa. African governments are ill equipped to broach a charge against MNCs operating adversely within their territories, since they lack the expertise, relevant legislation, and commercial confidence to take on entrenched corporations.

**TABLE 2.3 THE LOOTING OF AFRICA**

| |
|---|
| Slavery and dispossession of more than 12 million people |
| Net supplier of energy and raw minerals to the Global North |
| Loss of 20,000 skilled professionals in annual brain drain |
| 95% of cultural property trafficked (antiquities in British museums, private collections) |
| Unquantifiable losses due to stripping of animals and animal matter (e.g., endangering rhinoceroses, and elephants [especially for taking ivory]) |
| 30% of Africa's GDP moving offshore |
| US$30 billion–$148 billion moving to overseas financial institutions (Geneva, Monaco, Jersey, London, and other financial secrecy jurisdictions) |
| US$174 billion siphoned via corruption |
| US$10 billion leaving in multinational transfer mispricing |
| 50% shortfalls in national tax revenues due to tax evasion and avoidance measures |

*Sources:* Bond, 2006; Tax Justice Network-Africa, 2013.

Africa continues to be looted (Table 2.3). The corporate crime wave is global, stretching from Africa over to the Americas and to China, India, and almost everywhere else. The wealth, power, and illegality enabled by this hidden system are now so vast as to threaten the global economy's legitimacy. Outflows are a system driven from supply-side institutions (London, Washington, D.C., Delaware [for tax avoidance]) and accommodated by partners in Africa (Pretoria, Libreville, Abuja). The looting of Africa that began centuries ago continues today, entailing transfers of natural resources, human resources (slave labor in the past replaced by recruitment of skilled graduates today), antiquities, animals, and ivory; this trend is most evident today in the transfer of siphoned monies. This multibillion-dollar industry means African wealth is being subtracted from funds that could be otherwise deployed within Africa, for projects such as poverty alleviation. Particularly egregious about the outflows is that a minority of the elite is appropriating large amounts of the region's private assets while the general public, via their government, is bearing the excessive public debt.

The 2013 index of financial secrecy in Table 2.4 shows that only a few African countries register as the world's top 50 supply-side havens (Mauritius is ranked 19th, Liberia 27th, Seychelles 28th, and South Africa 36th); in sharp contrast, Europe and to a lesser extent the United States predominate in global financial secrecy activities. The United Kingdom and dependents under British control stand out as a major offshore network, accounting for one third of all offshore activities linking businesses and London that otherwise might not be related. The geography of UK financial networks within financial networks operates like a recharged version of the colonial empire where modems rather than gunboats orchestrate vast transfers. Few funds ever return to Africa. London is the epicenter of African outflows, determined by an apparatus that enabled the banking and legal systems to expand to provide an escape route for capital at the time those African countries were achieving independence. The French established a parallel system, whose veil was lifted by the Elf scandal (see Box 2.2).

Opponents of looting argue that policies designed to help Africa are nothing more than a facade and a charade because they end up benefiting Western institutions, multilateral institutions, and the global financial industry more than Africa itself. Mainstream development theorists need to better understand the destructive dynamic of looting. Supply-side curbs need to be set up. Africans need to engage in self-activity and in campaigns to ensure that declarations are taken seriously and accommodated with new curbs and more transparency.

## POLITICAL LEADERSHIP

The world's largest annual prize, the Ibrahim Prize, is awarded to improve leadership in Africa. It offers a monetary award that exceeds the Nobel Peace Prize. The prize is awarded to a democratically elected former African executive head of state of a government who demonstrated excellence in office, abided by constitutional term limits, and gracefully left office within the last three years. This incentive to improve leadership is

**TABLE 2.4 AFRICAN COUNTRIES IN THE FINANCIAL SECRECY INDEX (FSI) RANKINGS**

| Rank | Secrecy Jurisdiction | FSI Value | Secrecy Score | Global Scale Weight |
|---|---|---|---|---|
| 1 | Switzerland | 1,765.2 | 78 | 4.916 |
| 2 | Luxembourg | 1,454.4 | 67 | 12.049 |
| 3 | Hong Kong | 1,283.4 | 72 | 4.206 |
| **4** | Cayman Islands | 1,233.5 | **70** | 4.694 |
| 5 | Singapore | 1,216.8 | 70 | 4.280 |
| 6 | USA | 1,212.9 | 58 | 22.586 |
| 7 | Lebanon | 747.8 | 79 | 0.354 |
| 8 | Germany | 738.3 | 59 | 4.326 |
| **9** | Jersey | **591.7** | **75** | 0.263 |
| 10 | Japan | 513.1 | 61 | 1.185 |
| *19* | ***Mauritius*** | 397.8 | *80* | *0.047* |
| 27 | **Liberia** | 300.8 | 83 | 0.014 |
| 28 | **Seychelles** | 293.4 | 85 | 0.011 |
| 36 | **South Africa** | 209.7 | 53 | 0.260 |

*Source:* http://www.financialsecrecyindex.com/introduction/fsi-2013-results.

## BOX 2.2 ELF SCANDAL: CHASING FRANCE-AFRIQUE SHADOWS

The Elf affair was a jolt to the world. A dedicated magistrate in France, Eva Joly, investigated a corporate legal dispute, and her inquiry led her to dig deeper and deeper until she was able to peel off layers that eventually showed complex Africa–France entanglements. The story, as it unraveled, linked President Omar Bongo of Gabon to a gigantic system of corruption involving the French state-owned oil company Elf Aquitaine and French political, commercial, and intelligence establishments. A web of deceit contained details straight out of a spy novel.

It connected former lingerie model Christine Deviers-Joncour, who was paid over US$6 million to help persuade her lover—who happened to be the French foreign minister—to reverse his opposition to the sale of missile boats to Taiwan, among other things. This money originated from oil proceeds from the small African state of Gabon. A seven-year investigation uncovered a chain of transactions from Libreville to Paris via Geneva and Luxembourg and a host of other tax havens, as well as from Paris to French MNC affiliates via a slush fund for spending on oil bribes in African oil frontiers. This was made possible by selling African oil cargoes and splitting the proceeds among a range of bewildering accounts in myriad tax havens. The investigation put Deviers-Joncour in prison, and after she felt that the establishment had deserted her, she broke her code of silence and in 1999 wrote a best-selling book, *The Whore of the Republic* (English translation of the title). She revealed that she had used an Elf credit card to buy the French prime minister a pair of handmade ankle boots from a Paris designer shop so removed from reality that the shop owner offers to wash customers' shoes once a year in champagne.

The Elf affair turned out to be the biggest corruption case of the postwar era. This secret oil money was used to covertly finance French political parties, the intelligence services, other well-connected parts of French society, and other projects. This stash even provided a fund to pay bribes for large contract bids in Venezuela, Germany, and Taiwan. The out-of-sight Gabon origins meant that the money trail was invisible, and the slush fund was able to grow in size and complexity over time. Gabon does not appear on any list of tax havens, but the Elf system it funded was intertwined in an offshoring complex that may have comprised 4,000 channels around the world.

The secretive Elf pot of hundreds of millions of dollars enabled France to "punch above its weight" on the international scene. Shady dealings meant that Elf operated as a state within a state. Monies even reached the United States. A U.S. investigation showed that former president Bongo used loopholes in American laws to transfer millions to bank accounts of American lobbyists. U.S. and international banks facilitated these transactions without questions.

Nicolas Sarkozy's first phone call (after assuming the French presidency in 2007) was not to the president of Germany or of the United States but to Omar Bongo in Libreville. Bongo ruled Gabon, a small country on the coast of Central Africa, from 1967 until his death in 2009, making him the world's longest-serving head of government. Assuming office at age 31, his reign coincided with Gabon's oil boom. During his rule, Gabon became the fourth-largest oil producer in Africa. The country is richly endowed in other resources, containing one fifth of the world's known uranium supplies, iron ore, manganese, columbite, and talc deposits, and abundant dense forests with some of the world's most valuable woods. Gabon shares (with neighboring Equatorial Guinea) a world monopoly on the production of *okoumé*, a light wood used in the production of plywood. Bongo became France's point man in the region, and Sarkozy referred to this as a special relationship. By many accounts this special relationship is ongoing: France is reported to have been involved in rigged elections, enabling Bongo's son to take power after his father's death. One thousand French troops remain in Gabon today, connected by underground tunnels to the presidential place. Gabon is an oil-producing country, but with a small population of 1.7 million. During the Bongo years, the government built more pipelines than roads, and the country astonishingly registered among the highest infant mortality rates in the world. During Bongo senior's days his name and image adorned everything from stadiums to bottled water, and the Bongo system doled out cash to buy friends and to appease the opposition as needed. Bongo was said to be so averse to writing checks that when summoned to the presidential palace one traveled with a suitcase to collect payments. A local joke that made the rounds in Libreville was that the quickest way to become a millionaire was to set up an opposition party.

Bongo's presidential salary was €20,000 per month (US$26,322), yet he lived an outrageously extravagant lifestyle—dozens of luxurious properties in and around Paris and in southern France, a US$500 million presidential palace, 70 bank accounts, nine luxury vehicles worth about US$2 million, and many other cars (and these holdings are only what prosecutors knew about).

Bongo's rise is a full-blown snapshot of what went on behind the scenes when African countries gained independence and the French empire acted to set up new ways to stay in control. When Gabon's first president, Leon M'ba, died prematurely, Omar Bongo was the ideal replacement as far as French interests were concerned. Belonging to a tiny ethnic minority group and with no natural base of support, he would have to rely on France for protection. In exchange for French backing, Bongo granted French companies almost exclusive access to the country's resources, on preferential terms at the expense of the needs of the local population. Gabon acquired a taste for things French, and it is the world's largest consumer of champagne per capita, but most of its people are impoverished. Wealth is so unevenly distributed that half of the population lives below the poverty line—remarkable for a country with such a huge GDP and further corroborating the small elite's stranglehold on all power and economic resources. All this could occur because Bongo was a shrewd international operator and tapped into French Freemasonry networks and African secret societies to become one of the most important power brokers in France itself. Bongo facilitated France's relationship in such a way that it has been remarked that France left by the front door at independence but returned in the side, opened window.

The Elf affair illustrates shadowy dealings and the workings of cross-border corruption. The gigantic system of corruption affected ordinary people in both Africa and France in invisible but profound ways. Ordinary Africans have their resource monies siphoned off to the rich world through unfair contracts and general corruption, while French politicians helped prop up Gabon's leaders, making them less accountable to their citizens. At the same time, the Elf system made France's elites unaccountable to their own citizens. The Elf case symbolizes the problems with the global regime in international accounting standards. It turns out that the International Accounting Standards Board (IASB) is a privately funded company financed by the largest accountancy firms and global MNCs, headquartered in London and registered in Delaware.

the brainchild of African billionaire Mo Ibrahim, who made his vast fortune from the mobile phone industry. Prize money consists of US$5 million over 10 years and US$200,000 annually for life thereafter. The Mo Ibrahim Foundation also considers granting an additional US$200,000 per year, for 10 years, toward public interest activities and good causes espoused by the winner. The inaugural winner was the former South African leader Nelson Mandela in 2007 (Box 2.3); the 2011 recipient was Pedro Verona Pires, former Cape Verde president (Fig. 2.2). According to the awards committee, under the latter's 10 years as president, the nation became only the second African country to graduate from the United Nation's Least Developed category and has won international recognition for its record on human rights and good governance. The committee chose not to award a prize in 2009, 2010, and 2013, signifying that no leaders met the criteria for the award.

## BOX 2.3 NELSON ROLIHLAHLA MANDELA: AFRICA'S ICONIC LEADER

Nelson Mandela, who passed away on December 3, 2013, is widely regarded as the greatest African leader of the 20th century and a leader of global historical significance. A charismatic man of Africa and a worldwide symbol of resistance to the injustice of his country's apartheid system, Mandela evolved into a global iconic figure. He was jointly awarded the Nobel Prize in 1993 (with F. W. de Klerk) "for their work for the peaceful termination of the apartheid regime, and for laying the foundation for a new democratic South Africa" (Nobel Prize Organization 1993). As the first black president of South Africa and leader of the African National Congress (ANC) party, Mandela shepherded a difficult transformation from apartheid and racial conflict to democracy. He became the face of the "new South Africa" that re-engaged internationally and helped reimagine and turn South Africa toward Africa, considerable feats as the country had long been regarded as an international pariah and a settler colony run by Europeans, many of whom did all in their powers to be separate from Africans and Africa.

Mandela's global status is such that he is perhaps the closest the world has come to know as "a secular saint," but as one of his biographers emphasizes, he would be the first to admit that he was in essence something far more pedestrian: a quintessential politician who knew when to make the transition from warrior and martyr to diplomat and statesman (Stengel 2008:1). His autobiography *Long Walk to Freedom* (1995) is a very compelling account of his coming of age as a political leader, his immense human spirit, his education, and his deep reflection during his 27 years of imprisonment. In his memoir he notes, "I had no epiphany, no singular revelation, no moment of truth, but a steady accumulation of a thousand slights, a thousand indignities and a thousand unremembered moments produced in me an anger, a rebelliousness, a desire to fight the system that imprisoned my people. There was no particular day on which I said, henceforth I will devote myself to the liberation of my people; instead, I simply found myself doing so, and could not do otherwise" (Mandela 1995:95). Mandela was also deeply concerned with justice for all, and he maintained that "a nation should not be judged by how it treats its highest citizens, but its lowest ones." (Mandela 1995:201). His autobiography was made into a 2013 Hollywood biopic (breaking news of Mandela's death coincided with the London premiere), and Mandela's dedication to reconciliation is well chronicled in various texts and films. For example, the Academy Award-nominated *Invictus* film details Mandela getting behind the South African rugby team the Springboks (who personified the sport of his former oppressors); they were crowned world champions at the 1995 Rugby World Cup hosted by South Africa.

The Mandela legend storyline and his persona made him a pop cultural icon. From Mandela shirts to musical powerhouses

(*Continued*)

**BOX 2.3 (*Continued*)**

writing and performing songs in his honor (U2, Springsteen, Miles Davis, etc.) to Hollywood heavyweights portraying his life (Danny Glover, Sidney Poitier, and Idris Elba), Mandela is the most celebrated African.

Mandela was born into the Madiba clan in Mvezo, Transkei, in 1918 but grew up in Qunu, a small village in Eastern Cape Province. His father, Chief Henry Mandela, was a member of the Thembu people's royal lineage and his mother was one of his four wives. At age seven, he became the first member of his family to attend school, and he went on to complete a high school education and enter university. He studied for a BA at University College at Fort Hare (a higher education institution for the black population) but never completed his studies because he was expelled for joining a student protest. He studied law at the University of Witwatersrand (Johannesburg) but also never completed this program. By his own admission, he was a poor student, and he dedicated himself to political activism, which culminated in his joining the ANC in 1994. With a group of fellow ANC members (including Walter Sisulu and Oliver Tambo) he established South Africa's first black-run law firm as well as the ANC Youth League. Later he resumed his studies via a distance-learning program from the University of London, but he did not compete that degree and it was not until 1989, while in his last months of imprisonment, that he obtained a law degree from the University of South Africa.

Mandela spent his formative activist years as volunteer-in-chief of the ANC's Campaign for the Defiance of Unjust Laws, whereby volunteers defied selected laws in nonviolent resistance. Mandela changed tactics in 1961 when he decided to go underground to help create the ANC's paramilitary wing—*Umkhonto we Sizwe* (Spear of the Nation)—that engaged in sabotage against the government. Captured in 1962, he was accused of traveling outside of the country without a passport and of inciting workers' strikes. Subsequently, he was charged with treason in what became known as the Rivonia Trial and was imprisoned, along with seven other members of the top command of the ANC, for life without parole. Until 1982 he was imprisoned on Robben Island, South Africa's most notorious island prison, situated 11 km (7 miles) from Cape Town. There, he was confined to a 2.1 × 2.1-meter cell (7 × 7 feet), was made to do hard labor (breaking rocks), and was permitted to write and receive only one letter every six months. As international pressure mounted on the apartheid regime (through boycotts, divestments, the "Free Mandela" global media movement), Mandela was transferred to Pollsmoor Prison in 1982 and later to Victor Verster Prison, and from these locations negotiations with the apartheid regime ensued for over a decade.

Mandela's legacies are many. His most important accomplishments include engineering and overseeing South Africa's transformation from apartheid; championing the cause of forgiveness and reconciliation; dismantling the geographical lines of segregation and separation; promoting the "rainbow nation" (the coming together of peoples from different backgrounds, colors, and creeds); developing a new foreign policy, with special attention to the Southern African region through the promotion of the Southern African Development Community (SADC), which allowed South Africa to be both provider and recipient of migrant labor, transport services, and hydroelectricity; actively engaging with African institutions such as the Organization of African Unity (now AU) that rallied African leaders behind the vision of a unified Africa as well as an African Renaissance; demonstrating that public protest and boycotts can bring about change; and establishing a precedent that stepping down from the office of the presidency is important for the functioning of democracy.

Mandela, the "father of the nation," continued to engage in public life and philanthropy after leaving the presidency, establishing the Nelson Mandela Foundation, which is committed to rural development and education and combatting HIV/AIDS.

Of course, no leader can achieve everything, and Mandela has been criticized for not doing enough in the political and economic transformation of South Africa. He did not complete the transformation of the ANC party into a mature political organization that is not built around the cult of the leader. The hegemony of the ANC is such that a mature multiparty democracy has yet to emerge. His policies did not do enough in ending spatial and social disparities in South Africa. The gap between the rich and poor remains wide, and unemployment remains very high (officially it is almost 40% and it may be even higher in reality). Shockingly for a resource-rich country such as South Africa, nearly one third live on less than US$2 per day (World Bank 2013). South Africa has not flourished as well economically as the rest of Africa, and its economy is stumbling as a brain drain of South African professionals has occurred for many years.

Subsequent South African presidents (Thabo Mbeki and Jacob Zuma) have failed to live up to Mandela's legacy and have been rather flawed. Corruption and patronage of the ANC has become rampant, and the Zuma presidency continues to be embroiled in a series of scandals (sex, misappropriation of state funds, and the Gupta wedding jet scandal). Despite the overall economic presence of the country in the region, South Africa holds limited soft power in Africa. Even in its immediate region, South Africa has not provided leadership in managing the Zimbabwean crisis. However, it is important to emphasize that South Africa began its transformation under Mandela in 1994, and 20 years is not a lot of time for deep transformation. As Mandela noted, "we have not taken the final step of our journey, but the first step on a longer and even more difficult road" (Mandela 1995:624).

**FIGURE 2.2** Mo Ibrahim and African Leaders at Prize Ceremony, 2012. *Source:* © Sylvain Cherkaoui/AP/Corbis.

The prize is aimed at tackling the leadership deficit in African states. Business and intellectual African elites are convinced that improving government leadership is one of the most effective ways to transform Africa from within. Well established is the reality that many African leaders are motivated to cling to power rather than to put all of their energies into developing their economies. Retired presidents receive no benefits in many African countries, increasing the incentives for corruption as a means to provide for their own welfare when they leave office. Retirement brings an abrupt end to presidential perks; the mansions, cars, meals, wine, and social events are withdrawn. Renting a house in the capital can even become unaffordable to many, especially with regime changes, when the opposition wants to entrench its entourage and marginalize the old regime. Some attribute Africa's reverence for "big men" presidents as a leftover from precolonial reverence to traditional social systems of leadership, followed by a shift to colonial administrations that encouraged this kind of leadership (native administration), and then to postcolonial development policies that privileged top-down development until recently. Leadership is very important because it sets the priorities for development. In the past, African leaders were preoccupied with blaming their failures on external factors, scarcely reflecting on internal deficiencies. A legacy of past leadership failures is that international governments and their publics appear incapable of distinguishing between African leaders and Africans. Therefore, any efforts to improve Africa's leadership deficit can only be positive for all.

## ENVIRONMENTALISM OF THE POOR AND SUSTAINABLE DEVELOPMENT

Environmentalism was a preserve of the Global North until recently, and environmentalism of the poor in Africa has been overlooked. Wangari Maathai (2009) was a leader of the sustainable development and environmental conservation movement among the poor who was very influential in Africa. Through her work and advocacy, she showed that poor women often have intimate understandings of what is at stake with regard to poor management of the environment. By relying directly on land and its natural resources, many of Africa's poor understand the relationship between environmental degradation and poverty and are motivated by conservation. Their situations stand in sharp contrast with many in the Global North, where the relationship with the natural environment as a source of livelihood is largely lost.

Maathai, born in Kenya, excelled as a leader: she became the first woman in Eastern and Central Africa to be awarded a PhD, the first woman to be appointed professor at the University of Nairobi, and the first African woman to be awarded the Nobel Peace Prize (2004). She led an international campaign that prevented the Kenyan government from erecting the tallest skyscraper in Africa and extended her environmental activism into urban centers like Nairobi. Her career also encompassed crossing the political divide, from being an outsider activist to a member of Parliament insider; she even went on to hold a deputy ministerial appointment. She gained most international prominence for her leadership of the Green Belt Movement (GBM), a grassroots environmental conservation and community development NGO (Fig. 2.3).

Under Maathai's stewardship, the GBM became a vehicle for mobilizing community consciousness (it utilized tree planting and tree care as an entry point) and for facilitating self-determination, equity, improved livelihood and security, and environmental conservation, and in the process made progress toward invigorating self-confidence. Since 2006, the United Nations

**FIGURE 2.3** Wangari Maathai. *Source:* © RADU SIGHETI/X00255/Reuters/Corbis.

Environment Programme (UNEP), in Maathai's honor, has embarked on annual international campaigns to plant 1 billion trees. Remarkably successful, some 7 billion trees had been planted by 2011. GBM is well on its way to becoming a pan-African movement and is successful in scaling up its very local community advocacy to the highest level of society. For example, GBM successfully lobbied for the new constitution of Kenya (promulgated in 2010) to guarantee environmental rights: the fundamental right of all Kenyans to live in a healthy and clean environment and to have the environment protected for the benefit of present and future generations.

GBM built community, livelihoods, and sustainable development. Women's groups functioned as the conduit for the successful mobilization of a massive reforestation movement. GBM validated women's willingness to mobilize and work together to help meet their families' and communities' needs through sustainable strategies. Through activism, the marginalized poor demonstrated that the poor do provide leadership (as much as the elites). Creating healthier environments that sustain communities, GBM created spaces in which people affected by environmental degradation were provided with a voice in the negotiation, sharing, and management of resources. In this regard, GBM provides a vehicle for strengthening local cultures and economies, and an alternative to the continuation of past negative influences of colonial-enforced changes in land practices and external commodity prices. Maathai's most important legacy is her leadership in influencing people and communities to overcome the ingrained attitude that solutions to problems should come from the outside and from the top down.

Maathai contended that many of Africa's problems stem from a destructive dynamic where national resources turn into a curse and where effective resource leadership is absent, a dynamic rooted in colonialism. Colonists introduced commercial crops that were not native to the land; forests were cleared for export-oriented agriculture, and other areas were planted with exotic species of trees for the international lumber industry. Once harvested, exotic timber species were rarely replaced with seedlings, exotic or otherwise. Kenya, for example, lost 90% of its forest in 50 years (1950–2000). In time, deforestation leads to soil runoff and the degradation of water supplies. Scarce water in rural areas motivates men to migrate to cities for work, and rural communities and family networks unravel at the source.

Maathai advocated for an African leadership revolution. Current leadership and democracy deficits constitute a major development bottleneck. Africans remain stuck on a wrong-bus syndrome: behaving as travelers who have boarded the wrong bus, many people and communities head in the wrong direction or travel on the wrong path, while allowing their leaders to lead them farther from their desired destination. Poor education, naiveté, fear, pride, pack mentalities, powerlessness, and poor advice have led Africans on the wrong development path. The disempowerment of Africans is one of the most unrecognized problems, and it finds expression in a lack of self-confidence, apathy, fear, and inability to take charge of one's situation. Disempowered Africans find it acceptable to rely completely and without question on third parties (government, aid agencies, and God) rather than have faith in their own efforts. African leaders shoulder a large portion of the blame by not being transformational leaders. Their attitudes and policy decisions have supported the belief that Africans are helpless, unable to act on their own behalf. Values such as self-determination, embraced by the majority of Africans in the period after independence, have seriously eroded. Personal and collective uplift have been shattered recklessly. Africans must embrace the freedom of self-discovery and Africa-led solutions. Strong psychological, intellectual, and other rationales are advanced for not accepting the prevailing freedoms (i.e., the global marketplace and global financial system).

Counterintuitively, Maathai called for reimagining community and embracing micro-nations instead of wholly concentrating on artificially constructed countries. African states merely serve to issue necessary documents such as passports and identity cards. Anything less than embracing Africa's multitude of micro-nations perpetuates the cultural deracination that has left millions of modern Africans with the traits deriving from disempowerment discussed in the previous paragraph. She advocated for Africans reconstituting their nations with the support of grassroots civil society. Only through broad-based African ownership, foresight, and collective action will a positive and alternative future emerge. Africans cannot change the past, but they can manage the present differently to determine a better future.

## ENDING AID?

Within the last 50 years, US$1 trillion of foreign aid has been remitted to Africa. The continent ranks consistently as the largest aid-receiving region. Aid is given via loans and grants or materials, labor, and/or expertise. National governments, multilateral agencies (e.g., the United Nations), private organizations, and NGOs participate in a multilayered global aid regime. There are different types of aid: project aid (schools, ports), program aid (sectoral support, such as for agriculture or education), budget support (direct financial support), and various combinations, such as sector-wide aid (build and maintain schools). Aid has financially benefited African governments, NGOs, and community-based organizations, but the effectiveness of aid is always open to question and debate. Its impacts are difficult to measure and quantify. Establishing causality on many dimensions of the impact of international assistance is inherently difficult (e.g., aid and economic growth). Intangibles are not captured by standard assessments (e.g., transferring know-how and reverse transfers of African know-how). Nevertheless, it has not gone unnoticed that the terms upon which it is delivered are rarely defined by the people of Africa. African aid critics have lacked a public platform until recently.

The argument that aid is not working has been gathering momentum among academics and activists in Africa. There is a striking irony to current 21st-century aid realities. The capitalist West declares Africa to be in need of aid, while former communist and socialist countries such as China and India see Africa as an economic opportunity. There is an aid constituency in Africa that would like to continue the current aid regime with modifications (some radical), but none of the most passionate advocates for aid to Africa are Africans.

The perspective that aid may actually prevent African countries from developing by allowing their rulers to avoid long-term solutions is gaining ground. Aid may be destroying local initiatives and holding Africans back. As the Ugandans say, "begged water does not quench a thirst." One of the biggest opponents of aid is Dambisa Moyo, a Zambian-born economist. Her book *Dead Aid: Why Aid Is Not Working and How There Is a Better Way for Africa* (2009) has received much attention. Moyo gained recognition quickly as an articulate, African-born economist, with an academic and business pedigree (Harvard and Oxford University–educated and Goldman Sachs investment strategist). Her argument is not entirely new; others (e.g., William Easterly, a professor at New York University and a former World Bank economist) have come to similar conclusions. Moyo argues succinctly that aid is a disaster in Africa and that it is a disease pretending to be a cure. Like many others, she believes that aid fosters dependence, encourages corruption, and ultimately perpetuates poverty and bad governance. Aid is compared to oil in that it enables powerful elites to misuse and embezzle public monies. The siphoning of funds to replenish Swiss bank accounts happens all too often. She reasons that aid is too-easy money for governments; if governments were to rely on private investments, they would be accountable to private institutions, and if they relied on taxation, they would be accountable to their citizens. As it stands, African governments are hardly accountable, as the aid professionals are not predisposed to lift the lid on aid and undermine their own positions. In her writing, Moyo fervently calls on Africans to start representing themselves on the world stage rather than leaving it to Western rock stars (such as Bono and Bob Geldof). The core of her argument is that economic development in Africa can come about through borrowing on international capital markets, promoting Chinese investments, microfinance, and remittances. How these ingredients can combine and interrelate is left unexamined.

Moyo's arguments have been roundly criticized and contested for being based on weak analysis and for coming to mistaken conclusions about the life-and-death realities of Africa's poor. Jeffrey Sachs, director of the Earth Institute at Columbia University and one of the most influential economists in the world, is very vocal in asserting that "dead aid" is dead wrong. Sachs believes that ending aid would be irresponsible and would damage the poorest of Africans, some 400 million people living in extreme poverty. Lumping all aid into one undifferentiated mass program and project does not allow for the identification of the most successful aid programs. Examples of effective aid include support for farmers to grow more food, childhood vaccinations, deworming, school meals, training and salaries for community health workers, safe drinking water, antiretroviral medicine for AIDS sufferers, and clean low-cost cooking stoves to prevent respiratory diseases in young children. Some countries in Africa are on their way toward graduating from aid (e.g., Ghana, Rwanda, and Tanzania), but all of the fastest-growing African economies are still highly dependent on aid. Aid paved the way for growth in Asia (e.g., Taiwan, South Korea), but it is not clear why African states should be an exception. Moyo has been sharply criticized for giving too little consideration to the structural integration of African economies into the global economy. She ignores the impact of the debt crisis in undermining economic development, reinforcing poverty, and eroding health and education systems.

The future of foreign aid in Africa is an important debate. It may be easier to argue for ending aid than to demand a radical restructuring of the entire global architecture of international assistance. Aid can make a difference, but it should not distract us from considering other important mechanisms for promoting development (e.g., transparency, leadership, accountability, civil society, international finance). Africa's future will ultimately not depend on aid; it will depend far more on its people and its governments. The framing of Africa's development futures is now firmly shifting away from what the outside world owes Africa toward what Africans owe themselves.

## CIVIL SOCIETY

Civil society has come to play a significant role in Africa's democratization process and development initiatives. A plethora of civil society organizations and NGOs emerged in the 1990s, coinciding with the tilt toward democratization and more freedom of expression within particular African states. The term "civil society" pertains to the population of groups formed primarily outside the state and the marketplace for collective purposes. At its core, civil society holds ambitious aims of equality and justice, democracy and tolerance. It encompasses many forms of voluntary collective action—formal and informal, traditional and modern, secular and religious—extending well beyond high-profile formal NGOs with international support and funding. This landscape is diverse and complex, but poverty reduction is the common denominator for many civil society organizations in Africa. These organizations attract membership from the urban poor because state structures that affect their day-to-day lives seldom permit their active participation. Working at the grassroots permits a better opportunity to create local ownership of development (arm's-length government and international aid agencies often have difficulties in this respect). In theory, civil society constitutes a space for action outside the state, but in practice, distinctions and roles blur as mainstream development organizations co-opt civil society. Nevertheless, a critical feature of civil society organizations at the grassroots level is that they provide civic energy and, therefore, unleash hope and potential, even in the midst of unpromising circumstances.

Some scholars (e.g., Arjun Appadurai [2001]) hold very ambitious visions for civil society: a globalization from below to counter top-down globalization, members of the urban poor forming global alliances, and simultaneous activism at different geographical scales (local, urban, international). Intermingling the various scales of political action strengthens civil society by enhancing their knowledge bases and strategic arsenal. Cross-border activism is viewed as one of the most effective strategies for shaping a new political horizon of the urban poor and for pressuring local governments to make cities more inclusive of the poorest of the poor.

Some civil society organizations are registering successes. A very prominent civil society organization active in 15 African states is Shack/Slum Dwellers International (SDI). SDI is active in Angola, Ghana, Nigeria, Kenya, Liberia, Malawi, Mozambique, Namibia, Sierra Leone, South Africa, Swaziland, Tanzania, Uganda, Zambia, and Zimbabwe. The organization was founded in 1996 as a grassroots movement of the urban poor. Initially, intellectual leadership came from India and South Africa, but the organization has broadened its expert base over time. SDI has been successful in advancing a pro-poor agenda that aims to make cities more inclusive. Its central message is that slum dwellers need to organize into local groups (federations) in order for governments to change the way they operate toward the poor. SDI has a tried and tested methodology that is deployed in their member communities (Fig. 2.4).

Six interrelated components (saving and finance, community planning, exchange and learning, partnerships, slum upgrading, and women) allow for the poor to insert themselves (via the NGO) at the center of strategies to develop their communities. A central tenet is that ownership of the process leads to better development outcomes. This raises an important question: can international aid agencies better reduce poverty by direct relationships with the poor as opposed to operating at government levels? Unfortunately, donor agencies and their constituencies are more preoccupied with accountability upward in their own countries (national governments, media, and pressure groups) rather than with the people they serve. SDI's successes contribute to reassessment of these relationships.

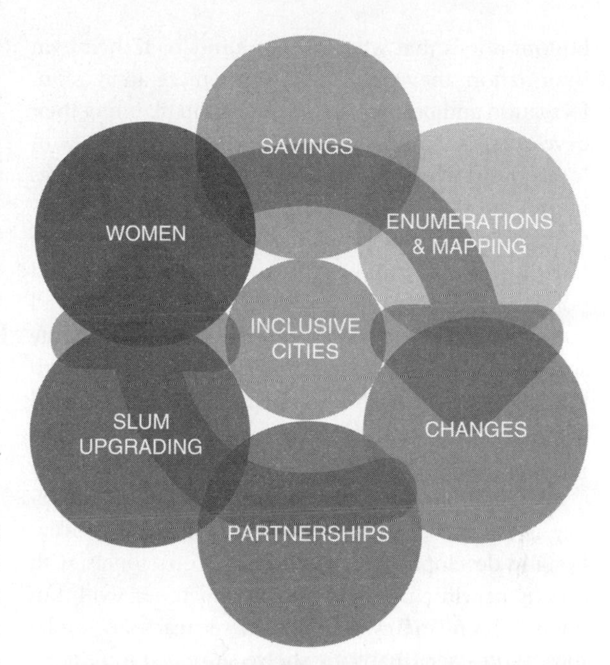


**FIGURE 2.4** Self-help: Shack/Slum Dwellers International (SDI) Methodology.

SDI federations excel in the initial stages of community planning. Their community-led enumerations have been vital in giving visibility and legitimacy to the poor, communities otherwise rendered invisible in national censuses (official censuses are notorious for undercounting slum dwellers) and in mainstream urban policies. Community-led enumerations count households, map settlements, and survey at the household level to provide detailed socioeconomic profiles of settlements. For example, the Ghana National Population and Housing Census of 2000 recorded 10,000 people in Old Fadama, a slum in the center of Accra, the capital of Ghana. A community-led enumeration in 2009 tallied 76,684 persons, providing a more exhaustive baseline. The settlement profile also revealed that almost all of the working-age population is employed in the informal economy and thus not recorded by government data. Given that the Old Fadama community is under an eviction threat and relocation is being discussed, these discrepancies have enormous implications for settlement provisions and their possible outcome. The bottom line is that when communities own their own information, they are able to gather more accurate information and become active partners in planning their development. Ultimately, as SDI federations work with local governments to verify and legitimize their findings in order to mainstream community-collected information for citywide planning purposes, everybody benefits from a bottom-up approach. Enumerations thus build political capital for communities both internally and externally. Activities such as community mapping create space for communities to identify development priorities, organize leadership at the street and household levels, build community consensus, facilitate local development, and envision an alternative future.

Another transformative aspect to the SDI methodology is that people learn through exchanges. Conventionally, development is planned by professionals, with experts providing solutions that sometimes work but often fail, and many solutions never reach an implementation stage (many are shelved, deemed too expensive and/or low priority). A deplorable outcome is that the communities are unable to advance their understanding of their own problems, which is undermined by professionals coming and going on project work with little regard for community-learning processes.

Two types of exchanges are practiced among federations of the urban poor. First, horizontal exchanges are the primary vehicle for learning and occur when one federation visits another to discuss communities' savings groups and/or toilet building, for example; learning takes place within the urban and national network. Exchanges encourage self-learning in that "doing is knowing," promoting the belief among the poor that they can become experts in their own development and can contribute toward a collective vision. This breaks the isolation of those in poverty, reclaims local learning, and prevents development professionals from repackaging ideas as their own. Second, vertical exchanges involve international exchange, for example, slum dwellers from Ghana can visit with slum dwellers in Nairobi, although in more recent times most of these exchanges take place within a proximate African region. International exchanges are instructive about successful outcomes and about failures, thus generating a knowledge bank that is a collective asset of the SDI network. The major advantage of international exchanges has been their ability to turn the tables on local development planners: slum dwellers now possess knowledge. When they meet with local officials and external actors to debate development policies, they can draw from international examples, forcing their government and other stakeholders to listen as well as to think. Certainly, the federations of the urban poor working with SDI dramatically increase their exposure to international organizations, the international press, academics, and a range of other domestic actors.

SDI has been remarkably successful in garnering support from academics, development practitioners, and funding agencies. It attracts millions of funding dollars from multilateral institutions (e.g., Cities Alliance), national development organizations (e.g., Department for International Development, UK [DFID], Swedish International Development Cooperation Agency [SIDA]), and philanthropic organizations (e.g., Gates Foundation). SDI intellectual supporters claim to have created a new optimistic development strategy of coproduction (Mitlin 2008), which purports to link the poor, their support professions, the state, and international funders in a win–win alliance in several domains, such as providing tenure and shelter to slum dwellers and incorporating the role of women in development.

As with every international organization, SDI has its critics. The most frequent criticism is that SDI simply mirrors outside interests and agendas—in essence, that activists operate within networked societies rather than focus on resistance. Radicals phrase this as co-opting or sleeping with the enemy. An ever-present money and power dynamic means that NGO opinion leaders are prone to look to the coordinators rather than to the constituents. A deeper complaint is that civil society, in general, is a dangerous distraction from the unfinished business of building national identities and developmental states, states that would have the authority and legitimacy to redistribute land and other assets and to provide coherent direction for economic development. Some observers cast doubt on whether civil society can actually be realized in African societies so fragmented along particularistic lines that public or common interest is little more than a pipe dream. In some instances SDI has monopolized the voice of the urban poor. SDI's successes may inadvertently reduce the role of smaller and alternative NGOs focused on poverty that simply cannot compete on the same playing field. There is also the charge that activists are no more than other urban elites; influence is restricted to cyberspace, international conferences of like-minded intellectuals, World Social Forums, and the corridors of Davos, London, Washington, and Pretoria. In spite of these criticisms, the voices of Africa's urban poor are now being heard.

## AFRICAN POSTCOLONIAL URBANISTS

Postcolonialism is a body of scholarly work that provides a postmodern reaction to, as well as an analysis of, the cultural legacy of colonialism. Unlike the previous perspectives, there is not one African individual who represents the perspective. Instead, there is a body of scholarly work that follows on from the work of literary theorists such as Edward Said and Frantz Fanon. African urbanists (e.g., AbdouMaliq Simone and Edgar Pieterse) and urban Africa geographers (e.g., Garth Myers and Jennifer Robinson) are examples of contemporary urban postcolonial scholars.

The postcolonial lens reverses the gaze of urban history and contemporary urbanity, invoking African perspectives. Emphasis is directed toward grassroots actors (and away from the axes of power) and toward the politics of struggle and opposition and understanding ordinary cities in their true state as opposed to comparing African cities with Global Northern cities. The postcolonial project aims to move beyond naive, false utopians and project-targeted development initiatives in order to liberate thinkers to imagine other, and different, future possibilities.

Oral histories, previously unused sources (sometimes in local languages), and community-based research are powerful methods for capturing varied and interlocking dynamics of lived urban experiences. The city is represented more as a product of indigenous, local agency even against the backdrop of powerful Western modern influences (e.g., urban planning, planning ordinances, and continuities in the built environment).

Postcolonial analysis recovers ordinary lives and illustrates diverse trajectories and complex interconnectivities (e.g., between planned and unplanned buildings, between formal and informal economies), intersections largely rendered invisible in mainstream urban analyses. Contemporary postcolonial scholars underscore informality as mode of urbanization in Africa, and in the process, they go a long way toward valorizing ordinary Africans' own agency (Myers 2011; Pieterse 2008; Simone 2010). However, there continues the perception that ordinary Africans' deliberations about present and future ways of living are still heavily dependent on global power dynamics.

Postcolonial theorists argue for switching the emphasis toward ordinary cities (away from the upper echelon of global cities such as London, Paris, and New York) to create a more cosmopolitan urban theory—a theory based on a diversity of cities in Africa and not grounded only on Global Northern experiences. Routinely, African researchers are expected to frame their contributions within the theoretical terms and concerns of leading Global Northern scholars and almost never frame their work on Africans' work. The comparative urban research tradition offers a middle ground by assessing what can be learned in different regional contexts—for example, Africa and the United States (Mitchell 1987; Robinson 2002). The comparative urban approach illustrates that African and North American cities in different times and different places (the United States at the turn of the 20th century and Central Africa in the 1960s) can inform each other in understanding urban social processes (Mitchell 1987). For example, Chicago of the 1860–1910s and Zambian towns in the 1960s display similar economic growth trajectories, with high demands for cheap male labor, which found geographical expression in patterns of poverty, overcrowding, racial segregation, and ethnic associations. The comparative lens moves beyond idiosyncratic and single-case studies toward understanding general sets of circumstances that arise in different times and spaces. To produce cosmopolitan urban theory is one of the most important challenges that the academy faces (Robinson 2002). To make truly cosmopolitan urban theory a reality, scholars in privileged academic environments have to find responsible and ethical ways to engage with, learn from, and promote the ideas of intellectuals in Africa and to share in the coproduction of urban knowledge.

The particular value of a postcolonial approach in an urban context is its sensitivity to diversities of experiences and its assertion of a global perspective that views different urban forms as integral to an understanding of the contemporary world and the rightful place of African urbanism within global urban knowledge. Building more adventurous as well as more nuanced urban accounts is an ambitious intellectual project. However, the perspective offers more in terms of critiquing mainstream approaches but less in the transfer of its insights into urban practice. This approach, which is gaining ground in the social sciences, is yet another advance against the traditionally overgeneralized and totalizing narrative.

## CONCLUSIONS

The Africa conversation is changing. It has moved on from the colonial dark days into contemporary splintering light. Africa knowledge is becoming more diverse, democratic, and informed. A wealth of up-to-date material is available online for those who want to be informed about Africa. Social media with an Africa focus is becoming ever more important. Still, the general public in the Global North's basis of Africa knowledge is low: ignorance, misunderstanding, and oversimplification are an enduring colonial legacy.

There is no shortage of ideas about how to shape Africa's future. Some of the leading ideas from African scholars/thinkers were presented in this chapter. Scholars such as Amin and Bond use Marxian reasoning to demonstrate how the capitalist system unfairly treats Africa and Africans. Others (e.g., Ibrahim and SDI) opt for more incremental transformation. Moyo, on the other hand, calls for an abrupt end to foreign aid and the cutting off of African "aid addicts." Postcolonialists argue that different modes of reasoning are needed to liberate people to imagine more sustainable and socially just futures. Lively discussions are taking place on genuine leadership, offshoring, civil society, recovering voices, and the possibilities of restructuring the international development architecture and about the United Nations' sustainable development agenda post-2105 (discussed in Chapter 13). More self-confident African leaders expect nothing less than to be partners in current and future development instead of the objects of development.

Africa knowledge is very splintered. It no longer makes sense to have two different academic planes of knowledge on Africa: that of Northern experts and that of African experts. In the twenty-first century it makes more sense to combine global and local knowledge; only together will our knowledge be sufficient to fully understand the complexity and diversity of the region. Coproduction of knowledge is an important path, and significant support is required. Educational approaches need to contextualize better and more creatively knowledge on Africa, and the academy needs to find ways to make theories more cosmopolitan, incorporating evidence and ideas from Africa.

On a practical level the current global financial crisis offers a rare moment in history where the interests of rich countries and those of people living in poverty in Africa are synonymous. At the core of the global financial crisis is the absence of transparency of the entire international system and burgeoning global inequality. The shadow financial system is bleeding Africa as well as putting the global economy on life support. It is particularly egregious that pundits, journalists, and politicians fawn over people who get rich by abusing the global system—getting around taxes and regulations and forcing everyone else to shoulder the associated risks and taxes. Much of the public in the Global North is too distracted by their own immediate financial challenges to notice that much is happening in the world and that Africa is not the place many imagine it to be.

## REFERENCES

Ake, C. 1991. "How Politics Underdevelops Africa." In *The Challenge of African Economic Recovery and Development*, eds. A. Adedeji, O. Teriba, and P. Bugembe, pp. 316–329. Portland: Frank Cass.

Amin, S. 1990. *Delinking: Towards a Polycentric World.* London: Zed Books.

Appadurai, A. 2001. "Deep Democracy: Urban Governmentality and the Horizon of Politics." *Environment and Urbanization* 13(2):23–43.

Bestman, C., and H. Gusterson, eds. 2005. *Why America's Top Pundits Are Wrong: Anthropologists Talk Back*. Berkeley: University of California Press.

Bond, P. 2006. *Looting Africa: The Economics of Exploitation*. New York: Zed Books.

CIA. 2012. *The World Factbook*. https://www.cia.gov/library/publications/the-world-factbook/rankorder/2172rank.html (accessed April 3, 2012).

Conrad, J. 2014. *Heart of Darkness* (reprinted). New York: Miller/Strauss Publishing.

Davis, M. 2007. *Planet of Slums*. London: Verso.

Deviers-Joncour, C. 1999. *La Putain de la Républic*. Paris: Mass Market Paperback.

Ferguson, J. 2007. *Global Shadows: Africa in the Neoliberal World*. Durham: Duke University Press.

Kagen, R. 2000. "The Return of Cheap Pessimism: Inside the Limo." *New Republic Online*, Apr. 10. Available at http://business.highbeam.com/4776/article-1G1-61410834/inside-limo-return-cheap-pessimism.

Kaplan, R. 1994. "The Coming Anarchy." *Atlantic Monthly* 273(2):44–76.

Kar, D., and D. Cartwright-Smith. 2010. "Illicit Financial Flows from Africa. Hidden Resource from Development." Global Financial Integrity. Available at http://www.gfintegrity.org/index.php?option=com_content&task=view&id=300&Itemid=75.

Maathai, W. 2009. *The Challenge for Africa*. New York: Pantheon Books.

Mandela, N. 1995. *Long Walk to Freedom: The Autobiography of Nelson Mandela*. New York: Little, Brown and Company.

Mitchell, J. 1987. *Cities, Society, and Social Perceptions: A Central African Perspective*. Oxford: Clarendon.

Mitlin, D. 2008. "Urban Poor Funds: Development by the People for the People. Poverty Reduction in Urban Areas Series: Working Paper 18." London: Institute for International Environment and Development.

Monga, C. 1996. *The Anthropology of Anger, Civil Society and Democracy in Africa*. Boulder: Lynn Rienner.

Moseley, W. 2011. "China's Farming History Misapplied in Africa." Available at http://aljazeera.com/indepth/opinion/2011/10/201110241502249406.html (accessed April 3, 2012).

Moyo, D. 2009. *Dead Aid: Why Aid Is Not Working and How There Is a Better Way for Africa*. London: Allen Lane.

Myers, G. 2011. *African Cities: Alternative Visions of Urban Theory and Practice*. London: Zed.

Nobel Prize Organization. 1993. "The Nobel Peace Prize 1993." Available at http://www.nobelprize.org/nobel_prizes/peace/laureates/1993/ (accessed December 18, 2013).

Owomoyela, O. 2010. "The myth and reality of Africa: a nudge towards a cultural revolution." In *Reframing Contemporary Africa: Politics, Economics and Culture in the Global Era*, eds. P. Soyinka-Airewele and R. Edozie, pp. 47–60, Washington DC: CQ-Press.

Palen, R., R. Murphy, and C. Chavagneux. 2010. *Tax Havens: How Globalization Really Works*. Ithaca, NY: Cornell University Press.

Pieterse, E. 2008. *City Futures. Confronting the Crisis of Urban Development*. New York: Zed Press.

Popke, J. 2001. "The Politics of the Mirror: On Geography and Afro-Pessimism." *African Geographical Review* 21(1):5–27.

Robinson, J. 2002. "Global and World Cities: A View from Off the Map." *International Journal of Urban and Regional Research* 26(3):531–544.

Samatar, A 2013. "African Union: Between Hope and Despair." Available at http://www.aljazeera.com/indepth/opinion/2013/05/201352762217612138.html (accessed September 26, 2013).

Shaxson, N. 2011. *Treasure Islands: Uncovering the Damage of Offshore Banking and Tax Havens*. New York: Palgrave Macmillan.

Simone, A. 2010. *City Life from Jakarta to Dakar. Movements at the Crossroads*. New York: Routledge.

Stengel, R. 2008. "Mandela: His 8 Lessons of Leadership." Available at http://www.leadershipnow.com/leadingblog/2008/07/mandela_his_8_lessons_of_leade.html (accessed December 18, 2013).

Tax Justice Network. 2013. "Financial Secrecy Index 2013." Available at http://www.financialsecrecyindex.com/introduction/fsi-2013-results (accessed November 11, 2013).

United Nations Economic Commission for Africa (UNECA). 2013. *The 2013 African Economic Outlook*. Addis Ababa: UNECA.

World Bank. 2013. "Poverty Headcount Ratio at $2 Per Day (PPP)." Available at http://data.worldbank.org/indicator/SI.POV.2DAY (accessed December 18, 2013).

## WEBSITES

African Arguments. http://africanarguments.org (accessed December 4, 2013). A multi-blogging site that covers unfolding, contemporary African events and develops debates on hot topics.

Africa Good News. http://www.africagoodnews.com (accessed December 4, 2013). South African website that focuses on positive news coming out of Africa.

Africa Journals OnLine. http://www.ajol.info (accessed December 4, 2013). Portal of 465 peer-reviewed African journals.

Devarajan, Shanta. *Africa Can End Poverty.* http://blogs.worldbank.org/africacan/archive/201310 (accessed December 4, 2013). Blog by former Africa Chief Economist at the World Bank, 2005–2013. Other World Bank staffers have since continued the blog.

Easterly, William. *AidWatchers.* http://aidwatchers.com/author/easterly (accessed December 4, 2013). Blog by prominent economist and expert on African aid 2009–2011.

Good News About Africa. http://www.goodnewsaboutafrica.com (accessed December 4, 2013). London-based site that focuses on positive news emanating from Africa.

Green Belt Movement. http://greenbeltmovement.org (accessed December 4, 2013). Environmental organization that empowers communities, especially women, to conserve the environment and improve livelihoods.

Global Financial Integrity. http://www.gfintegrity.org (accessed December 4, 2013). Organization that monitors and researches illegal cross-border flows of monies, and reports on the African region.

Moyo, Dambisa. 2009. TEDxBrussels. http://www.youtube.com/watch?v=_QjiiM4jhbk (accessed December 4, 2013).

Nelson Mandela Foundation. 2013. http://www.nelsonmandela.org (accessed December 18, 2013).

PBS. 2013. "Remembering South Africa's Nelson Mandela." Available at http://www.youtube.com/watch?v=LnGeMBNS9ZA#t=126 (accessed December 18, 2013).

Slum/Shack Dwellers International. 2013. http://www.sdinet.org (accessed December 4, 2013).

Tax Justice Network-Africa. http://www.taxjusticeafrica.net (accessed December 4, 2013). Not-for-profit organization that promotes awareness about tax issues and engages in advocacy and research on Africa tax issues.

Think Africa Press. http://thinkafricapress.com (accessed December 4, 2013). Online magazine that brings together African writers and international experts on a range of topics (e.g., health, gender, the environment, economy, politics, development).

WhiteAfrican. http://whiteafrican.com (accessed December 4, 2013). Leading Africa technology blogger.

CHAPTER 3

# AFRICA'S ENVIRONMENTS

## INTRODUCTION

Aspects of the environment are foregrounded in the majority of visual and written representations of Africa. There are two dominant, yet competing, popular narratives of Africa's environment. First, the more dominant one depicts the region as an increasingly fragile milieu subject to different and recurring environment crises: droughts in the Sahel, deforestation, declining large animal populations and fish stocks, more frequent floods, and dirty and dilapidated urban settlements. Some 30 African mammals (e.g., chimpanzee, African wild dog, and mountain gorilla) are endangered species, and several species are threatened with extinction (e.g., cheetah, African elephant, white rhinoceros). This narrative has colonial origins, whereby African environments came to be viewed as strange and defective by outsiders compared to Europe's familiar and productive environment (Davis 2011).

Many deeply engrained images of African environments can be incomplete and fixed in a particular crisis period. For example, most people hold an image of Ethiopia as a site of famine and human-environmental tragedy: starving children, adults in ragged clothes, all helplessly corralled in refugee camps waiting for outside assistance or death. Considerable environmental challenges remain, but fast-forward to 2012: Ethiopia has become the world's largest livestock producer and its economy grew 7.5% in 2011.

The second prevailing narrative advances Africa as a resource frontier where minerals (gold, diamonds, uranium, oil, natural gas, etc.) are abundant, and other resources (timber, land, and fisheries) are available for processing and export to wealthier countries. Private deals can be struck (with highly favorable and expedient contract terms) for immediate access to virtually all of Africa's resources. The continent's biophysical resources can be deployed to sustain distant industries and populations. This underexplored and underexploited African environment narrative is reminiscent of the colonial scramble for the region. Many label the current resource grab "the second scramble for Africa" (Carmody 2011).

Both narratives are highly selective and incomplete. They overlook the complex nexus among people, environment, economy, and domestic and international political contexts. Ironically, the resource frontier narrative feeds into the environmental crisis narrative, as environmental damage caused by mining, overfishing, and so forth eventually leads back to the older environmental orientalist viewpoint. Thus, the need to "improve," restore," and "repair" African environments provides powerful justifications for innumerable imperial and developmental projects, privileging external assistance to restore the environment that hitherto other external forces altered dramatically. This does not, however, negate the role that Africans have in destroying their own environments (by artisanal mining, exploration, logging ventures, etc.), and a small cadre of Africans have profited greatly from these and other externally driven resource activities.

## THE ENVIRONMENT AND INTERNATIONAL DEVELOPMENT AGENDAS

Beginning in the 1970s, and especially since 2000, environmental issues have risen on the global development agenda. International organizations (e.g., World Bank, the Intergovernmental Panel on Climate Change [IPCC], and the United Nations [UN]) and nongovernmental organizations (NGOs; e.g., Greenpeace, Wildlife Conservation Society, World Wide Fund for Nature) are spearheading a more internationally driven emphasis on the environment. The UN body, for instance, is composed of several prominent environment-focused organizations (e.g., Food and Agriculture Organization [FAO] and United Nations Environment Programme [UNEP]) that engage in high-profile initiatives in various policy arenas. The environment and sustainable development have risen to the top of the global agenda.

As mentioned in the previous chapter, the Millennium Development Goals (MDGs) became a key driver in development policy and practice after 2000. Particular MDGs create ambitious environmental sustainability targets. For example, MDG 7 seeks to advance environment sustainability by reversing the loss of environmental resources (e.g., forests), by preserving biological diversity (wildlife, plants, and fisheries), by enabling access to drinking water and basic sanitation, and by improving the lives of slum dwellers.

However, the UNEP (2008) progress report indicates that more African countries are failing than succeeding on each of the environmental targets, with the exception of water (Fig. 3.1). Many governments are acknowledging the environment more, but many African elites only pay lip service and instead focus on business as usual and political priorities.

The environment is critical to Africans' livelihood: over half of the population engages in agriculture, and forest activities and subsistence fisheries (both inland and coastal) support many others. Overharvesting, destructive methods by large, external industrial fisheries, coastal oil and gas exploration, and unsustainable urban and industrial development—to mention just a few environmental threats—are causing unknown (and to date unquantifiable) coastal and marine resource damage. Extreme wave events, sea-level rise, and land sources of pollution are growing threats to marine and coastal resource productivity.

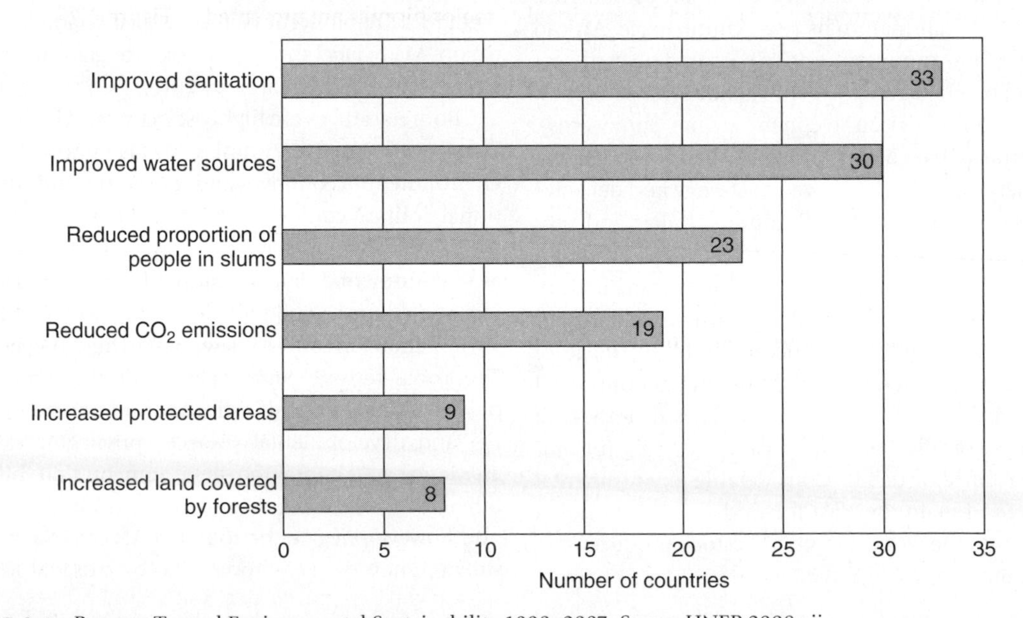


**FIGURE 3.1** Progress Toward Environmental Sustainability 1990–2007. *Source:* UNEP 2008:xii.

On land, deforestation, the conversion of more land to biofuel production and other export crops, and mining are inducing environmental change. Moreover, fluctuations in rainfall, the diffusion of plant diseases and pests (e.g., locusts), and some human diseases (especially water-borne ones) that are sensitive to environmental and climate change are becoming more problematic. Given the prevailing low levels of technology, weak environment regulations, and scarcity of funding for ecosystem preservation, African populations are highly vulnerable to environmental change. Human adaptation to environmental change is being emphasized as the most expedient response.

In Africa, there is a tendency to equate the term "environment" with wilderness and rural settings, overlooking the critical urban-human environment and downplaying other external drivers (e.g., global climate change and the role of foreign investors and multinational corporations [MNCs]). The environment in this chapter is conceptualized in a broad sense, encompassing the complex biophysical arena and urban-ecological and coastal and marine systems, and an attempt is made to provide examples of international environmental drivers. From a social science perspective, environmental issues need to be contextualized—socially, economically, and politically. For example, political instability, war, government policy, and international contractual arrangements related to resources have profound implications for Africa's environments.

## THE DIVERSITY OF AFRICAN ENVIRONMENTS

Africa is the second largest continent (after Asia) and covers one fifth of the planet's total land surface. It has many diverse and varied environments: deserts, rain forests, savanna grasslands, wetlands, and large urban settlements. The environments in Africa "span entire moisture and temperature gradients, from perhaps the most arid to among the most well-watered places on earth, from the coolness of the Cape to the furnace that is the Sahara" (Meadows 1999:161).

Several of the largest states in Africa exhibit considerable environmental diversity. Its largest state—Democratic Republic of the Congo (DRC)—comprises a plethora of different environments: rich forests, abundant minerals and fish, and plentiful agricultural land. The DRC is well endowed with water supplies, which have enormous potential for capturing for hydroelectric power generation and for trade/sale or diversion to water-insecure countries.

In Africa, broad biomes can be identified (Meadows 1999). A biome is defined as a large region whose interactions of plants and animals with the climate, geology, soil types, and water resources produce broad uniformity. These large-scale mapping units represent reasonably homogeneous areas of the land surface that capture the close ties among climate, vegetation, biota, and general environmental conditions. In Africa, plants tend to be used as the fundamental classification basis (rather than animals) because vegetation is the most reliable indicator of other environmental factors. However, natural vegetation is greatly modified by human activities (e.g., planting of species, fires, grazing). Thus, what is actually mapped is more properly termed anthropogenic vegetation: that influenced or disturbed by human beings. Emphasis is placed on the broad uniformity within the biome because significant variation occurs within the classification based on local changes in soil, wildlife, and human settlement patterns. Biomes provide a useful starting point for an overview of habitats at the macroregional scale in Africa. Six major biomes are presented in Figure 3.2.

### TROPICAL RAIN FOREST

A vast tropical rain forest (TRF) centers on the Congo Basin (almost 700,000 square miles [1.8 million km$^2$]); other rain forests are concentrated in coastal West Africa (extending westward and then northward from the Cameroon Highlands through the Niger Delta and along the Guinea coast as far as Sierra Leone) and eastern Madagascar. TRFs occur in close association with equatorial climates, which are characterized by consistently high temperatures, heavy rainfall (exceeding 55 inches [1,400 mm] per year and at least 2 inches [51 mm] per month), and a minimal dry season. These climatic conditions permit the development of a diverse biome and lush forest.

Biodiversity in TRFs is the greatest among all terrestrial biomes. However, African TRFs contain fewer numbers of plant species but more richness in primates (e.g., endemic forest gorillas) compared to the

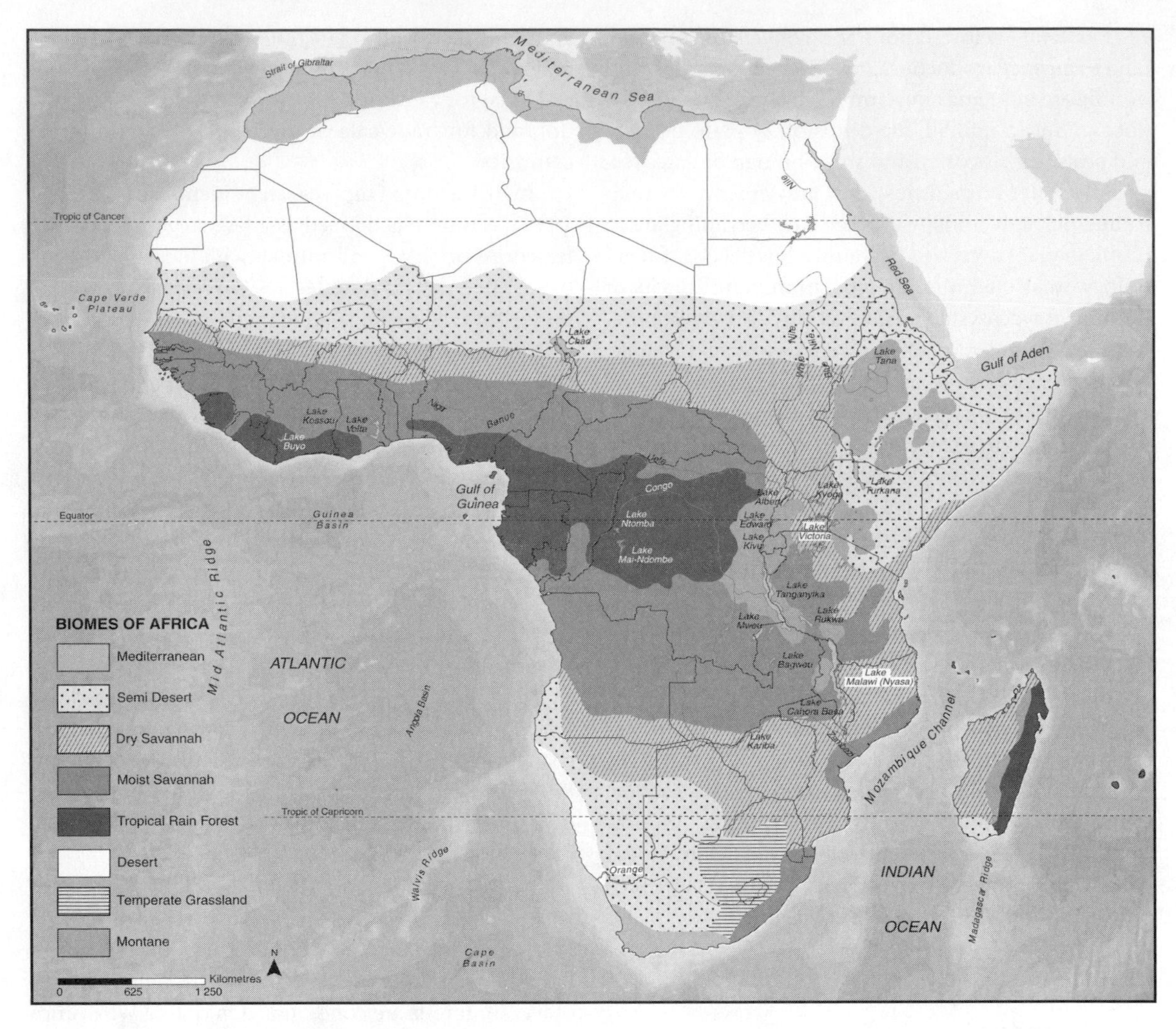


**FIGURE 3.2** Map of Biomes. *Source:* UNEP 2008:10.

Amazon Basin. An apparent impoverishment of plant biodiversity is still a matter of debate, partly reflecting the dearth of African TRF research compared to related research on Amazonia. Some posit that plant biodiversity relates to the breakup of the world landmass into continents before the development of seed flora; others argue that climate fluctuations in former ice ages triggered the retreat of rain forests into the current TRF distribution refuges, which represent a net loss of diversity; still others posit that refuge species richness is more a function of local climate and soil conditions than any geophysical epoch event (Meadows 1999). In this regard, soils in African TRFs are poorer in nutrients and more weathered and therefore less fertile than those in Amazonia.

TRF vegetation characteristically encompasses three layers: a ground cover of shrubs and ferns (6–10 feet [2–3 m]); a middle story of trees, palms, and woody climbers (59 feet [18 m]); and a dominant canopy of broad-leaved evergreens (66–98 feet [20–30 m]). Trees with commercial value include mahogany, rubber, and oil palm, and other trees are felled for plywood. Legal timber concessions cover extensive areas. An area the size of Sweden (173,731 miles² [449,964 km²]) has been cleared in West and Central Africa (Mercer et al. 2011). There can be no doubt that the corridors of African deforestation have widened in the last 30 years.

Other human activities have added to deforestation pressures: land conversion to agriculture and human settlements, road construction through rain forests, local lumbering and the cutting of timber for firewood and charcoal, and even civil conflict (where informal logging takes place unabated). According to the FAO (2010:xvi), 386,102 miles$^2$ (1 million km$^2$) of Africa's TRF have been cleared in the last three years alone. Clearing is still taking place at very high rates (averaging 8.4 million acres [3.4 million hectares] per year since 2000), even though it is slowing somewhat. For instance, five of the world's top 10 countries with the highest annual loss of forests are in Africa (FAO 2010:xv1). Human stewardship is still quite limited there: less than 14% of TRFs are under legal protection (UNEP 2008).

### TROPICAL SAVANNAS

Savanna vegetation covers an extensive area (approximately two thirds of the continent) and is largely located between TRFs and the deserts to the north and south. Savannas prevail in strongly seasonal climates, punctuated by a dry season (three to eight months' duration), with heavy precipitation at other times. African savannas have been studied more extensively than Africa's TRFs. Tree species and densities vary according to the amount of rainfall, soils, and anthropogenic influences. The savanna is an elongated belt stretching from east to west just under the Sahel, and it is also found in Central Africa and parts of Southern Africa. Two different subregions—dry savanna and moist savanna—are often delineated.

The dry savanna is located around Sudan, where scattered trees and short grasses predominate. The moist savanna is found closer to the Equator as well as the Guinea coast and includes a broad zone from Angola to the Indian Ocean and a zone in the Southern African region in the lowlands of Zimbabwe and South Africa. In wetter areas, the vegetation changes to a mixture of trees and tall grasses, extending to woodlands in a band adjoining TRFs. The vegetation range in this biome is enormous, varying from region to region.

The baobab tree (a cylindrical tree with a trunk (16–26 feet or 5–8 m wide) and the acacia tree (popular with grazing giraffes) are species typical to this biome. The savanna supports the bulk of Africa's cattle population and the world's most diverse wild animal populations (Meadows 1999). Animals such as lions, steenbok, antelopes, wildebeest, buffalos, zebras, and rhinoceros are found on this land, and protected game reserves such as the Serengeti (Tanzania) are known for abundant grazing herbivores.

Human influences on savanna vegetation are extensive. Farmers, herders, and hunters have disturbed the evolution of vegetation in the savanna by their activities, selectively eradicating less fire-resistant species and favoring grasses that regenerate more rapidly. Moreover, savannas have been commonly cleared for agriculture.

### DESERT

At the polar margins of the savanna, mean annual precipitation declines, and the length of the dry season becomes a major geophysical constraint (Meadows 1999). Dry savanna gives way to grass steppes, which transition to semidesert (e.g., the Kalahari and Karoo in southern Africa and the Sahel in northern Africa) and then eventually into desert. For example, the semidesert Sahel zone lies between the dry savanna of Sudan and the Sahara to the north. The area's expanding human and livestock populations have been affected by prolonged and recurring droughts. Anthropogenic influences (e.g., overgrazing, fire, collection of firewood) have had a negative impact on the environment. However, there is considerable ongoing debate about the main drivers of the Sahel's desertification and the precise role of human-induced changes, climate change, and various combinations (see Chapter 11).

At the outer edges of these arid areas lie two major deserts, the Sahara and Namib. The Sahara is the world's largest hot desert, extending over 3.6 million miles$^2$ (9.4 million km$^2$), occupying an area almost equivalent to that of the United States. Vegetation adapts to sparse and unpredictable rainfall (less than 4 inches or 100 mm per year), extremes in temperature (daily and seasonal), and poor soils. Saharan vegetation comprises fewer than 500 plant species, an extremely low number considering its size. It is generally sparse, with scattered concentrations of grasses, spiny shrubs, and trees in highlands (e.g., acacia tress and oasis depressions [often consisting of date palms and pomegranate and other fruit trees grown with irrigation]). The population of the Sahara is thin: 2.5 million people are scattered wherever there are available and reliable water sources and vegetation that supports grazing. Vast expanses of the Sahara are virtually empty of human population.

### TEMPERATE GRASSLAND (VELDT)

Temperate grasslands are concentrated in Southern Africa (known as prairie in the United States); this biome is characterized by a landscape dominated by grasses with some trees, largely a function of an interior area of higher elevation and moderate temperatures. Subsets of this biome are short-grass steppe (prairies) of humid climates and tall-grass steppe that can support forest. Soils, deep and rich in organic matter, and ample annual precipitation (18–28 inches or 450–700 mm, concentrated in summer months) produce very productive agricultural land. Sauer (1950) and others postulated that because its grasses are fire-resistant, it is of anthropogenic origin.

In South Africa, this area is known as the veldt (an Afrikaans word for field), but the area extends into Zimbabwe. The veldt is one of the world's oldest regions inhabited by humans: fossil evidence indicates that members of the hominid genus *Australopithecus* were present some 3 million years ago and Stone Age people some 300,000 years ago. The natural state of this environment has been dramatically transformed over time; most profound and most recent has been environmental change. Large tracts of land have been converted to dry land agriculture and livestock production.

### MEDITERRANEAN

The Mediterranean biome is found in both Northern and Southern African regions. For example, the South African Cape region is characterized by warm, dry summers, alternating with a cool, rainy period. The season during which temperatures are most suitable for growth coincides with the period of minimum available moisture, affecting the vegetation selection process (Meadows 1999). As a result, there is a lack of true woodland, otherwise referred to as the "vacant trees" niche (Meadows 1999). Nevertheless, tremendous biological diversity is evident: over 9,000 plant species grow in a 35,000-mile$^2$ (90,000 km$^2$) area, and 70% of these species do not grow anywhere else on earth.

The Cape region also contains a distinct fynbos ("fine bush" in Afrikaans) concentration, containing 7,000 species in an area stretching along the coastal belt of the southern Cape. The Cape Floral Kingdom in South Africa is the only floral kingdom (of six) concentrated in a single country. Covering an area of 1.36 million acres (553,000 hectares), it accounts for 0.05% of the African continent yet encompasses 20% of all of its plant species: such a high level of diversity is comparable to that of TRFs. Fynbos areas are threatened by the spread of alien species, in particular acacia species from Australia, and of pine plantations in the Cape foothills. Many species have become extinct, and more than 1,000 are endangered. Anthropogenic influences, especially urbanization and expanding viticulture, threaten much of this area (see the section on Ethical Wildflower Trade and Sustainability: Flower Valley, South Africa, in Chapter 10).

### MONTANE

This biome is found in relatively isolated mountain areas, such as the Ethiopian highlands, the Arc Mountains of East Africa, the Albertine Rift in central East Africa, Drakensberg, the Cameroon highlands, and the Atlas Mountains. In these discontinuous areas, altitude is a primary determinant of vegetation, soil, and climate. Accordingly, the vegetation is patterned in vertical zones with distinctive species graduation.

For example, the Albertine Rift (named after Lake Albert) is a 920-mile-long geologic valley of savannas, lake chains, and wetlands that rise to highland forests and snowcapped mountains, producing a luxuriant and biodiverse zone. The mountain chain making up the Albertine Rift straddles five countries (Uganda, Tanzania, Rwanda, Burundi, and Congo). Paradoxically, the Albertine Rift's richness (soils, animals, fish, and minerals) has led to human overexploitation of its resources. People crowded into this area because of fertile volcanic soil, plentiful rainfall, and vegetation biodiversity, and population numbers flourished in high

**FIGURE 3.3** Albertine Rift. *Source:* Image from National Geographic Society.

altitudes that were inhospitable to mosquitoes and tsetse flies and the diseases associated with these pests. The lack of management has put significant pressures on available resources. For instance, rural population densities in Burundi and Rwanda are among the highest in Africa. Population pressures require more forest clearing for farm and grazing lands, producing ecosystem changes evident in the sharp human-induced demarcations between land uses (Fig. 3.3).

Wars have also made the management of forest areas difficult, exacerbating encroachment pressures and other informal activities. For example, large numbers of refugees from the Rwanda/Burundi/DRC wars have led to the deforestation of some areas, and bands of rebels use other parts of the forest to hide in between periods of raiding and fighting. Despite high biodiversity importance, it is therefore not surprising that this forest montane is poorly studied.

## AFRICA'S RAINFALL AND CLIMATE CHANGE

Most of Africa's climate is tropical, and its latitudinal extent is between the Tropics of Cancer and Capricorn (i.e., the extreme southwestern tip of South Africa, the northwestern African coast, and the Sahara are not included). Temperatures above 70° F or 21°C for nine months of the year characterize the majority of the continent. The mean temperature in the hottest and coldest months of the year varies little for most of equatorial Africa, but away from the Equator and along the coast, seasonal variation is more dramatic. Still, there is considerable diversity within tropical climates due to (1) distance from the Equator, (2) winds and pressure systems, (3) maritime and continental influences, (4) ocean currents, (5) elevation, and (6) large lake effects in East/Central Africa.

Eight climatic zones related to precipitation and temperature characteristics have been identified (Goudie 1999), and there is considerable overlap between vegetation biomes and climatic zones (Fig. 3.4). Working from north to south, desert in North Africa and the Namib is the most extensive climate type. Africa's desert climates receive very little precipitation, and daytime temperatures are extremely high. The desert merges into a zone of tropical climate (the Sahel, the southeast of Africa) with long dry spells and low rainfall for at least six months (Goudie 1999).

The tropical savanna is located within a large area extending from the Equator in West and Central Africa and a small area on the east coast and in eastern Madagascar. This zone is characterized by precipitation and high temperatures throughout the year. Adjacent to this zone is a tropical wet climate with a short dry season and two rainfall maxima related to the intertropical convergence zone's movement back and forth: the first occurring in the months surrounding January as the zone moves poleward and the second arising when it retreats toward the Equator in the months surrounding August. The tropical summer-rain climate is located in the interior plateaus of Central and Southern Africa. Precipitation is concentrated in three summer months. The northern and southern extremities are characterized by hot, dry summers and mild wet winters, and precipitation can be quite low (e.g., Cape Town receives 24 inches or 615 mm annually).

The majority of scientific opinion recognizes that climate change is affecting Africa. Most scientists agree that anthropogenic influences (deforestation, pollution, industrial emissions) are altering climate worldwide, and human impacts on climate are projected to accelerate as the 21st century progresses. Human influences have been linked to increased concentrations of carbon dioxide, methane, nitrous oxide, and other greenhouse gases in the atmosphere. Anthropogenic $CO_2$ emissions have increased since the industrial revolution with the burning of fossil fuels, and more urban-oriented industrial development and emissions

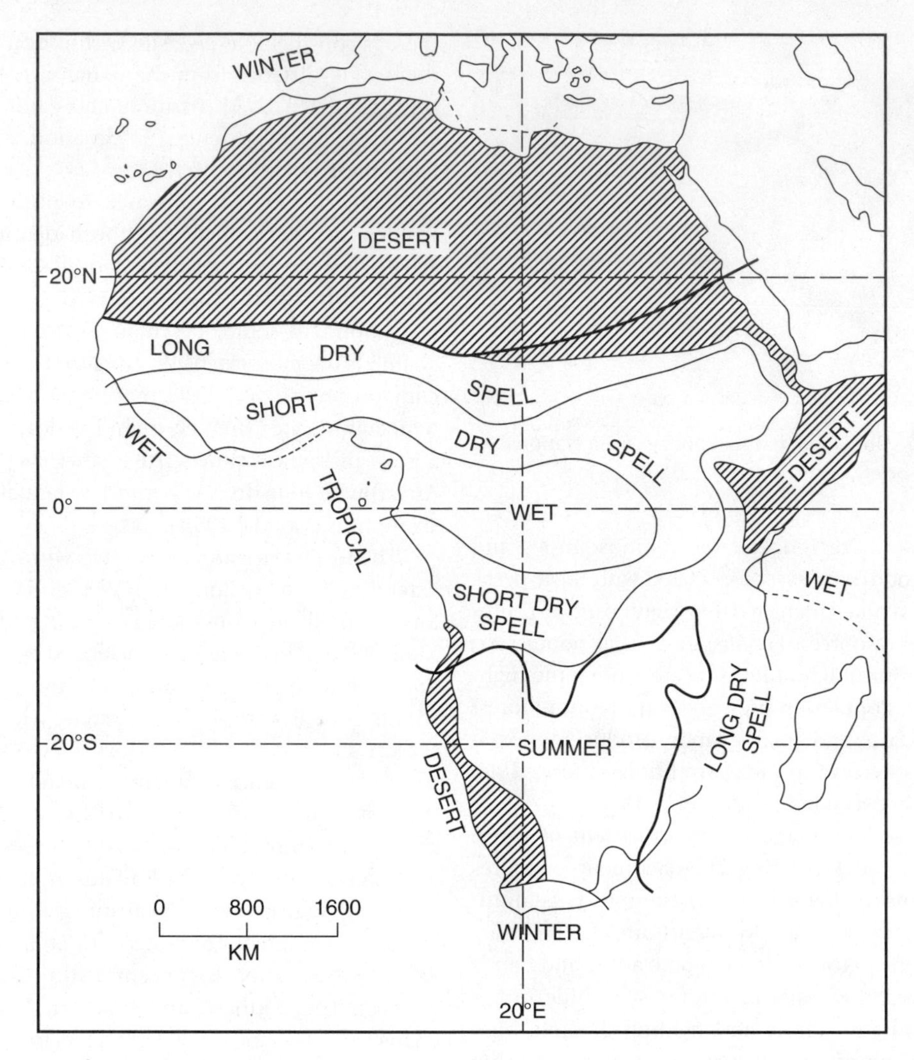


**FIGURE 3.4** Map of Main Climatic Divisions. *Source:* From Adams, Goudie, and Orme. 1999. *The Physical Geography of Africa*, p. 37.

have accelerated in recent years with the heavy reliance on carbon-based fuels (coal, petroleum, natural gas). Unfavorable and fluctuating climate conditions have been widespread in African since the 1960s, and the adverse and unpredictable climate contributes to poverty and underdevelopment. African environments are sensitive indicators of global climate change.

The authoritative IPCC forecasts that Africa will be hardest hit by climate change even though it is the region that contributes least to global carbon emissions. In countries and regions where poverty is endemic and vulnerability is ever present in everyday life, climate change will exacerbate existing conditions and may undermine MDGs and other development goals. Climate is forecasted to have a dramatic impact on the desert margins and the savanna regions, which are expected to become hotter and drier. While the wet climates will become warmer and wetter, flooding is expected in the Nile Delta and along the east and west coasts. The geographical distribution of diseases is likely to change, with the diffusion of malaria into previously malaria-free areas: the East African highlands and a southward shift into South Africa.

Rising sea levels and more frequent storm surges and flooding may affect the west and east coasts of Africa in particular. Many people live in coastal and delta regions in West Africa and along the Nile. The existence of informal settlements with poor building materials and without adequate infrastructure (drains, pipes, sewers, and waste collection systems) makes these areas particularly vulnerable to rising sea levels, storm surges, and flooding.

## AFRICA'S MINERAL RESOURCES

Africa contains an enormous wealth of mineral resources, including some of the world's largest reserves of fossil fuels, metallic ores, gems, and precious metals. Of the world's supply, Africa is thought to contain 42% of its bauxite, 35% of uranium, 42% of gold, 57% of cobalt, 39% of manganese, 73% of platinum, 88% of diamonds, 10% of oil, and 5% of copper (Carmody 2011:2; Custers and Matthysen 2009:20).

Africa in the 21st century is regarded, along with Siberia, as the largest remaining underexplored mining frontier. Africa extractive potential is now being courted by a growing list of suitors (China, India, Brazil, the European Union, the United States, and Japan), all seeking reliable supplies (Economic Commission for Africa 2011).

The mineral and metal deposits in Africa have various geographical distributions (Fig. 3.5). For example, several African countries have significant deposits of uranium: South Africa possesses 8% of global reserves,

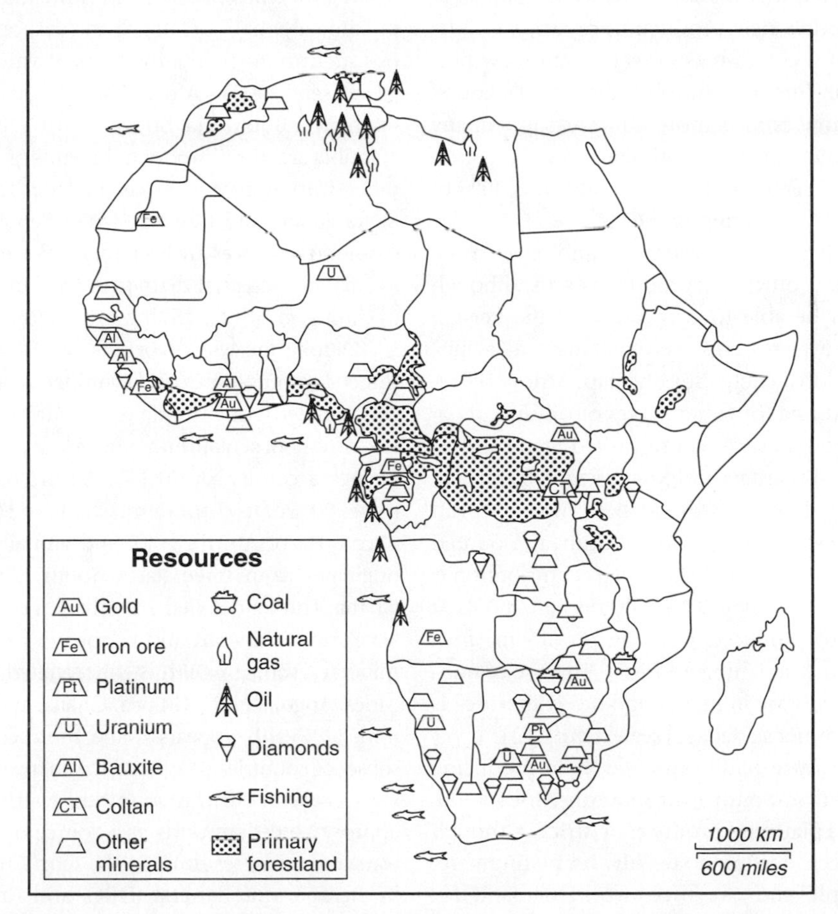


**FIGURE 3.5** Map of Africa's Resources. *Source:* Reprinted with the permission of Polity Press. Pádraig Carmody. 2011. *The New Scramble for Africa*, p. xii.

Niger 5%, and Gabon, DRC, Central African Republic, and Namibia 5% each. The uranium used for the atomic bomb dropped on Hiroshima originated from DRC.

Africa's proven oil reserves (including North Africa) in 2011 represent 10.4% of the world's total, amounting to 417.4 billion tons (British Petroleum 2012). Shares of global oil production among top African producers are Nigeria (2.9%), Angola (2.1%), Sudan (0.6%), Gabon (0.3%), and Equatorial Guinea (0.3%) (British Petroleum 2012). New emerging oil producers in the region (e.g., Ghana, Uganda, Kenya, and Tanzania) have experienced investment surges to the tune of billions of dollars of investment, and investors are widening their Africa foci to include other emerging producers (e.g., South Sudan, Somalia, Liberia, Mozambique, and Sierra Leone). Africa accounts for 20% of the world's new production capacity, and its contribution may increase as emergent and new producers come on line. ExxonMobil, the world's largest oil company, now extracts more oil in Africa than any other region: 25% of its oil originates in the region (Carmody 2011). French oil giant Total holds 30% of its reserves in Africa (Carmody 2011).

African oil is less expensive, safer, and more accessible than many other oil producers, and although Africa may not be able to compete with the Persian Gulf as far as proven reserves, it is emerging as a "swing" region on the global energy supply map. African oil exports are accounting for larger shares of oil imports of major global powers such as the United States (32%) and China (11%) (Custers and Matthysen 2009). There is also a regionalization to the global powers' African oil supplies. For example, the Gulf of Guinea accounted for 16% of U.S. imports in 2002, and its contribution is expected to rise to 25% by 2015 (Ghazvinian 2007). At the same time, oil produced in East and Southern Africa is exported mainly to China and Asia. The rise of non-Western contemporary influence and ties with Africa is the subject of immense debate (see Chapter 12).

The best-known platinum- and gold-producing area is the Rand in South Africa. South Africa is the largest gold and platinum producer in Africa. Although its gold has been declining steadily, its platinum reserves are ample and can meet world demand for many decades—up to a century using current mining techniques (Cawthorn 2010). Mined platinum is used mainly in automobile catalytic converters as well as in electronics and jewelry. Africa's second largest gold producer is Ghana. In colonial times, West Africa was known as the Gold Coast: outsiders were interested in gold from Ghana to Guinea. In recent times Tanzania and Mali have rapidly become Africa's newest gold producers.

Copper and associated metals are mined in South Africa, Botswana, Namibia, and Zimbabwe. DRC has the second largest copper reserve in the world (70 million tons) after Chile, and reserves in Africa's copper belt, stretching from Katanga in the DRC through northern Zambia to Angola, hold the world's highest copper grades.

Diamonds are the most important nonmetallic industrial minerals produced in Africa and account for approximately half of the world's supply in terms of production and 62% in terms of value (Custers and Matthysen 2009). Africa has 16 diamond-producing countries; Botswana, South Africa, DRC, Angola, and Namibia are most important in terms of carat output, and other important producers include Tanzania, Ghana, Sierra Leone, and Central African Republic. Included in diamond resources are both high-value gemstones as well as smaller industrial diamonds used in the manufacture of cutting, grinding, drilling, and polishing tools.

Despite the wide geographical dispersion of Africa's mineral resources, certain key minerals are geographically concentrated. For example, South Africa possesses 92% of the continent's platinum reserves and 90% of its coal. Guinea accounts for 90% of Africa's bauxite, and DRC mines 90% of its chromium. Ghana and South Africa produce 60% of Africa's gold, and virtually all manganese originates from three states: South Africa, Gabon, and Ghana. Thus, a detailed examination of relative shares of key African minerals and petroleum reserves shows that mineral resource wealth is concentrated heavily in South Africa, Angola, DRC, Guinea, Ghana, and Nigeria. Africa's mineral wealth appears to be more concentrated in a subset of countries compared to other world regions.

In several African countries, warring factions have appropriated diamonds as a source of income to fund wars and raids, resulting in the term "blood diamonds" in Sierra Leone, Liberia, DRC, and Angola. There are cases, however, where nonconflict gems have been mislabeled conflict diamonds (e.g., tanzanite in Tanzania)

(Schroeder 2012). Addressing the issue of blood/conflict diamonds, a major international effort in 2002 among industry, government, NGOs, and the UN led to the implementation of a system of certification known as the Kimberley Process Certification Scheme (KPCS). KPCS aims to stem the flow of conflict diamonds so that rebel movements cannot use these resources to finance wars against legitimate governments. Under KPCS, only rough diamonds accompanied by a government-issued certificate can be imported and exported, ensuring that diamonds originate from conflict-free dynamics.

The world's largest mining companies (e.g., BHP Billiton, Rio Tinto, Anglo American, Xtrata, Chinalco, and Vale) are converging on Africa at an unprecedented rate and scale. More than 1,500 industrial mining operations are scattered across the region; many are privately owned, but some involve African partners. For example, the government of DRC owns 17% of the Tenke Fungurume copper mine in central Congo. There are also thousands of artisanal or small-scale mines that use more labor-intensive techniques (and some deploy child and/or forced labor). Increased activity by Chinese migrant artisanal miners in Ghana is taking place, despite being banned by the 2006 Ghana Mineral and Mining Act. In total, artisanal mines account for 10% of Africa's gold output and 30% of its cobalt (Economic Commission for Africa 2011).

African fuels and minerals make important contributions to the global supply and account for over half of Africa's total exports. Opinion is divided as to whether the region's ample supply of resources is a development curse or a development windfall to propel African's growth and deepen it, particularly with regard to poverty reduction and gender equality (see Box 3.1). Some countries have resource-dependent economies: minerals account for at least 90% of exports of Nigeria, Angola, and Equatorial Guinea. Many other African countries depend on exports of two or three commodities, accounting for the lion's share of their trade output. African efforts to transform the extractive industrial sector away from colonially created enclaves have so far registered only partial success. South Africa is a partial success story in that mining returns facilitated the country's economic development as mining companies reinvested profits and diversified into other sectors, but miners' salaries have remained low and contested (e.g., a 2012 miners' strike at Marikana resulted in extreme violence and the deaths of 44 strikers).

### BOX 3.1 THE RESOURCE CURSE OR CURE?

Africa's mineral wealth typically does not result in broad-based development or trickle-down effects: half of Africa's population lives on the equivalent of US$1.25 per day (World Bank 2012). Indeed, Africa mineral-rich countries are most characterized by underdevelopment. This is an apparent paradox of plenty, often dubbed "the resource curse": countries and regions with an abundance of natural resources, specifically nonrenewable resource such as fuels and minerals, tend to grow more slowly than those in which natural resources are scarce (Sachs and Warner 1995). Furthermore, there is often an empirical correlation between resource richness and low investments in human capital (especially education), high levels of corruption, and failures in economic diversification. The resource curse has been identified in Angola, Chad, DRC, Equatorial Guinea, Guinea, Nigeria, Sudan and Zambia.

A complex array of sociopolitical factors lies behind resource-rich countries' inability to develop. Typically, mineral returns are siphoned off by a small, powerful wealthy elite (often members of the government); government institutions are weakened (non-transparent, corrupt, and tolerant of high levels of human rights violations); and resource volatility and excessive borrowing lead to an unstable macrofinancial situation (Le Billon 2012). Moreover, in countries highly dependent on mineral exports, the interests of the mining industry drive national priorities and often lead to the neglect of other sectors, such as agriculture. This often results in a vicious circle of agricultural decline, a greater reliance on food imports, and a corresponding requirement to bolster mining to pay for food and other imports. Countries dependent on nonrenewable resources also run risks of not conserving their resources and/or mining companies moving elsewhere when reserves run low or less expensive supplies become available.

The classic example of a natural resource-rich region prone to chronic underdevelopment is the Niger Delta (Nigeria). In this oil-abundant region the majority of Niger Deltans live on less than US$1 per day, and conditions of abject poverty have worsened rather than lessened since the discovery of oil; 2 million barrels of oil are pumped per day, which equals, on average, US$180 million per day in revenue (Watts 2008). Uneven development (caused by government neglect and poor profit-sharing mechanisms) and severe environmental degradation (oil pollution caused by oil operations with decayed, leaky infrastructure, gas leaks, flares, etc.) are everyday realties of the Niger Delta

*(Continued)*

**BOX 3.1 *(Continued)***

(Le Billon 2012). An average of two oil spills occur daily there. Flaring (the burning of natural gases that can neither be processed nor sold) releases pressured gases ($CO_2$ and other toxic gases) produced from oil extraction into the atmosphere and the immediate vicinity, causing health problems (especially respiratory ailments) and contributing to acid rain and the destruction of the ecosystem. Flaring continues in Nigeria despite a 1984 government law declaring it illegal. Moreover, the exclusion of the majority of the Delta's population from the benefits produced by the natural resource has led to conflict and violence in the region (see video images from *Curse of Black Gold* at http://talkingeyesmedia.org/curse-of-the-black-gold). In other cases, conflict over controlling the returns from resources is manifested in separatists' claims (e.g., Angola's oil-rich Cabinda province), but routinely it causes intra-government ministerial wrangling over access to budgetary allocations.

Oil-rich countries are not the only ones experiencing the resource curse. Guinea, in West Africa, contains almost half of the world's supply of bauxite (the raw material for aluminum), yet its government budget is 0.0005% that of its former colonial ruler (France) (Carmody 2011:2). Of course, a large part of the explanation of the adverse incorporation of Guinea and other commodity-producing African states into the global political economy lies in the colonial relationship established centuries earlier. African countries provide many minerals for export with little domestic consumption, resulting in their unfavorable incorporation into the global economy and failure to deploy their minerals in domestic industrialization and/or to specialize in more profitable parts of the mining value chain (e.g., refining and value additions).

Botswana is an important example of a resource exception. It is a shining example of a resource-rich country that ascended to middle-income status by 2007, largely on the basis of its diamond wealth and sustained economic growth since 1966. At independence, Botswana was one of the poorest states in the world, with a gross domestic product (GDP) per capita of US$283, and its landlocked status, poor soils and drought proneness did not bode well for its economic future. Despite these geographical constraints, the government has been able to increase GDP per capita to US$16,800 in 2012. The country has been progressive in deploying mineral wealth for societal development and avoiding the resource trap that other resource-rich countries have fallen into. Breaking the mold of exporting minerals in raw form, the government of Botswana has partnered with De Beers (South Africa) and begun diamond processing, marketing, and sales from Gaborone in 2012 in a highly ambitious effort to relocate diamond processing from London to Gaborone (Fig. 3.6). Gaborone is being turned into a major diamond hub that processes a US$6 billion share of the global diamond economy.

**FIGURE 3.6** Diamond Trading Company, Gaborone, Botswana.

## AFRICA'S FOREST RESOURCES

Africa's forest cover is estimated to constitute 17% of global forest resources and 1.6 million acres (650 million hectares) (FAO 2010). The major forest types are dry tropical forest in the Sahel and Eastern and Southern Africa, moist tropical forest in Western and Central Africa, and mangroves in the coastal zones. Only 1% of African forests have been planted, so the vast majority is natural. DRC, with 284,171 acres (155 million hectares), and Sudan, with 172,973 acres (70 million hectares), are ranked in the top 10 countries in the world with the largest forest areas. The majority of African forests are located in rural and often remote areas.

Africa's annual net loss of forest area is proceeding at the rate of 8.4 million acres (3.4 million hectares) per year (the second highest rate of regional loss after South America). Still, 20% of Africa is forested. Much of the deforestation (as is the case in other environmental degradation) is the result of large numbers of individuals engaging in decisions that are privately rational but collectively destructive. Each year Africa is losing forest cover the size of the Netherlands.

Still, there is a growing movement to place more forests under conservation and protection: since 1990, 12.3 million acres (5 million hectares) have been designated for conservation. Forests are important for storing a vast amount of carbon. When a forest is burned and cut down and converted to another use, such as agriculture, carbon is released back into the atmosphere. Forests are critically important in terms of both supporting biodiversity and mitigating climate change.

Deforestation is very controversial. Anthropogenic influences are most important, but naturally occurring physical events such as fires, landslides, diseases, pests, and floods also play a part. In Africa, the major human factors include clearing for agriculture (subsistence and increasingly commercial agriculture for palm oil plantations and biofuels), commercial logging, fuel wood consumption, and infrastructure expansion (especially transport infrastructure, human settlement, and mining in certain locations) (Osei 1993). Oil exploration in forest areas is a newer phenomenon. For example, oil concessions have been given out to cover 85% of Virunga National Park (Africa's first conservation park and world heritage site) in the northeast DRC, but the World Wide Fund for Nature (WWF) is leading a high-profile campaign to prevent a UK company from beginning drilling. More typically, the growth of towns and cities brings increased demand for charcoal and fuel wood and results in a decline in tree stocks (typically in a radius of 50–100 miles [80–160 km] of settlements). The construction of access roads (by forestry and mining companies) can also make the situation worse by opening closed forest areas. When resources are more easily accessible, this can enhance the profitability of the timber trade and thereby encourage more illegal activities and human activities in the area.

Loss of forest quality and biodiversity is also affected by different economic activities. For example, selective vegetation removal (during logging and fuel wood collection) affects the forest ecology. Indeed, when forests are cleared for agriculture and other activities, there is no guarantee that trees can even grow back, as the soil composition is altered in transformation. The building of roads not only affects vegetation but also accelerates the killing of animals for bush meat, the trade (prevalent in Central and West Africa) in which endangers forest-dwelling animals. Deforestation is largely responsible for Africa's losing one half to two thirds of its original wildlife habitat (World Bank 2000:195). Overharvesting of nontimber forest resources, such as medicinal plants, is also part of the deforestation process. Suffice it to say that the causes and drivers of deforestation cannot be reduced to a single explanation and vary from place to place.

Mismanagement of Africa's forest resources is indisputable, but considerable debate surrounds its main driver. For some time, it was often too easily assumed that deforestation was due to indigenous shifting cultivation practices, resulting in land degradation and desertification. Evidence can be found of local practices that have led to these results (e.g., Freetown), but there is also extensive evidence of indigenous management of forest resources (Binns 1995). In addition, there is evidence from Guinea of indigenous adaptive agroecological practices, working with and conserving the diversity of vegetation (Fairhead and Leach 1995). From a livelihood and sustainable development perspective, trees and shrubs can be considered as a savings to be drawn on in an emergency. Forests can serve as a source of food, a wild food bank, to be tapped in times of famine (Cline-Cole 1995).

Forests play an important role in sustaining livelihoods, especially for the rural poor, providing a range of products such as timber for construction, fuel wood for cooking, foods, medicines and other materials (reeds for weaving and wood for utensils/crafts). Forest foods (wild fruits, berries, roots, tubers, leafy vegetables, palm fruit and wines, cola and shea nuts, and mushrooms) are a regular part of rural diets. Wild meat contributes an important (and often main) source of protein in some rural communities in West and Central Africa. Local inhabitants can also derive benefits in terms of spiritual and aesthetic needs and employment. Tens of millions of rural households in Africa depend on forests to survive and to supplement cash incomes.

Many experts blame governments for their inability to manage forest resources and to curb illegal forest activities. The exploitation of forest resources can be traced to the colonial period, but the fundamental extraction system has remained intact: timber is a cash commodity and the basis of an extractive industry that pays low taxes and mostly profits external actors, with little trickling down to rural communities that depend on and legally own the resource (Owusu 2012; see also Al Jazeera's 2011 report on illegal logging in Sierra Leone). Typically, agreements are drawn up between governments and logging companies without local participation, and logging companies are not held accountable for paying taxes or for adverse environmental consequences. Corruption is rife in the logging business across Africa, but in 2011, the Malawi and Sierra Leone governments intervened to curb corruption by temporarily suspending timber exports.

On balance, most African communities are unable to balance the protection of forest resources with the pressures from population, urbanization, and poverty. State power can be misused and corrupted, resulting in "Faustian deals" whereby agreements are made with international and domestic interests for present gain without regard to future costs or consequences (Owusu 2012). Some NGOs have emerged to assist rural populations in the responsible use of forest resources—for example, Triple Forest Trust's (2011) activities in Cameroon and Congo. In addition, the Nile Basin Reforestation Project (Uganda), launched in 2009, became the first project approved under the UN's Clean Development Mechanism that linked increased reforestation and supporting the poor to adapt to climate change.

Africa is a major timber exporter: some of the strongest, hardest, and most sought-after timber is harvested in the region (e.g., mahogany is world-renowned for its use in fine furniture making, paneling in homes, and boat building). Around 45% of African timber is exported as unprocessed logs, but the largest share of its exported timber is processed (sawn wood, plywood, and veneer). The big timber companies in Africa are located in the Congo Basin (Congo, DRC, Gabon, and Central African Republic) and West Africa (Ghana, Nigeria, and Côte d'Ivoire). Commercial forestry provides an important source of national income in these countries. For instance,

## BOX 3.2 COLTAN

Fifteen years ago, few nongeologists and high-technology researchers had ever heard of coltan, an obscure mineral that is an essential ingredient in electronic components (Nest 2011). From mobile phones, laptops, Xboxes, PlayStations, and iPods to military weapons (and electric hybrid motor vehicles in the near future), the use of coltan has become commonplace.

Coltan is an abbreviation for columbite–tantalite, a mixture of two mineral ores (creating a dull, black-colored rare metal). When coltan is refined, it becomes a heat-resistant powder that can hold a high electric charge, vital for capacitors in a vast array of electronic devices and a key reason why devices can be miniaturized. Coltan has become a vital rare metal to the contemporary global economy (Carmody 2011). Coltan mined in eastern Kivu Province (DRC) ends up on the mobile phones used on U.S. college campuses (Fig. 3.7). Demand for the mineral is predicted to grow at 10% to 20% per annum (Lalji 2007).

Many industries, as well as the U.S. government, stockpile this strategic mineral. Outside of Africa, the mineral is known as tantalite, but the label "coltan" became associated with the artisanal mining of this ore in the Congo. Concerned activists and the general public now associate coltan with conflict minerals and the Congo. Coltan, however, properly refers only to the mineral found in the Central African region.

Coltan has recently become a prominent activist issue. Around 2,000 reports leaked out of DRC of mines deep in the jungle where coltan was extracted in brutal conditions under the control of various warlords. A UN team dispatched to the region to investigate the situation produced an exposé that claimed a relationship between violence and the exploitation of coltan. Since then, "conflict coltan" has become synonymous with a human disaster: forced labor, killings, rape, and environmental destruction in eastern Congo. UN Special Envoy of the Secretary General on Sexual Violence and Conflict, Margot Wallström, has referred to Congo as "the rape capital of the world," highlighting extraordinarily high incidences of sexual violence against women and the rape of the environment (UN News Center 2011).

The politics of coltan interweaves artisanal miners, local warlords, MNCs, and concerned activists. Many documentaries have detailed coltan mining in the area—for example, *Blood in the Mobile: Mining in the Congo* (2011), and ABC News (Australia) (2009) *The Congo Connection.* Conflict minerals in Congo are embedded within a complex geopolitical environment. At its height, conflict in Congo involved seven external African armies (Rwanda, Uganda, Burundi, Namibia, Chad, Angola, and Zimbabwe), and the regional scramble was partly fueled by coltan (Carmody 2011). A UN peacekeeping mission commenced in 2010, and current (2012) UN forces number 20,000 in DRC, representing the largest UN peacekeeping presence in the world.

Congolese rebel groups have ruthlessly exploited conflict minerals (not just coltan but also gold, diamonds, etc.) and extracted considerable wealth to fund their wars. Since the early 1990s, 15 to 20 different antigovernment armed forces have participated in conflict mining, and new groups enter the scene, but some have limited longevity (e.g., the March 23 Movement [M23] was active 2012–13) as the 3,000-strong UN peacekeeping forces are being

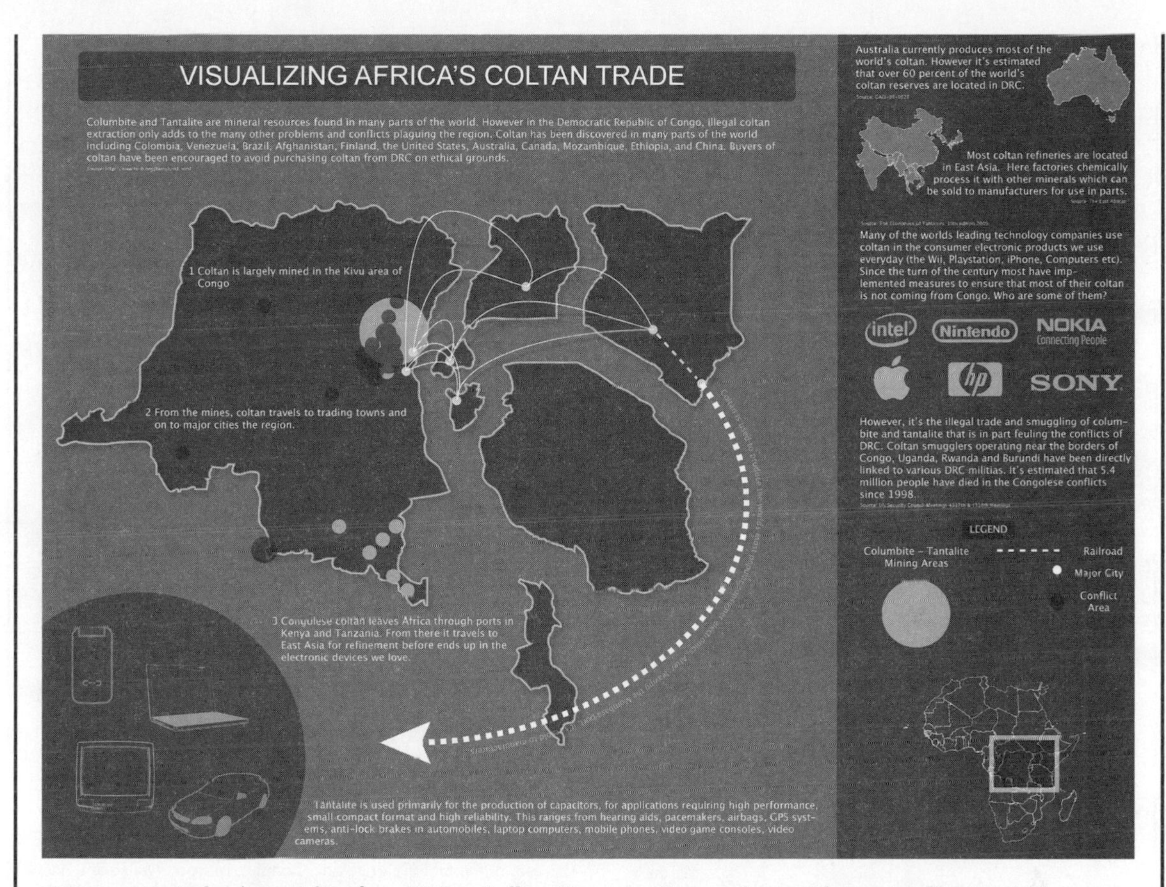


**FIGURE 3.7** Coltan's Spatiality: from DRC to College Campuses. *Source:* © Some rights reserved by Jon Gusier.

more aggressive in defeating and disarming local rebel groups. Militias control many of the mines and most of the trade routes within Congo as well as the cross-border smuggling routes. Coltan is eventually legally exported from neighboring countries (via airports such as Kigali in Rwanda and ports such as Mombasa in Kenya). Despite a UN presence, war lingers in the Congo.

Coltan is mined, with heavy impact on the environment, in eastern and southeastern Congo. In eastern Congo, mining takes place at two World Heritage Sites: Kahuzi-Biéga National Park and Okapi Wildlife Reserve. Coltan is found in high concentrations in Kahuzi-Biéga (75% within the boundaries of the Kahuzi-Biéga National Park in the Albertine Rift Valley) (Taka 2012). This area is home to the endangered lowland gorilla (whose population is estimated to be 250) and other threatened species (e.g., chimpanzee). The park does not have a delineated boundary, and thousands of displaced people and artisanal miners (of coltan and now oil and gold as well) have moved into area and practice intense artisanal mining (Fig. 3.8). To mine coltan, rebels have overrun Congo's national parks, clearing out large parcels of lush forests.

Furthermore, deforestation pressures are intense as people chop down forests for charcoal, kill gorillas as part of forest-clearing efforts, hunt other wild animals for food, and engage in poaching. Miners are far from food sources, and poverty and starvation caused by the war have driven some miners and rebels to hunt the parks' endangered elephants and gorillas for food. UNESCO has placed both the Kahuzi-Biéga National Park and Okapi Wildlife Reserve on the list of World Heritage Sites in Danger.

Campaigns to publicize the brutality of coltan processing in Congo (e.g., *Blood in the Mobile,* YouTube celebrity testimonies, and college student boycotts) have led to a recycling of facts about conflict coltan and some mythmaking. For example, Congo does not contain 80% of the world's tantalite, as is often claimed. Of the world's supply of tantalite, 40% is non-African, originating from modern industrial mines in Brazil, Australia, Canada, and China, where mining operates under long-term contracts and stable political conditions. Although more than half of the world's supply comes from Africa, approximately 13% originates from DRC. The actual number of artisanal miners in eastern Congo is unknown: estimates range widely from 20,000 to 400,000 (Global Witness 2012).

*(Continued)*

BOX 3.2 **(Continued)**

**FIGURE 3.8** Artisanal Coltan Mining. *Source:* © DAVID LEWIS/Reuters/Corbis.

Most of Congo's supply moves onto a spot market, 80% of which involves exchange for cash and/or supplies (e.g., flour, dried fish, beer) by "papas coltan" (i.e., wealthier brokers trade coltan). Several other intermediaries participate in trading and transporting coltan before it is packed into barrels and airfreighted out of the region. Coltan trade has led to a dollarization of eastern Congo's rural and urban economies, and the local economy is poorly insulated from coltan price fluctuations, which means that mining returns and prices for food and other commodities do not move in tandem (Jackson 2003).

Most mines are suitable only for artisanal mining because of the soft geology, requiring the use of basic tools such as picks, shovels, and buckets. Teams of miners claim a plot of about 33 × 33 feet (10 × 10 m); they dig down about 20 feet (6 m), make craters, and water the ore until the heavy coltan sinks to the bottom for sorting. Some mines are large, with thousands of miners, but others are small.

The coltan fever in eastern Congo has led to uncontrolled mining. Parcels of land are being eroded, and lakes and rivers are being polluted. Unregulated coltan mining destabilizes hillsides, leading to landslides, and further undermines agriculture potential. Half of the land seized for coltan mining is no longer suitable for farming. The war economy has boosted mining as the main survival mechanism for many poor rural communities, and it has become a lucrative endeavor compared to alternatives (e.g., farming and cattle ranching) deemed too dangerous because of marauding guerrillas. The mining push has drawn people away from farming, and agricultural production has declined, triggering a food crisis.

It has become unethical for major producers to source coltan from the Congo, but the coltan is rerouted to other producers and the supply line has become very complicated. Complicating the sourcing issue further is the confusion between legitimate mines in eastern Congo and illegal rebel operations, so the source of coltan is uncertain (Lalji 2007). According to Nest (2011), most Congolese coltan is routed to Australia and China for processing via neighboring countries such as Rwanda, Uganda, and Kenya.

There is a scientific effort by German scientists to "fingerprint" the geographical origins of coltan based on the geological dating of minerals. This research is an important first step in the development of a certified trading chain (similar to the KPCS for diamonds). Despite activists' call for a coltan certification scheme, it is unclear how such a system will improve the political conditions in the area and whether it will enhance or worsen local livelihoods.

wood exports represent 11% of total exports in Ghana (US$400 million) and in most years is the third largest export after cacao and minerals (Owusu 2012). Ghana's forests and wildlife resources provide direct employment to 100,000 people, and another 60% to 70% of the population depends on forest resources for livelihood and cultural purposes (Owusu 2012). A critical issue with timber is that once the most desired species are cleared and the commercially viable timber is gone, the export trade moves to a different country. In Africa, the DRC, with the most unexploited timber, is the timber frontier.

## ENVIRONMENTS OF AFRICA'S FISHERIES

Marine fishing off the coast of Africa accounts for 5% of the world's harvest, while inland fishing produces 8% of the world yield (Gates and Appiah 2012). African fisheries, in total, produce approximately 7.3 million tons of fish per year (Carmody 2011), and exports are accelerating. African marine resources are fully exploited and tending toward being overexploited. Inland fisheries, however, are generally underexploited (Kenyan and Ugandan inland resources are exceptions). In general, Africa's inland fisheries have not been developed for the international market and still serve only local consumers. The Institute for Security Studies (2007) warns that several African states are in danger of doing permanent damage to their marine resources. Overfishing not only depletes fish stocks, but it also causes severe and irreversible damage to marine and aquatic ecosystems.

Some of the world's most productive fisheries are located along Africa's Atlantic coast (from Morocco to South Africa) around a broad continental shelf and cold upwelling and currents. This area is becoming a fish basket for Western Europe: the European Union now imports 60% of its fish from Africa (Carmody 2011). The main near-surface fish include sardines, tuna, and mackerel, and the principal deep-sea fisheries bring in hake, octopus, snapper, and grouper. Africa's fishery exports produce US$2.7 billion in annual revenues (Carmody 2011).

European, Russian, North American, and Asian distant water fleets dominate Africa's marine fisheries. There are some direct benefits, such as the payment of fishing royalties (the European Union pays €600,000 to Tanzania for 79 fishing boats to catch 8,000 tons of tuna) (Carmody 2011), and large tuna canneries are located in Ghana and Mauritius. However, many fish access agreements (especially those involving Chinese interests) are much less transparent, and considerable illegal trawling by unregistered vessels occurs (Carmody 2011).

Total illegal, unreported, and unregulated (IUU) fishing in African could amount to US$1 billion per year. In Somali waters alone, up to US$300 million a year in tuna, shrimp, and lobster is being stolen by foreign illegal trawling (Carmody 2011). The unauthorized plundering of Somali waters partially explains the escalation in Somali piracy (see Box 3.3). Local communities are not compensated by royalties or other means for practices that undermine environmental conservation.

International trawling is efficient in harvesting fish but often to the detriment of local artisanal fishers. Local fishers often object to the large quantities of fish that are thrown away (the "by-catch") because of their limited international commercial value; the by-catch reduces fish reproduction rates and further disrupts the food chain, resulting in even more reduction of total fish stocks (Carmody 2011).

In many parts of Africa, fish provides 60% to 70% of animal protein consumption in the population's diet. Ironically, Africa is exporting more protein at a time when malnutrition is accelerating and population is increasing in the region. Generally, artisanal fisheries are reporting decreasing returns per fishing effort and reductions in the size of fish caught. Despite more evidence of scarcity from African fishers, governments continue to sign agreements with international fishing fleets and lack the capacity to monitor and enforce agreed-upon quotas. Under current conditions, African fisheries are unsustainable. One of the most critical issues is the lack of data: African governments have poor information about the extent of fish stocks and hence are making poorly informed decisions about marine and coastal environmental issues.

## HUMAN ENVIRONMENTS

The human ecological footprint in Africa has expanded especially in recent times, doubling between 1961 and 2008, and remains in an accelerated growth phase (World Wildlife Fund and African Development Bank 2012). As noted earlier, this expansion is largely the

## BOX 3.3 SOMALI PIRATES: RESPONDING TO AFRICA'S FISH RUSH?

Intense political fragmentation in Somali since the 1990s has meant that the country has been characterized by large-scale fighting among different regional clan groups until recently; consequently, the government had little control of the national territory. In the context of state collapse, international trawlers heavily harvest fish from Somali waters with impunity. Through the 1990s, approximately US$600 million of fish per year were poached from Somali waters. The UN estimated that some 700 foreign-owned vessels operated in Somali waters in 2005 alone. Closely paralleling illegal fishing, other opportunistic illicit activities, such as the dumping of toxic and nuclear waste off and on shore, occurred that also damaged marine and coastal environments.

Somali fishermen's livelihoods have been negatively affected by dwindling catches. Prior to the 1990s, some fishermen claimed that prohibited fishing methods were being deployed off the Somali coast: drift netting and use of underwater explosives, methods that accelerate the plundering of marine resources. Somali fishermen also maintained that IUU fishing fleets were attacking them: pouring boiling water on local defenseless fishermen, and in certain cases destroying Somali fishing nets. Beginning in the mid- to late 1990s, Somali fishermen began to respond. An early successful local defense group called themselves the Somali Coastguards, and their membership largely represented former fishermen. To this day, most Somali groups continue to label themselves as "saviors of the seas" (*badaadinta badah*), although this label masks various other agendas.

By 2003, the term "piracy" was commonly used to describe the ships and fishing vessels attacked by small, organized groups from Somalia (many of them initially operated from the northeast province of Puntland, but more recently groups have operated primarily from Galmudug in the central section of the country). Somali pirates approached ships (cargo, oil tankers, and chemical tankers) and fishing vessels (to a lesser extent) and demanded payment. When this strategy failed, hijacking became the strategy of choice. In the process, activities morphed from protection of Somali waters into an organized, lucrative criminal activity in its own right. The frequency of this activity slowly increased over the years, and Somali piracy attacks on ships peaked at 151 in 2011, with 1,118 hostages taken (One Earth Future Foundation 2012). Analysts report that Somali pirates earn on average US$5 million ransom per ship and crew via hijacking (One Earth Future Foundation 2012).

Of course, contemporary pirates need the security of a land base for conducting negotiations, accessing shipping and security information, and transferring the payments they receive from hijacking. Former rural fishermen need an extensive support system for piracy to function (muscle, technical and financial assistance). Analysts claim that wealthy external actors (possibly the Somali diaspora) play a command-and-control function in managing and financing these endeavors, although it remains to be established what level of organization exists between pirates and warlords, and between pirates and members of the diaspora. There is a dearth of academic research on the internal organization and structure of piracy, and many unsubstantiated claims are made.

Somali piracy cost various governments and the shipping industry US$5.7 billion to US$6.1 billion in 2012 (One Earth Future Foundation 2013). The coast of Somalia is a well-traversed shipping route (especially for oil tankers) due to its location near the Gulf of Aden and the Suez Canal: 20% of global trade passes by the Horn of Africa. Because it is a high-danger area for piracy, shipping companies pay US$2.7 billion in additional fuel costs to speed up vessels along the Somali coastline, making it more difficult to be hijacked. Other shipping companies avoid the area, rerouting tankers on a longer journey around the Cape of Good Hope. Somali piracy even affects Kenyan trade, as increased insurance premiums are levied on ships transiting the region. The 1,500 or so Somali pirates do untold damage to the international reputation of Somali and contribute to Somalia's international reputation as a pirate state.

Piracy activity off the coast of Somalia is now on the decline, however. Counter-piracy measures such as the use of private armed security guards by shipping companies, increased patrols by warships (with a 30-minute response time), and the imprisonment of more than 100 Somali pirates in Kenya has quelled the intensity of piracy attacks along the east coast of Africa. However, piracy is intensifying in West Africa and became a US$1 billion industry in 2012.

However, piracy appears to have had a positive effect on the overfishing problem. Local catches off the coast of Somalia seem to have recovered to 1990s levels, whereas overfishing seems to be a much greater problem farther down the coast in Tanzania.

Much remains to be researched about Somali piracy, and most of the information at this point comes from media reports and international maritime organizations. Nevertheless, the pattern of plundering Somali fisheries is illustrative of a broader system of looting African extractive resources. In a twist of argument, Carol Thompson (2009:300) claims that "piracy refers to the refusal to compensate or even acknowledge the original cultivators of the bioresource." Somali news media perhaps sum it up best by noting that the piracy of the rich has led to a piracy of the poor (Carmody 2011:153).

result of population increase and urbanization pressures, but extractive industrial expansion and infrastructural development (building of dams and some roads) also account for an enlarged footprint. Africa's current human ecological footprint is forecasted to double in size by 2040 (World Wildlife Fund and African Development Bank 2012).

Exploding population has become a driving force of environmental change on many fronts and at an unprecedented scale. According to some biogeographers, humans have emerged as a force of nature rivaling climatic and geological forces in sharing the biosphere and its process (Ellis and Ramankutty 2008).

Urbanization pressures, expanding population, and associated human activities are affecting vegetation, animals, land use, water, and air. Increased population to feed requires more space for agriculture, adding pressures on forests, wetlands, and other natural habitats. Urban concentrations of population necessitate a movement toward more intensive agriculture (irrigated crops), shifting land use away from pasture, forest, or other uses. Deforestation is a major reason for land degradation in Africa, especially when followed by overcultivation and overgrazing. Soil erosion and water pollution are other urgent rural problems. Agricultural runoff can also be manifested as an urban problem as cities draw from the water that flows through them, which can be polluted from toxic chemicals in fertilizers and biocides applied in urban hinterlands.

## URBAN ENVIRONMENTS AND ECOLOGICAL FOOTPRINTS

Africa's urban population growth rates are the fastest of any world region. Cities and towns are growing at twice the rate of rural population, and by 2030 more than half of Africa's population is expected to live in cities (projected to be 60% by 2050). Africa will have an urban population of more than 1.2 billion by 2050—more than the combined urban and rural population in the Western Hemisphere. Specific spectacular urban growth rates are being recorded: Abuja (Nigeria), 8.75%; Yamoussoukro (Côte d'Ivoire), 7.25%; Luanda (Angola), 6%; and Kinshasa (DRC), 5% (UN- HABITAT 2008). Growth of urban settlements is based on rural–urban migration, natural population growth, and especially in situ urbanization (the expansion of the urban footprint and the absorption of smaller satellite settlements within the spatial expansion of the larger city). Africa's population is clustered around resource-rich areas, and almost 25% of the population resides within 62 miles or 100 km of the coast. Much of this population is particularly vulnerable to climate change and future sea-level rise.

Urbanization directly transforms its immediate environment. Urban residential development incorporates various types of land use: hillsides and wooded areas can be cut and bulldozed; valleys, swamps, and lagoons can be filled in; and coastal areas can be developed with little regard for planning regulations. In the process, water and minerals are extracted beneath the city, and soil, groundwater, and local climate are altered in significant ways. The construction of roads, buildings, and factories changes the chemical, water, and energy balance. For example, the temperature in urban areas is affected by several factors, such as the way that walls and roofs of buildings conduct heat, which often results in an urban heat island effect (meaning that the urban area is significantly warmer than its surrounding rural areas).

African urban environments are poorly equipped to deal with rapid urbanization and its consequent inadequate infrastructure systems, phenomenal expansion of slums, and hazardous economic activities gravitating to poor urban environments (e.g., e-waste). Much of the economic dynamo of African cities resides in the informal economic realm outside of state regulation (Simon 2010). Negative environmental effects from both informal and formal economic activities are widespread as pollution concentrates in areas with the lowest regulation and the greatest gaps in detection.

Organic and unplanned urban expansion is proceeding without the provision of adequate health, shelter, water, and climate change adaptation (Sclar, Volavka-Close, and Brown 2012). The peripheries and fringes of urban settlements are mined for sand, gravel, and other construction materials such as wood for building as well as firewood/charcoal. Motor vehicle and industrial emissions, inadequate supplies of safe water, insufficient provision of sanitation and solid waste and toxic waste disposal, and water pollution (contaminated water bodies such as rivers, lagoons, lakes, oceans, and groundwater) are major contributing factors to disease and death in urban Africa. In addition, an array of common informal activities augment urban environmental pollution problems. For example, market women in the food niche of smoked fish (via burning fuel wood) sell their product for seven hours per day on busy streets and marketplaces to poorer urban dwellers, but the process of smoking fish amplifies the fuel wood/charcoal issue (beyond deforestation and its ecological and climate impacts). Collecting fuel wood is a means of livelihood for the urban and rural poor, but one with immense environmental implications.

Urban management of many institutional and industrial wastes is poorly developed, resulting in severe environmental health implications. Sewage sludge and hospital waste often contain bacteria, viruses, and cysts from parasites, all posing human health and safety risks. Many toxic wastewaters are either disposed of or channeled untreated into rivers, lagoons, and other nearby water sources. Urban toxicity is further augmented when toxic waste and e-waste dumping occurs at or near urban African sites. Several high-profile cases have been documented in recent years, but many continue to operate under the radar.

The Trafigura case of toxic dumping in Abidjan (Côte d'Ivoire) resulted in a court case and conviction. A Dutch court in 2010 found the Swiss MNC guilty of exporting toxic waste from Amsterdam and concealing the nature of the 2006 cargo that was unloaded at 12 dumpsites in Abidjan. Gases released from toxic chemicals in this waste led to 17 deaths and over 30,000 injuries. The Abidjan victims sued Trafigura but were unsuccessful in their request for financial compensation. Instead, Trafigura was forced to pay US$198 million for the cleanup of the waste and was further penalized €1 million for illegally transiting the waste through Amsterdam.

The ecological footprint also extends beyond the built-up area of cities into the surrounding regional environments. Urban centers and particularly primate cities (leading and disproportionately large cities within respective urban hierarchies) draw on water, minerals, and other resources from wider regional ecological environments (croplands, forests, grazing lands, and fisheries). In this process, the ecological environment of the hinterland is changed as it provides services and resources for urban dwellers. The ecological footprint of this relationship is estimated to be at least 10 times greater than that contained within the built-up area of a city (Mitlin and Satterthwaite 1996).

Although African ecological footprints are smaller than the global average, African citizens' footprints grew by 30% (1961–2008), leading some experts to conclude that African footprints are approaching the biocapacity available within Africa's borders, although it varies significant from country to country (World Wildlife Fund and Africa Development Bank 2012). For example, Mauritius, Mauritania, and Botswana have national footprints larger than the global average, but Eritrea, Rwanda, and Congo have footprints that are among the lowest in the world (World Wildlife Fund and Africa Development Bank 2012).

## CONCLUSIONS

Africa is well endowed with natural resources, yet it is characterized by low levels of economic development. In the last decade, external powers have intensified efforts to secure more African resources in a "second scramble for Africa" (Carmody 2011). External arrangements and local partnerships are reconfiguring Africa's resource geographies, and many extraction activities are having dramatic effects on the environment. Mineral and resource contracts appear to be unfavorable to African peoples and environments in the long run. Governments are taking a short-term view to obtain immediate returns from resources without giving adequate consideration to sustainable development. Current trends of resource depletion and ecosystem degradation are likely to accelerate in a future of increasing populations, urbanization, climate change, and further structural transformation of African economies.

Africa's human–environment nexus is highly complex. For a long time, Africans were viewed as passive victims of environmental change based on their particular incorporation into the global political economy, but others viewed Africans as responsible for their own environmental degradation. These ways of thinking still hold sway, but there is now a greater acknowledgment of the central role that government and other domestic institutions play in mediating the international and domestic environmental spheres. Globally, there are stronger voices advocating for a deeper appreciation of human–environment relationships. Development thinking and policy have moved in the direction of a more people-centered understanding of Africa's diverse environment, and sustainable development has risen to the top of many global agendas. There is a growing global recognition that environmental problems cannot be treated in isolation and that they need to be part of a comprehensive development agenda. MDGs are one effort, but, unfortunately, they will not be met by 2015. We urgently need a new, invigorated sustainable development agenda.

The concept of sustainable development implies that Africa needs to meet the needs of the present without compromising the ability of future generations to meet their own needs. Sustainable development requires a relative decoupling of natural resource use and adverse environmental impacts from economic growth processes. This, for example, requires African resources to be considered in sustainable ways (e.g., harvesting wood from forests at rates that maintain biomass and biodiversity), ensuring that extraction rates do not harm capacities to sustain future populations. Real sustainable development is going to require deliberate, concerted, and proactive measures to conserve and protect environments. In this context, it only makes sense for African countries to propel their economies away from the depletion of natural resources and toward sustainable development.

## REFERENCES


Adams, W., A. Goudie, and A. Orme. 1999. *The Physical Geography of Africa.* New York: Oxford University Press.

Binns, T., ed. 1995. *People and Environment in Africa.* New York: John Wiley.

British Petroleum (BP). 2012. *Statistical Review of World Energy.* London: British Petroleum.

Carmody, P. 2011. *The New Scramble for Africa.* Malden: Polity Press.

Cawthorn, R. 2010. "The Platinum Group Element Deposits of the Bushveld Complex in South Africa." *Platinum Metals Review* 54(4):205–215.

Cline-Cole, R. 1995. "Livelihood, Sustainable Development and Indigenous Forestry in Dryland Nigeria." In *People and Environment in Africa,* ed. Tony Binns, pp. 171–185. New York: John Wiley.

Custers, R., and K. Matthysen. 2009. *Africa's Natural Resources in a Global Context.* IPIS. Available at http://www.ipisresearch.be/att/20090812_Natural_Resources.pdf (accessed September 1, 2012).

Davis, D. 2011. "Imperialism, Orientalism, and the Environment in the Middle East: History, Policy Power and Practice." In *Environmental Imaginaries of the Middle East and North Africa,* eds. D. Davis and E. Burke III, pp. 1–22. Athens: Ohio University Press.

Economic Commission for Africa (ECA). 2011. *Minerals and Africa's Development: The International Study Group Report on Africa's Mineral Regimes.* Addis Ababa: ECA.

Ellis, E., and N. Ramankutty. 2008. "Putting People on the Map: Anthropogenic Biomes of the World." *Frontiers in Ecology and the Environment* 6(8):439–447.

Fairhead, J., and M. Leach. 1995. "Local Agro-Ecological Management and Forest-Savanna Transitions: The Case of Kissidougou, Guinea." In *People and Environment in Africa,* ed. Tony Binns, pp. 163–170. New York: John Wiley.

Food and Agricultural Organization (FAO). 2010. *Global Forest Resource Assessment 2010.* Rome: FAO.

Gates, H., and A. Appiah. 2012. *Encyclopedia of Africa.* New York: Oxford University Press.

Ghazvinian, J. 2007. *Untapped: The Scramble for Africa's Oil.* Orlando, FL: Harcourt Books.

Global Witness. 2012. "Artisanal Mining Communities in Eastern DRC: Seven Baselines Studied in the Kivus." Available at http://www.globalwitness.org/library/artisanal-mining-communities-eastern-drc-seven-baseline-studies-kivus (accessed October 15, 2013).

Goudie, A. 1999. "Climate: Past and Present." In *The Physical Geography of Africa,* eds. W. Adams, A. Goudie, and A. Orme, pp. 34–59. New York: Oxford University Press.

Institute for Security Studies. 2007. "The Crisis of Marine Plunder in Africa." Available at http://www.issafrica.org/pgcontent.php?UID=22399 (accessed September 17, 2012).

Jackson, S. 2003. "Fortunes of War? The Coltan Trade in the Kivus." Humanitarian Policy Group (HPG) Background Paper 13. Available at http://dspace.cigilibrary.org/jspui/bitstream/123456789/22614/1/Fortunes%20of%20war%20the%20coltan%20trade%20in%20the%20Kivus.pdf?1 (accessed September 6, 2012).

Lalji, N. 2007. "The Resource Curse Revised: Conflict and Coltan in the Congo." *Harvard International Review* (Fall), 34–37.

Le Billon, P. 2012. *Wars of Plunder: Conflicts, Profits, and the Politics of Resources*. London: Hurst & Co.

Meadows, M. 1999. "Biogeography." In *The Physical Geography of Africa*, eds. W. Adams, A. Goudie, and R. Orme, pp. 161–172. New York: Oxford University Press.

Mercer, B., J. Finighan, T. Sembres, and J. Schaefer. 2011. *Protecting and Restoring Forest Carbon in Tropical Africa*. Forest Philanthropy Action Network. Available at http://files.forestsnetwork.org/FPAN+Africa+report+chapter+1+HR.pdf (accessed August 27, 2012).

Mitlin, D., and D. Satterthwaite. 1996. "Sustainable Development and Crisis." In *Sustainability, the Environment and Urbanization*, ed. C. Pugh, pp. 23–35. London: Earthscan.

Nest, M. 2011. *Coltan*. Cambridge: Polity Press.

One Earth Future Foundation. 2012. "The Economic Cost of Somali Piracy." Available at http://oceansbeyondpiracy.org/sites/default/files/hcop_2011.pdf (accessed September 3, 2013).

———. 2013. "The Human Costs of Maritime Policy 2012." Available at http://oceansbeyondpiracy.org/sites/default/files/hcop2012forweb_6.pdf (accessed September 3, 2013).

Osei, W. 1993. "Woodfuel and Deforestation-Answers for a Sustainable Environment." *Journal of Environment Management* 37(1):51–62.

Owusu, H. 2012. *Africa, Tropical Timber, Turfs and Trade: Geographic Perspectives on Ghana*. Lanham: Lexington Books.

Sachs, J., and A. Warner. 1995. "Natural Resource Abundance and Economic Growth." Working Paper 5398. National Bureau of Economic Research (NBER).

Sauer, C. 1950. "Grassland, Climate, Fire and Man." *Journal of Range Management* 3(1):16–21.

Schroeder, R. 2012. "Tanzanite as a Conflict Gem: Certifying a Secure Commodity Chain in Tanzania." *Geoforum* 41(1):56–65.

Sclar, E., N. Volavka-Close, and P. Brown. 2012. *The Urban Transformation: Health, Shelter and Climate Change*. New York: Routledge.

Simon, D. 2010. "The Challenges of Global Environmental Change for Urban Africa." *Urban Forum* 21(3):235–248.

Taka, M. 2012. "Coltan Mining and Conflict in the Eastern Democratic of Congo (DRC)." In *New Perspectives on Human Security*, eds. M. McIntosh and A. Hunter, pp. 159–174. Sheffield: Greenleaf Publishing.

Thompson, C. 2009. "The Scramble for Genetic Resources." In *A New Scramble for Africa? Imperialism, Investment and Development*, eds. R. Southall and H. Melber, pp. 300–320. Durban: University of KwaZula-Natal Press.

UN-HABITAT. 2008. *The State of African Cities 2008: A Framework for Addressing Urban Challenge in Africa*. Nairobi: UN-HABITAT.

United Nations Environment Programme (UNEP). 2008. *Africa: Atlas of Our Changing Environment*. Nairobi: UNEP.

UN News Center. 2011. "Tackling Sexual Violence Must Include Prevention, Ending Impunity." Available at http://www.un.org/apps/news/story.asp?NewsID=34502#.UEoLFkK6W-8 (accessed September 6, 2012).

Watts, M., ed. 2008. *Curse of the Black Gold: 50 Years of Oil in the Niger Delta*. New York: Powerhouse.

World Bank. 2000. *Can Africa Claim the 21st Century?* Washington, D.C.: World Bank.

World Bank. 2012."Poverty." Available at http://web.worldbank.org/WBSITE/EXTERNAL/TOPICS/EXTPOVERTY/EXTPA/0,,contentMDK:20040961~menuPK:435040~pagePK:148956~piPK:216618~theSitePK:430367~isCURL:Y,00.html (accessed November 13, 2012).

World Wild Fund for Nature and African Development Bank. 2012. *Africa Ecological Footprint Report: Green Infrastructure for Africa's Ecological Security*. Geneva.

**WEBSITES**

ABC News Australia. 2009. *The Congo Connection*. Available at http://www.abc.net.au/foreign/content/2009/s2680172.htm (accessed December 5, 2013). News documentary on coltan, environmental destruction (eradication of gorillas and environmental pollution), and global connections.

Al-Jazeera. 2011. *Africa belongs to Africans. When will Sierra Leoneans be able to benefit from their own natural resources, instead of being cursed by them?* Africa Investigates documentary is available at http://.aljazeera.com/programmes/africainvestigates/2011/11/2011112313191424286l.html (accessed December 5, 2013). Documentary film on the resource curse in Sierra Leone.

*Blood in the Mobile: Mining in the Congo*, Documentary film excerpt. Available at http://www.youtube.com/watch?v=7z8nxxklDHE (accessed December 5, 2013). Film documents the travels of coltan from Kivu to handheld devices in Europe.

Triple Forest Trust. 2010. *Triple Forest Trust Congo, documentary film part 1*. Available at http://www.youtube.com/watch?v=qD0YGNbB5fQ (accessed December 5, 2013). Documentary film about environmental conservation.

CHAPTER 4

# THE "SCRAMBLE FOR AFRICA" AND THE STATE OF EUROPEAN GEOGRAPHICAL KNOWLEDGE ON THE REGION

## INTRODUCTION

The colonial experience created deep ruptures with Africa's past. In 1870, nine tenths of Africa was still under African control, but by 1914, nine tenths was under European control. Extraordinarily, within a six-year period between the Berlin West African Conference of 1884–85 and a series of tidying-up agreements in the 1890s, Europeans carved up Africa. These events are referred to as "the scramble for Africa" and are without historical or geographical precedent.

The Berlin proceedings were larded with expressions of good intent; however, no African leaders were invited to the Berlin conference. Amazingly, at no point in the proceedings were African voices heard. Instead, the conference transpired as a strictly European affair and consisted of European posturing, high-powered scheming, competition, and haggling over African territories (Fig. 4.1).

**FIGURE 4.1** · The Berlin Conference that Divided Africa. *Source:* © Bettmann/CORBIS.

According to historian John Reader (1998:551), "there was much to be won; although the cultured gentility of the gathering and the complexity of competing interests might suggest the proceedings were analogous to a game of chess, they were more akin to a backstreet game of poker—with marked cards . . . African leaders, however, appear to be extraordinarily trusting of these white men with guns with their haughty attitudes of superiority." Subsequently, leaders throughout Africa were persuaded or cajoled to put their marks on pieces of paper to agree to European delineations on African soil. Reader notes (1998:551–552) that "African leaders, no doubt, shared the European talent for diplomatic nullity or charlatanism where appropriate, but few managed to negotiate agreements with the Europeans that gained Africa more than a modicum of benefit; the Europeans got (or took) everything they wanted." The outcome of the Berlin conference was that Europeans marked out their "spheres of influence" (Fig. 4.2). Subsequently, each invaded the continent within its own "sphere," until foreign flags waved above nearly all African territories. The only territories to escape 19th-century invasion were Sierra Leone (a British colony established

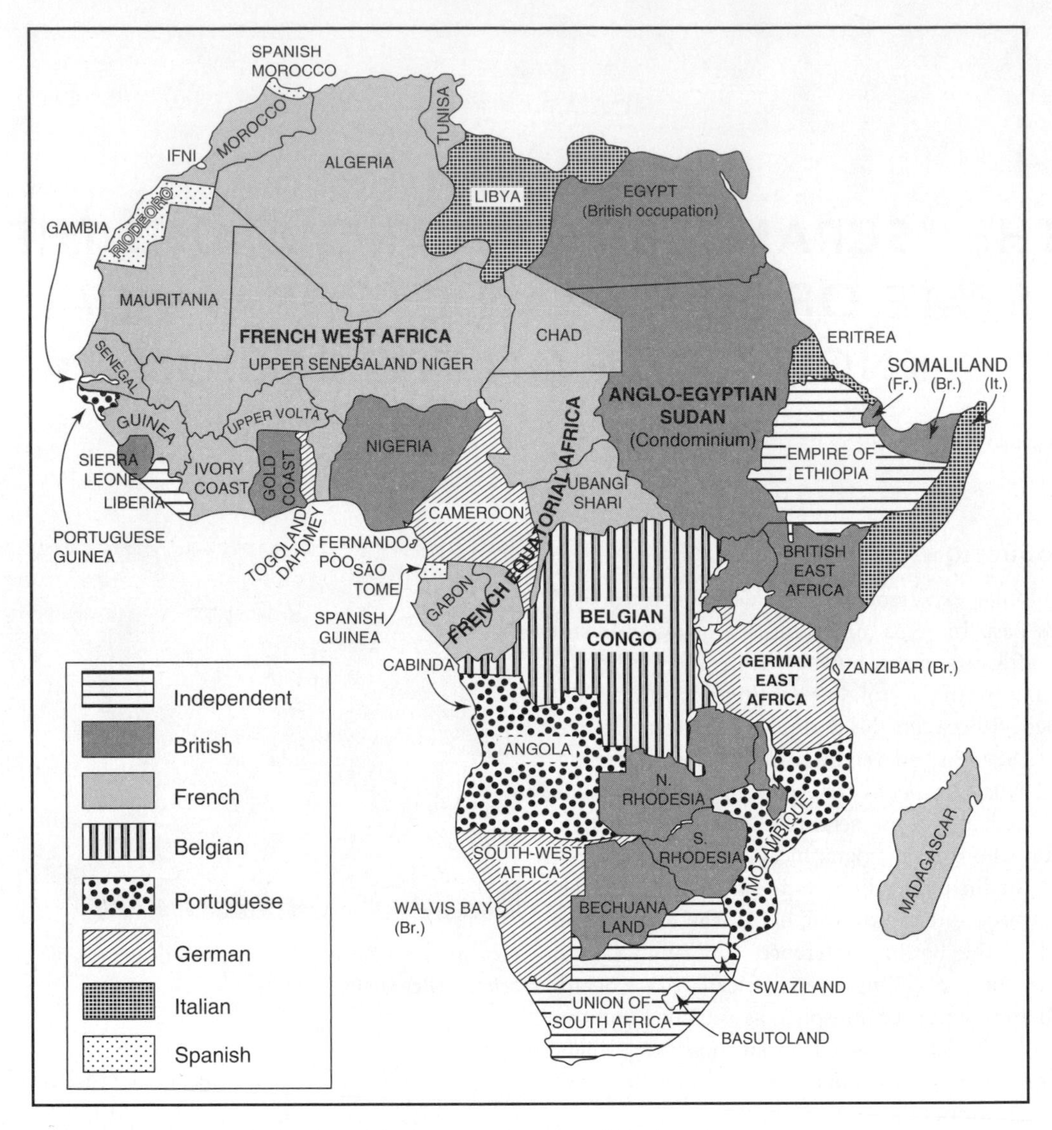


**FIGURE 4.2** Africa After the Scramble. *Source:* Cole and DeBlij, 2007:83.

in the 1790s with the special purpose of providing a home base for freed slaves), Liberia (under U.S. protection with an identical purpose but for American freed slaves), and Ethiopia (annexed by Italy but never colonialized).

All of this colonization took place in the context of Europeans and Africans not playing the same game, not following the same rules, and not speaking the same languages (Reader 1998). European translators drew up the treaties, translating among European languages, which provided "outsiders" the advantage of semantic subtlety when applying the treaties in their colonial territories. Many of the key terms written into treaties were untranslatable across cultures. For instance, rulings about land ownership had no direct counterpart in African understandings of land ownership: African traditions of land belonging to ancestors and spirits and being held by chiefs and families in communal ownership had no relation to Western land practices and law that privileged private ownership and town and country planning.

Through a 21st-century lens, it is unimaginable that an entire continent could be partitioned among external European powers, some of them not even great powers, without African consultation (Chamberlain 2010). But to 19th-century Europeans, it appeared natural, understandable, and justifiable.

Several factors may explain how this could transpire (Chamberlain 2010). First, Europe had undergone an industrial revolution and Africa had not. Thus, for the first time in history, an enormous economic, technological, and military gap opened between the two regions, with power skewed heavily toward Europe. The industrial revolution had transformed the cities of Europe, and urban Africa looked nothing like them.

Second, Europeans claimed that they were at the forefront of progress and civilization and that Africa was "backward" and primitive; Europeans garnered the resources and patronage to "civilize" the region. These views had been formed during the slave trade when humans were treated as merchandise. Although there were minor correctives to the general slave stereotypes (e.g., the noble savage) and although Africans were known to many Europeans (several thousand personal servants labored in Europe), a perception of "backward and inferior Africa" persisted.

Third, a moment of eclipse among several African empires (e.g., Savannah) in the mid- to late 19th century presented favorable conditions for European incursions. The opening up of Africa coincided with African political systems becoming more fragile and "Balkanized" (broken up into more micropolitical arrangements). Previously, Africans had supplied the goods that Europeans coveted, whether it was slaves, gold, or spices. Europeans knew that the gold mines lay just out of their sight—a bare 200 miles (322 Kms) from the coast—but West African societies were organized enough to keep Europeans at arm's length until the late 19th century.

Fourth, it was more than coincidental that Africa was opened up on a large scale when the North American frontier was closing. Africa become known as the last frontier, and this representation persists today.

Of course, lack of understanding and basic knowledge of Africa is another major consideration. A difficulty that bedeviled European understanding of Africa was that 19th-century Europe could not see beyond power embodied in the nation-state. Europeans found it almost impossible to recognize any other form of political organization; they regarded the absence of states as proof of chaos. Essential differences between European states and their underlying assumptions and those of African political entities were ignored and, even worse, misinterpreted when it was convenient. It is obvious that in this time Africans did not lack a sense of territoriality; an essential difference was that Africans mapped political space mentally and orally rather than cartographically. Europeans emphasized discrete spaces marked by boundaries, whereas Africans focused more of the arrangement of centers of power and influence surrounded by unappropriated territory. Europeans had a fixed sense of boundaries, whereas Africans maintained a fluid sense of boundaries that could oscillate and respond to ecological change, especially related to rainfall and drought. Europeans tended to view states and populations as constant and stable, whereas Africans were more open to including mobile and nomadic populations. Historian Basil Davidson (1992) concluded that this process resulted in "the curse of the nation-state": with the organization that was forced on Africa, these territorial containers turned out to be "the black man's burden" in Africa.

African historical and geographical knowledge was hidden from Europeans for a variety of reasons. Europeans were accustomed to studying and learning through written records. For large parts of Africa, no such records exist. Historical records of Arab travelers produced written accounts of parts of urban Africa (e.g., Western Sudan, Timbuktu, and Great Zimbabwe), but Europeans paid them no heed. From the seventh to the 14th centuries, Europe's written record on Africa is a lacuna; to Europeans Africa was a largely unknown and misunderstood region. Africa was often seen as one geographical and cultural mass, intermingled with tall tales of places of African splendor. Partly due to their limited understanding, Europeans developed a series of tantalizing mysteries about the interior of Africa in the early 19th century as they turned their attention to the wonders of Africa. There were stories of

Africa as an El Dorado—a land of fabled wealth (the legendary kingdom of Prester John and the city of Timbuktu were among the most famous).

Much European knowledge was restricted to the past two centuries, and information was geographically fragmentary and restricted to a few contact points because the Europeans (e.g., the Portuguese, Dutch, and Danes) were first restricted to a few port areas and relied on second-hand reports for information about the interior. When Europeans did engage Africans in arenas beyond trade, the former were always outsiders (e.g., explorers, missionaries, town planners, and colonial officers), and their documentary sources became the narrative about Africa. Their chronicles are those of non-natives rather than of natives. Outsiders did not always appreciate or understand what they observed; they were prone to bias as well as prejudice.

Many key European agents of change had complicated relationships with Africans and the colonial enterprise. For instance, missionaries portrayed their undertaking in Africa as humanitarian and saw Africans as capable of being uplifted, but they were also prone to denigrate indigenous culture. Historians continue to debate the relationships among mission activity, colonialism, and indigenous self-esteem. Missionaries opposed the harsher aspects of colonialism but backed European colonial expansion because it provided an opportunity to extend Christianity's reach. However, almost unwittingly, missionaries supported negative representations of Africans, largely to raise funds back home. Raising funds depended on public sympathy (many donations came from the working classes), so missionaries' literature was prone to selective representations of life in Africa: for the unconverted, it was brutal and barbarous. It is no surprise that the missionary debate is far from settled: humanitarian impulses produce ambiguous results.

Waves of explorers from several European countries traveled to Africa from the end of the 18th century until the late 19th century, the intense period of African exploration. Some explorers were captivated by discovery and fame, others were after souls to save, and the remainder were largely agents of European imperialism. Many of the most famous explorers were men, who became nearly legendary in Europe—David Livingstone, H. M. Stanley, René Caillié, and Pierre de Brazza (founder of the city of Brazzaville).

Explorers wrote vivid accounts of their travels and discoveries, fulfilling a 21st-century role as journalists/bloggers on the spot. Their writings generated great enthusiasm among the public interested in knowing more about the world as well as among industrialists and merchants interested in economic opportunities in Africa. Most explorers narrated as if they were the first people to penetrate an unknown region. For the most part, the peoples of Europe viewed the explorations as little more than entertainment and excitement set on an alien and exotic stage. In the eyes of the general public in Europe, exploration had a theatrical element to it. "The same men who saw their scientific missions endangered by African pomp and circumstance were in Africa as agents of an enterprise that was sold to the European public through pompous campaigns of propaganda, through exhibits and shows, and through drama and tragedy reported in travelogues and in the press. The drama has its heroes and villains and a grand plot called *oeuvre civilisatrice*" (Fabian 2000:121).

Most writing on colonial Africa focused on the contributions of men. There were, however, women travelers and explorers who contributed to European scientific knowledge of poorly known places in the 19th century and who continued to do so long after. English women travelers found greater freedom away from the socials norms of Victorian England, where they were confined to the home. Their journeys to distant places such as the African region were revolutionary. Women traveling during these times were mainly travel writers, missionaries, and colonial residents (typically the spouses of colonial administrators). An English woman, Mary Kingsley, stands out because of her achievements as a prominent travel writer, explorer, and amateur geographer and anthropologist (see Box 4.1).

In most accounts, Africans were reported as mere background, almost as if they were part of the local fauna. Some European explorations focused on incredible places like Timbuktu, a mythical "city with gold roofs," supposedly the hub of a fabulously wealthy trading system (Heffernan 2001) (see Box 4.2).

## BOX 4.1 MARY KINGSLEY (1862–1900)

Mary Kingsley had little formal education, and she got a late start in exploration (Fig. 4.3). Until the age of 30, she was a dutiful Victorian woman, caring for her sick mother. Her mother's death—within weeks of her father's unexpected passing—suddenly freed Kingsley to undertake her life's ambition to travel. She possessed a strong intellectual curiosity and was self-educated from reading books in the family's study. In a short time, she went on to become an amateur anthropologist, a famous writer, and an African explorer. She made frequent trips to Africa (e.g., Sierra Leone, Nigeria, Gabon, Cameroon, and South Africa); this was rare for a woman at that time. In West Africa, she collected valuable specimens of fish, she completed her father's study of world religions by researching fetish practices, and she became one of the first Europeans to reach the summit of the highest mountain in the region—Mount Cameroon (4,042 meters)—via a route not previously attempted by any other European. Kingsley challenged the ideas of gender, pushing the Victorian limits of acceptable behavior for women both in Africa and in Britain.

**FIGURE 4.3** Mary Kingsley.

Women faced credibility challenges in their pursuit of exploration. In sharp contrast to male explorers, who behaved as adventure heroes and authoritative experts, Kingsley issued disclaimers about her field observations, stressing the difficulties of African travel. Her approach enlivened her descriptions of people and places, and she became known for her anecdotal humorous travel tales (humor was not often demonstrated by male explorers at that time). Her subjectivity gave her a particular vantage point in observing Africa and Africans, and in writing about it. She was a single white woman with a lower-class accent, her parents came from different classes (her father came from a prominent literary family; her mother was a servant; and Mary came into the world four days after her parents were married). Her ambivalent position in society seemed to equip her with a sharp ability to reflect on her surroundings. Although her actions broke many barriers of the day, Kingsley was uncomfortable about being regarded as a feminist.

Kingsley's writings were widely disseminated. She wrote two books about her experiences: *Travels in West Africa* (1897) became an immediate bestseller, and *West African Studies* (1899) was positively received by academia. However, newspapers such as *The Times* refused to review her books because her views were seen as too radical. In her second book, she criticized colonial administrators and missionaries, writing disapprovingly of their "white man's burden" approach to Africa. She renamed the colonial enterprise "the black man's burden," and the Colonial Office responded by claiming she was "the most dangerous woman on the other side" (Davies 2012:1). She challenged the establishment with her actions as well as with her words.

Kingsley recorded a vast amount of data on the geography of West Africa, including information on local flora and fauna. In Victorian England, geography was considered a male-only science; thus, through her work, she challenged many accepted norms. However, when on expeditions her dress style was always that of a Victorian woman at home, massively unpractical for Africa travel. "She traveled through the bush and swamp in a long, tight-waisted skirt and a high-necked blouse, armed with an umbrella, paying her way as a trader" (Middleton 1991:103).

Kingsley described her primary purpose in traveling as publicizing West Africa, and she was motivated to inform the general public with different interpretations about indigenous people in an effort to develop more complex understandings of Africans and their culture. She said in one of her letters, "until the truth is known to the general public the GP will be content to let things slide there" (quoted in Blunt 1994:129).

Mary Kingsley's life ended prematurely at age 37. She had traveled to South Africa to care for Boer prisoners of war, and while there she died of typhoid. Kingsley's life was influential even if her public role was short-lived. Her writings and public lectures on life in Africa helped draw attention to Britain's imperial agenda as well as to the indigenous customs of Africans, both topics greatly misunderstood by the public. She had an

*(Continued)*

**BOX 4.1 (*Continued*)**

influence on public opinion and was instrumental in the development of the Fair Commerce Party, a colonial politics pressure group that fought for improved conditions for the natives of British colonies. Many of her accomplishments were recognized only after her death. For example, the Liverpool School of Tropical Medicine established an honorary medal in her name to be given for outstanding contributions in the field of tropical medicine. Her vision for a society focused on Africa that would bring together disparate interests (e.g., academics, friends, political groups, traders) to discuss alternatives to harmful colonial policies in Africa was realized when the Royal African Society (RAS) was established in Britain. The RAS has gone on to become an academic body with the mission to promote Africa in many spheres of UK public life (e.g., academia, business, politics. and international development).

## BOX 4.2 TIMBUKTU: COLONIAL ESCAPADES AND UNANSWERED QUESTIONS

Timbuktu is an important town in Mali (the current population is approximately 60,000) that was designated a United Nations Educational and Cultural Organization (UNESCO) world heritage site in 1988. In the 11th century, Timbuktu flourished as a hub within the caravan trade between Africa, the Mediterranean, and the Middle East. From the 13th through the 17th centuries, Timbuktu thrived as the world center for Islamic learning, instrumental to the spread of Islam in Africa, and a center of literary heritage (with libraries and intense scribe activity). During its heyday, a depository consisting of an estimated 200,000 manuscripts was produced. There were Islamic religious books, and other works on poetry, history, law, medicine, astronomy, and botany as well as stories from the region. Timbuktu functioned as an oasis of learning and literacy and a marketplace for books, where the trading of manuscripts was negotiated and books were circulated within the region.

Today we use the word "Timbuktu" to denote any distant or outlandish place; it even appears in Dr. Seuss' "Hop on Pop" book. But more importantly, the truth about the city dispels the Western narrative that Africa consists of people without a history. Timbuktu's learning, buildings, and urban lifestyle predated the arrival of the Europeans.

During the European colonial interlude, a frenzied competition ensued among European explorers to verify the mythical city's existence. The British government funded an expedition by explorer Mungo Park (UK £60,000 in 1805) and deployed a battalion of 40 well-armed soldiers to support the mission. Park failed in his efforts and was killed miles from Timbuktu in 1810, but the race to "discover" the city continued.

Rivaling the British, the Paris Geographical Society offered a prize in 1824 of 10,000 francs to the first person to provide a first-hand, verifiable, and scientifically valid description of Timbuktu. In 1828, French explorer René Caillié won the prize, but his claim sparked considerable controversy and raised ethical questions. Recounting how he was the first European to survive the journey to Timbuktu, Caillié highlighted the importance of disguise and deceit: he had assumed the identity of a Muslim (Abd Allahi) in light of his predecessors' poor survival rates.

A three-volume account of his journey published in 1830 (translated into English the same year) provided a surprising description of Timbuktu: "I now saw this capital of the Soudan, to reach which had so long been the object of my wishes. On entering this mysterious city, which is an object of curiosity and research to the civilized nations of Europe, I experienced an indescribable satisfaction. . . . [But] I looked around and found that the sight before me did not answer my expectations. I had formed a totally different idea of the grandeur and wealth of Timbuctoo. The city presented, at first view, nothing but a mass of ill-looking houses, built of earth. Nothing was to be seen in all directions but immense plains of quicksand of a yellowish white colour. The sky was pale red as far as the horizon: all nature wore a dreary aspect, and the most profound silence prevailed; not even the warbling of a bird was to be heard. . . . Timbuctoo and its environs present the most monotonous and barren scene I ever beheld" (Heffernan 2001:211, quoting a translation of Caillié 1830:49).

The idea of Timbuktu as a city of splendor and plenty was so deeply ingrained in the popular imagination that Caillié's discovery was greeted with disappointment, skepticism, and even indignation, especially in Britain. Doubt was cast upon the legitimacy of the messenger as well as of the message. In this and other explorations to the African continent, the search for scientific and geographical truth produced ambiguous results and raised fundamental questions. Sorely lacking were efforts at critical and deeper reflection on what constituted "real" geographical knowledge and truth. Europeans overlooked the real significance of the place: the African repository in the desert (Fig. 4.4). During the colonial period, the French imposed their language as the main language and its hegemony lessened the importance of Arabic and indirectly Arabic texts. As a result, many locals lost the ability to read and interpret the Timbuktu manuscripts in the languages in which they had been originally written.

In 2012, Timbuktu was added to the UNESCO list of World Heritage Sites in Danger. Climate conditions, time, and extremists pose different threats to the city. The building of the Ahmed Baba Center archive library in 2009, funded by Kuwait, Libya, UNESCO, and South Africa, to house a large collection of long-neglected historical works (20,000–30,000 works) is a contemporary effort in historical preservation of Africa's intellectual past and written tradition. However, in 2012, during a period of civil unrest, insurgents

**FIGURE 4.4** Manuscripts in the Desert.
*Source:* © Sebastien Cailleux/Corbis.

from Al-Qaeda in the Islamic Maghreb (AQUIM) entered Timbuktu and set fire to the library, destroying some works, although most of the collection had been moved to private homes for safekeeping prior to the attack. Few other great sites of learning in the world have faced such cultural wars to obliterate their meaning and historical significance.

During the colonial era, in many cases geographical information was misinterpreted, so misrepresentations continued. This happened with comparatively late European discoveries of some of Africa's great artworks and large urban ruins: for example, Great Zimbabwe, located in the southeast part of Zimbabwe (see Box 4.3). European misinterpretation of Great Zimbabwe eventually sparked one of the major controversies in the archaeological world. When Europeans stumbled on the remains of Great Zimbabwe in 1868, they supposed that they must have been relics from the Phoenicians (peoples originating from the Fertile Crescent, present-day Lebanon, Palestine, Syria, and Israel), who might have mined gold in the area. The European mindset precluded them from considering that the urban settlement had largely indigenous origins. It is a striking example of Europeans believing that Africans were passive in the making of their own histories and that Africans lacked their own urban creativity.

Geographical myths and legends became less prevalent in the late 19th century; in place of these, mapmakers recorded cities and towns, forests and savannas, snow-capped mountain ranges and deserts, and so forth. What the 19th century achieved for geographical knowledge on Africa, however, the 21st century has yet to achieve for a peoples' history of Africa. Closer detailed examinations can often reverse assumptions based on Eurocentric worldviews. The assumption that Africans lived in universal chaos and/or stagnation until the Europeans arrived was exceedingly convenient for colonial agents of all persuasions.

The story of the Timbuktu manuscripts is ongoing. One effort in 2013 involved an Indiegogo online crowd-funding campaign that raised over US$67,000 in one month from international individual donors to help preserve the historical manuscripts. Indiegogo funding will be used to purchase archival boxes and humidity-proof footlockers to safely store 600 manuscripts. The Ford Foundation is sponsoring larger efforts to preserve the rich heritage of Timbuktu and has provided grants totaling more US$3.3 million since 2000. An important initiative is the University of Cape Town's Tombouctu Manuscript Project, which provides expertise in archiving, preserving, and creating online database access to some of the manuscripts.

## COLONIAL MOTIVATIONS

Colonial motivations are complex, and the objectives of different European agents were not always in sync. The mélange of interests included an increasing awareness of the possibilities of Africa, some genuine scientific interest, and a modicum of sincere humanitarian interest. Beneath the surface, others had self-centered motivations such as (1) the acquisition of new scientific and geographical knowledge for Europe's benefit; (2) a desire to spread Europe's civilization and, for some, to extend Christianity; and (3) efforts to bolster Europe's grandeur and economic prowess by incorporating Africa into European-commanded circuits of trade and production, facilitating the international expansion of capitalism.

Many European attitudes toward Africa were formed during the long slaving and trading connection on the Guinea coast. Counterintuitively, negative images seemed to endure and even harden, rather than being modified or corrected, over the decades when the intensity of European explorations on the continent increased. Europe's civilizing mission was particularly blinkered. It rested on the assumption that indigenous people were inferior in mental, moral, and practical capacities compared to white Europeans. Rudyard Kipling's 1899 poem "The White Man's Burden" captures the essence of these feelings of racial superiority and responsibility. Europeans assumed that indigenous people needed training in basic arts and skills, which would occur with exposure to Western belief systems. Missions and educational systems were to become two potent forces for the dissemination of information.

Some Africans welcomed colonialism, many opposed it, but almost all were transformed by it to varying degrees. Europeans and Africans of all persuasions were at the mercy of processes of change that they neither comprehended nor fully controlled. Nevertheless, colonialism set the stage for a deep and inexorable misunderstanding between Africans and the rest of the world that has never been corrected.

Colonialism was objectionable and even evil on many grounds. It damaged societies economically, politically, psychologically, culturally, and in many other ways. Whether colonialism can be benign has long been debated by economic historians. Debating the pros and cons of colonialism, however, is akin to debating rape! Colonialism entailed coercive practices that conflicted sharply with basic human rights and democratic values. In both principle and practice, colonized people were considered subjects rather than citizens and were assigned duties but were never granted rights. The colonial years set the course of African states' integration into the world economy, a mode of incorporation that has lingered and continues to structure Africa's relations with the world. European rule in most parts of Africa lasted 60 to 80 years, a short period that produced profound effects. It shaped the contemporary urban, political, economic, and social geographies of Africa, most saliently by town planning and imported architectural designs (see Box 4.4 on the bungalow). The crises and difficulties that many African states experienced in the latter part of the 20th century as well as their current situations cannot be understood without reference to the colonial epoch.

## AFRICA'S URBAN PAST

Given the paucity of Western texts about Africa, as well as the biases many of them contain, we have to turn to archaeology as a major source of information about the region's urban past. According to urban archaeologist Graham Connah (2001:3), "archaeological evidence is like a jigsaw puzzle from which two-thirds of the pieces are lost, while the rest have the picture worn off or the corners missing." The goal of the archaeologist is to study the physical evidence of past human activities in order to reconstruct those activities.

Archaeological evidence is, however, not unbiased in contributing geographical information on ancient settlements in Africa. For instance, in Central Africa, archaeology has contributed much less information than in other areas, largely because many of the urban settlements were constructed of grass, wood, and other organic materials, making it much more difficult to excavate these structures than those made of stone. Moreover, the corpus of archaeological knowledge is based on human endeavor, and this is highly contingent on funding for excavation. It is easier to search for settlement sites in the open grasslands of the African savanna than in the tangled undergrowth of the rainforest. Funding agencies are motivated by finds and want the biggest bang for their funding monies. Therefore, the mapping of cities of ancient Africa is more likely to reflect the distribution of archaeological research and stone ruins rather than the complete ancient urban geography. Indeed, as Connah (2001:3) stresses, the word "unexplored" should be written across contemporary archeological maps, just as it was a couple of centuries ago with many early maps of Africa.

It is now recognized that cities existed in a wide range of African environments in the precolonial era. The broad pattern of the geography of African cities in precolonial times has been established by urban scholars and historians based on historical and ethnohistorical evidence. Some of the leading books on this topic are by Anderson and Rathbone (2000), Coquery-Vidrovitch (2005), and Freund (2007). Urban centers in Africa show considerable diversity: some were densely occupied for long periods, some were settled only briefly, a few were completely mobile (e.g., West African Savannah), some had transient residents from season to season and year to year, and still others were dispersed settlements covering immense areas. Such diversity indicates social complexity as well an indigenous urban tradition.

A key finding is that towns and cities are not somehow extraneous to the real Africa but rather an inherent part of developing Africa. Urbanization unfolded over a long time, and the process reflected indigenous origins rather than simply external beginnings derived from Arabs, Europeans, or others. There is no basis for the perspective that Africa lacked an urban tradition or that urban development in Africa was contingent on the colonial city. However, urban settlements did not meet all of the criteria used to define "true" cities (the Western industrial city). Similar arguments have surfaced in Middle America concerning Mayan cities in the North and South American contexts. True, Africa lacked industrial cities and, for the most part, still does; however, historical urban formations are evident, and these settlements were tailored to meet the needs of various inhabitants.

Figure 4.5 depicts the main areas of urban development in Africa: in West Africa, along the southern edge of the Sahara; in the West African forest, west of the

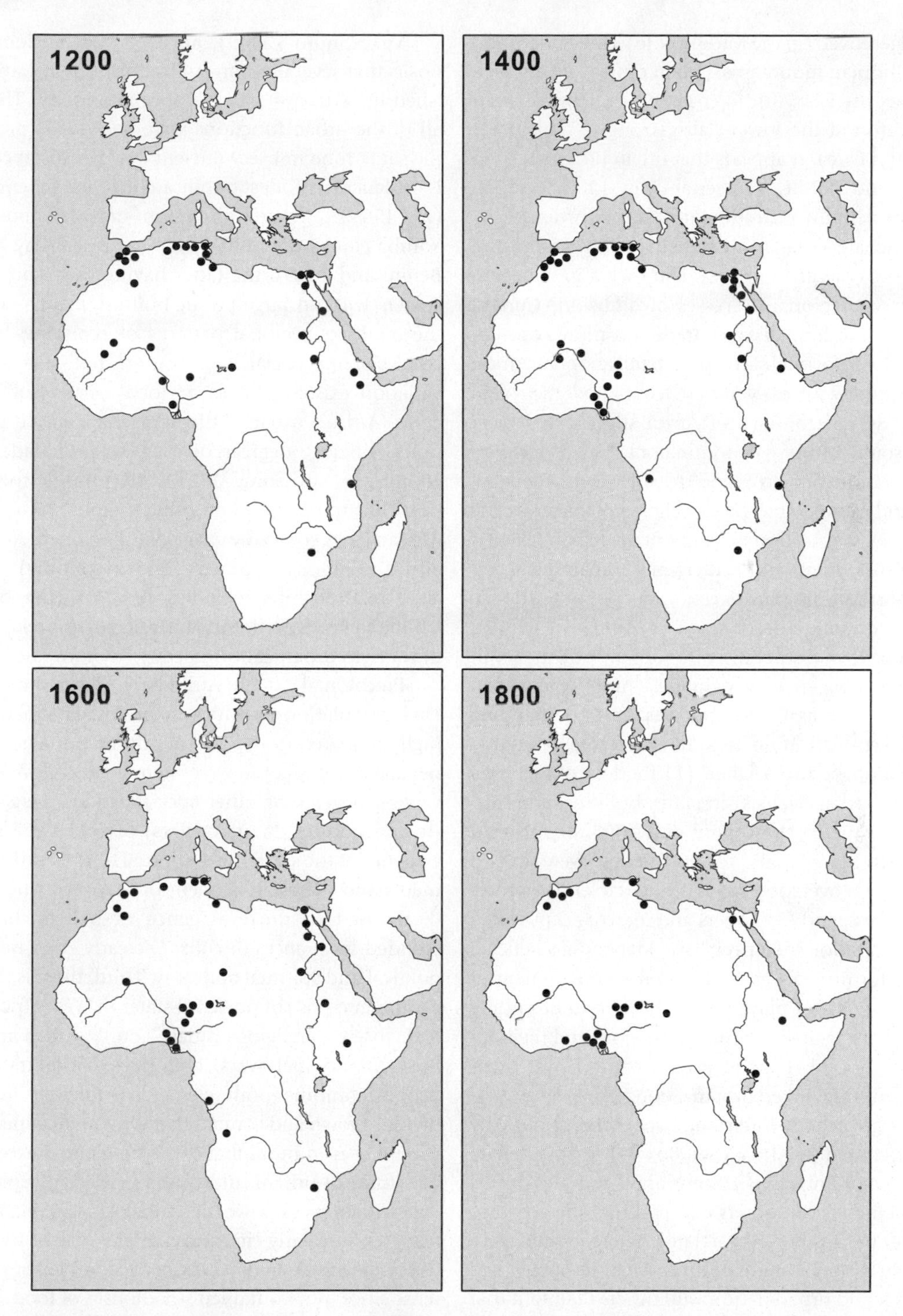


**FIGURE 4.5** Early Urban Settlements in Africa. *Sources:* Reprinted with the permission of Cambridge University Press; Connah 2001:13.

lower Niger River; on the Middle Nile in the Sudan; and in the Ethiopian mountains. Urban centers are also evident along the East African coast, on the Zimbabwean plateau, around the lower Zaire (Congo), and in the Lake Victoria area. It appears that urban life emerged in Africa around 750 BC, and certainly by 1200 BC urban life was evident throughout many parts of Africa.

The debate is far from settled on how, why, and where cities emerged in Africa. Did they arise independently or via regional processes of diffusion? Current thinking is that long-distance trade was more of an intensifier than an originator of urbanization in Africa. Several varieties of early urban forms and their geographical spread throughout Africa suggest that there must be some common denominators. There is evidence of ample population sizes and densities (settlements of 10,000 and more), which is especially notable given the sparseness of population throughout the region. There is also evidence of cities or settlements containing activities and institutions that affected wider political, administrative, economic, and/or religious realms.

We know more about some urban settlements in the region. Nigerian geographer Akin Mabogunje (1962) made major contributions in understanding functional specialization in Yoruba cities. He found that specialization occurred when (1) food surpluses were created to feed specialists (in crafts, building, administration) and when the central authority administered the surplus; (2) a small group of people were able to exercise power over food producers and to ensure peaceful conditions; and (3) traders and merchants provided raw materials for the specialists. Mabogunje believes that state formation was related to the development of urbanization: states played a pivotal role in defending urban centers against external aggression, and their apparatuses allowed them to extend control over other cities, permitting linked urbanization.

There are other contending hypotheses about why cities developed in Africa (See Box 4.3 for the debate about "Great Zimbabwe"). Some argue they developed through elite vision; others contend that the accumulation of political power related to economic gain from long-distance trade and/or scarce resources and minerals led to urbanization; still others maintain that they developed as a means to endure droughts or other environmental pressures (e.g., agricultural overuse of land and deforestation).

Mike Smith (2009), an expert on ancient cities, posits that several urban subtraditions may have flourished in Africa prior to European conquest. Thus, not all of the urban functions were necessarily present at the same time in every ancient city. For instance, there is evidence of modest public architecture in Jenne-Jeno (AD 450–110, located in present-day Mali) but not in Yoruba cities (AD 1400–1900, in present-day Nigeria, Benin, and Togo), the latter having busy commercial centers without large public buildings. Each city bore the marks of regional patterns of economy, political relations, and social organization, but high levels of variation existed. Put in a global context of ancient cities, Africa's urban settlements had ample populations, differing degrees of long-distance trade, social complexity, and some degree of economic specialization (but much less than other regions). But, overall, African cities showed weaker evidence of predictive science (arithmetic, geometry, and astronomy) and unclear relationships to political states (the basis of Childe's [1950] well-known thesis on the relationship of states to urbanization in early city formation).

Precolonial cities in Africa had some commonalties. First, population densities in urban settlements were high compared to the general, thinly populated continental region. People were mainly agriculturalists, but varying degrees of other specializations were evident (traders, administrators, crafters). Second, there is ample evidence of the sacred and spiritual roles of urban settlements and the establishment of places of worship, no doubt tied to the emergence of beliefs that transcended local ancestor cults. The early elites held both political and spiritual authority. Third, there is the well-documented rise of powerful states in West Africa, often tied closely to religious rituals. Fourth, urban and rural mesh in original ways, and large concentrations of people combine both elements (different from the divide of town and country that is often imagined). One should bear in mind that something rather exceptional (the force of urbanization) was needed to explain any concentrations of power in a tribal Africa. Fifth, another feature of the early cities was the presence of some type of fortifications. Early cities needed some form of defense since they contained storehouses of food, the seat of central authority, and concentrations of people.

Distinct periods of urbanization in Africa can be discerned (Fig. 4.6). First, there is the period of ancient

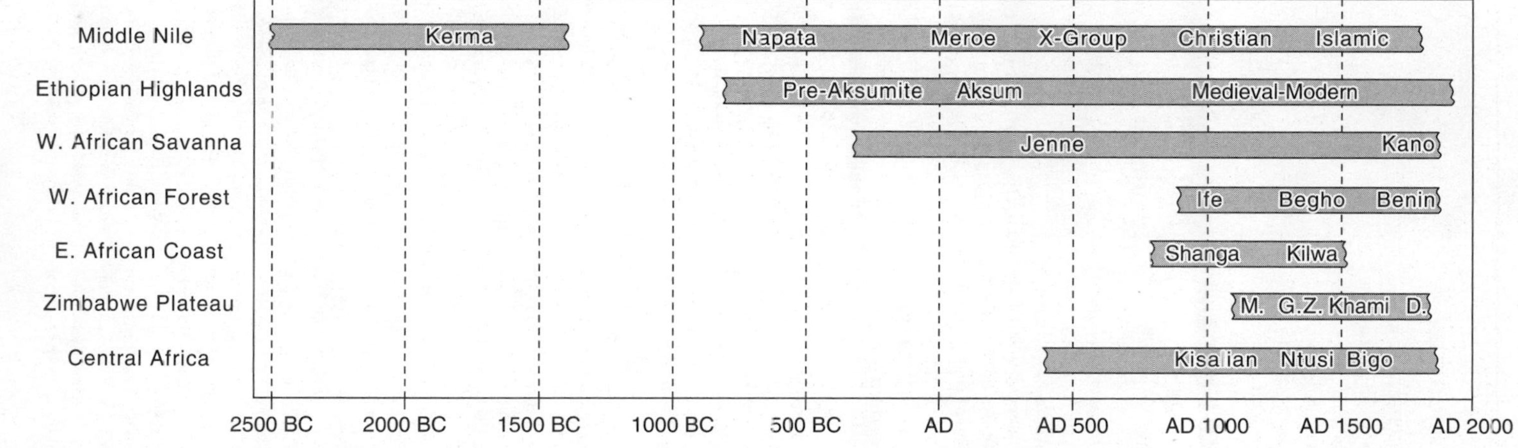


**FIGURE 4.6** Chronology of State/Urban Developments in Different Regions of Africa. *Source:* Reprinted with the permission of Cambridge University Press. Connah 2001:16.

## BOX 4.3 "GREAT ZIMBABWE": AN URBAN HISTORY ALMOST FORGOTTEN

Europeans largely justified colonialism on the basis that Africans were a people without history and incapable of indigenous development. One detrimental and damaging aspect of colonialism was Europe's role in denying African peoples their cultural heritage. In the southern African region, there was a well-documented effort to misinterpret evidence of an important indigenous urban civilization. Falsification became a handy tool to distort past African achievements and to support the representation of Africa as "uncivilized." This happened in the interpretation of Great Zimbabwe.

The word "Zimbabwe" means "stone dwellings" in the native Shona language, and remnants of stone buildings sparked one of the most important controversies in the archaeological world. Great Zimbabwe is known for its extensive stone complex and evidence of an ancient city. The central area of ruins extends about 200 acres (80 hectares), making Great Zimbabwe the largest of more than 150 major stone ruins scattered across Zimbabwe and Mozambique. The site is located in southeastern Zimbabwe, approximately 19 miles (30 km) southeast of Masvingo (Fig. 4.7). Great Zimbabwe was the heart of a trading empire that extended internationally from AD 1100 to AD 1400. At its height, the settlement was home to at least 18,000 people (Connah 2001:223), and gold in the vicinity was a major catalyst for long-distance trade and urban consolidation. The Great Zimbabwe complex is the largest precolonial ruin south of the Egyptian pyramids.

Cecil Rhodes, a well-known British imperialist and influential colonizer who conquered large parts of Southern Africa and named territories such as Rhodesia (modern Zimbabwe) and Northern Rhodesia (modern Zambia), played a devious role in concealing the truth about the ruins. The complex mesmerized Rhodes, like previous European explorers who had rushed to hasty judgments without a shred of evidence about the origins of the ruins, surmising nonindigenous origins (associating them with King Solomon and the Queen of Sheba). However, Rhodes saw an opportunity and a challenge in the authenticity of the site. He got involved on two levels: he established the Ancient Ruins company to prospect the ruins for treasure, and he financed a British archaeologist, Theodore Bent, to travel to the region and study the site. Bent published his findings in an 1892 book, *The Ruined Cities of Mashonaland* (Mashonaland is the home of the Shona peoples), and determined (conveniently for Rhodes) that the Great Zimbabwe complex "proved" that the civilization was not built by local Africans. Falsification continued, and subsequent archaeologists and explorers supported Bent's opinion: one archaeologist supported the nonindigenous thesis without ever visiting the site; another curator (unqualified for his post) at the site spent two and half years blundering recklessly by removing 12 feet of deposition and anything else that could be linked with the local population. Subsequent European excavators questioned the alien origins thesis, but the counterthesis barely saw the light of day. A minority of scholarly opinion has always supported the nonindigenous origins thesis. For example, Mallows (1985) claims the great enclosure was built by Arabs for storing slaves for shipment to India.

**FIGURE 4.7** Ruins of Great Zimbabwe.
*Source:* © Robert Harding World Imagery/Corbis.

The ruins were shrouded in controversy not only in the colonial period but also during the first years of independence. The new Rhodesian government put considerable political pressure on archaeologists to deny its construction by local, black people.

Eventually, radiocarbon testing and other archaeological evidence (e.g., close examination of dwellings, defensive sites, dams, and irrigation systems in addition to indications of trade, mining, and carts) enabled researchers to piece together the chronology of activities; this construction left no doubt that the ruins were of African origin and that the inhabitants had engaged in regional (within south-central Africa) and long-distance trade within a complex social system. Great Zimbabwe functioned as a capital city whose elite was tied to an extended economy of small towns on the Zimbabwean plateau. The settlement, in part, operated as a clearinghouse: goods were imported from the Islamic world and China (e.g., ceramics) and India (e.g., beads) for distribution to interior areas. These latter areas, in turn, exported primary products, especially gold, via the main center. The stone buildings even display evidence of considerable technological expertise in construction. According to Peter Garlake, a Zimbabwean archaeologist (1973:50), the indigenous dry-stone walling tradition in Great Zimbabwe is an "architecture that is unparalleled elsewhere in Africa or beyond."

European reaction to the discovery of Zimbabwe is just one striking example of the generally held belief that Africans had played a passive part in history, occasionally influenced by Phoenicians, Arabs, or other outsiders, but that they had lacked their own urban creativity. Europeans were more willing to jump to exotic theories of the origins of Great Zimbabwe than to accept the information that the site revealed. Eventually, incontrovertible evidence established that the urban settlement had emerged, developed, and then declined, and that the urban population was diffused throughout the region.

The historical interpretation of Great Zimbabwe became an important symbol for both sides in the struggle for majority rule after Rhodesia declared its independence in 1965, governed exclusively by whites. For black Africans, Great Zimbabwe was powerful evidence of an early African urban civilization, while for a white African minority, it became psychologically essential to detract from scenarios of Africans' organizing and controlling their own territories, both past and future. During the liberation struggles of the 1950s and 1960s, the ruins were presented as a potent icon of achievement and pride for the country's inhabitants and became a rallying symbol for unity. The national movement against colonial rule used the name "Zimbabwe" as a deliberate act of defiance. During white minority rule, the government suppressed information on the region's origin, censoring guidebooks, textbooks, museum exhibitions, etc.

When Southern Rhodesia eventually achieved majority rule, the new government chose the name "Zimbabwe" to enshrine African pride, achievements, and cultural values. The country's official name became Zimbabwe on April 17, 1980, the only country named in honor of a past urban civilization. In 1986, the site at Great Zimbabwe was declared a World Heritage Site. Unfortunately, despite its historical significance, Great Zimbabwe has received inadequate government funding for its preservation and scientific study, and the current political crisis in the country masks the world's understanding of its contribution to past urban life.

Great Zimbabwe (like Timbuktu and other places) is a mystery only when viewed in isolation. When we put the urban settlement into proper time and space contexts, it is evident that Africa had various urban traditions before the Europeans arrived.

cities (discussed earlier), which emerged with the expansion of agriculture, depending on location. Second, there are cities created through contact with Islam and the Arab world through long-distance trade. They appear along the eastern coast and in western Sahelian Sudan. Their function as commercial and cultural links is clear, but we know much less about the synergy between indigenous processes of urbanization and external impulses. The third period involves the beginning of European-directed urban development commencing in the second half of the 15th century; among the first was the Portuguese-directed urban model. Such European influences on urbanization predate the scramble for Africa and the building of colonial territories. Interactions between the Portuguese and the Congolese at this time showed some complementarity. For example, cordial relations developed between the two powers (Congolese came to Europe and Portuguese experts and printers went to Congo to instruct locals), and Lisbon's introduction of new crops like cassava and corn helped a Central African capital to emerge in M'banza-Kongo (in present-day Angola, near the border with the Democratic Republic of the Congo) and facilitated the centralization of power in other cities of the region. In much of Africa, Portuguese-style cities developed in accordance with local conditions, without the political and legal dependency on Western characteristics that punctuated the colonial period.

The fourth period is the colonial era, which brought about a divide but did not necessarily destroy the earlier urban centers. Colonial cities used, competed with, and, in some cases, completed previously existing settlements. The colonial period also determined the future trajectory of cities without regard for their history. Colonization and urban development were intimately related. Europeans were few in number and interested in speed and efficiency, so they used existing urban centers and dramatically changed transportation and trade networks to suit their purposes. All trade was reoriented toward port cities and tied to external interests. Colonizers created new port cities (e.g., Accra, Abidjan, Durban, and Port Harcourt), turned other strategic crossroads and small settlements into cities (Johannesburg, Brazzaville, Lusaka), and also created new cities from scratch (Nairobi, Cotonou, and Kampala). They had the choice of destroying or abandoning ancient cities or using them for their own purposes; they preferred the latter strategy whenever possible. Timbuktu, Jenne, and Kilwa are examples of cities that were not deemed pivotal for empire; they declined but did not disappear from the map.

## COLONIAL LEGACIES

The course of Africa's development was profoundly affected by colonialism. Each European power had its own complex motivations for colonizing African territory, but the imposed values, perceptions, and institutions left an indelible imprint on the people and territories in the form of changed economic orientation, language, education, law, urban planning, and so forth. Colonialism was far from a uniform process; it is better characterized by great heterogeneity because experiences varied from colonial ruler to colonial ruler and from place to place, but there were some commonalities. It unfolded in complex ways as European powers built their empires and faced various reactions from indigenous peoples.

A new elite with different value systems and beliefs was implanted on African soil. A small cadre of administrators and military officers operated to extend the European "sphere of influence" within the colonial territory. European colonial policy reflected European interests and bore no relationship to the needs of Africans. Decisions were made without any mechanism for African participation, consequently undermining and weakening traditional authority.

Colonial governments believed their primary responsibility was to maintain law and order at a minimum cost to the European taxpayer. The rationale was to extract revenue from resources and to have each colony (if possible) also supply the revenues to govern the colony. They anticipated it would take time to produce wealth from the territories, which was contingent on upfront investments toward the establishment of government machinery, immigration of the colonial apparatus and settlers, colonial settlement extensions, and other strategic apparatuses such as railways, roads, and ports to facilitate the development/exploitation of Africa's resources. No matter how resource-rich a colony was, the colonial government was starved of funds necessary for development. Colonial governments were scarcely able to provide basic infrastructure (e.g., roads and communication networks), and basic social services (e.g., education, health care, and housing) were severely and strategically limited. As a result, many colonies spent more funds on the military and policing than on education, housing, and health care combined.

Colonial powers differed in how they ruled. The French, Belgians, Portuguese, and Germans implemented direct rule. They functioned via centralized administrations in urban centers, and their policies emphasized assimilation. The motive of civilizing African societies so that they would be more like Europeans was often paramount. Divide-and-rule strategies and policy implementation that intentionally weakened indigenous power networks and institutions were common in direct rule. Europeans manipulated ethnic, cultural, linguistic, and religious differences to their own advantage. For example, Germans and then Belgians employed the same divide-and-rule strategy in Rwanda to elevate the minority Tutsi to positions of power in colonial administration. The level of animosity in Rwanda was so severe that it became a trigger to the 1994 Rwandan genocide. Colonial administrations recruited supporting cadres of Africans as lower-level bureaucrats. They tried to ensure loyalty by recruiting subordinate classes, such as former slave families, or by importing individuals from other regions, but they made good use of minority groups when it suited them.

In contrast, Britain used indirect rule. In principle, indirect rule was intended to prevent the destruction of indigenous culture and to be more cooperative than direct rule. This is not to say that its impacts were not as consequential as direct rule. Indirect rule incorporated traditional authorities into the colonial administration, with inferior roles. Some traditional rulers held on to power in particular areas and performed day-to-day government duties, gaining prestige and power but forfeiting local autonomy. Indirect rule allowed a small number of Europeans to govern effectively a large number of people over extensive territories.

## GEOGRAPHIES OF EMPIRE

Colonies were increasingly incorporated into global economic and resource systems and dominated by European metropolitan centers, far removed from the areas of production. Some plantation agriculture was introduced, though it tended to be on a much smaller scale than those of Southeast Asia and Central America and with stronger control. Crops such as cotton, groundnuts, coffee, and tea were planted in some areas. Mineral exploitation of copper, gold, tin, and diamonds took place on a larger scale because of European investments and use of technologies, largely driven by demands in industrial Europe. Europe's command and control of Africa's economies resulted in changes in the orientation and direction of trade; previously, the flow of mineral wealth was mainly via land, moving north to the trading cities of the savanna belt and across the Sahara, but with colonialism, it all went via the ports and the sea to Europe.

Colonial powers built the arteries of empire by establishing transport and communication networks. Africa had long-existing transport networks that included coastal navigable rivers, but navigation was much more difficult in the interior (e.g., the Zambezi River had rapids and waterfalls). Dirt paths, routes used by nomadic people, and foot tracks, especially in East Africa, were routine links in transportation systems. Traditional means of transportation by portage, cart, and river were extremely slow. In contrast, Europeans built new larger-scale roads to speed the movement of goods and people (especially troops), exemplified by the construction of frontier roads in South Africa. New roads were also built as feeder roads to railways, and at the beginning of the 20th century, better-quality roads for motor vehicles were built in colonial urban centers. Roads principally served European's interests; there were disproportionately fewer in border regions (away from the colonial capital), and very few actually crossed international boundaries.

The development of railways in Africa accelerated fragmentation of the region and at the same time consolidated European economic control. Constructed to facilitate the transportation of goods from inland areas to the coast, railways inevitably hastened the drainage of resources, agricultural products, and human power. Their economic effects were profound in the development of a colonial economy: transportation costs were reduced by "90 to 95 percent," trading systems were restructured, labor was released, and outlets for inland commodity production were created (Butlin 2009:481). Railways influenced the expansion of ports (e.g., Dar es Salaam, Mombasa, and Dakar), while interior areas not connected to railways declined. Railways were also influential in the development of European settlement and commercial agriculture in interior areas (e.g., the Kenyan highlands, Southern Rhodesia, and southwest South Africa).

Few railways cut across international frontiers in Africa; most do not even approach them. In French West Africa, this resulted in a series of almost parallel routes linking the hinterlands of French territories to coastal ports at Dakar, Abidjan, Lomé, and Cotonou, but these were never connected. In practice, this meant that many regions lay within 50 miles (80 km) of a railway, but the railways were not used within the regions because of their location on the "wrong side" of an international boundary. For example, eastern Ghana, separated from the rest of the country by the Volta River and Dam, would have benefited from a connection to the Togolese railway, but colonial partitioning made sure that this would never happen. The British and French considered building a trans-African railway. The main proposal was a railway from the Cape to Cairo, largely the idea of British explorer Cecil Rhodes (Fig. 4.8). The project commenced but was never completed. It was deemed too expensive, and colonial rivalry eventually meant the project was doomed.

**FIGURE 4.8** The Rhodes Colossus: Striding from Cape Town to Cairo. *Source: Punch*, December 10, 1892.

Demarcating colonial territories and boundaries was a critical feature of colonialism. According to Clapham (1996:31), "The previously fuzzy borderlands between indigenous centers of government, together with the large areas which possessed no formalized government structures at all, were replaced [at least on the map, though only much later and with more uncertainty on the ground] by precisely demarcated frontiers of the sort that European concepts of statehood deemed necessary."

Colonial boundary making was dubious and unscientific. Nugent (1996:41–42) records that 44% of African political boundaries were drawn according to astronomical lines (meridian parallels), 30% were mathematical lines (arcs and curves), and only 265 were related to geographical features. One obvious paradox was that Africans' local knowledge was infinitely more detailed than Europeans' knowledge, yet Europeans were infatuated with their sense of purpose, which propelled them to draw lines without second-guessing their legitimacy. It is a myth that the Berlin conference settled the boundaries once and for all. Some efforts at readjustment took place, and various boundary commissions worked to adjust lines in some instances to include known chiefdoms' boundaries. For instance, the boundary between Nigeria and current Niger was redrawn three times, the last effort making use of local trade routes in the precolonial period.

The work of boundary commissions varied enormously. Lamb (2000:11) presents a glimpse into a boundary commission in Northern Rhodesia tasked with boundary resolution pertaining to an early agreement between Britain and King Leopold of Belgium:

> The Anglo-Belgian Boundary Commission has been something of a disappointment, too, though [Gore-Brown, a member of the commission] has reveled in the outdoors life and the hunting. It was all far too bureaucratic and tied up with the petty egos of the other officers. Within three months Captain Everett, the second in command, was eaten by a lion, a most ill-omened start. Major Gilliam, who headed the commission, spent most of his time in a haze of whiskey and kept issuing and rescinding orders. . . . The work marking out the border has been damnably slow. The 1894 agreement between Britain and Belgium's King Leopold was vague, simply stating that the border runs southwards from Lake Bangweulu to its junction with the watershed separating the Congo and Zambezi rivers, following this line for 500 miles 'til it reached the Portuguese frontier.

International borders even bisected villages in particular cases. For example, Jassini, a village on the Tanzania–Kenya border, suffered because of neglect due to colonially imposed lines partitioning the village between two colonial entities. An Anglo-German treaty of 1900 demarcated the official international boundary via mangrove poles, anthills, and baobabs. Jassini's location around the mangrove swamps and shifting watercourses of the tidal estuary on which the village is situated illustrates the problems of drawing African national boundaries from desks in Europe and of arranging space to adhere to a line on a map. None of the original markers survived the course of time. Efforts at nation-state building in the independence period also sought to erase the past. For example, the government of Tanzania after 1975 withheld services to Jassini and actively discouraged people from living there, instructing the villagers to relocate some miles away, where services would be provided. They refused and have since been left to their own devices. Jassini to this day does not appear on any Tanzanian or Google map.

Colonial partition resulted in Africa becoming the most divided continent on Earth. It was Africa's misfortune not only to be plundered by Europe but also to have been colonized at a time when the concept of the nation-state was firmly entrenched as the determining principle in territorial organization. As Reader (1998:574) emphasizes, "Africa south of the Sahara contains 48 states, more than three times the number of states in Asia (whose land surface is almost 50% larger) and nearly four times that of South America." The sum of Africa's boundaries is more than 28,600 miles (46,000 km), compared to 26,100 miles (42,000 km) in Asia (Reader 1998:574). Not surprisingly, most African states have more than one neighbor—20 possess four or more—and the Congo has the most international borders, surrounded by nine states (making it seem more than a coincidence that it is afflicted by internal strife and neighboring interference). One of the severe criticisms of the Berlin conference is that the resulting boundaries divided cultural areas and tribes. One calculation is that "103 international boundaries in Africa cut through a total of 131 cultural areas, some of which are partitioned by more than one boundary" (Griffiths 1996:74, 81; cited in Butlin 2009:342). Boundaries are frequently ignored by people circulating within a cultural area and by nomadic peoples. Some borders are not clearly marked, so leaky state borders are common, and all sorts of cross-border transactions take place (e.g., migration, smuggling, and legal trade).

There are many examples where artificial boundaries severed established trading relationships and ripped apart communities. The boundary between Senegal and Gambia is a case in point. The Gambia, 311 miles (500 km) long and only 12.4 miles (20 km) wide in places, sits astride the navigable section of the Gambia River, which is a wormlike intrusion into the state of Senegal. Apart from some small sections, the boundary is entirely geometrical, consisting of arcs and straight lines that in some cases run directly through villages, severing trade routes and dividing populations between French- and English-speaking administrations. At the same time, the Gambia River is one of the easiest and most extensively navigable rivers in Africa. The international boundary has prevented the river from being developed as the principal artery of trade for Senegal, landlocked Mali, and the Gambia. Instead, French colonial authorities used the inferior Senegal River as a means to access their inland territories in Senegal, the Gambia, and Mali, resulting in the inability of all three states to access and use the river to the same degree; thus, each state pursued an independent and uncoordinated economic strategy. The Gambia is the most artificial state in Africa, an English-speaking country within a French-speaking subregion of West Africa.

## URBAN DEVELOPMENT AND PLANNING THE COLONIAL CITY

African urban development had deep roots, and urbanism had flourished in particular places at particular times, while in certain areas it had disappeared before the Europeans arrived. Early African urbanism experienced considerable discontinuities—and, it seems, fewer continuities. Colonization was a crucial turning point as urbanization was implanted over wider areas from the late 19th century onward. Prior to Europeans' arrival, the region was not highly urbanized and remained rural in orientation. Urbanization existed in pockets, and cities were only moderately integrated into regional urban systems and, in fewer cases, articulated into systems beyond the continent.

Whenever possible, colonial powers modified existing settlements because Europeans were few in number and colonizers were motivated by speed and efficiency. The selection of new sites for colonial headquarters—in other words, the implementation of urban development from scratch—was much less common than is supposed, although examples do exist (e.g., Lusaka, Pretoria, Nairobi, and, to a lesser extent, Accra) (See Figure 4.10). As can be expected, colonial-orchestrated urban development negatively affected some interior towns and urban settlements. For example, settlements formerly dependent on caravan routes could not operate within the context of colonial borders and withered (e.g., Timbuktu, Jenne). Colonial capitals, many of which were ports, experienced the most dramatic expansion as transport networks for the shipment of minerals; agricultural exports became focused on the capitals; and imports into colonial territories were circulated via the same networks. In most cases in Africa, port cities became the colonial capital as well as the economic and population center for the entire territory. Exceptions existed in landlocked African countries (Bamako, (Mali), Ouagadougou, (Burkina Faso), and N'Djamena, (Chad)) and in some places in East Africa (Kampala and Nairobi) where the capital was in the interior.

Although most colonial urban planning was grafted onto older settlements, its schemes overturned indigenous spatial and physical arrangements. Formal planning efforts concentrated in the central part of old towns and in new elite residential areas. Europeans took over the most desirable land: the strategic arteries of the city, the beaches and breezy hilltops, and so forth. Whenever more land was needed, colonial authorities enacted laws (e.g., zoning ordinances and/or powers of eminent domain) and rezoned and usurped land. The function of colonial headquarters was political, administrative, and commercial, and the built

## BOX 4.4 THE BUNGALOW

Another colonial legacy was the promotion of alien architectural design, which changed housing preferences. In Africa the colonists, as newcomers, did not consider adopting the dwellings and settlement form of the indigenous peoples; instead they opted for a radical change in the imported housing style of the bungalow. The bungalow (known as the "banggolo") originated from 17th-century India and was used to define a peasant's rural hut in Bengal, made out of mud, thatch, and bamboo. The model was transferred from India to Britain in the second half of the 19th century, and the bungalow was developed in new ways for other regions of the world. A well-known manual of that time, John Murray's 1895 *How to Live in Tropical Africa*, described how the bungalow was the best house for tropical Africa (King 1995:199–200). The popularity of the bungalow since its introduction into Africa has been remarkable: this housing model is now a dramatic and persistent element in urban Africa (and is now visible in rural Africa).

The British modified the standard design for Africa by raising the structure off the ground for greater dryness, ventilation, and freedom from dust and invasion of insects (Fig. 4.9). In contrast, the Dutch, Spanish, and Portuguese built brick and more solid structures firmly on the ground. Some prefabricated houses were even exported from Liverpool to British Africa. The dwelling unit was a technological device: a form of shelter for British colonial officials to provide protection against malaria and shelter from the tropical heat. At the turn of the 19th century, the bungalow design involved new material, science, and technology developed in industrial Britain. Building bungalows was expensive, ranging from £1,000 to £2000. For instance, in Lagos, the unit belonging to the general manager of the railways cost £1,500 and that of the principal medical officer cost £1,775. Adoption of foreign building design and the dependence on imported



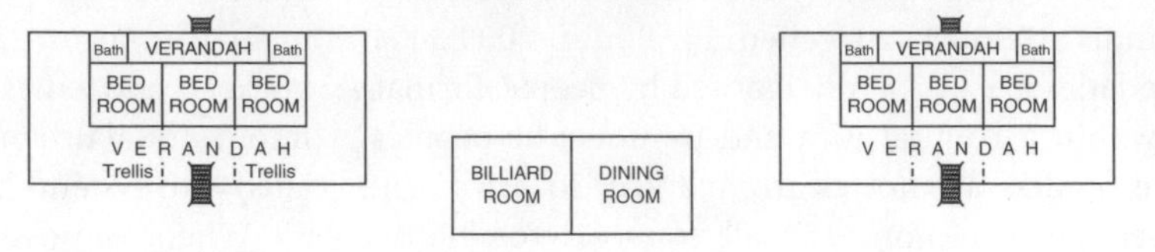

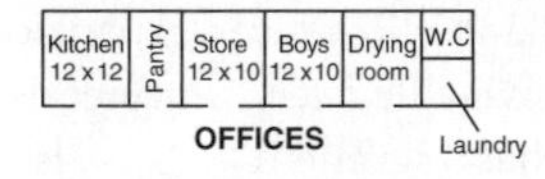

**FIGURE 4.9** The Bungalow Transferred: Official Colonial Housing in West Africa. *Source:* King, 1995:211.

construction technologies resulted in the high costs of units; bungalows were beyond the means of most of the population.

The bungalow design served European ideas of low-density living and privacy in which domestic activities such as eating, sleeping, and bathing occurred within the house and out of sight of indigenous inhabitants. The official residing in a bungalow was typically a single person or a nuclear family (with servants), isolated from their home community (some thousands of miles away) but socially distant from surrounding populations. Bungalows were diffused across colonial economies in the late 19th century. In the 1880s and 1890s, growing number of merchants, missionaries, doctors, engineers, and government officials came to West and East Africa, and bungalows were built to serve officials' needs.

Africans, too, adopted bungalow living in time. The bungalow preference gained momentum among Africans in the 20th and especially the 21st century. Many Africans switched their living arrangements from residing as extended families, farming on family-held land, and living in compound dwellings of grass or mud to modern urban living as nuclear families, working for wages, and residing in single-family bungalows built with concrete materials. This is in sharp contrast to traditional chiefs and families, all part of a collective, living in higher-density dwellings. Thus, the social and spatial organization of building design based on multifamily, tribal structures and on religious and indigenous political life was cast aside.

The bungalow is a good example of how foreign design and building aesthetics were imposed onto Africa, and it has social, political, and economic ramifications. It was a "tool of Empire" (like quinine, the railway, and the Enfield rifle), accommodating colonial officials who exercised direct control over expanded territorial economies. Its emergence as a sphere of (European) knowledge marks the expansion of Europe into a region where Europeans had not previously lived. The bungalow is a marker of the physical and spatial process of urbanization that incorporated modern Africa into a capitalist world economy. Bungalow construction relied on imported concrete and materials and was highly differentiated from traditional forms of shelter built with local materials. Housing construction practices during the colonial period became insensitive to local resources, and the practice has endured. It is obvious that the proliferation of bungalows has contributed greatly to sprawl throughout the region, and it is no coincidence that housing remains one of the most pressing problems in Africa.

environment was organized spatially to allow the colonial city to fulfill these roles. In other cases, such as the mineral areas of the copper belt in Zambia and the Democratic Republic of the Congo, the function of urban settlement was different, oriented toward housing and providing services for the mining industry rather than for colonial administration. Colonial cities never fully realized a holistic objective; economic control was limited to pockets within colonial territories (no doubt the more lucrative), and administrative and political control ebbed and flowed and was undermined by the indigenous people. Furthermore, neither the development of markets nor the conscious programs of social, political, and physical transformation succeeded in propelling the urban centers into a truly capitalist world economy. On balance, colonial planning produced uneven, patchwork outcomes, and colonial cities were fragmented places. In time, these fragmented colonial epicenters became new indigenous urban centers and brought forth the shift from a rural to an urban African orientation. Indeed, as the years passed, a new African leadership emerged in cities, where the most opportunities presented themselves for African social mobility, albeit limited to a small group of educated African elites. From the early days of colonialism, Africans became interpreters, clerks, police sergeants, and foremen, prospering from these positions as the "African face" of the new colonial order.

Colonial administrations operated, for the most part, on a shoestring budget, which limited the development of basic public infrastructure; the vast majority of urban funds was earmarked for the European city. Racial spatial segregation policies allowed colonial authorities to target European districts for the provision of limited supplies of scarce infrastructural services (e.g., electricity, tarred streets, pipe-borne water, and police patrols). Therefore, "the good life" was concentrated and overlooked the indigenous population.

The colonial push toward urbanization did not totally transform bonds between urban Africans and their rural home areas. It was quite a stretch for planners to assume that planned urban space and planned neighborhoods could break traditional tribal and kinship bonds. Many acts of resistance took subtle forms—everyday expressions of African identity superseding the colonial enterprise rather than submitting to the colonial city; for example, speaking indigenous languages or acting within ethnic, extended family, and religious memberships. Cultural opposition was manifested by individuals' holding on to spiritual beliefs and also found expression in poetry, songs, and

**FIGURE 4.10** Colonial Cities. Adapted from Binns, Dixon and Nel 2012:149.

storytelling. More salient forms of opposition involved organizing outside spheres of European control. There were acts of open defiance, such as nonviolent protests (e.g., opposition to the imposition of the Afrikaans language in education led to demonstrations and sparked the Soweto Uprising of 1976 in South Africa). Earlier resistance was suppressed heavy-handedly by the colonial authorities; for example, the Maji Maji Rebellion in German East Africa (Tanzania) in 1905, in which the locals opposed labor and tax laws, was put down by the Germans (the indigenous people sprinkled their bodies with holy water—*maji* in the Swahili language—to protect their bodies against bullets). Resistance also found expression in revolts against colonially imposed chiefs and colonial collaborators. One example is the Igbo women's revolt in 1929 in southeastern Nigeria: it was an entirely rural women's rebellion against an impending tax that threatened their livelihoods. Rather than rebel directly against the Europeans, they directed their opposition against Warrant Chief Okugo (a warrant chief is a recognized chief who served as tax collector for the British).

Overall, there was not a unified path of African resistance (in ethnic, class, gender, age, or other ways) against the colonial administration. Colonial opposition was expressed most often in subtle forms as African individuals and groups sought to remove themselves from the colonial sphere of influence rather than to challenge it. Acting in this way, Africans resisted the colonial order and undermined European authority.

Colonialism brought profound socioeconomic changes in African society. In many ways, the European elite and their African cadre were cultural pioneers. For example, colonialism was based on assumptions that the state should both promote development and provide particular services such as education, agricultural instruction, and some housing. According to Gann and Duignan (1978:347). "the very notion of the state operating as a territorial entity independent of ethnic or kinship ties, operating through impersonal rules, was one of the most revolutionary concepts bequeathed by colonialism to post-colonial precedent. . . . All of them have taken over, in some way or form or other, both the boundaries and the administrative institutions of their erstwhile Western lords." Colonialism also diffused different cultural values such as dress. Myers (2003:165) shows how dress was used to inculcate ordinary urban Africans: colonial residents of Zanzibar, were told by the British to dress like gentlemen, keep their hair cut, wear a sports coat, and press their trousers. They were made to feel dirty because they did not have good clothes. For a small African elite, colonial rule provided new opportunities (particularly for Western-educated natives), and they became indispensable to European rule and influence.

There was an absence of critical reflection on the part of the planners, who understood their mission as a constituent part of a larger project to realize the goals of the colonial enterprise. Physical planning was viewed as an exportable commodity: town planners, town planning legislation, building design, and the European experience of new towns were cut and pasted into a different landscapes without much deliberation and even less reflection. Colonial planners did not work in isolation or for themselves; ideas about the planning of urban form and function moved within and between imperial networks. Urban planning was not just a tool to impose order and juxtapose land-use activities; it was used for social control and to ensure that the colonial headquarters city was the focal point in colonial Africa (See Figure 4.11). Elsewhere, particularly in the interior and in areas where there was no previous urban settlement, Europeans established cities for the first time. Cities became the vehicle for the colonization enterprise.

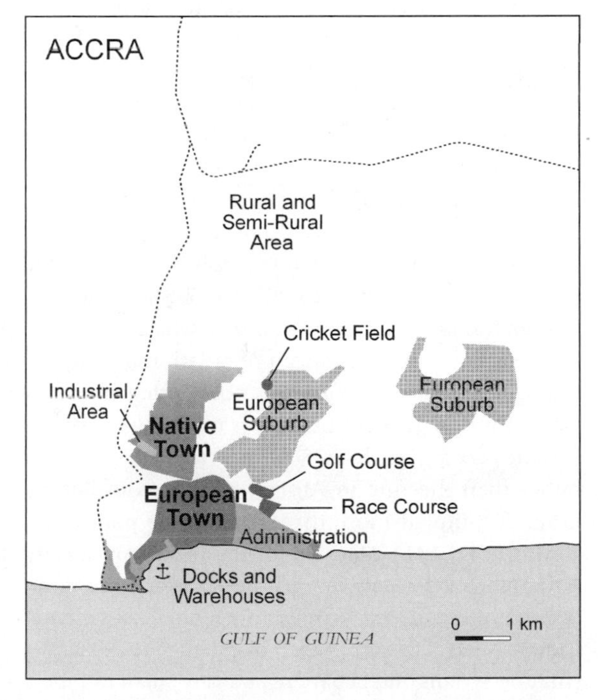


**FIGURE 4.11** Colonial Accra. *Source:* Grant and Nijman 2002:325.

Throughout colonial headquarters cities, the built environment was socially and spatially organized around a number of key zones: (1) a European town that functioned as a central business district (CBD) for the colonial headquarters of European companies as well as a warehouse center for storing and transshipping commodities; (2) an administrative center that contained the key buildings and institutions of colonialism with army barracks or cantonment, the police, hospital, jails, etc., in close proximity; (3) upscale planned European residential areas (on higher elevations whenever possible for security, climatic, and psychological reasons) for colonial administrators, consisting of bungalows and recreational spaces that were well serviced (sanitation, water, and road systems); (4) a native town for African workers, who were wedged into dense, mixed low-cost dwellings and indigenous commercial zones without planning, services, or amenities; and (5) an agricultural hinterland where food could be grown to feed expanding urban populations. There were large differences in population densities between the areas of the colonial elite and the areas of the indigenous population, influencing lifestyle and quality of life (Fig. 4.11). Indeed, expatriate life in many African cities today still reflects the colonial ordering of physical and social space, although new elements (shopping malls and supermarkets) have been added to the landscape of privilege (Smiley 2010).

The European population was concentrated in colonial cities, but not to the degree one might imagine based on most accounts. For example, in Nairobi (the capital of a settlement colony), only 7.5% of the total population in 1933 were Europeans, which amounted to about half of the European population in Kenya at that time. There is a common notion that the cities were primarily "white," but this is misleading. Even in a settler colony such as South Africa (an exception rather than the rule in Africa), people from Europe never constituted more than 20% of the population (Butlin 2009:141). Natives, for the most part, greatly outnumbered Europeans and the more modest numbers of international immigrants from Asia (mainly Indians and Chinese) and from the Middle East (mainly Syrians and Lebanese and very small numbers of Jews). Asian and Middle Eastern groups occupied a middle rung in the occupational ladder. Many engaged in import-export trade and construction, and they played an important intermediary role in the colonial enterprise.

Many Europeans went to Africa on a temporary basis as administrators, engineers, missionaries, health and educational specialists, and workers for infrastructural projects and in due course returned to their countries of origin. The French, Germans, and Italians were less successful than the British in encouraging the movement of people from the metropolis to the colony. Non-British colonies had very thin European populations in many territories: Equatorial Guinea (French) had a population of 3.19 million in 1931, and only about 4,657 were Europeans (Butlin 2009:157). Togoland (German) had a population of 1.03 million in 1914, and the white population numbered a mere 368 persons (Butlin 2009:160). In French colonies, there tended to be greater mixtures of other Europeans: Italians, Spaniards, and Maltese, all of whom jostled each other for status, with the French elite looking down on all of them. In addition, the white population was heavily stratified by occupation. Although the whites were primarily in control, colonial cities in Africa were far from "white" cities.

The colonial city became a quintessential expression of white power. Company headquarters, banks, administration structures, military organization, and, of course, jobs were centered in the city. Large numbers of workers were drawn into cities to serve the colonial enterprise, while the responsibility of reproduction was left to the countryside. Urban life acquired specific features that distinguished it from rural life—military parades, art, modern celebrations—but this is not to imply that indigenous traditions disappeared: they took place out of colonial sight in different parts of the city.

African communities had a life rich in association: participation in social activities such as funeral groups, hometown groups, drinking clubs, and drumming groups was common. Despite attempts at rigid segregation and occupational stratification, different cultures came into contact. Migrants refashioned the urban environment more quickly than the white population cared to acknowledge. For example, migration proceeded despite colonial governments' attempts to restrict labor to what they needed and could control. The flows into cities meant that Africans, out of necessity, opted for their own patterns of settlement and irregular employment. Even in well-laid-out native residential locations, Africans paid no heed to municipal building codes. Typically, dwelling structures exceeded the maximum as a ratio to plot sizes because builders were motivated to enlarge dwellings and to increase the numbers of rooms that could be used for subletting arrangements, a further violation of building codes.

Urban life in colonial cities entailed more blending and melting of cultures than is typically recognized, and cultural crystallization clearly took place. Colonial cites were, however, "colonized spaces" above all else, wide open to modernization influences from Europe and dominating Western values, but European urban ways of life underwent transformation in the process. In colonial cities, people combined the old and the new, integrating their heritage with the present as individuals adapted to different urban environments and cities in the making. Beneath the surface, colonial cities were far more complex arenas than simply European or white outposts in Africa. Instead of understanding African colonial cities as adaptations to the Western urban model, it is more appropriate to appreciate them as examples of synthesis and creativity.

Administrators used urban designs and the policy apparatus to make colonialism more attractive to Europeans and more tolerable for the colonized peoples. The spatial order (gridiron street systems combined with racial segregation, neighborhoods, etc.) and the architectural form of public buildings were reminiscent of Europe. European architectural forms were adapted to meet the climatic and resource constraints of colonial regions. Colonial officials supported the urban design of cities with policies that could dampen any possibilities of social unrest and this elevated the instruments of their authority, over and above the level of authority that existed in urban Europe, to deliver the colonial enterprise.

Obviously, the colonial city symbolized much more than the built physical environment. A crucial aspect was the social complex and social structure of the city and the colony. The colonial administrative center, the European town, and the European elite residential neighborhoods signaled power and control in grander ways than anything similar in earlier African cities. They played a vital role in impressing on the natives that colonial authorities controlled abundant resources that could be tapped for colonial government objectives. Therefore, the size, scale, and order of structures within these areas cannot be divorced from the discourses of domination and intimidation. The built environment epitomized a domineering and inviolable image of the imperial power and visibly embodied Western concepts of aesthetics and order. The environment signaled Europe's extensive power and ability to rule without a distance constraint, compared to African political power, which was territorially restricted as well as spatially confined and centered on the villages.

Urban living resulted in changes in economic activities and occupation, and in the way people lived. The city was not merely the incubator of new values and ideas, but it also became a vector for transmitting to the villages external values and ideas as well as new urban African ways of doing and being. As urban centers grew, they disrupted surrounding regions and often contributed to their reorganization. The colonial headquarters city was also instrumental in the organization of an urban hierarchy that greatly extended the reach of colonialism.

Despite differences in emphases on town planning and social control (the British adhered to racial segregation and the French to cultural and socioeconomic division), the outcome was segregated urban space (Njoh 2007). Efforts were made to segregate Africans by ethnic and religious origin, allegedly to reduce conflict. Africans' living and working spaces were placed in separate parts of the city. African migrants were essential to the working of the colonial city and economy but were denied any claim of citizenship; instead, they were regarded as impermanent urban dwellers. Segregation was often justified on the basis of security and public health, but an underlying motivation was racial division and containment. Poor Africans lived in precarious circumstances: low-income African neighborhoods could be flattened when it was convenient. For instance, the early years of planning Nairobi were marked by demolition episodes and destruction of impoverished settlements. Demolition was often said to be justified based on public health concerns, and this type of justification was used to explain the removal of black Africans from inner-city Johannesburg and their forced relocation to the South-West Townships (Soweto) on the outskirts of the city. Cultural conflicts were part of the tensions endemic in the character of colonial towns. Europeans adopted attitudes of superiority and almost contempt for local cultures. In colonial Brazzaville, the French administration went out of its way to impose cultural restrictions on the local population:

> the administration did issue a draconian decree which imposed restrictions on drumming, affecting not only recreational dancing but also funeral rights and the celebration for the end of mourning known as *matanga*. Not only was "drumming and noisy dancing" limited in time and space, that is to hours prescribed by the administration and on the outskirts of the town, but Africans had to give advance notice and pay a fee which was particularly resented, since it was set at a rate that was difficult for workers to meet. (Martin 1995:37)

In South Africa, the apartheid laws were an extreme form of control. For example, laws criminalized interracial marriages and interracial sexual intercourse, and restricted access to facilities such as beaches, toilets, and all entertainment venues for the black population.

An obsessive concern with "health" (a sort of sanitation syndrome) was another underlying consideration in colonial planning. The creation of "healthy" environments was defined in accordance with the cultural criteria of European powers with no appreciation of indigenous definitions of health and the means to achieve health. Culture- and class-specific perceptions of health hazards were more instrumental than an actual assessment of health risks in determining a specific health tilt in colonial urban planning policy. In the interests of health and the new economic and social order, new environments were created (King 1995:35)—"rows of minimal 'detached' housing units, surrounded by 'light air,' open space, gardens, and recreational areas in total disregard of the religious, social, symbolic and political meaning of the built environments."

Planning went ever further in terms of the criteria for preservation. Buildings of colonial architecture and history were to be preserved while the remnants of indigenous culture were allowed to disappear. For instance, in planning Kaduna in northern Nigeria, the Ministry of Overseas Development suggested the retention of a small iron bridge that had been erected by Lord Lugard, the previous colonial governor, for sentimental as opposed to practical reasons (King 1995:56).

Town planning expertise was increasingly transplanted to Africa. Town planning as a university discipline and profession emerged during the early years of the 20th century, and garden city ideas were incorporated into aspects of African town planning. The sphere of influence of the English Town and Country Planning Act of 1932, developed in the West Indies, was extended to Africa. For example, Uganda adopted the entire act in 1948. An urban planning apparatus with the locus of power outside of Africa emerged: external professional agents (appointed by the colonies) were sent, legislation was transplanted, professional publications on planning in tropical environments emerged, and reliance on the expertise of external consultants became routine. The predominance of international networks of communication and expertise was so extensive as to make local channels unimportant, and it set a precedent that has been hard to break. Colonial students studying, returning, and reinforcing European practices subsequently encouraged this kind of thinking. It was further strengthened by international organizations and professional entities communicating expert planning knowledge to the colonies and then to the former colonies. For example, the United Nations Human Settlement Programme (UN-HABITAT), UNESCO, and the World Health Organization (WHO) are among the prominent expert organizations that African governments engage today. The assumption that urban planning expertise has to lie outside of Africa and be imported has been challenged only in recent years.

## CONCLUSIONS

The promotion of colonial cities as centers of economic, administrative, and political power had profound implications for African society. The implantation of a colonial urban model in the region intensified urbanization processes that eventually led to society being anchored to urban centers in each African state. The gateway function of the colonial city was paramount: these cities served as conduits for the export of African rural, raw materials to Europe and as sites of import and dissemination of European manufactured goods and European ways of life that were heavily privileged above traditional and rural ways of life. African rural life and rural areas were of little interest to the Europeans except as sources of raw material and labor reserves.

An enduring legacy of the colonial era was that the mode of integration into the world economy was established. African economies concentrated on exporting minerals and agricultural primary commodities rather than focusing on production for domestic consumption. Extreme specialization took place in many colonial territories. There are examples of resource economies where one commodity dominates domestic production and exports (for example, coffee in Rwanda and Burundi); in most countries, three commodities or fewer account for most production and export activity (e.g., cocoa, coffee, timber in Côte d'Ivoire; tea and coffee in Kenya; copper and cobalt in Democratic Republic of Congo; cocoa and gold in Ghana). In most cases, colonial powers relied on African small-scale farmers to produce crops that were sold to state marketing boards and then exported. There are cases where colonial powers forced Africans into export agricultural production. For instance, in Gabon, the French coerced indigenous Gabonese to shift from staple food crops to export crops such as coffee or cocoa, and/or to harvest timber for export in order to pay colonial taxes. This meant that African farmers began to purchase a larger proportion of their food supplies, many of which were traded in urban centers.

Even though colonial cities grew and expanded, continuities and links with rural life remained, and rural affairs continued to have an impact on urban life, particularly in migrant and indigenous areas. In parts of the colonial city, idealized European norms were incorporated into the built environment. Spatial planning became a vehicle for organizing urban functions, and European concepts of aesthetics, order, and so forth were introduced. On the one hand, a small African educated urban elite helped reinforce the colonial city project. These Africans aspired to "modern" urban living and participated in urban lifestyles. Their urban social mobility was contingent on educational, occupational, and European-language achievement. On the other hand, the vast majority of urbanites were excluded from participation in European urban life, especially the "good life"; their urban existence revolved around low-status labor without urban citizenship. Rural Africans were confined to serving as peasants in the hinterlands, growing crops for export.

European economic and cultural values were disseminated from colonial cities outward into the region. European influence extended well beyond planning and organizing the built environment, and was manifested in colonial cultural life in terms of sports, the arts, and, in particular, eating habits and dress. Education became a crucial mechanism for inculcating in Africans Western practices of modern work: work revolved around regularity, discipline, and hierarchical structures based on work specialization and the organization of time and energy around work routines. According to Cooper (1980:70), education prepared workers for "internalizing cultural values and behavior patterns that would define their role in the economy and society." When Africans did not obey Western regimented time, their behavior was chastised as "Africa time." There were other more obvious ways that African urbanites engaged in Western modernization, such as housing preferences (e.g., bungalows) and food consumption. Consumption patterns changed at first because they were encouraged and convenient, but over time, taste preferences were altered. In due course, migrant laborers helped introduce these and other products (e.g., consumer durables) into their rural home villages.

The colonial conquest of the late 19th century ended Africa's relative isolation, and the slow pace of indigenous urban development strongly influenced the rise and fall of African societies that had prevailed for centuries. Colonialism increased the tempo and the spread of urbanization in Africa and rapidly dragged Africans into the 20th century, sometimes brutally. Not every European idea was flawed, and not everything the Europeans brought was bad. Medicine, water, urban drainage systems, education, and new methods of agriculture made positive contributions to societies. Furthermore, intra-African warfare was suppressed and internal slave trading was officially abolished. In Africa, the number of children attending school grew faster than did school enrollments in Europe during colonialism's boom years. Urban planning and political and economic ideas were more controversial but had the potential to be used for good or for bad, but mostly they served the privileged.

Ultimately, colonialism was imposed upon Africans, peoples who had long and vigorous historical experiences with their own societal and territorial organization. Europeans largely justified colonialism on the basis that Africans were a people without history and incapable of indigenous development. Colonialism forged patron-Client relationships where handpicked African elites became clients of colonial and overseas states (and they benefited accordingly). The political organization of African colonies had a total disregard for the political and territorial organizations that had existed prior. From the outset, European ideas were privileged and African ways of life were smothered, marginalized, and discredited. The development of colonial cities within externally determined entities (i.e., colonial territories) ruled the day, and subsequently Africa was divided into far more countries than would naturally occur with smaller and, as it turns out, more ethnically diverse populations, on average, than any other world region. These two colonial influences of small populations and ethnic diversity would have major impacts on the way that states, cities, and rural areas experienced development.

## REFERENCES

Anderson, D., and R. Rathbone. 2000. *Africa's Urban Past.* Portsmouth, NH: Heinemann.

Blunt, A. 1994. *Travel, Gender, and Imperialism: Mary Kingsley and West Africa.* New York: Guildford Press.

Butlin, R. 2009. *Geographies of Empire: European Empires and Colonies. c. 1880–1960.* Cambridge: Cambridge University Press.

Caillié, R. 1830. *Travels Through Central Africa Through Timbuctoo; and Across the Great Desert, to Morocco. Performed in the Years 1824–1828* (2 vols). London: Colburn & Bentley.

Chamberlain, M. 2010. *The Scramble for Africa.* 3rd ed. New York: Longman.

Childe, V. 1950. "The Urban Revolution." *The Town Planning Review* 21:3–17.

Clapham, C. 1996. *Africa and the International System: The Politics of State Survival.* New York: Cambridge University Press.

Cole, R., and H. DeBlij. 2007. *Survey of SubSaharan Africa. A Regional Geography.* New York: Oxford University Press.

Connah, G. 2001. *African Civilizations: An Archaeological Perspective.* 2nd ed. New York: Cambridge University Press.

Cooper, F. 1980. *From Slaves to Squatters: Plantation Labor and Agriculture in Zanzibar and Coastal Kenya, 1890–1925.* New York: Yale University Press.

Coquery-Vidrovitch, C. 2005. *The History of African Cities South of the Sahara.* Translated by Mary Baker. Princeton, NJ: Markus Wiener.

Davidson, B. 1992. *The Black Man's Burden. Africa and the Curse of the Nation-State.* New York: Times Books.

Davies, M. 2012. "Mary Kingsley." Available at http://www.royalafricansociety.org/index.php?option=com_content&task=view&id=169&Itemid=165 (accessed January 28, 2012).

Fabian, J. 2000. *Out of Our Minds: Reason and Madness in the Exploration of Central Africa.* Berkeley: University of California Press.

Freund, B. 2007. *The African City: A History.* New York: Cambridge University Press.

Gann, L., and P. Duignan. 1978. *The Rulers of British Africa, 1870–1914.* Stanford: Stanford University Press.

Garlake, P. 1973. *Great Zimbabwe.* London: Thames and Hudson.

Grant, R., and J. Nijman. 2002. "Globalization and the Corporate Geographies of Cities in the Lesser Developed World." *Annals of the Association of American Geographers* 92(2):320–340.

Griffiths, I. 1996. "Permeable Boundaries in Africa." In *African Boundaries, Barriers, Conduits and Opportunities,* eds. P. Nugent and A. Asiwaju, pp. 68–84. London: Pinter/Caswell.

Heffernan, M. 2001. "A Dream as Frail as Those of Ancient Times: The Incredible Geographies of Timbuctoo." *Environment and Planning* 19(2):203–25.

King, A. 1995. *The Bungalow: The Production of a Global Culture.* New York: Oxford University Press.

Lamb, C. 2000. *The Africa House.* London: Penguin.

Mabogunje, A. 1962. *Yoruba Towns.* Ibaden: Ibadan University Press.

Mallows, W. 1985. *The Mystery of the Great Zimbabwe.* London: Robert Hale.

Martin, P. 1995. *Leisure and Society in Colonial Brazzaville.* New York: Cambridge University Press.

Middleton, D. 1991. "Women in Africa." In *The Royal Geographical History of World Exploration,* ed. J. Keay, pp. 102–103. London: Paul Hamlyn.

Myers, G. 2003. *Verandahs of Power: Colonialism and Space in Urban Africa.* Syracuse: Syracuse University Press.

Njoh, A. 2007. *Planning Power: Town Planning and Social Control in Colonial Africa.* New York: University College London Press.

Nugent, P. 1996. "Arbitrary lines and the people's minds: a dissenting view on colonial boundaries in West Africa" In *African Boundaries: Barriers, Conduits and Opportunities*, eds, P. Nugent and A. Asiwaju, pp. 30–46, London: Pinter/Caswell.

Reader, J. 1998. *Africa. A Biography of the Continent.* New York: Alfred Knoff.

Smiley, S. 2010. "Expatriate Everyday Life in Dar es Salaam, Tanzania: Colonial Origins and Contemporary Legacies." *Social & Cultural Geography* 11(4):327–342.

Smith, M. 2009. "Ancient Cities." In *Encyclopedia of Urban Studies*, ed. R. Hutchinson, pp. 24–28. Thousand Oaks, CA: Sage.

## WEBSITES

BBC. 2010. "Lost Kingdoms of Africa: Great Zimbabwe." Available at http://www.youtube.com/watch?v=2be1gO36Fs4 (accessed December 8, 2013).

Indiegogo. "Timbuktu Libraries in Exile Campaign." Available at http://t160k.org (accessed December 8, 2013).

Royal African Society. http://www.royalafricansociety.org/who-we-are/27.html?Reference=61) (accessed December 5, 2013). A UK African organization that promotes better understanding of the region.

University of Cape Town's http://www.tombouctoumanuscripts.org (accessed December 5, 2013). A web resource on Timbuktu's manuscripts.

CHAPTER 5

# RURAL AFRICA

## INTRODUCTION

Most Africans still live in rural areas, despite recent accelerated urbanization trends. Most remain either physically located in or dependent on rural environments (Binns, Dixon, and Nel 2012). In 2011, 60% of the population (some 632 million people) lived in rural areas (United Nations [UN] 2012). While the urban transition is well under way, not until some time after 2030 will urbanization be fully representative of modern Africa.

Rural Africa is largely characterized by underdevelopment and widespread persistent poverty, and the latter is deepening in many instances. The expansiveness and isolation of many rural areas, colonial neglect, failure of national rural development strategies, infrastructure deficiencies, projected rural population growth, and the widening urban–rural development gap—combined with complex ethnic, poverty, and climate contexts—all magnify contemporary rural development challenges (Binns, Dixon, and Nel 2012).

Agriculture is the backbone of Africa's rural economy. Employing 60% to 70% of the population and accounting for 15% of Africa's gross domestic product (McKinsey 2010), agriculture encompasses crops, livestock, and fisheries. It continues to be central to national development despite massive flight from the countryside and the failure of rural development policies to stem this migration. In the 21st century, most experts target agricultural transformation as the core of rural development initiatives. Broadly defined, the different types of agriculture consist of large-scale commercial activities, smaller-scale commercial activities, and subsistence activities. Across Africa, most are smallholder subsistence farmers, engaged in agriculture in regions characterized by underdevelopment with poor infrastructure and services. Smallholders and their families are highly vulnerable populations with very high incidences of poverty, especially among children and women.

Rural Africa is in a nascent state of deep transformation. This chapter focuses on three major topics of rural development, the outcomes of which will deeply affect the kind of rural transformation that will take place. First, there is the debate about the push for an African green revolution (GR) and the role of genetically modified (GM) crops to increase agricultural productivity, enhance food security, and bolster rural livelihoods or whether such an experiment is misguided. Second, there is the contentious debate over the multifaceted development initiative of the Millennium Villages Project (MVP; see Box 5.1) and whether spending on a set of interventions (to reach Millennium Development Goals [MDG] targets) has resulted in measurable successes or whether the funding could be better spent on alternative emphases. The third major topic is the identification of neglected development foci in rural Africa such as (1) the gender gap, which brings into focus female farmers (as opposed to male farmers) and adolescent girls (as opposed to contemporary emphases on children and adults) and (2) diversified rural livelihoods (rural "informals" [i.e., those earning a living from informal economic activities or petty trade] as opposed to an exclusive focus on agricultural) and alternative rural industries (e.g., tourism). The outcomes of debates on what, whom, and where is prioritized in rural Africa to bring about sustainable development will be critical to determining how the rural revolution unfolds. Rural Africa is characterized as "globalization's shoreline," and once the tide comes, particular futures (out of many possibilities) will be determined (Carr 2011).

## RURAL DEVELOPMENT THEMES AND POLICY EMPHASES: THE ELUSIVE QUEST FOR SUSTAINED DEVELOPMENT

Perspectives on rural Africans have come a long way in the last half-century. Various decades are associated with heydays in particular rural development paradigms, but several competing paradigms operate during most of the time. Nonetheless, a basic understanding of the chronology of rural development emphases is helpful in understanding how policy switched from concentrating on large commercial farms to small farmers to eventually embracing integrated and diversified rural development.

An agriculture-centered development strategy has remained central, but ideas about the roles of agriculture and of African farmers in development have ebbed and flowed. Early 20th-century observers considered Africans lazy and "backward" farmers, employing unproductive, rigid, and unscientific agricultural methods. The modernization paradigm (1950–70) aimed to bring regressive African agriculture into the modern era by promoting plantation estates and large commercial farming, but the rural emphasis was secondary to the core idea of promoting industrialization in urban settlements. Small-scale agriculture was allocated a more passive role, contributing to economic development when possible by supplying resources to the modern urban economy.

In time, the modernization paradigm was widely refuted and counterbalanced by other perspectives highlighting the cultural dimension of rural development and underscoring that indigenous knowledge and practices were unrecognized and misunderstood by modernist proponents. Indeed, in many instances, small-scale farming could be dynamic and productive (Warren, Slikkerveer, and Brokensha 1995). Toward the mid-1960s, the rural development paradigm switched to considering small farms as the very engine of growth, although it did not defeat a set of ideas about large farms. Indeed, many agricultural advisers believed that the faster small farms expanded, the faster their eventual demise would be (Ellis and Biggs 2001). However, the inverse relationship between farm size and economic efficiency gained currency. Small farmers were proven to be more efficient than large farmers, making use of abundant labor, small plots, and scarce capital. In spite of resource deficits, small farmers generated higher yields per unit area (Berry and Cline 1979).

The mid-1980s, with the emphasis on structural adjustment policies (free market and liberalization), led to governmental withdrawal of large-scale management of the agricultural sector, and development policies concentrated on urban development projects that aimed to facilitate more linkages to global economic circuits. The liberalization era of the 1980s to 2000 led to retreat of the state from rural Africa, and although small-farm efficiency ideas were not debunked, rural development policy moved into a phase of neglect.

At the same time, nongovernmental organizations (NGOs) for rural development filled a gap created by the retreat of big government. This led to another rural paradigm shift (mid-1980s–2000) from a top-down or blueprint approach to rural development, characterized by external technologies and national-level policies, to a bottom-up grassroots or process approach (Ellis and Biggs 2001). Rural development was approached as a participatory process that empowered rural dwellers to set their own priorities for development and change. It was also accompanied by a growing acknowledgment of the validity of indigenous technical knowledge and of the agency of the rural population. Researchers and farmers moved toward rebuilding agricultural practices based on farmers' knowledge and local resources, and some of these efforts led to reductions in the use of pesticides. For the first time, Africa's rural poor were viewed as capable of contributing solutions to the problems they encountered and interdependencies with urban areas were recognized (e.g., circular and permanent migration, food, remittance and durable commodity flows, and so forth).

A sustainable livelihoods approach from the late 1990s onward has aimed to broaden the rural development debate by deepening understandings of the dynamics of rural poverty and survival. This people-centered approach focuses on how individuals and households make livelihoods for themselves and how the assets that they draw on (skills, social networks, nature, and financial capital) are affected by their vulnerability contexts, which take into consideration such events as shocks (e.g., environmental change), seasonality, and rural and urban trends. This approach extends thinking about ordinary rural Africans, and it moves the center of attention away from farming and farmers and toward broader survival mechanisms among the rural populace. Most sectoral programs of major international organizations now incorporate sustainable livelihoods thinking and analysis, acknowledging the multidimensionality and interrelationships among a range of factors that affect rural livelihoods and the survival choices made.

The MDGs of 2000 to the present day center on a multifaceted approach to development, emphasizing key development targets. MDGs renewed investment in agriculture, which had been abandoned by international investors in the 1990s as they short-circuited rural development initiatives that they had heavily promoted in earlier decades. MDGs prioritize social investments as a requirement for rural development (especially human capital through education and health), alongside infrastructure and agricultural investment. Several important initiatives have been launched on the ground to bolster the MDGs. The MVP (2005–present) aims to promote sustainable and complementary investments in health, education, family planning, infrastructure, gender equality, water, sanitation, and agriculture to move people out of poverty. Subsidized fertilizers and seeds are provided to all farmers in these communities to increase crop yields and to diversify crops (to improve household nutrition), and various training programs based on best practices have been introduced (e.g., agronomics).

A separate initiative launched by the Alliance for an African Green Revolution (AGRA) (2008–present) aims to accelerate agricultural productivity in rural areas. This major initiative provides a different and separate road map to realize the MDGs. Both approaches have revitalized the 21st-century rural development path after previous efforts have lost momentum. Science-informed strategies have gained more broad-based support than the narrower economic and planning initiatives of earlier times.

Large international organizations (e.g., the World Bank, African Development Bank, and Food and Agriculture Organization of the UN [FAO]) continue to emphasize the centrality of agriculture in economic and rural life but recognize differences among countries: for instance, agriculture dominates the gross domestic product in states such as Central African Republic and Sierra Leone but makes only minor contributions in states such as South Africa and Botswana (World Bank 2011). Indeed, the World Development Report of 2008 emphasized agriculture's role in development and called for more investment in African agriculture (World Bank 2008). For example, only 4% of public investment and 4% of official development assistance goes into agriculture (Asian countries averaged 12% in the same budget categories during the 1970s) (World Bank 2008). Nevertheless, the future of rural Africa remains contentious, with widely different views about what should be prioritized, how rural and urban Africa development gaps can be narrowed, and what role foreign direct investment should play (if any) in the transformation process.

There are four key debates in rural Africa: (1) increasing agricultural productivity and promoting food security; (2) diversifying rural livelihoods and supporting alternative, nonfarm employment opportunities; (3) closing the gender gap in agriculture; and (4) improving education and access to education, particularly making education equal and relevant, especially for girls. Although these issues are interrelated and contemporary development efforts take an integrated approach, there remain considerable differences of opinion on how to prioritize and proceed. Of course, ideological and sectoral debates are always present in the development arena, but the stakes could not be higher as rural population numbers continue to climb and as a decade of strong growth rates at the national level have resulted in little improvement in rural livelihoods.

## THE GREEN REVOLUTION

William Gaud, a U.S. Agency for International Development (USAID) official, coined the term "green revolution" in 1968. This term refers to a series of research, development, and technology initiatives commencing in the mid-20th century that accelerated agriculture productivity in many regions of the world (but not Africa). Relying on hybrid crops, improved fertilizers, herbicides, and much irrigation, this modern scientific approach signaled a large-scale historical agricultural transformation of traditional farming methods. It further demonstrated that the same scientific revolution that occurred in Global Northern agriculture could be transferred and adapted to the Global South. Proponents credited the GR with helping move large numbers of poor people out of poverty and with aiding many nonpoor people to avoid poverty and hunger traps that they would have otherwise experienced if agricultural output had stayed on the traditional agricultural productivity trajectory.

The achievements and consequences of the GR are hotly debated and contested (see Table 5.1).

There are strong views on both sides of the debate. Leading environmental activist Vandana Shiva (2000:7) emphasizes that "small farmers are pushed to extinction, as monocultures replace biodiverse crops, as farming is transformed from the production of nourishing foods and diverse foods into the creation for genetically engineered seeds, herbicides, and pesticides. As farmers are transformed from producers into consumers of corporate-patented agricultural products, as markets are destroyed locally and nationally but expanded globally, the myth of 'free trade' and the global economy become a means for the rich to rob the poor of their right to food and even their right to life."

GR proponents counter that environmentalists miss the central issue of feeding more people as the population expands and resources become scarcer. Norman Bourlag, an agronomist instrumental behind the GR, responds that environmentalists are not the ones going hungry. "They do their lobbying from comfortable office suites in Washington or Brussels. If they lived just one month amid the misery of the developing world, as I have for fifty years, they'd be crying out for tractors and fertilizer and irrigation canals and be outraged that fashionable elitists back home were trying to deny them these things" (quoted in Tierney 2008:1).

**TABLE 5.1 ADVANTAGES AND DISADVANTAGES OF THE GREEN REVOLUTION**

***Advantages***

- Production and yields increase.
- Extra production can be used to feed animals, and more animal products can be incorporated into human diets.
- National and economic security is enhanced with reliable food supplies.
- Smarter agriculture and new concentrations on vegetables, fruits, and dairy can position farmers to be integrated into urban markets (e.g., supermarkets).
- Standards of living improve for some farmers.
- Improvements in rural infrastructure (e.g., irrigation systems) bring more land under cultivation.
- Credit facilities and access to inputs enable farmers to be more entrepreneurial and productive.
- The rural economy becomes diversified.

***Disadvantages***

- Pollution, contamination, and degradation of land and water occur from fertilizers and pesticide runoff.
- The burning of soils by heavy use of chemical fertilizers leads to overreliance on chemical inputs to compensate for deteriorating soil quality.
- Biodiversity changes; for example, the genetic diversity of crops diminishes as monohybrid crops are introduced into regions previously characterized by crop diversity.
- The erosion of biodiversity may impair the ability to improve crops in the future.
- Poor farmers cannot afford inputs and machinery, and many run up debts.
- High-yielding crop varieties require more water and fertilizers, which are expensive.
- Cash cropping for a global market does not always improve local food security.
- Labor is displaced, increasing rural–urban migration.
- Cultivation of marginal land can lead to soil erosion.

## AFRICA'S FAILED GREEN REVOLUTION

Attempts to implement a GR in Africa in the 1960s and 1970s failed for several reasons. First, African diets are based primarily on grains (e.g., millet and sorghum) and tropical root crops (e.g., cassava, yams, and sweet potatoes), all considered peripheral crops in the GR framework. Instead, the GR concentrated on rice, wheat, and maize; only maize could be considered an African staple, but not throughout the region. Second, African agriculture is predominantly rain-fed rather than irrigated, and fears prevailed that irrigated monoculture would enhance disease and pest risks. Third, Africa's poor transportation and commercial infrastructures were serious impediments to the distribution of agricultural inputs and outputs; transportation costs alone doubled the input price for smallholders and raised costs to a level beyond their means. Fourth, government support was generally lacking, and agricultural science and research in the region was not sufficiently developed to play a supporting role. Indeed, the African region still maintains the world's lowest agricultural research capacity (i.e., 70 African agricultural researchers per million population in 2013 compared to 2,640 in the United States). Fifth, most of Africa's farmers were subsistence smallholders deficient in training and skills to harness new technologies even when they could afford them. Sixth, women, the bedrock of African agriculture (responsible for 70% of stable food production and contributing 60%–80% of the labor used to produce household crops), lacked access to institutional credit and were generally excluded from participation (Negin et al. 2009).

Africa's heterogeneity and diversity meant that instituting the GR's innovations was particularly challenging. However, a few African countries (e.g., Kenya, Nigeria, and Zimbabwe) registered increased maize yields from introduced hybrids. Other GR improvements included the introduction of a breakthrough rice variety (weed, drought, pest, and disease resistance) in the late 1990s (termed the "New Rice for Africa" or "Nerica") to upland areas. Nerica's cultivation now extends to 300,000 acres (1.2 million hectares) (Rockefeller Foundation 2006). Despite these few modest successes, the GR largely bypassed Africa.

**FIGURE 5.1** Kofi Annan, Chair for Alliance for a Green Revolution in Africa at its Headquarters, Nairobi, Kenya.
*Source:* © ANTONY NJUGUNA/X90056/Reuters/Corbis. Corbis image 42-18662346.

## AFRICA'S NEW GREEN REVOLUTION

Former UN Secretary-General Kofi Annan called for a new African GR in 2004 (Fig. 5.1). Advances in agricultural research as well as a political commitment to harness these advances in Africa underpinned the renewed attention (Sanchez, Denning, and Nziguheba 2009). In the first GR, influential donors maintained that a market-based approach would be sufficient to support Africa's agricultural transformation; in the second GR, the emphasis switched to a capital-intensive approach.

The case for an African GR is supported by the need to feed Africa's hungry (200 million Africans, one third of them children, are undernourished) and expanding population. It seeks to address the stagnating and downward spiral of agricultural productivity and inertia of farming methods that rely on traditional practices with few inputs (on average, African farmers apply one-quarter the fertilizers that farmers in other regions use). Evidence shows that average yields are 2.6 metric tons per hectare—less than half that of comparable world regions—and farms are producing 20% less (on a per capita basis in 2005) than they did in 1970. Small-scale farming has been proven to be ineffective in nourishing Africa's population, and it cannot manage projected growth and environmental changes.

Although 60% to 70% of Africans engage in agriculture, Africa has become a net food and agricultural importer (the food trade deficit is climbing; it hovers in excess of US$26 billion per year and is anticipated to increase in coming decades), and food insecurity has become a major African and global concern since the food price hike of 2007–08 (Rakotoarisoa, Iafrate, and Paschali 2011). Thus, a switch to more intelligent farming methods and the production of higher-value crops such as vegetables and fruit would boost farmers' incomes and contribute more food.

There are also arguments that Africa's farming future is going to depend on rigorous research and a multifaceted technological approach, and that the region's farmers need to be pushed toward using proper science- and evidence-based methods. Such an approach is essential to build resilience to the effects of climate change on crops, growing seasons, rainfall/droughts, and pests. It promises to deliver results well beyond those of traditional agricultural adaptation.

Africa's underperforming agricultural sector is to be transformed through smallholders gaining access to agricultural inputs—primarily fertilizers, high-response seeds, and small-scale water management technologies—all within a comprehensive rural development strategy. The African GR has been endorsed by the UN as well as by African heads of states on several occasions. It fits within their agreed target of 10% public expenditure on agriculture and their annul target of an 8% agricultural productivity growth rate, as outlined in the 2002 Comprehensive Africa Agricultural Development Program of the New Partnership for African Development. African leaders endorsed the GR again at the 2006 Abuja Africa Fertilizer Summit, and they continue to support the effort.

Private philanthropy, largely spearheaded by the Bill and Melinda Gates Foundation and the Rockefeller Foundation, was instrumental in launching and funding AGRA in 2007. By 2011, it was mobilizing US$76.4 million to support the African GR (AGRA 2012:32). AGRA's mission "is to transform African agriculture into a highly productive, efficient, competitive, and sustainable system that assures food security, lifts millions out of poverty, and protects the environment" (AGRA 2012:11). This is to be achieved by improving seed quality (750 new varieties are targeted for introduction) and soil fertility, strengthening farmers' access to agricultural inputs and markets, ensuring better access to water, establishing innovative financing methods, improving agricultural education, and creating an environment in which governments and international organizations can support farmers and introduce incentives to adopt new technologies (AGRA 2012).

The AGRA concentrates on four "breadbasket" areas—Ghana, Mali, Mozambique, and Tanzania—but its initiatives have been extended to 12 other countries. The breadbasket strategy aims for quick wins in achieving food security, so breadbasket regions were selected on the basis of good soils, dependable rainfall, preexisting basic rural infrastructure, and supportive national political environments.

AGRA programs have recorded some successes: 40,000 metric tons of 322 improved varieties of seeds (e.g., maize, wheat, beans, sweet potato, cassava, sorghum, millet, cowpea, and rice) have been introduced (183 of which have been commercialized) (AGRA 2012). In Burkina Faso, seed companies and input dealers have supplied improved seeds to 20% of farmers. Building capacity through postgraduate training has produced 2,500 new agro-certified, trained people to work with smallholders. Smallholders have received training ranging from planting to storage to making more informed market decisions (e.g., use mobile phones to access real-time commodity market prices).

Africa's new GR has staunch critics, and many impediments persist. The single magical technological bullet has been criticized for failing to address diversity and heterogeneity throughout the region's farming systems (Thompson 2007). There is a danger that African farmers will lose control over seeds as a major farming input: 80% use saved seeds, enabling them not to purchase seeds every season. Africans' wealth of indigenous ecological knowledge, reflecting centuries of adaptation (e.g., interspersing different plants to enrich the soil and to deter pests from food crops), could be lost forever. High-tech solutions to Africa's food crisis are seen by many as incorrect answers because they pollute the environment with fertilizers and pesticides, destroy small farming, and transform the genetic wealth of Africa into cash profits for a few global corporations.

Another viewpoint holds that political solutions more than technical ones are needed to end hunger. Anti-GR proponents claim the silent revolution is not primarily about helping peasants to produce more food but rather about creating a food system in which smallholder agriculture, widely regarded as backward and unproductive, is subordinated to a global commercial and capital-intensive mode of production. Thus, the African GR is set to reshape social relations, transforming rural production by incorporating small farmers into the global marketplace. In the meantime, the issue of who is responsible for protecting the rights of smallholders has been overlooked.

Other critics (e.g., Moseley 2011) claim that the successes of the GR elsewhere (in India and China, in particular) are being misapplied to Africa. For example, in the Punjab region of India, yields increased fivefold within 20 years, but groundwater dropped by 12 inches (30 cm) and one quarter of small farmers were driven off the land, unable to continue purchasing seeds, fertilizers, and pesticides (Patel 2009). Suicide rates among small farmers escalated as many fell into serious debt. The social and environmental costs of the Indian GR were very high. The international lesson was that smallholders were easily seduced into a new system that was detrimental to their survival: they participated in a higher-yielding system unaware that productivity costs would rise and that new crops would be subject to unpredictable international markets. Therefore, many (e.g., Patel 2009) contend that Africans should follow a different path by sustaining plant and food biodiversity rather than eliminating it by industrial monoculture. It should be noted that African-grown food for human consumption is derived from 2,000 plants, while the U.S. food base is dependent on 12 plants (Thompson 2007).

## GENETICALLY MODIFIED CROPS

GM agricultural crops are laboratory-designed or engineered crops that increase or modify a gene containing the desirable trait in the living DNA of a host plant. The most common desirable traits include resistance to pests, diseases, and drought, and/or tolerance of herbicides. Sometimes additional modifications can enhance the taste, appearance, and color of crops. Large-scale commercialization of biotechnology for the African market has so far delivered only two traits—pest-resistant (known as Bt) crops and herbicide-tolerant crops—while other modifications remain in the pipeline. Leading-edge GM crops include maize, soybean, cotton, and canola.

Current GM research aims to improve the shelf life of harvested products, whereas other frontier research focuses on increasing nutritional value, although in both these areas scientific breakthroughs have been slow. For example, the development and introduction of golden rice, an enhanced (biofortified) staple crop enriched with provitamin A, is well under way but not yet in the African marketplace.

GM crops were first officially commercialized in 1996, when six countries planted them on 1.7 million hectares. By 2011, GM planting had expanded to 29 countries with 148 million hectares in production, roughly equivalent to 2% of the world's agricultural land (Africa Center for Biosafety 2012). Globally, an 87-fold growth makes GM the crop technology to be most quickly adopted in the history of modern agriculture (Adenle 2011). Uptake in Africa has been slow: only three countries south of the Sahara have commercialized biotech crops—South Africa (1997–present), Burkina Faso (2008–present) and Sudan (2012-present). Nevertheless, the percentage of total area allocated to GM crops is increasing (2.2% of total area in South Africa and 0.03% in Burkina Faso [African Center for Biosafety 2012]), and research is under way in another 18 African countries (mostly confined trials). Some (e.g., Kenya, Ghana, Nigeria) have introduced forward-leaning biotechnology laws and frameworks, paving the way for large-scale introductions.

South Africa is the leading and largest African adopter of GM crop technologies: GM maize, GM soybean, and GM cotton have been introduced, and a GM potato is under development. Since 2010, South Africa has exported GM crops to other African countries. A few African countries (e.g., Kenya and Ghana) have facilitative biotechnological frameworks in place, expected to pave the way for the introduction of GM crops. However, despite Kenya's 2009 Biosafety Act (regarded as the more facilitative biotechnology framework in Africa) and Ghana's 2011 Biosafety Act, public confidence in national regulatory systems of GM technologies is low. In 2010, protesters at Mombasa's port blocked a GM maize import shipment from South Africa. Protesters criticized the Kenyan government for failing to preform relevant safety checks and for not alerting Kenyans about GM maize imports (Cooke and Downie 2010). The Kenyan government responded by establishing a temporary ban on GM imports in 2010, but it has since been lifted. In Ghana, the Food Sovereignty movement has been a vocal opponent of GM imports.

Kenya has joined an increasing number of African countries developing their own GM research and development capacity: 20 countries had commenced R&D by 2012 (African Center for Biosafety 2012). Research cooperation between Africa-based institutions and companies and institutions based in the United States and/or Europe is driving this research effort. The private sector primarily determines the research themes (with public sector collaboration), and profit is a dominant motive, as opposed to needs fulfillment (e.g., focusing on herbicide-tolerant crops rather than on employment generation). Declining public-sector African agricultural research, combined with the privatization of agricultural research, has led to a focus on providing high-tech solutions, including transgenics, over other agricultural options.

There are an increasing number of Africa-focused research efforts. Significant attention is being directed to develop "water-efficient maize for Africa." Field trials are under way in South Africa, Uganda, Kenya, Tanzania, and Mozambique, and the project involves local agricultural research institutes, with significant financial backing (US$47 million) from the Gates Foundation, the Buffett Foundation, USAID, and Monsanto. Proponents of water-efficient maize for Africa argue that a drought-tolerant crop could increase yields by 30% and, importantly, safeguard food security in drought years. Similar but less resourced initiatives target the introduction of other crops: GM soybean pilot projects are proceeding in Mozambique and Zambia; a virus-resistant sweet potato and a stem borer-resistant maize (a pest that destroys, on average, 20% of the crop) are being developed in Kenya; and GM banana and GM cassava development is continuing in Uganda.

GM technology is a revolutionary, young science spawning a powerful industry and support basis. It is changing agriculture, the food system, and environments. Only time will tell if it proves to be the biggest innovation, or the biggest mistake, of our times (Makoni and Mohamed-Katerere 2006). Risks and benefits of GM technologies are hard to quantify at these early stages of development. Resistance to new technologies and new food has historical precedents (e.g., coffee was restricted in 17th-century Cairo, Stockholm, and London, and tomatoes were viewed as poisonous in the United States until the 1830s). As in all breakthrough sciences, securing acceptance is difficult upstream and virtually impossible downstream. In general, there is a lack of quality information to support policymakers and the public in evaluating their options. Much of the available information is either highly specialized (incomprehensible to non-GM experts) or overly simplified and polemic, focusing on one aspect of the debate without considering a balance of effects. Nevertheless, the stakes are very high: a major miscalculation could lead to major changes in the planet's ecosystem with significant repercussions for humankind. It is hardly surprising that GM technologies are controversial.

There are three key debates in Africa: (1) the interpretation of science and whether genetically modified organisms (GMOs) are safe for human health, the environment, and biodiversity; (2) whether GMOs are a sustainable food security option for Africa; and (3) what the capacity of Africans is to participate in GM research, monitoring, and assessment of GM technologies.

Arguments in support of GM crops for Africa include that the technology is available and can be refined and transferred to African contexts. GM crops have been demonstrated to be safe in the United States, Brazil, Argentina, India, Canada, and China. The United States is the leading producer of GM crops, and its government (especially the Food & Drug Administration [FDA], agroindustry, and mainstream scientists) has tested GMOs and declared them safe. GMOs are well embedded in the U.S. food chain, and some foods now contain 70% GM content. They have not been proven to be harmful: neither death nor illnesses have been attributed to North Americans consuming food with GM content.

GM support for Africa has garnered momentum among major philanthropic and international organizations and global corporations. GM proponents claim that it is only a matter of time before a global scientific consensus emerges (as was the case in the climate change debate). Advocates are armed with scientific evidence showing that the introduction of GM crops will increase crop yields, elevate farm incomes, and reduce herbicide spraying by 50% (Adenle 2011). Major international regulatory bodies, including the World Health Organization and the FAO, reaffirm that the application of GM technology has not had negative effects on human health.

There is also an argument that GMOs can support biodiversity. By preventing more land from being brought into agricultural production, GM technology helps conserve land. GM technology can even be applied to safeguard plants under threat of disease and pests. Furthermore, GM planting can reduce the amount of pesticides and fertilizers used and limit the amount of harmful runoff into the environment.

Arguments against GMOs in Africa center on this biotech solution as being yet another one-size-fits-all initiative for rural development. Moreover, the fast-track approach privileges crops developed elsewhere over African varieties, with little regard for the traditional African diet, the diversity of ecological zones, and farming systems. There is the specter of fostering another unhealthy dependency, this time on a handful of powerful companies (e.g., hybrid seed companies such as DuPont and Syngenta and agrochemical companies such as Bayer, BASF, and Monsanto). Fears of global corporate dependency resonate with many Africans, deeply suspicious of further Western intrusion into African ways of life. Global corporations dominate seed and chemical sales worldwide, and Monsanto has secured patents (on the basis of a World Trade Organization 1995 ruling) that allow companies to claim private ownership of GM seeds. Critics charge that GM powerhouses are purchasing patents on nature and depriving Africa's small farmers in the process. As a result, corporate control, profits, and agenda setting of agricultural practices take priority over small farmers' interests. This means that African farmers will not longer be in control of seeds and of saving and storing them (a very common practice); instead, they will be forced to buy seeds each year. Participants in the anti-GM lobby perceive themselves as waging a David-versus-Goliath struggle. Their campaigns are spearheaded by environmental NGOs such as Greenpeace and Oxfam, but these organizations lack the resources to conduct research on communities experimenting with GM technologies. Published scientific evidence is skewed heavily toward positive evidence-based reports, and the anti-GM lobby is restricted to moving testimonies from local environmental activists. Since seed companies wield patent power over their products, independent researchers are not provided access to GM seeds for testing. The GMO lobby is powerful and has deep pockets and research clout; its research results inform FDA approval processes and global GM policy. For the most part, regulations in the United States, as elsewhere, have been adapted for GMOs rather than developed from the outset to test their safety. The inability of the anti-GM lobby to compete in research has resulted in a concentration on advocacy, but even in this domain they compete against well-financed pro-GM campaigns.

Peer-reviewed scientific evidence of the harmful effects of GMOs is limited to testing on animals. In a study that assessed how Monsanto's Roundup-tolerant maize affected rats' health, a group of French researchers (Séralini et al. 2012) found severe adverse health effects, including tumors and kidney and liver damage, that led to premature death. This article findings were explosive and led to a Europe-wide ban on GMO maize and a Kenyan ban on all GMOs. However, this article became mired in controversy, and the journal *Food and Chemical Toxicology* subsequently retracted the article in 2013, noting that the rats used in the study were cancer-prone and the experiment did not adequately distinguish tumors that might have been caused by GM from those that were spontaneous. Skeptics charged that the French researchers were blinkered by their anti-GMO predispositions; for example, Séralini had already published a popular anti-GMO book (Séralini 2012). However, the retraction of the article has not dulled the campaign against all GM foods.

Civil society organizations in the region claim they have been excluded from discussions about the safety of biotechnologies in agriculture and that their governments lack the scientific, legal, and administrative capacity to monitor GMOs. This means that most of the regulating is within industry (self-regulation) with very little information sharing (African Center for Biosafety 2012). Civil society is adamant that the GM revolution is not a farmer revolution and that indigenous agricultural knowledge could be lost forever. Anti-GM protests have occurred throughout Africa,

**FIGURE 5.2** Anti-Monsanto Protest, Cape Town, South Africa. *Source:* © NIC BOTHMA/epa/Corbis. Stock Photo 42-47027259.

and Africans have joined a global movement against large agrocorporations such as Monsanto (Fig. 5.2).

Fears have been raised about the promotion of "superweeds" and "superpests" as well as the substitution of an enhanced secondary pest in place of a primary pest (as occurred with Bt cotton in India) (Makoni and Mohamed-Katerere 2006). Superweeds and superpests require farmers to use more and more herbicides. Crops engineered for pest resistance are also potentially problematic, as pests themselves are capable of developing resistance.

Potential risks of GMOs continue to be raised, particularly in Europe, and the European media are calling attention to the question of whether GM products are safe. Largely based on a precautionary principle, the European Union (EU) decided not to introduce GMOs. The European stance needs to be contextualized, however: EU agriculture is already highly productive and therefore not urgently in need of a productivity boost. Based on close historical, political, and trade ties, African governments are very mindful of the EU's stance and fear damaging their traditional agricultural export market.

There are grave concerns about harming Africa's biodiversity. Small farmers and diverse farming practices are viewed as nurturing biodiversity. A movement toward large-scale GM agricultural farming will undoubtedly lead to a decrease in farm biodiversity and perhaps overall biodiversity. For example, GM seeds can contaminate neighboring farms by drift (GM canola pollination occurs up to 1.9 miles [3 km] away). These situations raise patent-infringement issues, as the presence of GMO on a non-GM farm means that farmers can be charged with infringement of intellectual property rights of seeds. In the United States and Canada, lawsuits have resulted in favorable judgments for GMO producers. (Also, the blowing of GM pesticides onto adjacent farms can damage non-GM farming, so the issue of a buffer zone comes into play.) According to La Vía Campesina, an international peasants' movement, the UN estimates that 75% of world's plant genetic diversity has been lost as farmers have abandoned native seeds for the genetically uniform varieties offered by corporations (quoted in Morvaridi 2012:253).

Throughout most of Africa, some governments, most media, and many people maintain hostile views toward GM crops, largely based on perceived negative environmental and human health consequences. In fact, even in crisis times, several African countries have rejected GM food aid (e.g., Namibia, Zambia, Benin), while others (Malawi, Zimbabwe) accepted it only on the condition that the maize was milled prior to distribution in the country (Makoni and Mohamed-Koterere 2006). It is understandable that African governments are taking a go-slow approach toward the adoption of GM technology. The South African and Tanzanian governments are the most pro-GM in the region. Within these states, however, there is strong resistance to GM crops by urbanities, farmers' groups, civil society, and rights-based organizations. For their part, African farmers are slow to embrace GM technologies because of significant local variations in climate, soil, water, and disease. Biotechnology may present long-term promise, but farmers are preoccupied with their immediate needs and are risk-averse. GM monocropping is, therefore, the antithesis of biodiverse (and often organic) farming, the mainstay of African agriculture.

## RURAL LIVELIHOODS: A CHALLENGE TO THE PRIMACY OF SMALL FARMS?

In Africa, farmers have always diversified their sources of income, but in the past it was a seasonal rather than a permanent phenomenon. Farmers were known to pursue two or more livelihoods simultaneously or to

## BOX 5.1 AFRICAN MILLENNIUM VILLAGES

The Millennium Villages Project is a 10-year, UN-backed initiative that promotes an integrated approach to rural development and aims at achieving the MDGs by 2015. It is led by Jeffrey Sachs at the Earth Institute of Columbia University and implemented by the Millennium Promise (an NGO founded in 2005). The MVP is well publicized, backed by celebrities such as U2's Bono and Angelina Jolie, and funded by high-profile individuals (e.g., the Lenfest family), international foundations (e.g., Open Society), and international development organizations (e.g., the UK's Department of International Development [DfID] and the Islamic Development Bank [IsDB]) (Carr 2008; Tran 2013). In 2013, the IsDB agreed to provide US$104 million in loans (provided in the form of Islamic finance of long-term repayments at a 0% interest rate) to extend program efforts. Described by its proponents as a "bottom-up approach to lifting African villages out of the poverty trap" (Millennium Villages Organization 2012), the MVP aims to combine proven science-based interventions with local knowledge to create a new, practical approach to alleviate rural poverty at the village level. Its goal is to halve rural poverty by 2015.

The inaugural MV was established in Sauri (Kenya) in December 2004. Rather than targeting an individual village, the project encompasses clusters of several nearby villages. Phase One (2005–12) comprised 14 villages, involving more than 80 communities (Fig. 5.3). MV sites span a range of population sizes (5,000–80,000 villagers). The villages are located in 10 countries (Ethiopia, Ghana, Kenya, Malawi, Mali, Nigeria, Rwanda, Senegal, Tanzania, and Uganda), which are all stable countries whose governments have functioning service-delivery structures (although in most cases they are far from optimal). Carefully selected in close coordination with national and local governments, this initiative targets areas prone to chronic hunger and disease and lacking basic services and infrastructure. In total, the effort delivers targeted development assistance to approximately 500,000 villagers in diverse agroecological environments.

**FIGURE 5.3** Jeffrey Sachs Dancing with Rural Women at Ruhiira Millennium Village, Uganda. *Source:* © Guillaume Bonn/Corbis. Corbis 42-50223430.

In Phase Two (2013–present), the project has expanded in two ways. First, the UN Development Program (UNDP) and partner governments are cooperating in 17 independent MVP efforts in 13 countries, in addition to the original 10 MVP sites. Second, the IsDB, the Earth Institute, and its partner the Millennium Promise are extending the MVs by launching a new initiative, the Sustainable Village Project, a flagship program that focuses on rural villages in Sudan, Chad, and Mozambique in an effort to fight rural poverty by improving agriculture, health, education, and standards of living. In total, more than 20 African countries in 2013 are implementing rural strategies based on the MVP's coordinated development approach (Fig. 5.4). The interventions are deemed so successful in Uganda that the government is scaling them up.

The main principles of the MVP are:

- proven science- and evidence-based technologies and practices
- community-based strategy, with a participatory approach to planning, implementation, and monitoring in which the interventions are adapted to the context of each village
- enhanced by the development of local capacity in technical, managerial, and participatory skills
- multisectoral and integrated interventions based on the MDGs
- links to district, national, and global strategies
- costs shared by the community, government, and donors
- supported by increased national financing of public goods in line with increased official development assistance to the respective African governments (Millennium Villages Organization 2011).

The MVP has three basic principles: (1) to demonstrate that integrated, community-based, low-cost interventions can speed progress towards the MDGs in these rural settings; (2) to identify processes whereby MVP interventions can be scaled up to support rural, regional, and national development strategies based on the MDGs; and (3) to empower villagers to frame their development concerns in the context of integrated development (as opposed to isolated policy interventions), further promoting cross-village shared understandings while enabling village-level issues to be intelligible to national policymakers applying the MDGs (Carr 2008).

At the investment level, the MVs invested on average US$99 per capita per year in multiple sectors during the first five years of implementation. Interventions are multisectoral and typically encompass fertilizers, insecticide-treated bed nets, schools, HIV testing, microfinance, electric lines, road construction, piped water, irrigation, and so forth, the precise mix tailored to each village cluster's requirements. Multiple buy-ins are secured by the sharing of funding contributions among stakeholders in the

*(Continued)*

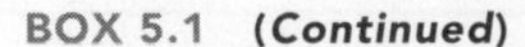

**BOX 5.1 (*Continued*)**

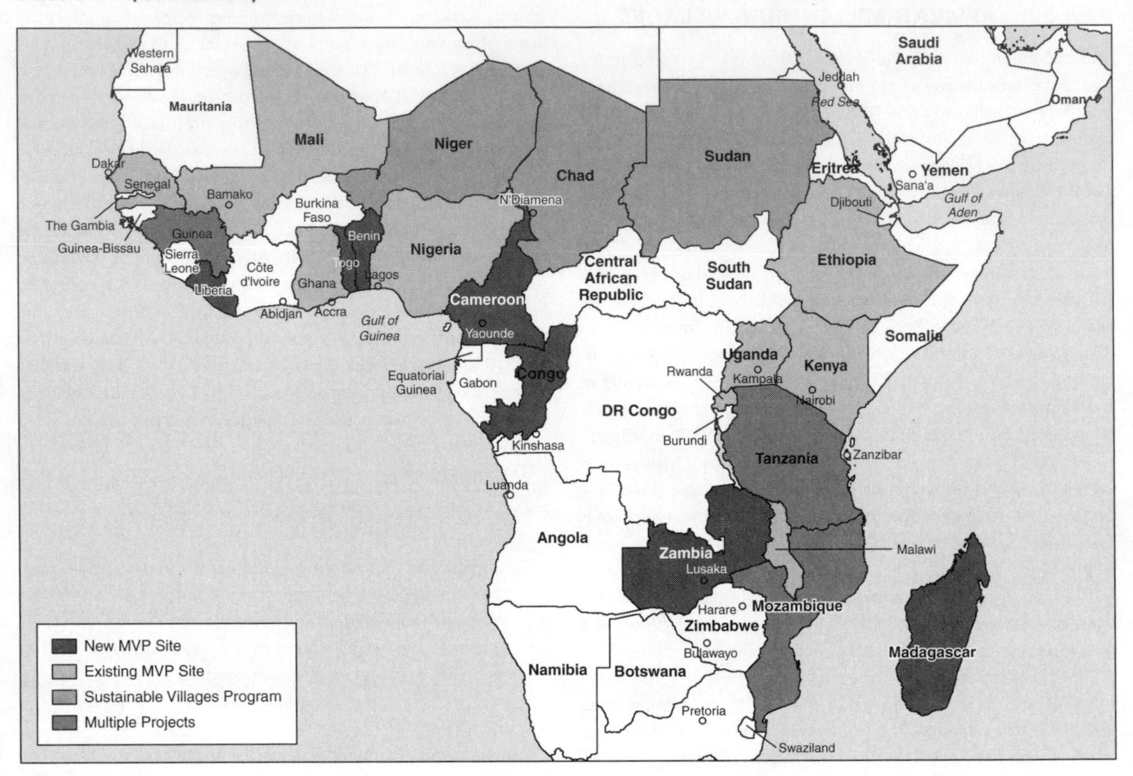


**FIGURE 5.4** Location of Millennium Villages.

approximate ratios: MVP and core partners (49%), external donors and NGOs (10%), local and national governments (33%), and the village community (8%) (labor/service commitments count as contributions).

According to the Millennium Villages Organization (2011:5), substantial improvements have been recorded in MVs (Fig. 5.5).

The MV group has published peer-reviewed results. One paper highlighted the multisectoral package of interventions improving levels of child undernourishment and lowering rates of stunting (low height for age in a given population) by 33% (Remans et al. 2011). Another paper reported preliminary reductions in poverty, food insecurity, stunting, and malaria after three years of project implementation (Pronyk et al. 2012). The same study also initially claimed a decline in child mortality at three times the rate of the general rural trend; this finding embroiled the MVP in deep controversy due to an error in the calculations, however.

A cadre of researchers, bloggers, and policy analysts remain skeptical of the effectiveness of MVs and MVP assessment (Bump et al. 2012; Carr 2008 and 2011; Clements and Demombynes 2010; Wanjala and Muradian 2011). Debates have been extensive and heated both online and in the print media (a summary of these debates is available on Tom Murphy's [2011] "A View from the Cave" blog).

The four main criticisms of the MVP are (1) there is a lack of rigorous assessment, particularly by independent researchers; (2) MV team assessments rely on results that are based on a before-and-after project and the scaling of a number of indicators, but fail to disaggregate general positive development trends such as economic growth rates from MVP results; (3) the initiative is not sensitive enough to understand (and change) village power dynamics (e.g., the structure of gender relations); and (4) agricultural productivity growth may not result in a significant improvement in recipients' incomes compared to rural villagers who were switched out of agriculture. Carr (2011:209) argues that the MVP "is predicated on problems and solutions designed in advanced economies and implemented in villages along globalization's shoreline in a manner that marginalizes community participation." Carr (2011) contends that globalization and development are much more unruly and defy easy conceptualization in any predetermined set of policy interventions. Instead, Carr (2011) advocates for more locally informed sets of community interventions, new and different data metrics, and more open-ended possibilities.

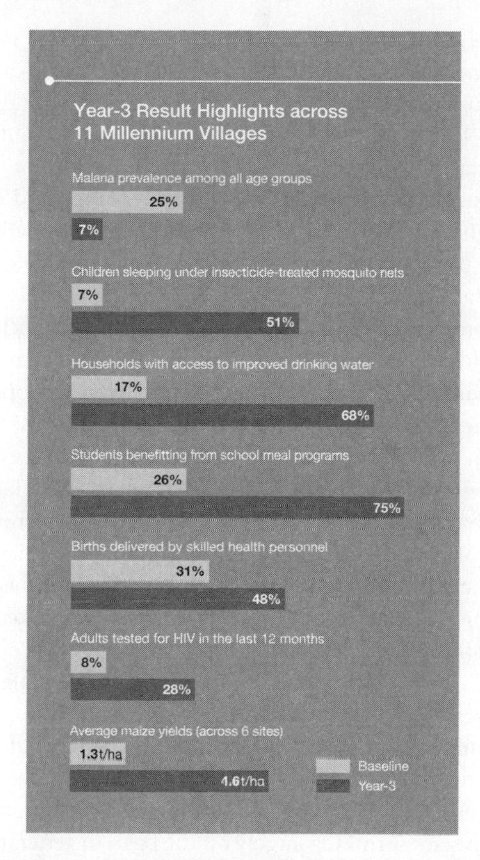


**FIGURE 5.5** Year-3 Result Highlights Across 11 Millennium Villages.

Different pieces of evidence are marshaled to criticize or support the MVP. For example, critics often claim that MVP-"imposed" interventions preclude farmers from diversifying into more profitable nonfarm employment (Wanjala and Muradian 2013). Proponents argue that MV interventions have diverse livelihood impacts and farmers actually receive training to diversify their livelihoods; some go on to become fisherman, barbers, and small kiosk owners.

The debate about the MVP results reached a fever pitch in the pages of the prestigious medical journal the *Lancet* in 2012, when a letter by prominent critics (Bump et al. 2012) prompted the researchers to withdraw one of the study claims—that child mortality declined at three times the rate of the general rural trend (on the basis of conceptual measurement difficulties). After the withdrawal of this finding from the overall results, the *Lancet*'s editors emphasized that a correction did not detract from the study's overall merits. However, the MVP was challenged and put on the defensive, and this led to a new commitment to independent evaluation.

In June 2012, DfID announced a US$18.1 million project to launch a new MV in northern Ghana. DfID also allocated US$3 million for a 10-year evaluation of the project to measure its impact, sustainability, and value for money. In addition, Robert Black of the Johns Hopkins School of Public Health is leading an independent expert group to advise MVP on its final evaluation. Institute of Development Studies researchers (DfID 2013) designed a sophisticated instrument to conduct an independent evaluation of the MVPs in northern Ghana that will be completed in 2016. It is impossible to select a control village, so the DfID methodology selects matching villages and matching households based on observable village-level characteristics to compare the effects of the MVP interventions with those of nonparticipants in the vicinity. The independent assessment will measure the impact of the MVPs and determine whether benefits diffuse from MVs to nearby villages and also to assess whether demonstration effects occur whereby other villages adopt packages similar to those of the MV districts.

The range of development ideas for rural Africa is large, with few and hard-won victories. Funding for rural Africa is always finite, so it matters greatly how and where money is spent. Particular interventions will always produce winners and losers within the villages as well as in wider rural development contexts. The effectiveness and success of the MVP will undoubtedly attract more funding and will extend this rural development initiative in scope and scale, indirectly diverting funds from alternative paradigms.

switch from one activity to another in a process of experimentation, trying to offset losses in one area with gains in another (Bryceson 2009). Now, rural Africans are engaging in more diverse economic activities. This transformation began in the mid-1980s and coincided with the withdrawal of big government from the countryside and the introduction of liberalization policies, but it has gained considerable traction over the past decade (Bryceson 2009).

Labeled the "deagranization thesis," this phenomenon captures processes of occupational adjustment, income-earning reorientations, and social identification with nonfarming activities. Simply put, rural dwellers are shifting away from exclusively agriculture-based livelihoods. Approximately 30% to 50% of rural households' income has been found to be derived from nonfarm sources, and the share is as high as 90% in some rural parts of southern Africa (Ellis 1999).

Some income is generated by engaging in multiple endeavors, farming being just one of them. More people are deriving incomes from nonfarm activities, such as (1) petty trade (food selling, brewing, handicrafts, mobile phone airtime selling and selling to urban periodic markets); (2) employment in construction, operations, and maintenance generated by village and town consolidation with accompanying infrastructure development (houses, roads, mills, transport depots) (Bryceson 1996) and mobile phone network expansion; (3) transfer payments (remittances and pensions from family members); (4) selling their labor to others; and (5) the development of new consumer markets for products formerly freely available in rural areas (e.g., fuel wood and water supplies). Also, rising urban demand has encouraged rural dwellers to "mine" tree stocks for supplies, whereas in earlier periods, rural people obtained fuel wood communally rather than through market processes.

Typically, rural households engage in several income diversification strategies at the same time, reflecting different opportunities at various times of the year and in different places (Binns, Dixon, and Nel 2012). Most of the activities are highly opportunistic, involving quick responses to market demand and supply (Bryceson 2005). Different informal economy niches are also populated by different genders: women gravitate to all-year-round petty and curios trade, beer brewing, hair plaiting, knitting, tailoring, and soap making, while men dominate in businesses based outside of the home, such as construction and transport (Bryceson 2005). There is inconclusive evidence of whether women benefit—by being empowered as entrepreneurs—or are doubly burdened—by being straddled with petty trade responsibilities on top of home responsibilities. Access to diversification opportunities varies by income group. Wealthier rural entrepreneurs are able to diversify into more lucrative nonfarm businesses (brick making, trade, transport, etc.), while rural dwellers in poverty are much more prone to casual labor diversification (Ellis 2004).

The rise in rural engagement in informal economic activities and the participation of women and youths are changing age-old patterns of a predominant agrarian economy and undermining the traditional transfer of farming skills from one generation to the next. However, not all nonagricultural activities are specialized, and considerable occupational churning (movement back and forth among occupations) takes place. Even at the household level, many families still strive to maintain a foothold in farming in order to contribute subsistence food. However, there is more individualization of economic activity than ever before, suggesting a break from the cultural tradition of pooling income among household members (Bryceson 2005).

Declining agricultural returns and uncertainty are deterring younger generations from agriculture. Continued subdivisions of land at inheritance and other highly varied land tenure arrangements impose limits on participation, making it impossible for youths to obtain enough land to take up farming as their main economic activity. Poor farm performance, declining yields due to declining soil fertility, a degraded natural environment, and increased climatic variations are other factors. In areas heavily afflicted by HIV infection, the shortage of able-bodied labor to take on the physically onerous tasks can compound these factors. Urban residents' investment in commercial farming is displacing undercapitalized small farmers, particularly in areas close to urban centers. Inputs have become more expensive, and many agricultural markets are declining for small producers. Two diverging processes are taking place simultaneously. First, large-scale land investments are consolidating farms into larger, more commercially oriented units, and the appropriation of land by foreign investors is disadvantaging and even dispossessing many smallholders. Second, family farms are shrinking largely on the basis of generational subdivisions, especially in agricultural fertile areas.

Rural tourism and mining activities have expanded in different areas, offering alternative employment options. Often mining is the most lucrative alternative to agriculture, but mining cannot absorb all rural Africans, and mining practices are highly problematic and unsustainable. There are alternative policy efforts that target rural livelihood diversification as a means of lifting people out of poverty (e.g., pro-poor tourism) (see Box 5.2). However, the orthodoxy of agriculture first is difficult to dislodge, and contributions of nonfarm activities tend to be shrugged off by proponents of mainstream views.

Rural areas do not exist in isolation and are connected to urban areas in diverse ways other than through agriculture. In fact, rural development and urbanization are highly interwoven. For example, there is a rising urban demand for all kinds of rural resources. As cities and towns grow, they create additional demand for resources (land, labor, food, water, energy, and other

## BOX 5.2 PRO-POOR TOURISM BEST PRACTICES FROM RURAL SOUTH AFRICA: MAKING AFRICAN GAME PARKS WORK FOR THE LOCAL POOR

Pro-poor tourism is an initiative first supported by DfID that generated significant scholarship on rural (and urban) best practices to build partnerships between tourism businesses and local residents to unlock opportunities for economic gains for the poor. The UN's World Tourism Organization has endorsed pro-poor tourism, and rural South Africa is a center of significant experimentation and innovation.

Four potential advantages of pro-poor rural tourism have been identified (Rogerson 2006). First, the community can be engaged directly as a partner in environment resource management. Second, the community can be positively incorporated into global supply chains, altering historical physical, social, and economic isolation. Third, rural residents are enabled to draw on assets of natural (wildlife and scenery) and cultural capital and to participate in viable employment alternatives to subsistence farming. Fourth, bringing a wealthy customer base to target destinations (e.g., remote rural areas) creates economic multiplier effects.

Game lodge tourism generally occurs in remote pristine rural areas also characterized by rural poverty. In the past, it functioned as an enclave economic space in rural Africa with weak local linkages and few benefits accruing to the rural poor. An alternative tourism model has emerged combining market-based sustainability with robust poverty alleviation. New leasing concessions have enabled new high-end safari lodges to be developed in such a way that the local community forges partnerships with the state and the private sector (African Safari Lodge Foundation 2012). This model allows the community to benefit as a shareholder (from lease fees and profit-sharing arrangements), directly from employment creation and indirectly from small business development as providers of auxiliary services (e.g., managing gate access, horticulture farming). This pro-poor ecotourism venture is also based on an ethos that conservation is most effective when local communities participate as co-owners and realize direct economic benefits.

Madikwe Game Reserve in the North West Province of South Africa (bordering Botswana) is based on a best practices pro-poor rural tourism development model. The reserve was established in 1991 on 75,000 acres (30,351 hectares) of land formerly degraded with abandoned cattle farms and derelict buildings; the entire area had been emptied of natural wildlife by extensive poaching. The project commenced with Operation Phoenix, which represented the largest transfer of game to have ever taken place in the world (1991–97). More than 8,000 animals (of approximately 28 species) were released into the reserve, enclosed by a 150-km perimeter fence (Madikwe Game Reserve 2014). Importantly, the Madikwe project aimed to put local people first, promoting livelihoods before wildlife and conservation (Rogerson 2006).

**FIGURE 5.6** Madikwe River Lodge, South Africa. *Source:* © Atlantide Phototravel/Corbis.

Locals obtained the lease rights for a prime tourism concession in the reserve and brokered these rights to raise capital, develop two luxury lodges (Fig. 5.6), and develop partnership agreements with private firms to operate and maintain the lodges. These lodges became anchors for generating local employment (in hospitality, field guide, and cultural interpretation areas). By 2007, the Madikwe Reserve employed 773 workers, three fourths of whom were local people (Relly 2008).

Another best practice in community-led tourism development is the Makuleke model in the northernmost portion of Kruger National Park, in an area known as the Pafuri Triangle, a 54,363-acre (22,000 hectare) reserve with abundant natural resources. Historically, the area belonged to the Makuleke people (a Tsonga tribe that settled the area around 1700), but the apartheid government forcibly removed the community in 1969 in an effort to secure international borders.

Its development as a tourism initiative grew out of a landmark community restitution case in 1998 that returned the land to the Makuleke community. The Makuleke opted not to resettle the area but rather to maintain the land within the national park. They initiated a new relationship with the land by engaging with South African National Parks and leasing their tourism rights to the private sector, convinced that conservation and tourism would generate more revenues than cattle farming or other agricultural activities. Under an agreement with South African National Parks, the community participated in joint management of the area to maintain the same standards of the national park and retained the right to access the natural resources (e.g., firewood, stone gravel, medicinal plants) (Shehab 2011).

The Makuleke community engaged in several private-sector partnerships. Lease fees (10% of turnover) and other profit-sharing mechanisms generated approximately zar 6 million (US$688,374) (Shehab 2011). Two private partners, Outpost and Wilderness Safaris, built luxury tourist lodges, and the latter also built and operated the EcoTraining facility. Since the beginning, 31 locals

(Continued)

**BOX 5.2 (*Continued*)**

have been trained as field guides, 61 are employed in hospitality services, and additional members of the community are employed as rangers (engaged in security and antipoaching monitoring) (Shehab 2011:175).

Community profit sharing from tourism underwrote the electrification of two Makuleke villages (outside the reserve) and paid for some school improvements. Nevertheless, community opinion remains divided on whether income generated from the project was distributed equitably: many perceive that well-connected community members usurped the profits (Shehab 2011). In addition, Wilderness Safaris participated in several small joint-venture community projects; for example, a hydroponic farm (tomatoes, spinach, and lettuce) to produce food for the lodges and a village-based bed and breakfast. Unfortunately, neither of these enterprises turned a profit, so the private-sector partner withdrew from both. However, another project—a children's life skills camp (Children in the Wilderness)—was sustained.

Clearly, pro-poor tourism initiatives cannot solve most rural poverty issues, but they can result in modest improvement in local livelihoods and serve as another economic anchor in depressed, remote rural areas. High-end, low-volume safari lodge tourism can be developed to maximize the engagement of locals and to allow the community to participate in corporate profit sharing, elevating local people as active participants in the economic development of their own communities, even though this will never be an uncontested process.

commodities). Urban demand generates rural income and creates opportunities for nonfarm income. It also increases pressures on rural areas to produce and contribute more to urban and national economies. Moreover, in regions where the town can be brought to the countryside, enhanced possibilities for rural development exist (water and sanitation systems, road infrastructure, regional markets, etc.).

Diversification points to different paths in rural development. However, the rural livelihood perspective has waned in terms of its impact on changing development thinking (Scoones 2009). The approach was more useful through its descriptions (showing complexity and diversity and assessing good and bad livelihoods) than its prescriptions of solutions (although pro-poor tourism is an example of how practitioners are grappling with diversification of livelihoods). For the most part, proponents failed to articulate a vision of what future livelihoods should be like. Nevertheless, the policy implications are that livelihood diversification trends point to endowing rural people with the skills they need to escape farming and even the countryside, if they choose. At this juncture, Africa's rural future looks uncertain, but it is clear that substantial displacement of smallholder agriculture is highly likely. However, the rural livelihood perspective needs to be reinvigorated to face contemporary and future challenges.

## GENDER IN RURAL AFRICA

Scholars and development planners paid scant attention to the role of gender in development in Africa until the 1970s. Ester Boserup's (1970) groundbreaking scholarship demonstrated that women's work in the household and in subsistence economic activities was discounted in official statistics. As a consequence, development interventions failed to consider women's full economic roles and their social and political constraints. Boserup's work inspired the Women in Development movement, which gained momentum during the UN Decade for Women (1975–85) and opened a space for engendering development. In 2007, the UN declared October 15 as the International Day of Rural Women in Africa (Fig. 5.7). The establishment of the Association of African Women for Research and Development in 1977 marked the beginning of efforts to

**FIGURE 5.7** International Day of Rural Women in Africa, Dogon, Mali. *Source:* © Jose Luis Cuesta/Demotix/Corbis. Corbis image 42-37181127.

institutionalize gender and women's studies in the region. Notable individuals from the region (e.g., Wangari Maathai) also championed the cause of rural gender issues as women began to raise their voices individually and collectively (Mama 1996).

Combined, these efforts prodded the development community to question gender-neutral cost–benefit assumptions in various development interventions. A critical development benchmark is the MDGs that underscore gender equality; MDG 3 explicitly targets gender equality and women's empowerment.

Feminists contend that African women had to confront "two colonialisms"—that of the former European colonialists and that of contemporary men (Urdang 1979). Colonialism produced a patriarchal dividend, although gendered power relations existed prior to the arrival of the Europeans. Colonialism was responsible for elevating the Victorian and missionary ideal of women's spatial confinement and entrapment to the home to concentrate on childrearing and domestic duties. Certain colonial policies reinforced patriarchal dominance. For example, taxes were paid by men, which encouraged them to assume their "appropriate" breadwinner/head of household role. In addition, employment and migration practices targeted male labor, leaving rural women to fend for themselves without the traditional African economy support structure. Overall, colonialism led to a decline in women's economic independence and social status. However, the decades following political independence brought little to halt gender inequalities.

While the disadvantages of rural women are unequivocal, they were not passive victims. Women have been active individually as well as collectively in exercising control over their own destinies, and many participate in several spheres. However, their contributions may go unnoticed because they do not fit the way that Western minds consider home and work as separate domains. In rural Africa, much of women's work straddles the home–work continuum. Indeed, rural women are active on several fronts, assuming responsibility for domestic and productive responsibilities (childrearing, educational duties, household subsistence activities such as food preparation, farming, and water and firewood collection) and engaging in various income-earning activities (selling food, wood, alcohol, and provisioning services). Some women combine these activities with civil society activism, while a few concentrate on activism. Some household duties can be very time-consuming. For example, in Malawi, women can spend up to nine hours per week collecting water, eight times more than men spend getting water (UNDP 2011). Traditional stereotypes, however, may not allow men to dedicate more time to fetching water without reflecting negatively on their social standing. In general, women's jobs in the informal economy concentrate at the lower end of labor markets (in more crowded micro-activities that are less remunerative and with low levels of productivity). Beer brewing, very lucrative in some communities, can be an exception.

In other regions of the world, men dominate in agricultural employment, which is not true in Africa. Women in Africa have relatively high participation rates and the highest average agricultural participation rate in the world. Cultural norms have long encouraged women to be economically self-reliant, and since colonial times, women have assumed responsibility for subsistence agriculture. Significant geographical variation occurs: women are more heavily engaged as agriculturalists in the poorest countries and in the most deprived rural regions (FAO 2011). The share of women in agricultural employment ranges from 35% in Côte d'Ivoire to over 60% in Mozambique and Sierra Leone (FAO 2011:8). (Added to this, time-use studies in agricultural employment reveal that women work longer hours than men.) Official statistics show the share of women in African agriculture to be approximately 50%, but it could be higher as women underreport some of their efforts as work (FAO 2011).

Gender inequality is highly prevalent in African agriculture (FAO 2011). The gender gap refers to the fact that women have less access to productive resources and opportunities. The gender disparity is evident in terms of land, livestock, labor, education, communication and learning activities, financial services, and technology.

Land ownership, the most important household asset in rural Africa, is highly unequally distributed. Women, in general, account for 15% of all agricultural holders, but there is significant geographical variation among countries (women own less than 5% of holdings in Mali but 30% in Botswana and Malawi) (FAO 2011). Augmenting landownership inequalities, farms operated by women are smaller, more isolated, and/or closer to home, with poorer soils than those cultivated by men (Momsen 2004).

Livestock represent a valuable agricultural asset and a guarantee of income in hard times. Livestock ownership shows large gender differences: men own larger livestock holdings and larger animals (cattle, horses, and camels), while women tend to the smaller animals (chickens and pigs) that typically feed on household scraps. These differences also carry societal and symbolic meanings. Horvaka (2012:875) notes that "cattle are admired and respected, reflecting high social status and economic wealth of individuals and the nation; they drive the economy, feature in government development programs, reside in reserved, privileged physical spaces of cattle posts and ranches, and mark important social occasions through their exchange. Chickens garner much less attention, wield little status and power, and feature in low-valued domestic subsistence or impersonal industrial agriculture realms."

Women's jobs in rural agriculture tend to be more precarious and less protected than men's. They participate more as subsistence farmers, unpaid workers on family farms, and temporary and seasonal workers on other farms. Far more likely to be the lowest-paid workers, women typically are paid less than men for the same work, even when they have more experience. In several contexts (e.g., Ethiopia), the gender division of agricultural labor allows men to specialize in plowing and herbicide spraying and women to weed, and social norms oblige female landholders to recruit male labor to undertake activities traditionally done by men rather than to do them themselves.

The historical bias against girls in education has resulted in the widest gaps (between males and females) in attendance and educational attainment in Africa among all the world regions, especially in rural areas. In Ghana, for example, male heads of households attain more than double the education of their female counterparts, and the gender gap is even larger in many rural communities. MDG 3 targeted gender parity in education but has narrowed the gap only at the primary levels. A consequence is that the education pyramid remains unchanged for African women at higher levels. Women are almost totally unrepresented in professional agricultural management and decision-making positions.

Access to financial services illustrates another dimension of the gender gap. These services are vital for providing credit and insurance and for enhancing opportunities to improve agricultural output, to manage risk, and to accumulate and retain other assets. Cultural and sometimes legal barriers prevent women from entering financial contracts and from holding bank accounts. Women typically have less control over the types of fixed assets that can be used as collateral for loans. Even though microcredit agencies target women, in most countries, the share of female smallholders obtaining credit is 5% to 10% lower than that of males, and women are able to tap less than 10% of available credit (FAO 2011).

The use of agricultural technologies (e.g., machines and tools, improved plant varieties, animal breeds, fertilizers, and herbicides) also illustrates the gender gap. African women are only half as likely as men to use fertilizers. The net result is smaller yields (20%–30% less than men). However, this does not mean that women are less productive farmers than men (FAO 2011), but rather that their productivity levels are a function of their weaker structural positions.

Women-operated farms concentrate on different crop/livestock mixes, with an emphasis on production for home consumption rather than commercial sale. For the most part, women grow staple foods, while men focus on cash crops. This is a function of women's exclusion from modern contract farming arrangements. For example, in Kenya, women make up less than 10% of farmers engaged in smallholder contract farming integrated into fresh fruit and vegetable export markets (FAO 2011:13).

Closing the gender gap in agriculture would generate significant gains for women, the agricultural economy, and society. If women had the same access to productive resources as men, they could increase yields by 20% to 30% (FAO 2011). This could raise total agricultural output by as much as 4%, which could lift millions of women and their families out severe poverty and greatly contribute to food security (FAO 2011). Moreover, research shows that when women control additional income, they spend more of it on food, health, clothing, and education for their children.

Women play important roles in agriculture. Beyond laboring, women are reservoirs of plant knowledge, especially about seed saving and medicinal applications. They dominate rural retail trade in small-scale agriculture, and in some regions (e.g., West Africa), rural women traders are a major link and source of information between rural and urban economies and serve intermediary roles in some urban food supplies.

## GIRLS IN RURAL AFRICA: ALMOST INVISIBLE MEMBERS OF SOCIETY

Children under five have received increased attention in the MDGs, and significant progress has been made in increasing gender parity in enrollments (but not equality) in primary school education. However, girls in their second decade remain invisible in the global development agenda. Rural adolescent girls (aged 10–19) face triple disadvantages based on their age, gender, and location in rural poverty contexts (Chicago Council on Global Affairs 2011). Rural girls are one of the most marginalized and discriminated-against groups in Africa.

Development scholars have only recently noted adolescent girls' lack of human and educational rights. To raise the profile of this issue, the UN launched its first-ever International Day of the Girl on October 11, 2012, coinciding with Plan International's campaign to publicize the denial of education to one third of girls globally (half of the girls in Nigeria and Mali do not complete primary education) due to poverty, violence, and discrimination (UN Education, Scientific and Cultural Organization [UNESCO] 2012). These campaigns coincided with the UN Secretary-General's launch of an Education First initiative, positioning education at the center of the Africa development debate.

In contemporary African rural environments, adolescent girls' lives remain hard, insecure, and fragile. Farm activities are divided by gender: men handle the plow, women guide the oxen, girls weed the fields. Adolescent girls in virtually every rural household make significant economic and social contributions, but they are generally undervalued and uncounted as productive labor. Working alongside their mothers and family members, they participate in a wide range of economic activities. Their specific farm duties include weeding, sorting and pounding grain, collecting milk and eggs, and gathering wild foods. Their domestic duties routinely involve helping with food preparation and household chores (e.g., cleaning, collecting fuel wood, water, and trash). Girls engage in female-dominated informal activities such as petty trade, drinks/food production, daily wage labor, and informal construction (Population Council 2010). Some rural girls move to cities to earn income in informal urban economies, and moving to a city almost ensures that the girl will drop out of school.

Adolescent girls' duties are time-consuming, leaving little time and energy to focus on school. In extreme cases, girls spend five to eight hours per day fetching household water. On top of their many duties, adolescent girls also care for younger siblings, the sick (especially HIV-infected family members), and the elderly. Rural adolescent girls therefore carry heavy burdens, working more than urban girls and rural boys.

The UN's Universal Declaration of Human Rights, adopted in 1945, states that education is a basic human right for all children. MDG 3 promotes gender equality and women's empowerment, with specific targets set to eliminate gender disparity in primary, secondary, and tertiary education. As of the early 2010s, millions of girls are denied the right to education, unable to access the knowledge, skills, and capabilities necessary to take an empowered and equal role in society. In most African countries, girls have less than a 50% chance of attending secondary school, and only 5% of girls enroll in tertiary education institutions. A girl in North America is more than 15 times more likely to have a chance to attend college than a girl in Africa (Fig. 5.8).

Throughout Africa, girls' drop out rates to grade 5 are very high. A number of factors explain girls' limited

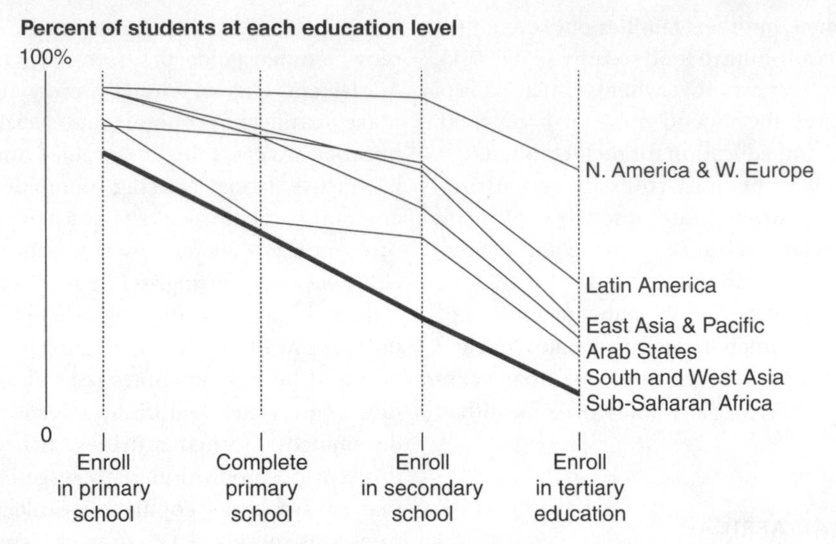


**FIGURE 5.8** Percentage of African Girl Students at Each Educational Level.

access to education and their low retention rates. First, poverty poses a severe challenge to educating girls. Education costs remain prohibitive for many of the rural poor, especially large families. Parents must decide how to use their limited resources and how best to provide a secure future for their families. While most African countries have national policies mandating free primary education, additional costs (examination fees, supplies, and transport) must be borne by the family. Families in poverty are highly vulnerable to poor health, and family illness can force a girl to drop out of school. Unfortunately, tough choices have to be made, and sacrificing girls' educational opportunities and ultimately their chances in life is all too common.

Second, social and cultural norms (within households, communities, and societies) that disadvantage girls are entrenched. Villages, ethnic groups, religious orders, and national communities bolster gender norms. Within households, boys are often valued as future family heads and providers, whereas girls are typically regarded as temporary family members because they are expected to join another family at marriage. For many parents, the choice not to spend money on girls' education is supported based on how far education supports or threatens traditional roles for girls. For example, perceptions that men feel threatened by and are reluctant to marry educated females are powerful disincentives against breaking the gendered norm.

Third, early pregnancy and/or early marriage can end girls' already slim chances of education. The pressures of motherhood, its stigma for young girls, and the lack of support mechanisms make it highly unlikely that girls will remain in school after they give birth. Some countries (e.g., Malawi, Kenya, Ghana, and Liberia) have amended legislation to safeguard the right of young mothers and pregnant girls to remain in school, but this legal right is unenforceable in remote, deprived rural school districts.

Fourth, sexual behaviors can hinder girls' education. Sexual violence at school can force girls to drop out. Reported sexual abuse rates among girls are alarmingly high: case studies report rates as high as 30% to 50% in some rural African contexts (Africa Child Policy Forum 2011). Transactional sex, found to be common in some severe poverty contexts where sex is used to raise money to pay for schools as well as luxury items (clothes and perfumes), has been shown to be negatively correlated with school completion.

A number of African governments have implemented child-friendly policies. For example, Uganda has introduced universal secondary education polices that target parity in secondary school enrollment. However, an enormous education challenge exists throughout the region. Nine out of ten countries of the world with the worst gender disparities in education are African countries—Niger, Mali, Ethiopia, Burkina Faso, Eritrea, Guinea, Togo, Mozambique, and Benin (UNESCO 2010). In these countries, life is especially tough for rural girls (Table 5.2).

Girls' obligations outside of school have high opportunity costs for themselves, their households, their communities, and their national economies. Whether they drop out of school, become victims of sexual violence, and/or marry early, these factors have major repercussions on their ability to realize their full potential as contributing, empowered members of society. Girls' adolescent experiences reverberate for generations. Unequal education means unequal job opportunities, made worse by social and cultural norms that hinder women's entrepreneurship and limit their participation (outside of prostitution) to the most crowded and least profitable economic activities. Enabling girls to gain an equal position in society begins with equal educational opportunities and attainment.

**TABLE 5.2 RURAL GIRLS' LIVES IN ETHIOPIA**

| |
|---|
| 1% are registered at birth |
| 71% have less than five years of education |
| 47% never attend school |
| 40% work for pay before age 15 |
| 91% fetch water |
| Over 50% did not know about menstruation before it happened; 25% managed menstruation by sequestering themselves in remote place (fields, forests, and deserts) |
| 58% report having experienced clitoridectomy (female genital mutilation) |
| 41% are married and 5% are divorced/widowed |
| 81% have an arranged marriage |
| Misconceptions about HIV are common |
| Health workers only reach 19% |

*Source:* Chicago Council on Global Affairs (2011:39).

## CONCLUSIONS

Rural development is very important for Africa's future. Several decades of large-scale, one-size-fits-all initiatives have failed to produce rural development. The bottom-up local approaches of the last 15 years have also failed to reduce poverty at the macro scale. There are a plethora of ideas and approaches on how to meet the rural development challenge. There is considerable debate and some controversy about whom, what, and where to prioritize. The most controversial of all is the introduction of an African GR and GM crops, which nevertheless is gaining traction.

While there is broad support for the MDGs, there is a need to sharpen and extend targets and to be more geographically sensitive to place and culture. Compelling arguments can be made that there should be more attention to adolescent girls, and a better focus on education outcomes at all levels is necessary (rather than a sole focus on primary education). Of course, the MDG approach is not the only game in town, and rural development initiatives are also needed in tourism, livelihood diversification, and other sectors (see Chapter10 for a discussion of food, and land challenges).

Education may be the policy arena where the broadest agreement can be reached. A focus on improving education and making it more relevant could greatly add to national economic growth by making rural residents more productive. However, the state of education in rural Africa is appalling and is the worst in the world. Some 31 million children are not in school, and the numbers are projected to increase (van Fleet and Watkins 2012). African countries that have conducted education assessments show that many schoolchildren are not learning reading and math skills to a fourth- and fifth-grade level (van Fleet and Watkins 2012). Current economic and financial optimism at the national level needs to be tempered by the reality of underlying structural problems in education. Investing in education and in making education more relevant is important so that the skills of the rural population can be upgraded and future workers can be better prepared.

African governments and their development partners should not underestimate the long-term consequences of continued poverty and gender inequalities in rural areas. Tacking the educational crisis is going to require a holistic and integrated approach: African governments must implement policies that reduce poverty in rural areas, such as improving infrastructure and health and sanitation conditions, developing the agricultural sector, and increasing spending on education. Africa's recent economic growth was propelled by increased foreign development investment and resource extraction, but sustained growth over the longer term will require an educated workforce; skills shortages will put a brake on any future growth. Development requires a better synergy between rural and urban education and rural and urban development trajectories.

Few dispute the need to develop rural Africa in a sustainable way. Rural development cannot be achieved using a single strategy or a single paradigm. It requires creative and integrated solutions that have buys-ins from local populations and longstanding international commitments rather than serial flirtations with the latest external development fad. Given the history of neglect and the imposition of external ideas, it is time that locals are fully included in the development process as true partners, both in practice and in transparent agreements, as opposed to being partners on paper only. Rural development in Africa must be people-centered and as equitable as is humanly possible, and it makes sense to acknowledge the diversity of rural environments (Binns, Dixon, and Nel 2012) and to link rural and urban development in a sustainable way.

## REFERENCES

Adenle, A. 2011. "Response to Issues on GM Agriculture in Africa: Are Transgenic Crops Safe?" *MBC Research Notes* 4, 388. Available at http://www.biomedcentral.com/content/pdf/1756-0500-4-388.pdf (accessed December 5, 2012).

Africa Child Policy Forum. 2011. "Violence against Children in Africa." Available at http://www.africanchildforum.org/site/images/stories/ACPF_violence_against_children.pdf (accessed October 30, 2012).

African Center for Biosafety. 2012. "Understanding the Impact of Genetically Modified Crops in Africa." Available at http://www.acbio.org.za/index.php/publications/rest-of-africa/371-understanding-the-impact-of-genetically-modified-crops-in-africa (accessed October 16, 2012).

African Safari Lodge Foundation. 2012. "Transforming Rural South Africa." Available at http://www.asl-foundation.org (accessed October 26, 2012).

Alliance for a Green Revolution in Africa (AGRA). 2012. "AGRA in 2011." *Investing in Sustainable Agricultural Growth—A Five-Year Status Report.* Available at http://www.agra-alliance.org/AGRA/en/our-results/annual-reports/ (accessed October 10, 2012).

Berry R., and W. Cline. 1979. *Agrarian Structure and Productivity in Developing Countries*. Baltimore: Johns Hopkins University Press.

Binns, T., A. Dixon, and E. Nel. 2012. *Africa: Diversity and Development*. New York: Routledge.

Boserup, E. 1970. *Women's Role in Economic Development.* London: Earthscan.

Bryceson, D. 1996. "Deagrarianization and Rural Employment in Sub-Saharan African: A Sector Perspective." *World Development* 24(2):97–111.

———. 2005. "Rural Livelihoods and Agrarian Change in Sub-Saharan Africa: Process and Policies." In *Rural Livelihoods and Poverty Reduction Policies*, eds. F. Ellis and H. Freeman, pp. 48–61. New York: Routledge.

———. 2009. "Sub-Saharan Africa's Vanishing Peasantries and the Specter of a Global Food Crisis." *Monthly Review* 61(3):48–62.

Bump, J., M. Clemens, G. Demombynes, and L. Haddad. 2012. "Concerns about the Millennium Villages Project Report." *Lancet* 26(379):1945.

Carr, E. 2008. "The Millennium Village and African Development: Problems and Potentials." *Progress in Development Studies* 8(4):333–343.

Carr, E. 2011. *Delivering Development: Globalization's Shoreline and the Road to a Sustainable Future.* New York: Palgrave Macmillan.

Chicago Council on Global Affairs. 2011. *Girls Grow: A Vital Force on Rural Economies.* Chicago: Chicago Council on Global Affairs.

Clements, M., and G. Demombynes. 2010. "When Does Rigorous Impact Evaluation Make a Difference? The Case of the Millennium Villages." Center for Global Development Working Paper 225.

Available at http://www.cgdev.org/files/1424496_file_Clemens_Demombynes_Evaluation_FINAL.pdf (accessed October 8, 2012).

Cooke, J., and R. Downie. 2010. "African Perspectives on Genetically Modified Crops: Assessing the Debate in Zambia, Kenya, and South Africa." Center for Strategic and International Studies, Report of the CSIS Global Food Security Project. Available at http://csis.org/files/publication/100701_Cooke_AfricaGMOs_WEB.pdf (accessed October 17, 2012).

DfID (2013). Impact evaluation of a New Millennium Village in Northern Ghana. Initial Design Document, 25th March 2013. Available at http://www.ids.ac.uk/files/dmfile/IDDFinal25Mar132.pdf (accessed October 23, 2013).

Ellis, F. 1999. *Rural Livelihood Diversity in Developing Countries: Evidence and Policy Implications.* ODI Natural Resources Perspectives No. 40. London: ODI.

———. 2004. "Occupational Diversification in Developing Countries and Implications for Agricultural Policy." Available at http://www.oecd.org/development/povertyreduction/36562879.pdf (accessed November 2, 2012).

Ellis, F., and S. Biggs. 2001. "Evolving Themes in Rural Development, 1950–2000s." *Development Policy Review* 19(4):437–448.

Food and Agriculture Organization (FAO). 2011. *The State of Food and Agriculture, 2010–2011: Women in Agriculture: Closing the Gender Gap for Development.* Rome: FAO.

Horvaka, A. 2012. "Women/Chickens vs. Men/Cattle: Insights on Gender-Species Intersectionality." *Geoforum* 43(4):875–884.

Madikwe Game Reserve. 2014. "Madikwe Game Reserve." Available at http://www.madikwe-game-reserve.co.za (accessed March 6, 2014).

Makoni, N., and J. Mohamed-Katerere. 2006. "Genetically Modified Crops." In *Africa Environmental Outlook 2: Our Environment, Our Wealth*, pp. 300–330. Nairobi: UNEP.

Mama, A. 1996. *Women's Studies and Studies of Women in Africa during the 1990s.* Dakar: Codesria.

McKinsey Global Institute. 2010. *Lions on the Move: The Progress and Potential of African Economies.* Washington, D.C.: McKinsey Global Institute.

Millennium Villages Organization (MVO). 2011. "Millennium Villages Project: The Next Five Years: 2011–2015." Available at http://www.millenniumvillages.org/uploads/ReportPaper/MVP_Next5yrs_2011.pdf (accessed November 13, 2012).

———. 2012. "Millennium Villages Overview." Available at http://www.millenniumvillages.org/the-villages (accessed November 13, 2012).

Momsen, J. 2004. *Gender and Development.* New York: Routledge.

Morvaridi, B. 2012. "Capitalist Philanthropy and the New Green Revolution for Food Security." *International Journal of Agriculture and Food* 19(2):243–256.

Moseley, W. 2011. "China's Farming History Misapplied to Africa." *Al Jazeera*, Oct. 27. Available at http://www.aljazeera.com/indepth/opinion/2011/10/201110241502249406.html (accessed December 5, 2012).

Murphy, Tom. 2011. A View from the Cave. Reporting on International Aid and Development. Available at http://www.aviewfromthecave.com/2011/10/empire-strikes-back-sachs-vs-world.html (accessed December 4, 2013).

Negin, J., R. Remans, S. Karuti, and J. Fanzo. 2009. "Integrating a Broader Notion of Food Security and Gender Empowerment into the African Green Revolution." *Food Security* 1(3):351–360.

Patel, R. 2009. *Stuffed and Starved: The Hidden Battle for the World Food System.* New York: Melville House Publishing.

Population Council. 2010. "Ethiopia Young Adult Survey: A Study in Seven Regions." Available at http://www.africanchildforum.org/site/images/stories/ACPF_violence_against_children.pdf (accessed October 30, 2012).

Pronyk, P., M. Muniz, B. Nemser, M. Somers, L. McClellan, C. Palm, U. Huynh, Y. Ben Amor, B. Begashaw, J. McArthur, A. Niang, S. Sachs, P. Singh, A. Teklehaimanot, and J. Sachs, for the Millennium Villages Study Group. 2012. "The Effect of an Integrated Multisector Model for Achieving the Millennium Development Goals and Improving Child Survival in Rural Sub-Saharan Africa: A Non-Randomized Controlled Assessment." *Lancet* 379(9832): 2179–2188.

Rakotoarisoa, M., M. Iafrate and M. Paschali 2011. *Explaining Africa's Agricultural and Food Trade Deficits.* Rome: FAO.

Relly, P. 2008. "Madikwe Game Reserve, South Africa—Investment and Employment." In *Responsible Tourism: Critical Issues for Conservation and Development,* ed. A. Spenceley, pp. 240–267. London: Earthscan.

Remans, R., P. Pronyk, J. Fanzo, J. Chen, C. Palm, B. Nemser, M. Muniz, A. Radunsky, A. Hadera Abay, M. Coulibaly, J. Mensah-Homiah, M. Wagah, X. An, C. Mwaura, E. Quintana, M.-A. Somers, P. Sanchez, S. Sachs, J. McArthur, and J. Sachs. 2011. "Multi-Sector Intervention to Accelerate Reductions in Child Stunting: An Observational Study from 9 Sub-Saharan African Countries." *American Journal of Clinical Nutrition* 94(12):1632–1642.

Rockefeller Foundation. 2006. "Africa's Turn: The New Green Revolution for the 21st Century." Available at http://www.rockefellerfoundation.org/uploads/files/dc8aefda-bc49-4246-9e92-9026bc0eed04-africas_turn.pdf (accessed October 10, 2012).

Rogerson, C. 2006. "Pro-Poor Local Economic Development in South Africa: The Role of Pro-Poor Tourism." *Local Environment* 11(1):37–60.

Sanchez, P, G. Denning, and G. Nziguheba. 2009. "The African Green Revolution Moves Forward." *Food Security* 1:37–44.

Scoones, I. 2009. "Livelihoods Perspectives and Rural Development." *Journal of Peasant Studies* 36(1):171–196.

Séralini, G. 2012. *Ces OGM Qui Changent Le Monde.* (How GMOs are changing the World). Paris: Flammarion.

Séralini, G., E. Clair, R. Mesnage, S. Gress, N. Defarge, M. Malatesta, D. Hennequin, and J. de Vendômois. 2012. "Long Term Toxicity of a Roundup Herbicide and a Roundup Tolerant Genetically Modified Maize." *Food and Chemical Toxicology* 50(11): 4221–4231.

Shehab, M. 2011. "Tourism-Led Development in South Africa: A Case Study of the Makuleke Partnership with Wilderness Safaris." PhD diss., University of Witwatersrand, Johannesburg, South Africa.

Shiva, V. 2000. *Stolen Harvest: the Hijacking of the Global Food Supply*. Cambridge: South End Press.

Thompson, C. 2007 Africa: Green Revolution or Rainbow Evolution?" *Review of African Political Economy* 34(113):562–565.

Tierney, J. 2008. "Greens and Hunger." *New York Times.* Available at tierneylab.blogs.nytimes.com/2008/05/19/greens-and-hunger/ (accessed January 1, 2013).

Tran, M. 2013. "Jeffrey Sachs' Millennium Villages to expand with £67m loan." Available at http://www.theguardian.com/global-development/2013/aug/13/millennium-villages-project-islamic-development-bank (accessed September 15, 2013).

United Nations (UN). 2012. *World Urbanization Prospects: The 2011 Revision.* New York: Population Division, Department of Economic and Social Affairs, United Nations. Available at http://esa.un.org/unup/pdf/WUP2011_Highlights.pdf (accessed October 1, 2012).

United Nations Development Program (UNDP). 2011. *Human Development Report, 2011. Sustainability and Equity: A Better Future for All.* New York: UNDP.

United Nations Education, Scientific and Cultural Organization (UNESO). 2010. *Education for All Global Monitoring Report*. Rome: UNESCO.

———. 2012. *Youth and Skills: Putting Education to Work*. Rome: UNESCO.

Urdang, S. 1979. *Fighting Two Colonialisms: Women in Guinea-Bissau*. New York: Monthly Review Press.

Van Fleet, J., and K. Watkins. 2012. "State of Africa's Education: Learning Barometer." Center for Universal Education, Brookings Institution. Available at http://www.thisisafricaonline.com/Access/Barometer (accessed January 1, 2013).

Wanjala, B., and R. Muradian. 2011. "Can Big Push Interventions Take Small-Scale Farmers Out of Poverty? Insights from the Asauri Millennium Village in Kenya." CISIN Working Paper 2011-1. Available at http://www.ru.nl/cidin/@831810/pagina (accessed October 4, 2012).

Warren, D., J. Slikkerveer, and D. Brokensha, eds. 1995. *The Cultural Dimension of Development: Indigenous Knowledge Systems.* London: Intermediate Technology Publications.

World Bank. 2008. *World Development Report 2008: Agriculture for Development.* Washington, D.C.: The World Bank.

———. 2011. "Africa Development Indicators." Available at http://data.worldbank.org/data-catalog/africa-development-indicators (accessed October 1, 2012).

#### WEBSITES

The Gaia Foundation and the African Biodiversity Network. *Seeds of Freedom.* Available at http://vimeo.com (accessed December 5, 2013). Documentary narrated by Jeremy Irons about global agriculture, seeds, and conflict of interest between agroindustry and farmers.

Millennium Villages Project (MVP). http://www.millenniumvillages.org (accessed December 5, 2013).

Plan International. *Because I Am a Girl.* Available at http://plan-international.org/girls/stories-and-videos-video-detail-16.php (accessed December 5, 2013). Video shot in Malawi about an international campaign to transform the lives of girls from impoverished backgrounds.

CHAPTER 6

# AFRICA'S MOBILE PHONE REVOLUTION

## Informal Economy, Creativity, and Informal Spaces

### INTRODUCTION

Informality looms large throughout Africa. Informality captures different ways of social life, forms of living, working, and operating outside the mainstream. The vast majority of Africans work outside of the formal economy, live in informal housing, and conduct business without using banks. Outside of governments' regulatory purview, informal Africans rely heavily on friends, family, and favors. Making do, being entrepreneurial and creative, and tapping into new technologies that have been developed for the base of the economy are essential to their everyday lives.

For a long time, informality had negative connotations of being small-scale, oriented toward mere survival, and devoid of innovation and creativity. Many still hold this perspective and maintain that the informal economy is in dire need of formalization as well as transformation by expert ideas (often originating from well-paid economists with international reputations and/or business consultants who advise on how to eradicate it). More recently, however, the perspective that aspects of the informal economy are a crucible for innovation, entrepreneurship, and creativity has gained ground. It has taken 40 years since Africa's informal economy was "discovered" by Keith Hart (1973) (in his research on informality in Accra) for a balance of narratives on informality to emerge.

African's massive uptake of information technology has caught the world by surprise. Africa's mobile phone revolution has been nothing short of spectacular in transforming economic, social, and political activities. Though less spectacular, a rise in computer imports, especially used devices that are repurposed, has been noteworthy in reducing the digital divide and sparking a plethora of informal information technology businesses (e.g., email, remittance, Web-based tourism enterprises such as slum tours) and other spin-offs (e.g., harvesting precious metals from used computers and operating "419" Internet scams). In societies where infrastructure is thin and land lines are sparse (and often not working) and where governments have controlled the flow of information as well as its content, the contemporary technological revolution is unprecedented.

Technological innovation is emanating from Africa's informal economy. Various prototypes are under development (e.g., ishacks, solar-powered toilets, devices attached to bicycles that can recharge mobile phone batteries), and these contributions are increasingly being recognized, adapted, and supported internationally. For example, the Ushahidi platform is an innovative solution born in Africa that was introduced to monitor election violence in Kenya and but has developed into a platform for grassroots activism, citizen journalism, and geospatial information that is easily accessible by an Internet-connected public. Since its introduction in 2008, the Ushahidi platform has been refined, improved, and deployed over 40,000 times in 159 different countries (Ushahidi 2013). Like everywhere else in the world, technology is embraced by young people, who use the technology to bypass the inefficiencies in existing systems. Some of their output (e.g., slum mapping) is used to challenge aspects of the status quo and municipal failure to count every citizen. The technological uptake has gone some way toward digital inclusion by universalizing the use of communication technologies to boost autonomous and continuous learning, enabling users to perform basic Internet research and to tell their stories first-hand. Technology is lowering the barriers of entry for everyone to get involved and to be heard and connected. For example, many of Nairobi's Matatu buses (informal privately operated transport vehicles) are equipped with free Wi-Fi. Technology has become a very important vehicle to advance the inclusion of African informals.

In the 21st century, informality encompasses activities that cut across many realms of social, economic, and political life. However, it is the economic arena where scholarship is most developed. Various terms (e.g., "shadow economy," "black market," "underground economy," and "system D") have been employed to capture this enlarged scope, and instances where a local informal activity is part of an international network. The term "system D" has a French African/Caribbean origin as a slang phrase (originating from the word *débrouillard*) that describes the resourcefulness and ingenuity of informal entrepreneurs engaging in business and livelihood strategies.

Subsisting, resourcefulness, improvisation, self-reliance, and using social networks to substitute for capital and other deficiencies are essential components of the informal life. In general, the informal economy is enlarging more than the formal economy. Portions of the informal economy are among the most dynamic sectors of the global economy: for instance, the growth rate of some businesses in the informal economy in South Africa is estimated to be higher than that of the world's top-performing national economy (Neuwirth 2011). Major international organizations now recognize the informal economy as contributing to the global economic output, although its measurement is very difficult. The informal economy is expected to expand, and the Organisation for Economic Co-operation and Development (OECD) projected in 2009 that by 2020 it will encompass two thirds of the world's workforce.

## THE PAST 40 YEARS OF INFORMAL ECONOMY EVOLUTION AND SOPHISTICATION IN AFRICA

Most experts recognize that informal activity has always been around, and it was a major characteristic in the development of all early and medieval cities. In colonial Africa, informal economic activity was tolerated but frowned upon, and two separate economies and cultures were recognized (Myers 2011). In independent Africa, policymakers focused on enlarging the formal economy and neglected the rest in the hope that informal activity would eventually disappear. Industrialization in Africa was expected to lead to the demise of traditional/marginal activities. However, the opposite transpired: African industrialization failed to took root, and instead informal work expanded in scale and scope and became deeply entrenched. Eighty percent of Africans now find employment in the informal economy (OECD 2009).

Study of the informal realm in Africa begun in earnest in the early 1970s. At that time, scholars and policymakers believed that the informal realm was a separate sector. Hart's (1973) research showed that the urban poor engaged in petty capitalism as a substitute for wage employment, from which they were excluded, but their work was articulated to economic activities beyond their place of residence. As failures of modernization became more apparent, it became evident that informality was not a temporary phenomenon but rather a permanent fixture that needed to be reckoned with. The informal arena came to be recognized as a wider sphere that provided coping and survival mechanisms for the poor in terms of housing, health care, food, employment, and social life.

A massive expansion of the informal economy coincided with the introduction of liberalization policies and the tilt toward markets from the mid-1980s onward and the corresponding retreat of the state from providing housing, employment, and social protection. During this era, governments' over-emphasis on the private sector and free markets resulted in a neglect of the informal economy, culminating in a colossal "elephant in the room."

In the 1990s, an expanded definition of the informal realm was adopted. The informal economy was defined as encompassing all forms of informal employment that are (in law or in practice) not covered, or insufficiently covered, by formal contracts. The extended definition moves beyond simply describing unregistered individuals/enterprises to shedding light on their invisibility by an acknowledgment that many workers are not protected by labor and/or social legislation. There is a continuum from the informal to the formal ends of the economy and thus the two sides are interdependent; the informal economy is not an isolated sector (as it was considered in earlier eras), and it affects all aspects of people's daily lives. Many workers operate with confidence in the informal economy but lack government protection (Meagher 2003).

In the 21st century the informal economy continues to be a hot topic for policymakers, activists, and researchers. The large share of the workforce that remains outside the world of full-time, stable, and protected employment makes it an important and timely issue (Rogerson 2007). The last decade has seen a fresh interest in the informal economy worldwide and in its social and technological dimensions. Internationally, this renewed interest is also attributed to the fact that contrary to the predictions and intentions of many economists' policies of regularizing the informal economy (by registering, taxing, and enforcing municipal codes such as health and safety), implementation of these policies has not resulted in its disappearance. Instead, it has expanded massively on a global basis, sheltering retrenched workers and permanently accommodating and providing livelihoods for the masses. Informal employment is growing faster than formal employment, and livelihoods within the economy are highly differentiated: there are survivalists, entrepreneurs, exploiters and the exploited, and all shades of in between that operate from discrete spaces (Figs. 6.1 and 6.2).

**FIGURE 6.1** Informal Street Selling in Soweto, South Africa.

**FIGURE 6.2** Informal Production in Soweto, South Africa.

The informal economy has emerged in new guises and in unexpected places, even on the streets of developed countries. Conventional wisdom that street commerce belonged to a bygone era, and was only a characteristic of an early stage of urbanization, has been overturned. Some (e.g., Davis 2006) even suggest that the opposite is transpiring and that the formal world now depends on the informal one. According to Davis (2006), our urban future is one of a "planet of slums" with only islands of formal spaces. There are claims that the formal economy would grind to a halt without the contribution of and reliance on informal workers and their entrepreneurial activities. World-famous architect Rem Koolhass (2003) offers an optimistic evaluation of informality as the genesis of a city of the future, whereby informal agents demonstrate large-scale efficiencies in constructing living spaces and creating livelihoods. Through a lens of living sustainably, informals use resources more efficiently and their activities incorporate more sustainable use of resources, with heavier use of recycled and repurposed materials. In the words of Koolhass: "Lagos [Nigeria] is not catching up with us. Rather, we may be catching up with Lagos." There is an emerging consensus that informalization is a mode of urbanization as well as a constituent part of rural development rather than an exception (Lindell 2010; Roy 2005).

## MAJOR INFORMAL ECONOMY DEBATES

Vigorous debates about the informal economy have been going on since the 1980s. As with all fundamental social questions, one's ideology affects the question and points to particular answers; thus, early on there was little consensus and much more confusion on what to do with informality. Four main perspectives on the causes of informality and the best course of action have been identified.

First, dualists (e.g., Hart 1973; Tokman 1978) emphasize the mismatch between people's skills and the needs of the formal economy. Subscribing to a regressive view of the informal economy, they view informal enterprises as operating on the basis of involution rather than entrepreneurship (lacking solid business practices, particularly the separation of personal and business finances). Dualists emphasize "up-skilling" and participating in a regularized market structure ("learning by doing"), as well as government support, as critical mechanisms to bridge the gap between the two economic worlds.

The legalist perspective (De Soto 1989, 2000) underscores the inadequacies of regulatory systems: informal entrepreneurs find it necessary to operate outside the official system because of its excessive red tape, the burdensome costs of registration, and their lack of know-how about operating an enterprise in the formal sphere. Strong arguments are made that the poor hold assets tied up in informal property, not legally recognized and so not leveraged: "dead capital" (De Soto 2000). Uncertainties surrounding ownership of such assets prevent individuals from lending or borrowing against their real value. De Soto's logic advocates converting these assets into formal property so that entrepreneurs (in waiting) can be unshackled.

A third perspective, the voluntarist school, sees micro-entrepreneurs choosing to operate informally to avoid taxes, commercial regulations, electricity and rental fees, and other costs of operating formally (Maloney 2004). Inadequate human capital consigns workers to the informal economy. Microfinance and mentorship and training programs are proposed to promote broader and more extensive engagement with the formal economy.

The fourth perspective, the structuralist one, argues that informal workers and informal enterprises are subordinate units to formal entities, and informal workers serve to reduce the costs of labor and represent a labor reservoir that more powerful entities can draw on (Breman 1996; Castells, Portes, and Benton 1989; Moser 1978). Horizontal networks rather than efficient bureaucracies are seen as more important in shaping the organization of life of the working poor in the developing world. Structuralists propose that there should be more equitable linkages between the formal and informal economy and social and labor protection for informal workers. Collective organizing by informal workers to engage in policy discussion is deemed essential (Lindell 2010).

Debates among the different perspectives have generated more heat than light, and each perspective focuses on only certain aspects of the informal economy rather than taking a holistic perspective. Dualists focus on those engaged in traditional and survival activities; structuralists emphasize petty traders and producers and subcontracted workers; and both the legalists and voluntarists concentrate on informal enterprises and entrepreneurs. The informal economy is, however, more heterogeneous and complex than any individual perspective can shed light on. Under close scrutiny, the mosaic of each livelihood is a complex one.

Informality is best conceptualized as encompassing a series of transactions that connect different economies and spaces to one another. However, without a clear consensus on how to engage with informality, policymakers and planners risk dealing with only a portion of informality at a surface level. This can protect the status quo as well as the interests of a small minority (slumlords, labor brokers, subcontractors, and exploitative firms) that benefit most from the current impasse.

## THE AFRICAN INFORMAL ECONOMY ON THE GROUND

African economies are different from economies in the Global North in that the informal economy is much more significant in Africa. African labor markets are much less segregated, and it is common for people to hold multiple jobs and work in different sectors and different locations at different parts of the day. Some workers even switch between the informal and formal economies in the same workday. It is, however, more common for a formal worker to participate in the informal economy as an income-boosting strategy. For example, a health professional can work at a national health facility during the day but make house calls at night; or a secretary at a formal firm can make jewelry at home in her off time (Owusu 2007; Portes and Castells 1989). In Africa, people routinely have multiple livelihoods conducted in different spaces (Owusu 2007).

The informal economy encompasses heterogeneous activities and spaces. The informal economy comprises a range of workers, from self-employed persons (street traders are the largest group in Africa, home-based workers the second largest group) to informal enterprises (often with less than five to 10 full-time workers) and other informal wageworkers (e.g., casual, part-time, unpaid family members and subcontracted homeworkers or industrial outworkers). Formal firms can reach inside informal spaces to employ homeworkers in manufacturing, retailing, transportation, and construction (Grant 2010). The informal economy is located everywhere: in markets, on the streets, in informal industrial areas, in parts of formal dwellings (so-called "boys' quarters") and in both upscale and low-income residential areas (in informal additions or backyard shacks that accommodate home-based enterprises or renters) (Oosterbaan, Arku, and Asiedu 2012).

The informal economy is characterized by:

- low entry requirements in terms of capital and professional qualifications
- small scale of operations (often family-organized)
- skills often acquired outside of formal education
- labor-intensive methods of production and adapted technology
- highly varied hours of operation and wages
- heavy reliance on mobile phone technology
- lack of regulation (for the most part)
- frequent reuse and repurposing of materials
- clientele based on personal contacts
- prices that are generally negotiable between the seller and buyer
- financial support and credit based on personal networks as opposed to bank or government-backed credit
- many work-related injuries.

Traditionally, the informal economy was perceived as comprising mainly survival activities. Its negative aspects were highlighted, ranging from undeclared labor, tax evasion, and unregulated enterprises, to more serious illegal activities. Some of this indeed takes place, but the vast majority of activities in the informal economy provide goods and services whose production and distribution are quasi-legal. Not all informal activities are for survival purposes.

The informal economy includes legal and irregular operators. The informal minibus taxi or "tro-tro" system of public transportation is a good example of the mix of legality and irregularity: regulations pertain to the vehicle (vehicles are registered and drivers are licensed but not always insured), but the mode of operation and the operator are unregulated. Some government officials (e.g., police officers) often enforce informal arrangements or "daily tickets" (as they are called in Nigeria) for the non-harassment of informal street hawkers and to ensure that their goods are not confiscated. It is reasonable not to include criminal activities in the standard conceptualization of the informal economy to avoid stereotyping the informal economy and conflating it with the criminal economy. Obviously, criminal activity occurs throughout the informal–formal continuum.

The bifurcation of the economy finds its spatial equivalent in the neat separation of informal from formal residential areas (Myers 2011). However, the informal economy does not operate just in slums: it is ubiquitous. Although informal settlements can be hives of informal economic activities, some of which are confined to the local slum economy, other activities are integrated into a wider space. The phenomenal growth of slums in Africa since the 1980s—they now house 62% of Africa's population (UN-HABITAT 2003—is a striking development. UN-HABITAT (2003:12) describes the physical characteristics of slums as having inadequate access to safe water, inadequate access to sanitation and other infrastructure, poor structural quality of housing, overcrowding, and insecure residential status (Fig. 6.3).

According to Becker (2004), 80% of non-agricultural employment, over 60% of urban employment, and over 90% of new jobs reside in the informal economy. There is an absence of non-farm employment opportunities in rural Africa. Women, in particular, are overrepresented

**FIGURE 6.3** New Shack Construction, Soweto, South Africa.

in the informal economy. The poor construct their own houses (often using recycled materials) and often invent their own jobs as self-employed workers in a wide range of production and service activities. The uptake of mobile technologies has been unprecedented. On average, African informal economies contribute 42% of gross national product (Schneider 2005).

Informality as a mode of organizing society was certainly around in 19th-century Europe, and there are some parallels with contemporary African informality. But contemporary African informality dominates societal relations to a degree that it never reached in industrializing and urbanizing Europe. The European past does not reflect the African present: informality in the region is now manifested in new forms and new geographies, both at the rural–urban interface and in terms of developments that may serve as a principal avenue to property ownership and livelihoods. Perceptions of informality have changed from being a marginal activity performed by destitute people to acknowledging African informals as articulated but differentially integrated into society. The informal economy simultaneously encompasses flexibility and exploitation, productivity and abuse, aggressive entrepreneurs and defenseless workers (Castells, Portes, and Benton 1989). Some contend that politically underrepresented, socially stigmatized, and culturally repressed people are overrepresented in the informal economy (AlSayadd 2004).

Engagement with informality is difficult for planners and policymakers. Informal spaces seem to be the exception to planning, lying outside its realm of control. In a very real sense, informality is the object of development, a seemingly natural phenomenon that is external and alien to those studying it and managing it. Official attitudes to the informal economy vary from being benign or mildly positive to showing open hostility and antagonism (Potts 2008). Street traders have been harassed in many African countries, with political leaders claiming that itinerant street activities are untidy, unhygienic, and not compatible with a modern city. Slum dwellers have been evicted under the guise of nonpayment of rent by unscrupulous politicians who are in cahoots with land investors (Otiso 2002).

To complicate matters, informal workers have little in common (e.g., street hawker, scrap worker, construction worker), and there are many different labor arrangements within the informal economy. Breman (1996) underscores that poverty can be endlessly franchised and the poor can be very exploitative of each other; some of the exploitative relations are even conducted at arm's length by agents and nonvisible owners, who can even reside in different countries or regions. The informal e-waste international economy is a good example of the web of distant and specialized players who engage in the commodity chain (recycling, shipping, refurbishment, repurposing, local collection, burning, metal extraction and re-exporting), with huge differences in work conditions and reward structures. Some parts of the chain are never in contact with other parts, and no one entity controls the entire chain (Fig. 6.4). Many Africans are not living in worlds "unto themselves" or leading lives divorced from wider global, national, and urban processes.

**FIGURE 6.4** E-waste in Accra, Ghana. Photograph by Martin Oteng-Ababio.

## INTERNATIONAL INFORMAL ECONOMY ENTANGLEMENTS: E-WASTE IN AFRICA

Electronic waste (e-waste) is now a US$7 billion global industry (Grant and Oteng-Ababio 2012). E-waste generated in the United States and Western Europe is increasingly being dumped in Africa. E-waste is defined as discarded, surplus, obsolete, or broken electronic devices that enter the waste stream but are subsequently reused, resold, salvaged, recycled, or disposed of (Grant and Oteng-Ababio 2012). E-waste encompasses obsolete computers, refrigerators, televisions, mobile phones, and other devices that are discarded in one country and shipped to unsupervised dumpsites in other parts of the world. The United Nations Environment Programme (UNEP 2005) estimates that 50 million tons of e-waste is generated globally per year, an amount equivalent to a line of dump trucks stretching halfway around the globe. Large e-waste sites are found in several African countries, including Ghana, Nigeria, Senegal, Côte d'Ivoire, Kenya, and South Africa. South Africa is the only country in the region with formal e-waste recycling capacity.

E-waste became a visible environmental issue in the 1990s as governments in Western Europe, Japan, and the United States set up e-waste recycling systems but did not have the capacity to deal with the vastly expanding quantities of e-waste. Global Northern players began the process of exporting the problem to developing countries, where laws to protect workers as well as the environment are inadequate and not enforced. The economics of e-waste recycling means that the cost of glass-to-glass recycling of computer monitors in the United States is 10 times more than the cost in Africa, making the business of dumping very profitable as well as expedient (even after factoring in shipping and agent fees). At first international e-waste exports converged on China and other locations in Asia, but by the early 2000s e-waste was also being routed to urban "wastelands" in Africa.

African policymakers facilitated the region's entry into the e-waste business. For example, the government of Ghana implemented a policy in 2004 to "bridge the digital divide" by reducing the import duty to zero on used computers; the country thus became a popular site for e-waste shipments. The national policy encouraging one laptop per child/household yielded unintended consequences: such as digital dumping and may have contributed to an escalation in cyber crimes.

Monthly estimates of e-waste shipments to Ghana range from 300 to 600 40-foot-long containers (each with 2,390 cubic feet of storage space) arriving at the port of Tema, just outside the capital, Accra (Grant and Oteng-Ababio 2012). Government import policies were developed with working computers/electronics in mind and without adequate consideration of nonworking devices. The 1992 Basel Convention was developed to help regulate trans-boundary movement of hazardous wastes, making it illegal to ship hazardous materials from the developed world to the developing world. There are ongoing efforts to strengthen the Basel Convention; under consideration is the Ban Amendment, which aims to close a loophole by restricting countries from exporting hazardous wastes intended for recovery, recycling, or final disposal. The geography of e-waste dumping is disguised by the comingling of working devices and nonworking devices in shipments, making it virtually impossible to monitor shipments and the movement of e-waste through different jurisdictions.

One of the most high-profile e-waste sites in Africa is Agbogbloshie, a large slum in the center of Accra, the capital and largest city in Ghana. This e-waste site emerged after 2005, and e-waste has expanded to dominate and engulf the slum, providing employment for more than 5,000 workers. The international media have produced exposés of e-dumping at this site (e.g., *Frontline*'s "Ghana Digital Dumping" [2009]) and local social media are playing an active role in monitoring illegal shipments and dumping (e.g., E-Waste Watch Ghana on Facebook).

Greenpeace researchers (Brigden et al. 2008) emphasize the local health and environmental consequences of e-waste toxicants wafting into the immediate atmosphere and leaching into soils, the lagoon, and the nearby sea. E-waste generates significant toxicity risks (posing serious threats to human and environment health), especially in low-technology informal backyard processing and the unregulated dumping of toxic components. Exposure to hazardous chemicals can cause acute damage to the lungs, kidneys, and other vital organs as well as increase cancer risks and lead to cognitive impairment in children. Illegal dumping also poses a security risk at the end of a commodity chain, with potential lack of security of data drives and even cyber-crime threats. It is no coincidence that Ghana is recognized as a major hub of cyber-crime activity, especially Internet scams. Numerous Internet scams (e.g., lotto, inheritance windfalls, business or export ventures, employment, and invitations to visit Ghana) are disseminated by young, tech-savvy "entrepreneurs" based in the region.

E-waste processing and dumping are not isolated to the Agbogbloshie slum. E-waste processing, open storage, dumping, burning, and extracting scrap metals take place at the site, but these activities are part of a wider space economy with hierarchies as well as specialized niches (Fig. 6.5). An informal e-waste economy centering on the settlement has expanded in recent years. As of 2012, some 30 electronic device repair facilities are present (computer specialization is clearly evident, but mobile phones, refrigerators, and other devices are also repaired). Thirty-six individual stationary scrap buyers (with weighing scales) participate as intermediaries between the collectors and the main scrap dealers. Many collectors hire pushcarts on the site on a daily basis for US$1.39. Trucks are available for hire from particular Old Fadama vendors for transporting larger loads (too heavy to be lugged by human power), but upfront rental costs are prohibitive for many collectors. E-waste burning takes place on the outer fringes of the settlement, furthest away from human settlement. A range of ancillary services has grown up around the e-waste and scrap market to serve workers ("chop bars" [local informal eating places], mosques, and various enterprises for renting carts and trucks for picking up and transporting e-waste).

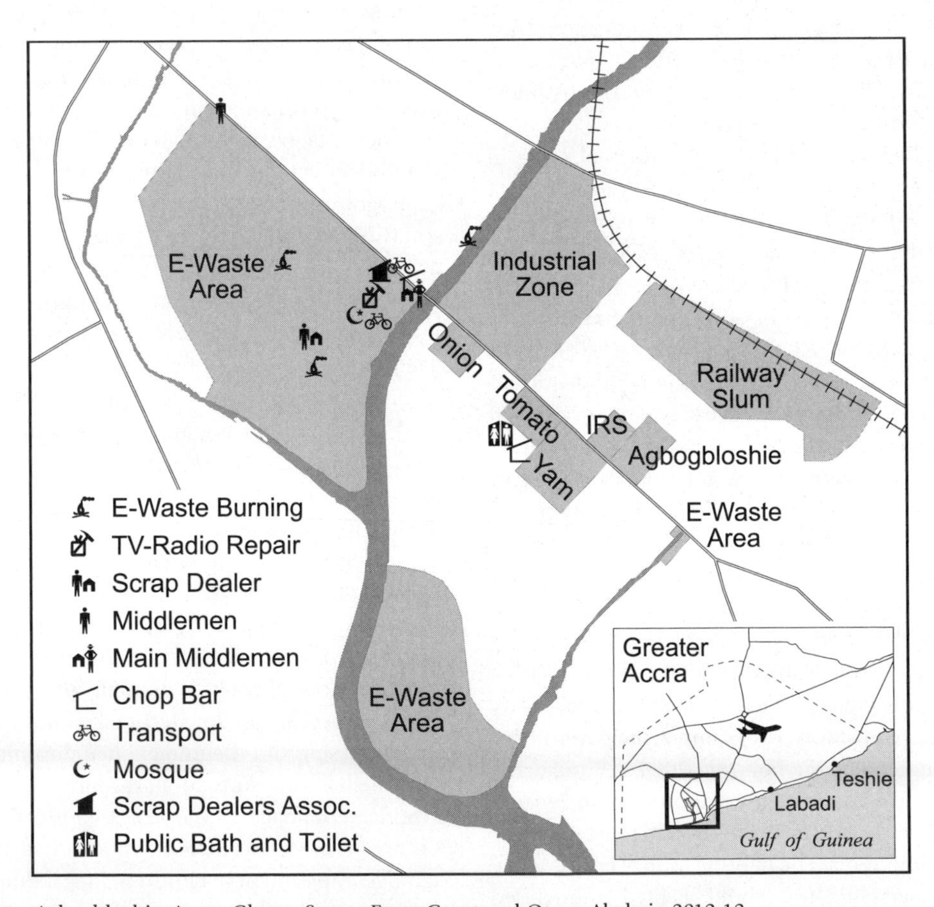


**FIGURE 6.5** Agbogbloshie, Accra, Ghana. *Source:* From Grant and Oteng-Ababaio 2012:12.

Warehousing and scavenging take place throughout Accra as well as other urban centers in Ghana. Importing of e-waste and exporting of scrap metals derived mainly from burning computers both occur via Tema; these scrap metals are sent to companies in Asia and Europe. A web of activities constitutes the e-waste economy: global agents and brokers, importers, warehousers, backyard processors, waste and metal intermediaries, itinerant buyers, waste pickers, small-waste sellers, second-hand electronics markets, formal/informal industrial representatives, and exporters (Grant and Oteng-Ababio 2012). As such, materials disposed of as e-waste in one location (e.g., the United States) become sources of value in Ghana when commodities are transformed through reuse, repurposing, and/or breakdown as raw materials for primary inputs for the production of new commodities (Lepawsky and McNabb 2010). For example, the minerals extracted from e-waste burning can be bundled and sold as inputs to make iron rods in formal industry or ground down as powder and sent to plants in Germany and China, where traces of precious metals can be extracted from the residue, and this metal can be sold to formal firms. Agbogbloshie thus functions as a site of informal work activity within the international e-waste exchange.

E-waste circulates through a hierarchically organized space economy with very different occupational and wage structures. In livelihood terms, e-waste collectors are daily-wage laborers who earn, on average, US$3.50 per day, about two and half times the average income of informal economic workers in Ghana. Collectors involved in dismantling and recovery earn an average of

**BOX 6.1 THE DIVERSITY OF INFORMAL WORKERS, SOWETO TOWNSHIP, SOUTH AFRICA**

Soweto is the best-known and largest township in South Africa. Townships were created under colonial/apartheid policies that separated groups on the basis of race, and they became infamous during the apartheid years as icons of poverty, oppression, and despair. Townships largely performed residential dormitory functions (without an internal business and commercial role, although under-the-radar businesses always operated) and housed workers (not firms) who performed jobs in the "white city." Everyday township life under apartheid included broken families, flimsy shacks on dusty streets, lack of infrastructure, and acute poverty.

Soweto is located in southwest Johannesburg, some 14 miles (22 km) from Johannesburg's central business district. It has become the largest "black city" in South Africa. In a concerted effort to begin to normalize townships, it was officially incorporated into the greater Johannesburg metropolitan area in 2001. The township comprises 43% of Johannesburg's population (approximately 1.7 million people) but contributes only 4% of total economic activity (largely based on estimates of formal activities) (Grant 2010). Soweto has grown to become a diverse residential area, ranging from upper-middle-income areas to many more low-income areas, and it also contains several informal settlements or slums. Shacks are still a ubiquitous feature in many residential areas (200,000 people live in backyard accommodation in Soweto; Crankshaw, Gilbert, and Morris 2000:845).

Informal work in Soweto operates in the context of historical factors such as the legacy of apartheid policies and South Africa's more developed economy status and townships as marginalized spaces accommodating a marginalized labor force. Unemployment is high in Soweto and hovers (unofficially) at 50%, with limited local employment opportunities. High township population growth and influxes of international migrants, some of them more skilled than the local people and willing to work for lower wages, add additional pressure. Often the complexities of spatial arrangements are underappreciated, such as the way the proximity and density promote social networks and township milieus that enable informal entrepreneurial endeavors (e.g., sourcing recycled materials for local suppliers for production processes, labor pooling arrangements among township firms, and securing family and friends' help as needed).

Despite the challenging context, workers find diverse employment opportunities inside the township as well as outside of Soweto. There are three general patterns. First, there are workers who participate in the township's informal retail economy as street sellers and hawkers, selling everything from curios to pirated DVDs to bottled water. A smaller subset engages in informal production in a number in township industrial estates (e.g., making furniture and clothing, constructing fences and gates for burglar-proofing). Many have no option but to become self-employed workers who use their home as a base (known as home-based enterprises) (Gough, Tipple, and Napier 2003). Home-based workers concentrate on retailing, preparing food, building shacks and offering services, running hair salons and *shebeens* (informal drinking establishments), and doing auto repairs; there are also traditional healers selling medicine and treating patients at home.

Second, the dominant spatial pattern is for informal workers to leave the township to obtain employment in the city beyond

*(Continued)*

**BOX 6.1 (*Continued*)**

as domestic help, gardeners, security guards, and low-skilled workers. These informals remain at the beck and call of more powerful individuals who demand chores and favors, and this activity necessitates extensive travel to work.

A third group comprises individuals and families excluded from formal and informal opportunities. To cope they emphasize self-help and mutual aid and their ability "to work" at making friends and contacts (i.e., create social capital). In South Africa (in contrast to the rest of Africa) the state allocates grants for child support, pensions for the elderly, and disability grants. Aid is nominally targeted to individuals (based on disability or parental status), but it is widely known that funds end up supporting more than the recipient. Typically larger, multi-generational households pool their resources (e.g., shelter, state grants). In some instances links can extend to rural and international households in support chains, and relatives with HIV/AIDS can hide away and receive some support in urban homes. Relationships are cemented for better (and sometimes for worse) when families devise several strategies to amplify their income, such as renting rooms and space (when available).

Figure 6.6 illustrates the diverse engagement of Sowetan informal workers. It is clear that the informal economy encompasses diverse places, workers, and sectors, and the township workers are active in the informal economy. Informal work is done at home, in the worker's block or neighborhood in Soweto, elsewhere in Johannesburg, or at a variety of locations; workers engage the urban economy at all spatial scales. The working spaces of the urban poor are more extensive than can be identified in any uniform spatial narrative or single perspective.

Work involving the retail and the food and beverage industry is heavily home-oriented. Small-scale manufacturing activities (furniture, arts and crafts, shack construction, brick making) also take place in the vicinity of the home. Individuals in domestic services, tourism, and transport work mainly outside of their neighborhoods—particularly in other parts of the township or elsewhere in Johannesburg. Construction and transport workers are the most mobile. Some skilled workers are moved from site to site by contractors, but other daily construction workers have more unstable work. Multisite work is also common: for about 20% of workers, work is split between two sites (e.g., making crafts at home but also at a fixed market site) (Grant 2010).

Soweto is the best-established township near Johannesburg's central business district (compared with newer and more remote urban informal settlements), but the Sowetan evidence demonstrates that even historically marginalized poor residents are connected to a wider space economy (albeit incompletely incorporated).

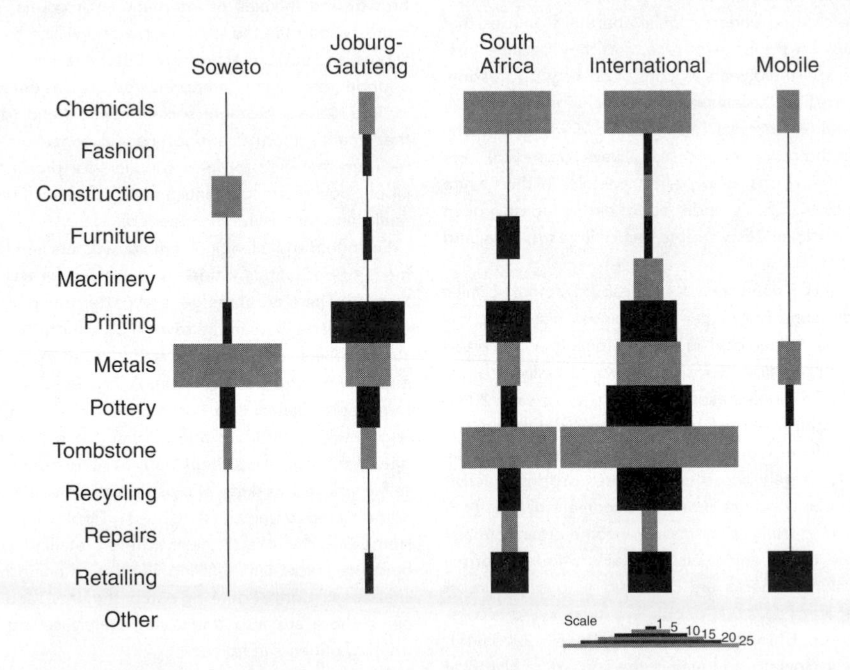


**FIGURE 6.6** Spatiality of Work by Sector in Soweto, South Africa.
*Source:* Grant 2010, p. 605.

US$8.00 per day, but incomes vary significantly. Another factor in the different income levels is the fact that precious metal recovery rates vary widely: recovery rates can be very low for gold but as high as 85% for copper. Youths under 15 years of age earn as little as US$20 per month. At the top of the informal income ladder are scrap dealers specializing in used computers, who earn around US$50 per day. Just below them are "the middlemen," making US$20 to $35 per day (but whose earnings can rise to US$80 on a good day). These people purchase scrap (e.g., copper, brass aluminum, iron) in bits from the recyclers and subsequently sell it in bulk to the dealers, who in turn sell it to refinery industries and exporters. Refurbishment produces a steadier but smaller income stream; earnings average US$7 per day. Workers at the upper end of the informal hierarchy have more regular working hours as well.

## DYNAMISM, CREATIVITY, AND INFORMALITY

Poverty is widespread in Africa: 51% of the population—763 million people—exist on US$1.25 per day (World Hunger Organization 2012). Population pressure and the spectacular growth of slums is leading to speculation that large mega-slums (slums housing 1 million people) are emerging in cities such as Lagos and Kinshasa (Davis 2006). Life is bleak for many people operating in precarious circumstances, and resourcefulness is a key strategy for survival.

Researchers identify the "wage puzzle" (the gap between what is reported in official statistics and what individuals have at their disposal) in Africa to account for how workers could possible survive on such meager sums. The wage puzzle grapples with the complexity and unpredictability of income streams and sheds light on fluid household budgets, especially relevant to extended family situations (e.g., on the one hand, heads of household can pay he school fees for relatives and allow extended-family members to live rent-free; but on the other hand, in different households extended-family members can be charged rent and even expected to pay to use household toilet facilities or contribute to household budgets). Household finances are highly complex and politicized.

Given the complexity of the informal economy, it is hardly surprising that there is a dearth of reliable statistics. Most informal economy data are derived from (1) surveys that sample a portion of the population; (2) government estimates that rely on incomplete employment data, enterprise registrations, and tax records (countries do not collect data on informal enterprise and informal workers); and (3) other proxies. Proxies such as electricity consumption and membership in informal worker organizations (e.g., scrap dealers' organizations) provide additional limited information (restricted by common phenomena such as illegal electricity taps and non-affiliation with member organizations). Some of the most reliable and insightful observations are derived from household and individual interviews.

Research on household finances, such as paying for the costs of housing, routinely necessitates separate interviews with different household members to obtain accurate data on income and spending: agreed-on budgets have been shown to differ widely among family members (Grant 2009). Of course, besides employment income, other income sources come into play, such as credit given by informal shopkeepers; loans from family members as well as loan sharks; inheritances; remittances; community support; charity; "gifts from the grace of god" or from "an aunty"; income from irregular work; income brought into the household from working children; and payments in kind (e.g., neighbors sharing food or food derived from urban agriculture). Household budgets are very fluid and show significant variability in time and space, making it hard to calculate what is actually circulating within the informal economy.

The sum of money circulating in the informal economy is immense. For example, the daily cash turnover of 5,000 to 8,800 informal traders in a Durban market ranks just below that of the largest shopping center in the city (Brown 2006). Migrant remittances to Africa rose to US$40 billion in 2012, surpassing foreign aid flows to the region (see Chapter 7 for a detailed discussion of remittances). Most of these flows are targeted at keeping families afloat (e.g., paying for food, housing, school, and health care), and some funds are used for business start-ups. Commonly, family members earning aboard set up a relative in a bootstrap enterprise and participate in profit-sharing.

Simone (2010:3–5) provides a rich description of the rhythms of the city and the multiple use of space, 24 hours a day. He describes the Oju-Elegba neighborhood and market in Lagos where, when the day's commerce recedes, different kinds of night operations are visible. Local government, trade associations, police and security companies, and sanitation crews are officially in charge of the market, but their oversight withers during the night hours. At 3 a.m. there are many activities going on, unlike most Global Northern urban spaces at this hour (with the exception of fast-food outlets and entertainment venues). Simone (2010:4) paints a vivid picture:

> Small stalls sell huge marijuana cigarettes next to those that sell votive candles and batteries throughout the night, and then there are sales of rice, cigarettes, laundry soap, and batteries throughout the night as well as cooked meals, many stalls specializing in regional cuisines. Pharmaceuticals, charms, and local medicines are hawked both by ambulatory sellers and various forms of makeshift stalls. Set back further along the streets are stores whose histories have known hundreds of functions and whose identities even now fluctuate according to the time of the day. At this hour, hardware stores become outcall services for sex workers delivered to almost any location in the surrounding vicinity; a small business center takes calls and sends out a fleet of young repairmen on motorcycles for various domestic emergencies such as broken water pipes or shorts in overtaxed electricity distributors that often occur when too many households try to connect to a branch line. Schools that teach computer classes during the day host all-night prayer meetings next to bars and small discos, next to a small law office which at this hour serves as a kind of floating "design workshop," where a local politician holds court soliciting ideas, plans, drawings, and models for new housing developments, roadways, drainage systems, and a host of small improvements in every aspect of life in the area.

Informals demonstrate a remarkable inventiveness in creating places that support a wide range of endeavors and aspirations, even in the face of dilapidated and poor conditions and government neglect. These pursuits have continued for decades. Most informals strive to "do the right thing" under harsh conditions: children are nurtured, clothed, and fed and sent to school (many of them charge fees in slums, as public schools are located elsewhere).

## MOBILE PHONES: "THE NEW TALKING DRUMS" OF EVERYDAY AFRICA

Following the first mobile call in Africa in 1987, the uptake of mobile phone technology in the region is taking place at an amazing speed and scale, leading many to describe the adoption and adaptation as "a mobile phone revolution" (Carmody 2010; Porter 2012). From slums to remote villages, Africans use mobile phones in hundreds of ways beyond voice communication: to email, bank, text, send money, tweet, send and receive public health or emergency messages, monitor elections, check market prices for agricultural commodities, search the Internet, and access educational content (Etzo and Collender 2010) (Fig. 6.7). Africans also perform everyday activities such as taking photos, making films, watching television, and accessing Facebook. Activists can take advantage of the range of functions on mobile phones to plan campaigns and respond quickly to unfolding events.

The mobile phone market in Africa is thriving, with approximately 642 million phones and a 65% penetration rate (penetration rate is the number of phone subscribers per 100 people) (Biosca 2012). Subscription rates range from 70% in Reunion to 1% in Burundi (Carmody 2010). Kenya, South Africa, Nigeria, and Tanzania are the early and principal adopters of mobile phone in the region.

**FIGURE 6.7** Bedouins using a Laptop and a Mobile Phone in the Sahara. *Source:* © Philippe Lissac/Godong/Corbis. Corbis image 42-46297569.

This expansion of technology represents a dramatic reversal of Africa's "black hole" in terms of information networks that prevailed in the 1990s, when Manhattan accounted for more landlines than the entire African region (Castells 1998). However, the use of mobile phones differs from that in the Global North: many users operate within a mobile footprint, accessing phones on an as-needed basis (at mobile phone kiosks or village phone desks), and many subscribers share phones, some possessing their own or even several SIM cards, and others rent SIM cards to use in phones borrowed from family members or friends or at commercial phone points. Therefore, there are many more users than subscribers in Africa. In many contexts in the region, mobile phones are more a communal instrument than a personal instrument, but this is changing as intense competition among providers is driving prices down. Mobile phones are progressing from simple communication tools into service delivery platforms. Mobile coverage is increasing and mobile platforms are enlarging. The rollout of 4G services and the introduction of affordable smartphones (under US$50–$90) are expected to have a further transformative effect in the coming years. There are projected to be 160 million mobile broadband connections by 2016 (Deloitte 2012).

Most Africans still use prepaid services with low-end phones, but a high cost per unit of time means that people have developed creative ways of communicating. Africans have devised "beeping" or "flashing" as a means of communicating via phones: the caller dials but hangs up before the call is answered so as to avoid a phone unit charge. Typically a beep is used to send a message such as "call me back," but callers and receivers can established a predetermined message such as "pick me up" or "I have arrived at my final destination." "Flashing" is a way to promote relationships at a low cost: it reminds distant call recipients of ties and of obligations with minimal effort (Porter 2012). For example, rural dwellers have a way of communicating with urban migrants or relatives (who are assumed to have more resources) (Porter 2012). By giving phones to rural relatives, urban dwellers can change extended-family dynamics. It can discharge some obligations to rural kin and reduce the number of journeys to rural areas. It can also strengthen family ties across space and can encourage more rural dwellers to come to the city in search of jobs (Porter 2012). Information technology can promote communication among a vast Africa network—not only those located in the region, but increasingly incorporating the Diaspora.

Economically empowered individuals such as family heads and business owners are often "beeped," signifying their centrality in social and financial networks and their ability to place calls. Mobile companies do not approve of beepers jamming networks, given the high network priority for voice, however. Telecoms such as Vodacom Tanzania responded by rolling out a beeping replacement with a free text message, and subscribers can send three "please call me" messages per day. Despite such creative ways to minimize phone charges, these costs remain a burden on poor people: poor South Africans spend 10% to 15% of their incomes on mobile phones, and it is unclear whether their livelihood strategies benefit in a proportionate way (Carmody 2010).

People have even devised creative ways to deploy phone and prepaid airtime vouchers. Sending prepaid vouchers between an urban and rural location has become a vehicle for transferring money. It works this way: an individual wanting to remit money to the village buys airtime, but rather than loading it into the phone, he or she calls the shared village phone operator and reads the code. The operator purchases the airtime for the village phone, and the transfer is complete when the operator hands over the money, minus a commission, to a specified recipient.

The informal economy is a crucible for mobile phone innovation and entrepreneurship. Long-distance trade networks have always been important in many African contexts, and mobile phones create new possibilities for facilitating communication within supply chains. They are very beneficial in situations where long-distance travel is onerous and time-consuming and road accidents are common. Face-to-face contact is particularly significant in African social and business contexts, so phone communication serves as a substitute, in part, for travel, but it has made a greater impact on eroding the role of intermediaries in conveying messages. Overall, it is taking time to consolidate trust in the technology in African business contexts where legal contacts do not exist and social capital is based on face-to-face contact. However, major industrial players are converging on this space and, in time, industry and consumers may feel assured that their data are secure, guaranteed, and well administered. Importantly, the technology allows for great efficiencies in exchanging all kinds of information (e.g., market information for agricultural commodities and between rural herdsmen and urban livestock traders, feedback about unscrupulous operators, updates from key buyers, and customer promotions), it reduces travel (and the carbon footprint), and it leads to a boost in productivity. Deloitte (2012:45) reports that an increase of 10% in mobile penetration increases gross domestic product growth by 1.4% in Africa.

The ubiquity of mobile phones is being matched by an expansion of value-added services and m-commerce. M-commerce is helping Africans to make the transition from an exclusively cash economy in the informal sphere and one in which savings took place under the mattress, in jars, and in secret hiding places to an m-money economy, where funds can be accessed, saved, and transferred in digital space. Mobile money systems are being introduced to Africa's unbanked population (roughly 80% in 2013) and offer great potential for financial inclusion (Ericsson 2012). One of the most innovative developments is the mobile wallet, which creates a smartphone-based equivalent of a physical wallet, a cloud or SIM-based collection of financial, personal, and identification information that an individual wants to carry based on his or her budget and technical abilities. Mobile wallets use digital currency or bitcoins, and they are emerging as the leading edge of grassroots financial revolution. Uptake has been fast in Kenya and Tanzania (Fig. 6.8).

A large informal economy has emerged to support the mobile sector, with individuals selling airtime; charging, fixing, unlocking, and renting devices; and selling accessories. Mobile services that travel to customer bases by bike with phones and spare batteries have emerged in rural areas. In addition to adding vibrancy to the informal economy, a critical mass of technologists, entrepreneurs, and activists experimenting with new ideas in the mobile sector is emerging. Mobile operators also invest in civil works, extending their networks to rural areas, building roads, and bringing electricity to remote areas.

**FIGURE 6.8** Mobile Banking, Nairobi, Kenya.
*Source:* Corbis image 42-46179058.

Mobile money transfers (m-banking) has put Eastern and Southern Africa within the forefront of the global mobile money industry. The M-Pesa ("m" stands for mobile and "pesa" is the Swahili word for money) mobile banking system in Kenya is at the vanguard. Open to people who do not have bank accounts, it has developed into a system that is accepted by over 100 partnering organizations (e.g., banks, media houses, government agencies, microfinance institutions, and insurance companies) (Fig. 6.9).

M-Pesa allows its customers to access account information, buy airtime, save money, receive payments, pay electricity bills, receive/pay salaries, purchase goods, make donations, and withdraw money at ATMs, and UK citizens can use the system to send remittances (up to UK£1,000 per month) to relatives in Kenya. From 2007 to 2012 the M-Pesa customer base has expanded to cover 17 million users (more subscribers than bank-account holders in Kenya).

Mobile phones have become deeply embedded in many aspects of Africa society, and the swift uptake has led to the description "the new talking drums of everyday Africa" (De Bruijn, Nyamnjoh, and Brinkman 2009). Upwardly mobile money is turning parts of Africa's informal economy into a cashless society.

There are some creative examples of rural Africans using mobile phones to link with distant clientele. Van Beck (2009) documents an indigenous healer from rural northern Cameroon who is using the technology to expand his business operations beyond his rural

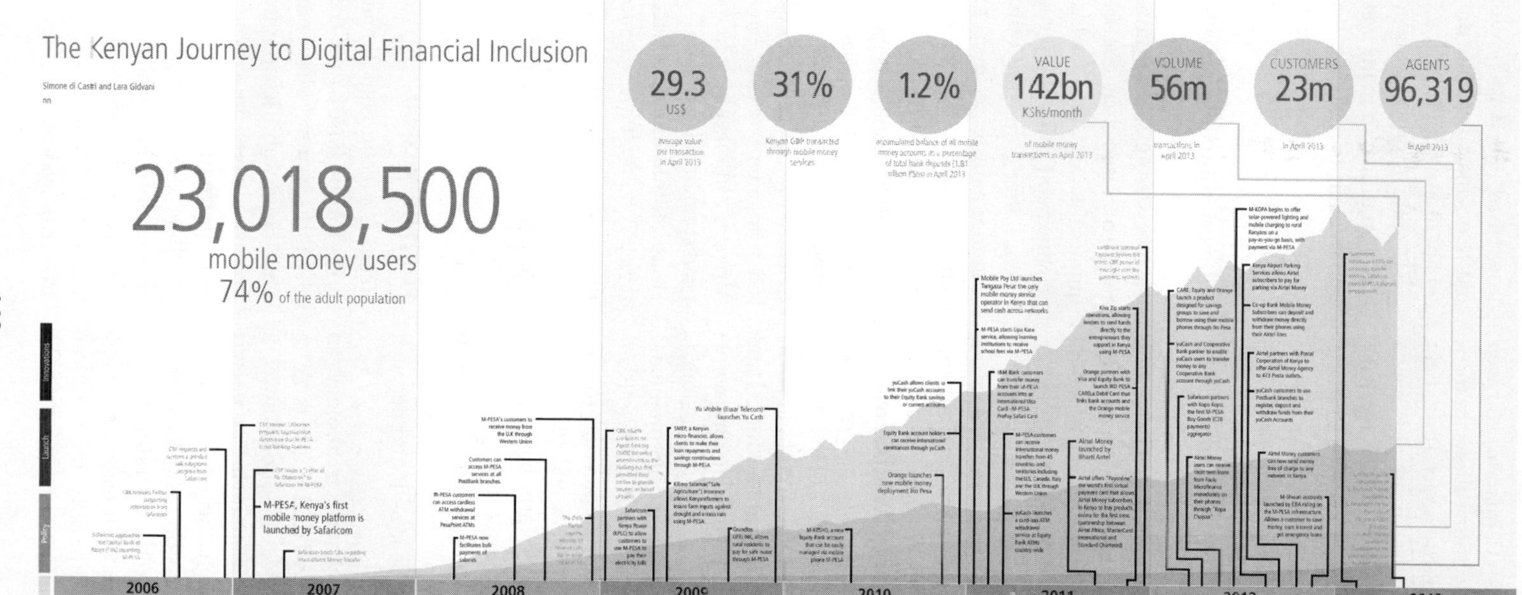


**FIGURE 6.9** Kenya's Path to Digital Financial Inclusion.

locality to Douala and Yaoundé (Cameroon) and even as far afield as Paris. The mobile phone allows for treatment sessions without face-to-face contact. The patient describes the ailment and after the consultation pays a fee that is transferred via the phone. Upon receiving the payment, the healer mails the products; each package is labeled with a different number. When the patient receives the packages, he or she is instructed to phone the healer; the patient is informed what the numbers stand for and is given detailed instructions on how to administer the products. The client is expected to memorize the instructions so that the secret healing formula is not written down. The healer checks back with the patient in two weeks; if the patient is feeling better, he or she is expected to send a final "gesture of appreciation" based on what he or she can afford.

This is still early days to assess the full social impact of mobile phones and whether they are truly benefiting the poor. In rural Africa mobile phone services represented the first modern telecommunications infrastructure of any kind. Mobile phone use has become part of everyday life in many African countries, even in poor households (Porter 2012). Without doubt mobile phones connect rural and urban Africans and extend family responsibilities by facilitating remittance flows to relations like never before. Obviously, mobile phones make people feel more important and connected, reduce rural isolation, offer numerous advantages to workers and the self-employed, promote digital inclusion, allow autonomous and continuous learning, and enable users to perform basic Internet research and use common applications (e.g., spreadsheets, text editors). Mobile services allow workers to search for employment with online job postings instead of depending on word of mouth and personal networks.

Pilot projects have shown that the technology can be applied to enhancing literacy. For example, the Yoza Project (South Africa) developed an episodic novel in English and isiXhosa. Readers were invited to interact with an unfolding story, through discussions about the evolving plot, and vote on the story, as well as submit a written sequel for a competition. During the first seven months, two stories had been read over 34,000 times on mobile phones, surpassing South African bestseller sales criteria of 4,000 units. By 2013 the project had added 28 novels, five Shakespeare plays, and 11 poems, suggesting the potential of technology-enhanced learning in Africa.

Despite improvements in information flow, access is still stratified, in many ways mirroring the way that employment is stratified. Employers and middlemen own smartphones, while many non–full-time employees, apprentices, and family laborers cannot afford a basic handset, let alone the running cost. In addition, this greater use of information technology appears to be enabling larger businesses and players to capture customers from smaller firms (Carmody 2010). Fears are also raised that the bitcoin currency is more untraceable and may facilitate more under-the-radar activities and illicit transactions. On balance, mobile phones offer great transformation potential, but challenges remain to scale up many of the initiatives in banking, health, education, agriculture information, and to provide more locally based content.

M-money goes a long way toward enhancing financial inclusion by assisting Africa's unbanked population to engage in convenient and affordable money transfers (the transfer fee in Kenya is US$0.04). It provides an alternative to bank transfers and couriers. Africans no longer have to walk long distances to their nearest financial institution or fear being turned away. Using mobile money services means that traders and travelers do not have to conceal their cash and run the risk of robbery and theft. It is clear that m-commerce is being primed to occupy an even bigger part of everyday lives.

## USHAHIDI: BUILT IN AFRICA FOR THE WORLD

Kenya is a technological and creative hub for applications for the base of the economy. It is home to the Ushahidi system (which means "testimony" or "witness" in Swahili), an open-sourced platform for grassroots activism, citizen journalism, and geospatial information. Developed by an ad hoc group to facilitate grassroots reporting of violence following the elections in Kenya in 2008, it collected eyewitness reports of violence reported by email and text messages and placed them on a Google map. The basic, low-cost system was developed by four Kenyans, experts in either software development or blogging (Erick Hersman, Juliana Rotich, Ory Okolloh, and David Kobia), and their tech-savvy networks during a very short period (see the WhiteAfrican [2010] blog on the three days of intense collaboration). Since 2010 iHub (Nairobi) has provided physical office space for the Ushahidi organization.

Today Ushahidi is a nonprofit tech company, specializing in developing free and open-source software for information collection, geovisualization, and interactive mapping. Ushahidi builds tools for democratizing information, increasing transparency, and lowering the barriers for individuals to share their stories. Through Crowdmap.com, Swiftly.org, and accompanying mobile applications, Ushahidi is expanding its global footprint and making crowd-sourcing tools available at no cost (Fig. 6.10).

The platform is capable of accepting input from SMS text messages and/or Internet-based Web forms while simultaneously creating a temporal and geospatial archive of events. It is now the most widely used platform for local-level geospatial reporting.

Ushahidi has been deployed throughout Africa and the world in 40,000 projects in 157 countries (Ushahidi 2013). It has been deployed within Africa to monitor elections and xenophobic outbursts, to track the availability of medicines, and as part of emergency response efforts. It has given rise to a new phenomenon, the African citizen-journalist, who is both an observer and an active participant in events. Outside of Africa, it played an important role in emergency response efforts in Haiti, Afghanistan, and Chile. For example, after the 2010 Haitian earthquake, individuals sent text messages reporting the locations of survivors; these locations were mapped and used in search-and-rescue efforts. Global corporations have also deployed the technology. For instance, Al-Jazeera used the platform to monitor elections in Kenya, obtain citizen-journalist reports from Somalia, and to gather Ugandans' reactions to the Kony 2012 campaign.

**FIGURE 6.10** Ushahidi Logo.

Ushahidi is planning on releasing a BRCK device (a backup generator for the Internet) in 2013. This US$215 device (which includes postage to anywhere in Africa) caters to the unpredictable state of Internet connectivity in much of Africa. This smart, rugged device (which can withstand humidity and dust) allows continuous Internet connectivity (switching seamlessly between Ethernet, Wi-Fi, and 3G/4G connections, allowing users to switch from one network to another) and creates a hotspot for multiple devices (up to 20) while plugged in or running on battery power. Over US$172,107 was raised on the crowdfunding site Kickstarter to develop the device, showing how global funding can be raised for technological developments for the base of the African informal economy.

## DESIGNING SUSTAINABLE PROTOTYPES FOR AFRICA'S INFORMALS

Considerable effort is being devoted to designing sustainable or green innovations for Africa's extensive informal environments. From toilets to solar panels, from fire-retardant building materials to the integration of urban gardening into dwelling units, diverse efforts are under way to improve the lives of the urban poor, reduce their dependence on costly inputs (building materials, long-distance fuel, and food imports), and ease negative environmental impacts (e.g., pollution of air, water, and soil). International competitions to champion new innovations are becoming common, and donors are committing support for these pilot projects. For instance, the Bill and Melinda Gates Foundation funded several seed projects to reinvent the toilet and to develop a new sanitation solution for the world's poor (typically harnessing solar power and clean water solutions) (Gates Foundation 2014).

Housing models for the poor have been prioritized (in much the same way that housing has been the focus of traditional urban development policies). There are increased efforts to mainstream sustainable building designs (including methods of construction and the incorporation of local materials) to address various aspects of slum living conditions. Three green prototypes are the US$300 shacks, earthships, and ishacks.

The US$300 housing challenge—to design and build more sustainable dwelling units for the poor—received international acclaim. The challenge, launched by *Harvard Business Review* in 2010, was the brainchild of Vijay Govidarajan, a Dartmouth business professor, and Christina Sarkata, a marketing consultant. The *Harvard Business Review* set up a competition "to design, build, and deploy a simple dwelling which keeps a family safe from the weather, allows them to sleep at night, and gives them a little bit of dignity" (Govindarajan 2010:1). The objective was to replace unsafe and unsanitary dwelling units with a sustainable prototype solution that could be mass-produced and built from local materials. Students, architects, and businesses competed to design the best prototype of a one-room prefabricated dwelling, equipped with solar panels, water filters, etc. The concept won the Thinkers50 Breakthrough Idea Award (2011), an Oscar-like award for innovative thinkers' ideas with potential to change the world.

A second effort is the earthship model. Largely the brainchild of environmental activist Michael Reynolds, the prototype is being introduced into Africa. Built by sweat equity, earthships promote living within an enclosure that integrates the environment. Using solar energy (no heating or air-conditioning is required) and local harvesting of water and recycled materials (old tires packed with dirt, aluminum cans, discarded appliances), these off-the-grid structures eliminate reliance on public utilities (electricity and water) and fossil fuels. African examples include the Sierra Leone Earthship School outside of Freetown (established in 2011) and the Kapita Earthship Community Center in northern Malawi (established in 2013). The Malawi project is the most ambitious African undertaking, with plans to build a rural community center (incorporating a community bank, a library, a health facility, offices, and a food bank) to serve 38 villages (5,000 people) in an area lacking basic infrastructure, off the development grid, and with a low presence of nongovernmental organizations (NGOs).

A third example is the ishack ("improved shack") initiative developed by the Sustainability Institute of Stellenbosch University in collaboration with the City of Stellenbosch (South Africa) (Fig. 6.11). Ishack models have been introduced in Enkanini, an informal settlement in Stellenbosch. The Bill and Melinda Gates Foundation has provided funding to retrofit 100 shacks, and there are plans to build ishacks in Ghana and Tanzania.

**FIGURE 6.11** An Ishack. Photograph: Sustainability Institute, Stellenbosch University.

The ishack model is a sustainable modification of the basic corrugated-iron shack. The design strives to protect occupants from extreme temperatures, and a central feature is a roof solar panel to provide basic energy needs (with capacity for two interior lights, a motion-sensitive exterior security light, and a cell phone charger). Windows are strategically placed for optimal air circulation and sunlight heating, while the roof is sloped so that rainwater can be harvested during winter months. Recycled cardboard boxes and "Tetra Pak" containers are used for insulation between the exterior zinc surface and the interior, while a flame-retardant paint is used to lessen the risk of fire. Inside the shack, rows of recycled bricks create a sturdy flooring base, serving as a "thermal mass" and protection against extreme temperature change. Excluding the solar power system, an ishack costs approximately US$660.

However, critics argue that the techno-housing-fix green experiments are divorced from the realities of social process, culture, and power (Swilling and Annecke 2012). They may provide a straightforward business solution, but they fail miserably as social solutions. They promote the misconception that a ready-made standardized housing strategy "fits all" of the poor in Africa. Housing prototypes cannot be inserted in a cut-and-paste logic, and the complexities of property demarcation and land ownership rights are not addressed. In many slums, some dwellings accommodate extended families, and in older slums many buildings have been in families for generations. Importantly, informal dwellings, for the most part, are not self-contained residential units: many function as industrial workshops, classrooms, and shops, and various additions provide accommodation for paying tenants. Such structures facilitate trade, production, service, and residential activities—often simultaneously. Indeed, informal construction is an important neighborhood economic activity, providing employment to shack builders, providers of recycled building materials, electricity installers (informal), and informal real estate brokers.

The availability of mass-produced houses may well turn the poor into consumers and away from being self-builders. Such a change would transfer the opportunity of coming up with creative solutions away from the poor to large corporations, elite universities, external entrepreneurs, and global innovators. Replacing incrementally self-built houses with ready-made prefabricated models could be likened to the introduction of the mass-produced British bungalows into colonial Africa. Indeed, the 20th-century experience with public housing experiments in the United States and Europe provides a telling lesson of good ideas and good intentions going awry. Housing experts since the classic work of John Turner (1976) have advocated that the poor should be deeply involved in building their own houses (deciding what to build and how) and that the state, NGOs, and the private sector should be enablers rather than direct providers of houses (Yeboah 2005).

### BOX 6.2 COMMUNITY-LED ENUMERATIONS AND MAPPING: COUNTING LIVES AND BUILDING COMMUNITY IN KIBERA, KENYA

There is a growing trend of civil society organizations working with the urban poor to engage in community-led enumerations, mapping, and other efforts to make informal life and practices more visible to mainstream society. Obviously, this is a complex relational topography that encompasses relationships among NGOs, local branch affiliates, and members of the urban poor, who constitute a community of diverse identities, interests and goals.

There is, however, some remarkable evidence of particular slum communities organizing within this context to use the city as a platform to connect their housing issues with the global world beyond. An important example of community-led mapping initiative took place in the Kibera slum in Nairobi.

Map Kibera is an open-source community map that has developed into a comprehensive interactive community information effort. For many years Kibera, located on 550 acres (223 hectares) of government-owned land, 3.1 miles (5 km) southwest of Nairobi's central business district and minutes away from UN-HABITAT's headquarters, was an illegible slum (missing from official maps) and therefore an invisible area to urban planners and the public. Population estimates varied wildly, ranging from 200,000 to more than 1 million. The dense and overcrowded settlement narrative dissuaded any attempt to tackle slum upgrading on the basis of its large population.

Historically, the land was given to members of the Nubian tribe (the word *kibera* in the Nubian language means forest or jungle) during the colonial era in exchange for their service in the British Army, but land tenure was never formalized. Over the years other tribes moved into the area and paid rent to Nubian landlords, and over time Kibera became ethnically diverse. It developed into a vibrant community often thriving with numerous informal enterprises, health clinics, schools, churches, mosques, movie theaters, water vending points, and pay-showers and latrines. Until recently, the area was not classified accurately on official maps: its land-use designation—as forest—was completely out of date.

Although it was heavily studied and represented in fiction (e.g., Michael Holman's [2007] *Last Orders at Harrods: An African Tale*) and film (e.g., John le Carré's [2005] *The Constant Gardener* included several scenes filmed on location in Kibera), public information was sparse and few research findings were shared directly with the community. The media typically reported only violence and evictions in Kibera, so that internal community resources remained largely unknown to the public as well as to the residents. The dynamism of the community meant that any preexisting map of the community's layout was outdated: the adding of new structures and the tearing down of older ones meant that the spatial layout was always evolving.

In 2009 13 trained members of the community initiated Map Kibera to produce an open-source community map in real time. Initial funding was secured from Jumpstart International, a U.S. NGO, and from local development partners, Social Development Network, Carolina for Kibera, and Kibera Community Development Agenda. Surveying with GPS devices enabled the community team to record spatial data and produce a free and open digital map of the community. Recording the settlement's footprint in terms of buildings and infrastructure using handheld GPS units uncovered an intricate community of buildings,

(*Continued*)

## BOX 6.2 (*Continued*)

**FIGURE 6.12** Map Kibera. *Source:* © OpenStreetMap.

firms, and services. Once that reference network was created, the mapping process was divided into four key themes: health, security, education, and water. Subsequently the map incorporated initiatives reflecting community concerns such as mapping safe and unsafe pedestrian routes for girls (bringing attention to lighted/unlighted spots and safe places for girls to access gender-targeted resources, such as literacy, counseling, and vocational training). This community initiative also contributed to other goals, such as raising awareness about women's safety (violence against women and AIDS/HIV vulnerability). The girls' security layer of Map Kibera illustrates the kind of practical information that can be produced by members of the community to enhance safety (Fig. 6.12).

Map Kibera has stimulated the development of offshoots such as Kibera News Network and SMS monitoring of services and incidents. Kibera News Network is a citizen video journalism program where citizen-reporters use hand-held video cameras to cover stories not covered by the mainstream media (e.g., local electricity disruptions). These Kibera projects are not just about the products: most importantly, they empower slum dwellers by allowing them to participate and create local knowledge that begins to change the broader political context that surrounds this community. Participatory and open technologies are an important tool to help marginalized communities to amplify their voice. The success of Map Kibera has stimulated grassroots efforts to map other neighborhoods in Nairobi, Dar es Salaam, and Kampala, and an ambitious community project is under way to map the entire country of Swaziland.

The first Map Kibera recorded a population of 170,070, but the 2012 version shows that the number of residents has increased to 250,000. The initial population number was dramatically lower than the figure assumed by the City Council of Nairobi and international development organizations, who commonly portrayed Kibera as one of the largest slums in Africa. Map Kibera's major contributions were to make the community more legible to residents, planners, civil society, and other organizations and to democratize spatial information and to allow the community to create their own content. The Wiki platform offers a mechanism for initiating a discussion about more inclusive spatial planning.

## BOX 6.3 SLUM TOURISM AND SHACK CHIC: ARE THEY GOOD IDEAS?

Slums are not typically thought of as tourist sites nor shacks as a fashion and design style icon. Nevertheless, slum and township tours (also referred to as "reality and poverty tours") have emerged as niche tourism in recent years. From community tours in Kibera to showcasing township eating establishments such as Mzoli's Butchery in Gugulethu (Cape Town) to historical walking tours in Ga Mashie (Accra) to overnight stays (e.g., township B&Bs in Soweto) to bicycle tours and adventure sports (e.g., bungee jumping in Soweto), outsiders have a growing number of opportunities to observe and engage slum communities at graduated levels, depending on the program. Kibera, for example, is marketed as "the friendliest slum in the world" by a local tour operator.

These mainly for-profit programs seek to provide learning excursions that offer some form of "meaningful" encounter between the tourist and the host. Some programs explicitly aim to move beyond tourists and visitors toward a delegation model, mindful of providing greater sensitivity in the encounter so that there can be greater transformative possibilities. A deeper level of engagement is provided in "volun-tourism," which seeks reciprocal benefit for both tourists (e.g., gaining cultural experiences) and host community residents (e.g., receiving assistance to repair or construct housing or preserve vital resources). Slum-dwelling hosts are recruited (and compensated) to provide delegates with the opportunity for social gatherings, to present their domestic and exterior spaces in their own voice, and to showcase general life and community initiatives (which often involve microenterprise projects); some programs also include short home-stays. Specialized programs organize micro-finance excursions for matching donors and recipients in socially responsible investments. More commonly, donors connect with communities in virtual space. For example, Kiva is an U.S.-based NGO that solicits loans of US$25 and above and brings together lenders and needy participants to alleviate poverty. Socially responsible investors have supported Kiva-sponsored projects in many African countries (e.g., Benin, Cameroon, The Republic of Congo, DRC, Kenya, Mali, Rwanda, South Sudan, Togo, Uganda, and Zimbabwe.)

Along with poverty tourism, an interior design style emanating from South African slum communities called "shack chic" is being promoted. Craig Fraser (2002), a freelance interior design photographer, captured artful images of Cape Town shack interiors in a book entitled *Shack Chic*, which has become a coffee-table best-seller and an inspiration for curio and clothing designers (Fig. 6.13) (Grant 2012). Shack chic curios (e.g., radios made from recycled materials such as wire and Coca-Cola bottle tops) are sold in various tourist shops and markets across Africa. This styling provides an illustration of how the patched-together life of slum dwellers finds artistic expression within the domestic interiors of some of their homes. The display of color and vibrancy by using recycled materials with an Andy Warhol flair provides a strong visual. The most striking examples of shack chic have become staging points on township tours in South Africa, whereby tourists facilitate and reproduce this style in the global imagination. Shack chic is a new culture-scape involving the spatial reimagining of shack living, leisure, and tourism that provides an alternative to conventional depictions of dirt, poverty and suffering.

However, there are serious debates about slum tourism and the repackaging of slum interiors as shack chic. Poverty chic treads the finest of lines and can easily be perceived as insensitive and complicit in doubly marginalizing informal subjects. There are strong arguments against poverty safaris, the aestheticization of the domestic spaces of poverty, and unleashing informals into a branded world where their communities become yet another site to be consumed. A counter-argument is that supporting poor-inspired entrepreneurial and tourism-focused projects creates badly needed jobs, allows money to flow to marginalized communities, emphasizes positive and creative practices, and begins to deflect entrenched negative stereotypes.

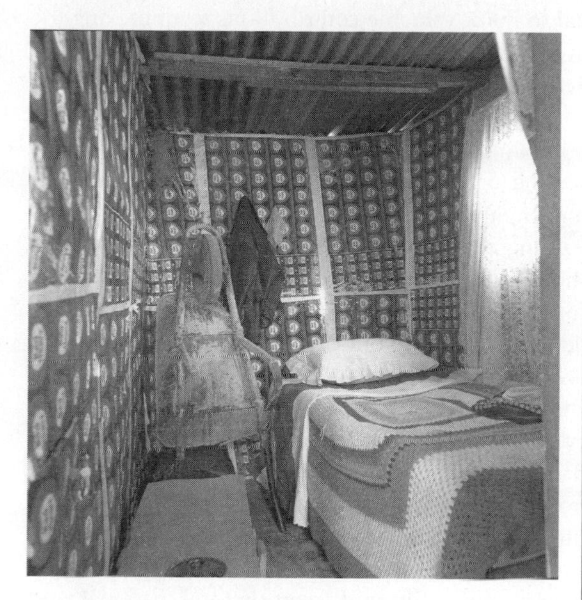

**FIGURE 6.13** "Shack Chic". Photograph: Craig Fraser.

## CONCLUSIONS

The informal economy can no longer be considered a temporary or abnormal phenomenon. It provides employment to almost half of all workers in the world (Neuwirth 2011), and it supports most Africans. Far from being a disorganized residue of petty capitalism, parts of the informal economy are vibrant, creative, and innovative. There has been an unrealistic expectation that economic development would be highly organized and subject to bureaucratic regulation and measurement, with local, national, and state government at the center. Instead, economic development in Africa has taken place outside of these conditions, and informality remains misunderstood by governments, donors, and the general public.

Mobile phone providers and mobile phone application entrepreneurs are embracing the business prospects of deploying their technologies in the informal economy. The uptake in mobile phones in Africa has been unprecedented. It has enhanced freedom but has also changed the way social networks operate (most visibly expressed by "a lot of more talking and a lot less walking"). Several Africa countries are at the vanguard of embracing a mobile technological future. There are more than 50 mobile tech hubs in Africa where entrepreneurs, innovators, and investors are focusing on finding local solutions for the most pressing regional issues. Kenya stands out as a center of mobile innovation; it represents the origin of the Ushahidi platform, which has been globally adopted. The uptake of mobile wallets in East Africa has been so fast that some speculate about a bitcoin (cashless) society emerging, with informals at the leading edge. In addition, Africans are using available technologies to engage in community mapping, blogging, and first-hand reporting. However, despite the hype about a grassroots democratic transformation, the jury is still out on whether mobile phone technology can facilitate an alternative type of politics and social change. To date, Africans seem to use the technology more as consumers and learners than political activists—but one monumental political occurrence could change everything.

The informal economy is not the problem it was made out to be in the past. Instead, it offers the potential to make valuable contributions to improving the lives of poor Africans. Given the economic contributions of the informal economy, governments should be developing policies that recognize its importance, restricting and regulating it when necessary but mostly seeking to increase productivity and improve the working conditions of informal workers. Credit and skills training are badly needed supports, but governments should not neglect their roles in providing land, workspace, and infrastructure to their poorer citizens. The development of mobile money and mobile wallets for the poor has the potential to be very transformative and to bring badly needed financial inclusion. However, not all parts of the informal economy are dynamic, and it is not clear how to separate dynamic from survival mechanisms and how to bolster the livelihoods of the poorest while also improving the most dynamic parts. Informal livelihoods are complex, dynamic, and highly contingent on local, national, and global processes.

At the same time, the informal economy should neither be romanticized as an entrepreneurial and creative hub nor accepted as a necessary reservoir of surplus labor. We need to be careful not to construct a myth of survival whereby the poor are expected to implement survival strategies, based on their endless capacity to work, to consume less, and to depend on self-help networks and/or a myth of entrepreneurship whereby the poor earn their way out of poverty by succeeding in business ownership. Policymakers and business elites who show a strong preference for concentrating on big businesses, privatization, and foreign investment can misuse such myths. We must not perpetuate a duality between the formal and informal economies where a minority enjoy disproportionate access to resources while the vast majority, though eager to participate, are excluded by virtue of decades-old policy biases and entrenched negative and misinformed stereotypes. Moving toward a more bottom-up process and linking more with community-based initiatives and away from top-down, overbearing control is urgently needed.

## REFERENCES


AlSayadd, N. 2004. "Urban Informality as a 'New' Way of Life." In *Urban Informality: Transnational Perspectives from the Middle East, Latin America and South Asia*, eds. A, Roy and N. AlSayyad, pp. 7–30. Lanham: Lexington Books.

Becker, K. 2004. Informal Economy. Fact-Finding Mission. SIDA. Department for Infrastructure and Economic Co-Operation. Available at http://rru.worldbank.org/Documents/PapersLinks/Sida.pdf (accessed August 9, 2012).

Biosca, M. 2012. Opportunities in Africa and the Middle East. Available at http://www.gsma.com/publicpolicy/wp-content/uploads/2012/03/mea12atkearney-africapresentation.pdf (accessed August 15, 2012).

Breman, J. 1996. *Footloose Labour: Working in India's Informal Economy*. New York: Cambridge University Press.

Brigden, K., Labunska, I., Santillo, D., and P. Johnston. 2008. *Chemical Contamination at E-Waste Recycling and Disposal Sites in Accra and Korforidua, Ghana*. Greenpeace report. Available at http://www.greenpeace.org/international/en/publications/reports/chemical-contamination-at-e-wa/ (accessed August 15, 2012).

Brown, A., ed. 2006. *Contested Space. Street Trading, Public Space and Livelihoods in Developing Cities*. Warwickshire: Intermediate Technology Publications.

Carmody, P. 2010. *Globalization in Africa. Recolonization or Renaissance?* Boulder, CO: Lynn Rienner.

Castells, I. 1998. *End of Millennium*. Malden: Blackwell.

Castells, M., A. Portes, and L. Benton. 1989. *World Underneath: the Origins and Effects of the Informal Economy.* Baltimore: Johns Hopkins University Press.

Crankshaw, O., A. Gilbert and A. Morris. 2000. "Backyard Soweto." *International Journal of Urban and Regional Research* 24(4):841–857.

Davis, M. 2006. *Planet of Slums.* New York: Verso.

De Bruijn, I., Nyamnjoh, F. and I. Brinkman (eds). 2009. Mobile Phones: The New Talking Drums of Everyday Life. Langaa, Cameroon and Leiden, Netherlands: African Studies Centre.

Deloitte. 2012. *Sub-Saharan Africa Mobile Observatory 2012*. Available at http://www.gsma.com/publicpolicy/wp-content/uploads/2012/03/gsma_ssa_obs_exec_web_10_12.pdf (accessed August 26, 2013).

De Soto, H. 1989. *The Invisible Revolution in the Third World*. New York: Basic Books.

———. 2000. *The Mystery of Capital: Why Capitalism Triumphs in the West and Fails Everywhere Else*. New York: Basic Books.

Ericsson. 2012. M-commerce in Sub-Saharan Africa. An Ericsson consumer insight summary report. Available at http://allafrica.com/download/resource/main/main/idatcs/00041763:cb6d314bacfd94ea5457b8169c3e6d8c.pdf (accessed August 26, 2013).

Etzo, S., and G. Collander. 2010. "The Mobile Phone 'Revolution' in Africa: Rhetoric or Reality?" *African Affairs* 109(437):659–668.

Fraser, C. 2002. *Shack Chic: Art and Innovation in South African Shack-Lands*. Cape Town: Quivertree Publications.

Gates Foundation. 2014. Water, Sanitation & Hygiene: Reinvent the Toilet Challenge. Available http://www.gatesfoundation.org/What-We-Do/Global-Development/Water-Sanitation-and-Hygiene (accessed March 7, 2014).

Gough, K., G. Tipple, and M. Napier. 2003. "Making a Living in African Cities: The Role of Home-Based Businesses in Accra and Pretoria." *International Planning Studies* 8(4):253–277.

Govindarajan, V. 2010. "The $300 House: A Hands-on Lab For Reverse Innovation? *Harvard Business Review Blog Network*." Available at http://blogs.hbr.org/2010/08/the-300-house-a-hands-on-lab-f/ (accessed March 7, 2014).

Grant, R. 2009. *Globalizing City. The Urban and Economic Transformation of Accra, Ghana*. Syracuse: Syracuse University Press.

———. 2010. "Working It Out. Labour Geographies of the Poor in Soweto, South Africa." *Development Southern Africa* 27(4):595–612.

———. 2012. "Material Cultures of Domestic Residential Spaces in Urban Africa." In *International Encyclopedia of Housing and Home,* editor in chief, S. Smith, pp. 200–205. London: Elsevier.

Grant, R., and M. Oteng-Ababio. 2012. "Mapping the Invisible and Real 'African' Economy: Urban E-waste Circuitry." *Urban Geography* 33(1):1–22.

Hart, K. 1973. "Informal Income Opportunities and Urban Employment in Ghana." *Journal of Modern African Studies* 11(1):62–89.

Holman, M. 2007. *Last Orders at Harrods: An African Tale*. London: Abacus.

Koolhass, R. 2003. *Lagos.* New York: Icarus films.

Lepawsky, J., and C. McNabb. 2010. "Mapping International Flows of Electronic Waste." *The Canadian Geographer* 54(2):177–195.

Lindell, I. 2010. *Africa's Informal Workers: Collective Agency, Alliances and Transnational Organizing*. New York: Zed Books.

Maloney, W. 2004. "Informality Revisited." *World Development* 32(7):1159–1178.

Meagher, K. 2003. "A Back Door to Globalization? Structural Adjustment, Globalization & Transborder Trade in West Africa." *Review of African Political Economy* 30(95):57–75.

Moser, C. 1978. "Informal Sector or Petty Commodity Production: Dualism or Dependence in Urban Development?" *World Development* 6(9):1041–1064.

Myers, G. 2011. *African Cities. Alternative Visions of Urban Theory and Practice*. New York: Zed Books.

Neuwirth, R. 2011. *Stealth of Nations. The Global Rise of the Informal Economy*. New York: Pantheon Books.

Oosterbaan, C., G. Arku, and A. Asiedu. 2012. "Conversion of Residential Units to Commercial Space in Accra, Ghana: A Policy Dilemma." *International Planning Studies* 17(1):45–66.

Organisation for Economic Co-operation and Development. 2009. *Is Informal Normal? Towards More and Better Jobs in Developing Countries*. Paris: OECD.

Otiso, K. 2002. "Forced Evictions in Kenyan Cities." *Singapore Journal of Tropical Geography* 23(3):252–267.

Owusu, F. 2007. "Conceptualizing Livelihood Strategies in African Cities." *Journal of Planning Education and Research* 26(4):450–465.

Porter, G. 2012. "Mobile Phones, Livelihoods and the Por in Sub-Saharan Africa: Review and Prospect." *Geography Compass* 6(5):241–259.

Portes, A., and M. Castells. 1989. "World Underneath: Origins, Dynamics and Effects of the Informal Economy." In *World Underneath: the Origins and Effects of the Informal Economy,* eds. M. Castells, A. Portes, and L. Benton, pp. 11–37. Baltimore: Johns Hopkins University Press.

Potts, D. 2008. "The Urban Informal Sector in Sub-Saharan Africa: From Bad to Good (and Back Again?)." *Development Southern Africa* 25(2):137–152.

Rogerson, C. 2007. "'Second Economy' Versus Informal Economy: A South African Affair." *Geoforum* 38(5):1053–1057.

Roy, A. 2005. "Urban Informality. Toward an Epistemology of Planning." *Journal of American Planning Association* 71(2):147–158.

Schneider, F. 2005. "Shadow Economies Around the World. What Do We Really Know?" *European Journal of Political Economy* 21(3):598–642.

———. 2010. *City Life from Jakarta to Dakar. Movement at the Crossroads*. New York: Routledge.

Swilling, M. and E. Annecke. 2012. *Just Transitions. Explorations of Sustainability in an Unfair World.* New York: United Nations University Press.

Tokman, V. 1978. "An Exploration Into the Nature of Informal–Formal Sector Relationships." *World Development* 6(9–10):1065–1075.

Turner, J. 1976. *Housing By People: Towards Autonomy in Building Environments.* London: Martin Boyars.

UN-HABITAT. 2003. *The Challenge of Slums. Global Report on Human Settlement 2003*. Sterling: Earthscan.

United Nations Environmental Programme (UNEP). 2005. *Ewaste: The Hidden Side of IT Equipment's Manufacturing and Use*. Geneva, Switzerland: UNEP.

Ushadidi. 2013. http://www.ushahidi.com (accessed September 9, 2013).

Van Beck, W. 2009. "The Healer and his Phone: Medicinal Dynamics Among the Kapsiki/Higi of Northern Cameroon." In *Mobile Phones: The New Talking Drums of Everyday Africa*, eds. M. de Bruijin, F. Nyamnjoh, and I. Brinkman, pp. 125–133. Leiden: African Studies Center.

WhiteAfrican. 2010. "Making Ushahidi." Available at http://whiteafrican.com/2010/08/12/making-ushahidi (accessed September 8, 2013).

World Hunger Organization. 2012. World Hunger and Poverty Facts and Statistics. Available at http://www.worldhunger.org/articles/Learn/world%20hunger%20facts%202002.htm (accessed August 9, 2012).

Yeboah, I. 2005. "Housing the Urban Poor in Twenty-first Century Sub-Saharan Africa: Policy Mismatch and a Way Forward for Ghana." *GeoJournal*, 62 (1): 147–161.

## WEBSITES

E-Waste Watch Ghana. Facebook. http://www.facebook.com/pages/E-Waste-Watch-Ghana/128902477153239 (accessed December 5, 2013).

Frontline. *Ghana: Digital Dumping Ground.* http://www.pbs.org/frontlineworld/stories/ghana804/video/video_index.html (accessed December 5, 2013). Documentary about the e-waste trail and Agbogbloshie dumpsite in Accra, Ghana.

Kibera Tours. http://kiberatours.com (accessed December 5, 2013).

Kiva. http://www.kiva.org (accessed December 5, 2013).

Map Kibera. http://mapkibera.org (accessed December 5, 2013). Citizen mapping project in Kibera, Nairobi.

The $300 House. http://www.300house.com (accessed December 5, 2013).

CHAPTER 7

# MIGRATION

## INTRODUCTION

Africans have always demonstrated a high propensity for geographical movement, a trend that appears to be intensifying in the contemporary era. Archaeological, genetic, and historical records reveal evidence of significant mobility and movement across as well as out of Africa. The immense human consequences of 100 or so Africans walking from the region and beginning the processes of human settlement some 130,000 years ago, leading to the emergence of regional civilizations, is the most fantastic consequence of mobile Africans (Reader 1999). Africans' oral histories are embedded with narratives of mobility. The rapid urbanization that is well under way in the region is just one recent manifestation of African people on the move.

Different types of movements and different mobility rhythms occur. Nomadic pastoralists, healers, traders, artisans, artists, adventurous individuals, and others move back and forth between two places or among several locations. Some movements are linked to seasonal rhythms in hunting, fishing, market commerce, and, especially, agriculture (e.g., pastoralists relocate for water and fodder, and agricultural workers move to keep up with harvests and migrate to towns in the off season); some are for survival (e.g., during droughts) or related to a life-cycle process (e.g., students) or family income diversification (e.g., male laborers moving to work in mining). Other circular movements are linked to religious obligations and pilgrimages. For example, a well-defined overland route of West African Muslims traveling to Mecca has been important for a long time.

Mobility and migration go beyond the movement of people: there are nonhuman and nonmaterial aspects. Ideas and values also circulate and migrate. Migration dynamics have affected greatly the development of societies throughout the region, influencing economic, cultural, religious, and political systems. Migration has added and subtracted to, and divided, places depending on a range of other factors.

Many migrants are drawn to cities to take advantage of economic opportunities; often family members will join the migrant once he or she has secured an urban livelihood. The character of a region is an important consideration in the decision to move: the presence of affiliated ethnic/religious or political groups, relative safety from persecution, and tolerance are important pull factors. Although flows to the largest cities are slowing, flows to middle-sized and small settlements are gathering speed, and some migration back to rural areas is taking place. Still, the largest cities are particularly attractive to younger, ambitious migrants, those wanting less oversight from tribal chiefs or family heads, and those seeking alternative living options. Patterns of mobility appear to be increasing and undergoing some change: some older patterns are decreasing and newer ones are emerging. For example, over the last two decades, women have been participating more in urban migration flows, and human trafficking has become a massive illicit global industry: Africa accounts for approximately one third of all global trafficking cases.

A good proportion of migrants are pushed to move because of the lack of opportunities in rural areas; many have no other option to escape poverty. Conflict, political instability, and human rights violations are also important drivers. Environmental change such as drought and floods are other factors forcing people to migrate. Droughts in the Sahel, for example, have led to significant movements from this region. Climate change scenarios envision greater unpredictability in climatic patterns, although it is still unclear whether climate change will empty certain rural areas, trigger reverse rural migrations, or vary from one region to the next region. As such, the climate/environment/migration nexus is a critical unknown for the future.

Most Africans move within their country or within the continent. Even though migration to the Global North captures most international headlines, it constitutes a much smaller flow than internal and intraregional migration. Movement of people from rural to urban areas (a classic urbanization trend) is the dominant pattern of recent decades. Two additional movements stand out in terms of the numbers and scale—internally displaced persons (IDPs) and refugees. Africans account for almost half of the world's IDPs, and the African region is home to the world's second largest concentration of refugees (after Asia) (Forced Migration Online 2013).

Efforts to collect data on migrants center on international flows and miss many internal flows or what are termed "irregular flows" (movement that eludes statistical coverage) Official statistics focus on national boundaries, even though many national boundaries have little relevance for Africans, who consider them porous. As a result, there is much weaker systematic knowledge on international movements: data on rural-to-urban movement, IDPs, and smuggled/trafficked persons are estimates—at best.

Flows to the Global North consist of four categories of migrants. First, there are refugees and asylum seekers. Second, there are smaller outflows of highly qualified Africans (e.g., health professionals, accountants, and educators) drawn by upward mobility, professional development, stable careers, and higher salaries. There is an ongoing debate as to whether this outflow represents "brain drain" (loss of bright, able-bodied, educated people) or "brain gain" (added benefit of return migrants with improved skills and members of the diaspora who engage in development from afar) for the region—or elements of both. Third, there is a smaller, but consequential, flow of athletes (Box 7.1). Fourth, there are flows of undocumented and trafficked individuals, mainly to Europe but, of late, to Asian and Middle East destinations as well.

## BOX 7.1 AFRICAN ATHLETES AND GLOBAL SPORTS RECRUITMENT

A growing exodus of Africa's top athletes abroad has been taking place, primarily to elite leagues in Europe and North America. This talent migration is termed the "muscle exodus" (Darby 2007). Kenyan, Ethiopian, Sudanese, and other track-and-field athletes migrate to the United States and the UK. Some of the basketball players going to U.S. college teams are moving on to become National Basketball Association (NBA) stars (e.g., Dikembe Mutombo of the Democratic Republic of Congo [DRC]). Ghana, Côte d'Ivoire, Mali, and Nigeria lead the way in exporting soccer players to European, Asian, and Middle Eastern leagues. Several top African athletes even opted to represent their host countries (e.g., Poland, Qatar, United States, Denmark, the UK, and Bahrain) in the 2012 London Olympics. One of the most high-profile international athletes is Mo Farah (a Somali-born UK citizen), who won two gold medals (for the men's 5,000 and 10,000 meters) at the 2012 Olympics (Fig. 7.1).

There is a smaller stream going the other way. A few soccer players are opting to represent their father's home country (e.g., Frédéric Kanouté chose to represent Mali after representing France in under-21 football, and Togo's only Olympic medal was won by French-born Benjamin Boukpeti, who placed third in the canoe individual slalom in the 2008 Olympics).

The largest exodus of elite African athletes is to Europe's professional soccer leagues, but flows are increasing to Australia, Russia, New Zealand, and China. European recruitment of African players extends back to the colonial period. Following the European Court of Justice's Bosman Ruling in 1995, however, soccer players' migration surged as limits on the number of international players on European team rosters were abolished. Indeed, France's recent international successes are, in part, attributable to the contributions of talented players who were born in Africa but became naturalized French citizens. Squad rosters of teams participating in the 2013 African Cup of Nations reveal the degree to which Africa's elite soccer players have been integrated into European markets: 188 of the 368 players were Europe-based. The Côte d'Ivoire team was composed almost exclusively of European club players (21 of 23 national squad members earn their living in Europe). Several African players are global soccer superstars (e.g., Didier Drogba [Galatasaray, Turkey], Samuel Eto'o [Chelsea, England], and the Touré brothers, Yaya [Manchester City, England] and Kolo [Liverpool, England]). Eto'o in former years was the top earner in the

**FIGURE 7.1** Mo Farah Wins Gold at the London Olympics. *Source:* © Tim Clayton/Corbis.

world soccer circuit (US$25.1 million per year) (Sports Business Daily 2012). Another African player, George Weah (Liberia; A.C. Milan, Italy), received the world's most coveted soccer player of the year award—the Ballon d'Or—in 1995. Nevertheless, the majority of African players are concentrated in the lower leagues, where there is a constant renewal and circulation of players (Poli 2007).

The European soccer industry's perspective on African player migration (i.e., scouts, agents, clubs, broadcasters, and promoters) is that participation in European leagues contributes to the development of the African game, provides global exposure to top African talent, and should be allowed to continue unfettered (Darby 2007). An opposing perspective considers the loss of Africa's soccer resources to Europe to be a neocolonial activity that results in a dependent relationship between African clubs and academies of the European soccer industry that drains talent and leaves less-accomplished players to populate African leagues, undermining its grassroots structure and ultimately restricting possibilities for developing a domestic soccer industry in Africa.

African soccer fans are drawn to follow elite European teams and African players who ply their trade at top clubs. An analogy can be drawn with the colonial economies that sourced, refined, and exported raw materials for European markets, where most profits accrue upstream. This time around, raw football talent is being exported at a low initial purchase cost, and players realize their highest values after one or more trades within the global football market. Most profits accrue to European clubs, agents, and the players themselves. Much lesser sums are distributed among scouts and members of the African network involved in the initial transfer of players (often outside of official channels). Still, the grassroots actors remain enticed to nurture the next "great talent" for export, anticipating a windfall. The lure of Europe and global football stardom and financial rewards is irresistible to young African players, who perceive the sport as their only escape from poverty. Unscrupulous local agents and officials participate in the associated underground network, falsifying documents such as age to obtain contracts outside of the region. While thousands of African players are given apprenticeship contracts outside of Africa, only about 20% move on to obtain professional contracts (Scherrens 2007). The majority are released after a few years, and numerous African players become stranded (some entered countries on improper student/tourist visas and, after playing football, gravitate to informal employment in their host countries), whereas others return home with little financial gain for the time invested in honing their athletic skills. The soccer conveyor belt has been likened to youth trafficking, with well-documented cases of abuse by unscrupulous scouts, agents, cutthroat clubs, and local African networks. FIFA president Sepp Blatter (2003:17) strongly condemns this labor recruitment system: "Europe's leading clubs conduct themselves increasingly as neo-colonialists who do not give a damn about heritage and culture but engage in social and economic rape by robbing the developing world of its best players."

The global soccer industry's great wealth inscribes itself within the African grassroots infrastructure. Its hegemony greatly accentuates the underdevelopment of the sport in Africa at all levels (apart from national teams), and the domestic sports infrastructure remains impoverished. Governments are the main sponsors of sport, but due to pressing budget concerns they cannot allocate adequate spending. Most playing fields and stadiums are owned by governments and managed by the Ministry of Sports or by local government administrations and therefore cannot be used by local clubs to generate income, further hindering the development of a professional infrastructure. Players are highly motivated to seek external opportunities for developing their talent in developed countries with superior infrastructure, education, remuneration, and social welfare systems. Overexposure by the media of the elite European teams and unbalanced accounts of a handful of successful African footballers further contribute to the African football dream.

Since the 1990s, football academies have become established throughout Africa to nurture talent for export. Some are government-supported; others involve partnerships between local academies and European clubs (e.g., Ajax [Amsterdam], Ajax [Cape Town], Feyenoord [Rotterdam], and the Feyenoord Fetteh Football Academy [Accra]); some are privately owned or corporate-sponsored. Most, however, are nonaffiliated academies. Academy networks frame the aspirations of young African players and operate within of broader systems of talent identification, recruitment, and export. Given the global financial clout of elite clubs, sponsors, and global TV rights, the draining of Africa's talented footballers will continue.

Migration has moved to center stage in development policy after decades of waiting in the wings. During the 20th century, migration was portrayed largely in negative terms—an African development problem area. Since 2005, the positive role of migration to the development process has been acknowledged beyond the migration expert community, for example by the central banks of various African states, nongovernmental organization (NGOs), and local communities. The discovery of the enormous amount of funds transferred worldwide by international migrants to the "home front" and the development implications of such remittances, skill transfers, and international networks (from trading networks to hometown associations) have generated excitement and anticipation (Zoomers and Nijenhuis 2012) (See Table 7.1 for a discussion of the positive and negative effects of the migration of educated Africans).

## COLONIAL POLICIES AND MIGRATION

Colonialism controlled the movement of people to serve European economic development. The development of the trans-Atlantic slave trade and the export of Africans to serve on overseas plantations and produce commodities controlled by Europeans for international markets represented the most brutal and inhumane system. Forced slave migration brought 10 million Africans to the Western Hemisphere between 1519 and 1867: 360,000 landed in what is today the United States (Capps, McCabe, and Fix 2011:2).

As Europeans established colonies in Africa, the emphasis shifted to the deployment of migrant labor to serve the colonial endeavor (in mines, commercial ventures, farms, and the lower echelons of colonial administration). The functioning and profitability of the colonial state required Africans to migrate to work, and it limited the mobility rights of other Africans.

Colonial migrants were enticed to "islands of economic development" (Prothero 1965) such as mining areas of the Witwatersrand (South Africa), copper belts in present-day Zambia and DRC, fertile agricultural areas in the Kenyan highlands and the southern part of present-day Zimbabwe and Eastern Cape, and major cash crop-producing areas (e.g., cocoa in Ghana, Nigeria, Côte d'Ivoire, and Senegal; tea and coffee in Tanzania, Uganda, and Kenya; rubber in DRC). The economic

**TABLE 7.1 POSITIVE AND NEGATIVE EFFECTS OF THE MIGRATION OF EDUCATED AFRICANS**

*Positive*

- Migrants are better off (professionally and financially).
- Individual freedom is enhanced by allowing individuals to migrate.
- The migrants' families benefit (e.g., from remittances and from the satisfaction of family members succeeding abroad).
- Migration may lessen the pressures on other less-skilled family members to migrate.
- Successful African migrants abroad provide an incentive for the youth to acquire as many skills as possible to succeed as migrants (the "Kofi Annan factor").
- Migrants may have a positive effect on their home politics and/or institutions from abroad.
- The threat of migration may be an incentive for governments to pay higher salaries to people in various professions and to ensure that the tax system treats them favorably.
- Migrants facilitate international networks.
- Some migrants return home and contribute positively to their societies.

*Negative*

- There is a loss of skilled professionals, which translates into poor return on government spending on public education.
- Migration of educated people undermines the Millennium Development Goals (MDGs) and other development efforts.
- Family separation due to migration causes suffering in nonmonetary ways.
- Tensions develop among families who receiving remittances and those who do not (remittances help families to afford housing and consumer items).
- Tensions arise within families over access to remittance funds.
- Migration of the educated population impedes the development of the middle class, limits the tax base, and contributes to the development of highly polarized societies consisting of an elite wealthy class and a large underclass of very poor people.
- Not all migrants abroad are successful, and those who are not successful return home and represent an added burden.

geography of colonies was punctuated by isolated modern economic nodes with requisite indigenous laborers and neglected peripheries that warehoused a reserve army of workers at the beck and call of colonial rulers. The implementation of poll and hut taxes, the expropriation of the best land for settler agriculture, and the provision of services for wage laborers compelled Africans to offer their labor to earn income. As infrastructure for Africans was nonexistent, Africans undertook travel on foot, routinely walking hundreds of kilometers.

Colonial administrators did not want the migrants to settle permanently in their new work locations, however. Laborers were welcomed but were expected to return home after their contracts ended to make way for new workers. This established a circular labor migration system, which was reinforced by the fact that migrants were encouraged to maintain their traditional ways of life and village households so that they had homes to return to. Above all else, African migrants were to be kept in their place.

In time, more and more unplanned migration took place to urban centers. For colonial rulers, the growing urban population of poor and underemployed workers presented serious social, economic, and moral challenges. Planners were faced with a dilemma—to allow families to migrate and create permanent settlements or to control more rigidly migrant settlements and social tensions.

After African states' independence, governments and development agencies emphasized rural development, with the goal of enabling people to achieve a better quality of life in their home areas. There were massive government investments in rural development programs, agricultural extension programs, and rural infrastructure (e.g., schools, health care facilities, and roads). Some observers contend that the agricultural/rural development emphasis had an underlying aim of controlling population movements (Bakewell 2007). Certainly within development thinking, migration was viewed as an indicator of failure. Thus, until the 21st century, development thinking on migration aimed to keep people rooted in rural areas.

## CIRCULAR MIGRATION

Circular migration is the movement of people, including temporary and permanent migration, within and between countries that is often multidirectional and involves different destinations (Grant 1995; Potts 2011b) (See Box 7.2 for a detailed case study of circular migration to and from Burkina Faso). No formal legal or administrative definition of circular migration has been adopted; its legality varies by country and region. Protocols on the freedom of movement within Southern Africa have

### BOX 7.2 BURKINA FASO: CIRCULAR MIGRATION IN TIME AND SPACE

Burkina Faso is a landlocked country in West Africa surrounded by six countries (Mali to the north, Niger to the east, Benin to the southwest, Togo and Ghana to the south, and Côte d'Ivoire to the southwest). Gaining independence from France in 1960, the country has struggled to improve its agricultural-based economy. Consistently rated among the poorest countries of the world, Burkina Faso ranks in the bottom seven least-developed countries according to the UN's human development index (UN 2013).

More than 80% of the population (17.2 million in 2012) is engaged in subsistence agriculture, and remittances provide a major source of national income. In the 2000s, prior to conflict in neighboring Côte d'Ivoire, remittances accounted for 6% of gross domestic product, but the level fell during the years of instability. However, outflows from Côte d'Ivoire recovered US$726 million in 2012 (World Bank 2013). Importantly, Burkina Faso's migration is a regional phenomenon: 90% of migrants leave for another African country (Wouterse 2011). Consequently, Burkina Faso is unlike many countries in that it has neither a large diaspora in the West nor high levels of within-country migration (less than 30% of total migratory flow) (Konseiga 2005). Despite its former colonial status, migration to France plays a very minor role, and only in the last decade has Italy emerged as a new European destination for Burkinabè, many of whom work as agricultural laborers.

Burkina Faso is representative of a number of poor African countries (e.g., migration from Lesotho and Mozambique is concentrated on South Africa/Botswana, and wars forced populations from Liberia, Sierra Leone, Burundi, and the DRC to seek refuge in neighboring countries) in that intraregional migration predominates. Burkinabè have traditionally migrated within the West African region to Côte d'Ivoire, Ghana, Senegal, and Nigeria. Two thirds of Burkinabè migrate to Côte d'Ivoire, traditionally the strongest French-speaking economy in West Africa, and this is a very important intra-African migration corridor.

Burkina Faso ranks very high in migration indices, simultaneously as a sending, receiving, and transit destination for migrants.

*(Continued)*

## BOX 7.2 (*Continued*)

Despite its economic problems, the country attracts a range of immigrants, including Syrian and Lebanese merchants, French professionals, and Malian workers and refugees. In 2012, 40,000 additional Malian refugees fled to Burkina Faso following the outbreak and intensification of conflict in Mali (UNHCR 2013a).

Circular migration is part of a long tradition, a way of life. For over 1,000 years, people have followed livestock, crops, and trading markets on a seasonal basis (Kress 2006). Foundations of circular migration were laid by the region's various ethnic groups, such as the Mossi (50% of Burkina Faso's population), the Lobi, and the Fulani, all operating within a fluid zone crisscrossing the subregion.

These groups and other Sahelians maintain traditions of circular regional migration, many passing through Burkina. Some transit migrants remain in Burkina Faso, replacing Burkinabè labor that has circulated elsewhere. In 2011, Burkina Faso's immigrant population climbed to 1 million, while 1.6 million Burkinabè circulated (often for durations of two or more years). By the mid-2000s, the long-established migration tradition had resulted in one in five Burkinabè residing outside their home state (Kress 2006).

French colonial rule sought to restructure forcibly the region's economy and developed a different logic for circular migration. During the colonial period, Upper Volta, as Burkina Faso was then known, unsuccessfully experimented with producing cotton for export, and colonial administrators decided that the area was unsuitable for plantation agriculture. Its distance from markets and ports reinforced their decision. As a result, inland West Africa was largely disregarded in colonial economic development initiatives, fostering a dependent relationship between the interior labor reserve areas and exterior colonial islands of economic development.

Many laborers were coerced to work on plantations and in factories in Côte d'Ivoire, while others were forced to work in construction and communications projects in other French colonial areas. A colonial edict set a target for Upper Volta to contribute half a million workers to Ivorian plantations. Colonial policies were not devised to meet the demands of seasonal agricultural work in Upper Volta but instead were driven by global market demands for cacao, groundnuts, and coffee. Heavy colonial taxation and the creation of a waged workforce ensured that migration was necessary. Some 60% of the adult male population migrated in the early 20th century, and the exodus was so large that it undermined farming in Upper Volta. Voltanique migrants also sought employment as colonial clerks and construction workers and gravitated to Ghana, Mali, and Senegal to take up these positions. Ghana, then known as the Gold Coast, was the preferred destination for Upper Volta migrants because of higher wages, better working conditions, and an alternative to forced labor in Côte d'Ivoire. In essence, the colonial era set in motion a pattern of labor mobility that wove a complex grid of relations and interdependencies across West African borders, overlaid upon traditional circulation patterns.

In the 1960s and 1970s, West African corridor migration system revolved around major immigration hubs, with Côte d'Ivoire, Ghana, and Nigeria as receiving countries and Burkina Faso, Mali, and Niger as sending countries. During recurring economic crises, host countries tended to adopt restrictive migration policies and to expel migrants (e.g., Nigeria in 1983–85 and Côte d'Ivoire in 1986), and migrants responded by altering their migration strategies and shifting to neighboring countries (Konseiga 2005). For the most part, periods of decline and episodes of restrictive policies in one country coincided with more liberal admission into another (Konseiga 2005). Therefore, despite disruption, population movements within the region were sustained by readjustment to alternative destinations as opposed to returning home.

The signing of the treaty creating the Economic Community of West African States strengthened the legal basis for migration, facilitating freedom of movement among member states. At the same time, possibilities were created for unscrupulous agents to engage in human trafficking, especially children, who are trafficked from Burkina Faso into Ghana, Côte d'Ivoire, Benin, Nigeria, and Mali. Girls are used as domestic helpers, beggars, and sex workers and boys are put to work in mines, farms, and houses.

In 2010, some 11.2% of Côte d'Ivoire's population was foreign-born, a decrease from more than 26% in 1982 (UN 2013). Prior to the civil war (2002–11), the country was among the top 12 destinations for migrants in the world. Almost a decade of civil war and political strife has shrunk the numbers of Burkinabè, but they still represent the largest foreign-born group. Xenophobia has become significant, and politicians have manipulated the rhetoric of "authenticity" (i.e., "Ivorite") to mobilize "real Ivoirians" against "other Ivoirians"; identity has become a key issue in presidential elections.

As a result, many Burkinabè (including children born abroad) have returned home, but even in the face of repatriation, emigration still proceeds, indicating the durability of circular migration. Returnees typically choose not to settle in their places of birth and instead to gravitate to where the jobs are: urban centers and rural areas undergoing economic development (e.g., western Burkina Faso). Recent migration to western Burkina Faso has been accelerated by urbanization, the eradication of river blindness, and successes in cotton production. Urban returnees tend to be self-employed entrepreneurs, and rural returnees engage more with agriculture. Of note, there are tensions between Burkinabè and the repatriated population, who find themselves marginalized at the national and local levels; they, in turn, are organizing to advocate for a new kind of Burkinabè citizen, whose identity is shaped by transnational and forced migration.

Burkina Faso illustrates a migration dynamic where regional mobility predominates. Labor has always been particularly mobile in West Africa, but environmental drivers are now adding to longstanding economic and social factors. The Burkina Faso government recognizes that migration of many of its people is a problem and that it also has to grapple with displaced people from neighboring countries.

been implemented (e.g., Southern African Development Cooperation allows for 90-day visa-free entry, permits employment, and includes provisions that allow visa extensions for citizens of Angola, Botswana, DRC, Lesotho, Malawi, Mauritius, Mozambique, Namibia, Swaziland, Tanzania, Zambia, Zimbabwe, South Africa, Seychelles, and Madagascar [though the latter's membership is currently suspended]).

Circular migration involves continuous engagement between a migrant's home and his or her adopted residence. Migrants become active in both spaces, creating connected economic spaces straddling separate locations. This type of migration ebbs and flows according to the strength of urban and rural economies, shifting to longer-term family migration when economies are vibrant and to shorter stays/return migration in hard times or the off season.

Circular migration affects those remaining in the home area as well. Migrants are typically young, able-bodied males and, increasingly, young women, whose exit means that they cannot contribute to their home rural economy. Losses of substantial numbers of young people can endanger the viability of a village economy, creating gaps in the physical and social infrastructure. However, migrants' absence can be compensated by the remittances they send back home, which are critical to sustaining families and to avoiding agricultural shortfalls. Moreover, the timing of circular migration, with family members moving during the unproductive season and returning before the onset of the rains, allows for the continuity of the agriculture cycle in home villages. According to Potts (2011b:593), wives in rural Africa "operate a rural string to the urban livelihood bow." In addition, urban migration relieves the pressure on feeding all rural family members during difficult parts of the year. In African contexts, migration is fundamentally a family affair rather than an individual activity, so maintaining connections and moving back and forth is often the norm.

Rural links are important for urban migrants. Maintaining a stream of rural food production for consumption or for sale in town to boost household incomes is part of the economics of rural–urban linkages. Rural families provide a safety valve and a welfare option for urbanites during hard times (Potts 2011a). Some urban dwellers become partly dependent on rural sources of food and/or income, evidence of a reverse flow illustrating a different symbiotic relationship.

## FORCED MIGRANTS: REFUGEES AND IDPs

A refugee is defined as a person who, owing to a well-founded fear of being persecuted for reasons of race, religion, nationality, membership in a particular social group, or political opinion, is outside the country of his or her nationality and is granted protection (United Nations High Commission for Refugees [UNHCR] 2012). IDPs are people who are forced to leave their homes as a result of or to avoid the effects of armed conflicts, violence, violations of human rights, or natural or human-made disasters but seek refuge in their home countries (not crossing international borders) (UNHCR 2012) (Fig. 7.2).

Many Africans in conflict zones encounter the destruction of land, homes, and livelihoods. Shattered lives and livelihoods compel many to move in search of safety and security. In 2011, Africa had 2.7 million refugees, or one quarter of all refugees globally, primarily from Somalia (760,800), Sudan (462,100), and DRC (457,900) (UNHCR 2012). Two principal subregions of displacement are evident: a West African axis covering Côte d'Ivoire, Guinea, Guinea-Bissau, Liberia, and Sierra Leone, and a Central/East African axis encompassing DRC, Congo, Burundi, Ethiopia, Kenya, Rwanda, Somalia, Sudan, Tanzania, Uganda,

**FIGURE 7.2** IDP Camp, Rutshuru, DRC. *Source:* © Yannick Tylle/Corbis. Corbis 42-28620393.

and Zambia. Adverse environmental factors (e.g., droughts, degradation, deforestation) may be compelling people to move, although many refute the degree to which the environment/climate is the ultimate trigger of refugee movement (Gill 2010) (Fig. 7.3).

Various types of violence, armed conflict, and oppression are driving most of the movement. Intercommunal violence can range from political oppression to interethnic cattle rustling for financial gain, assertion of dominance over neighboring tribes or groups (prevalent in northern Kenya and areas of South Sudan, Ethiopia, and Uganda), and armed conflict. Armed conflict itself takes various forms, such as governments and their armed forces being pitted against armed opposition groups, governments crossing borders into neighboring countries (e.g., Kenyan army attacking Al-Shabaab forces in southern Somalia), and, other times, armed groups crossing borders into neighboring countries (e.g., armed groups from Ethiopia and Somalia triggering displacement in northern Kenya).

Five of the top ten refugee-producing countries in the world are located in Africa—Sudan, Burundi, DRC, Somalia, and Liberia (Crisp 2010:2). Several of the world's top hosting countries are in the region—Tanzania, Chad, and Uganda. Refugee concentrations, or hubs, arise in areas where conflict has been endemic; most refugees settle in immediate neighboring countries or stable countries within the subregion. Few African refugees have the resources to travel outside their region. Many African states have refugee populations in excess of 50,000 (Fig. 7.3). According to Crisp (2010:2), eight of the 20 countries with the highest

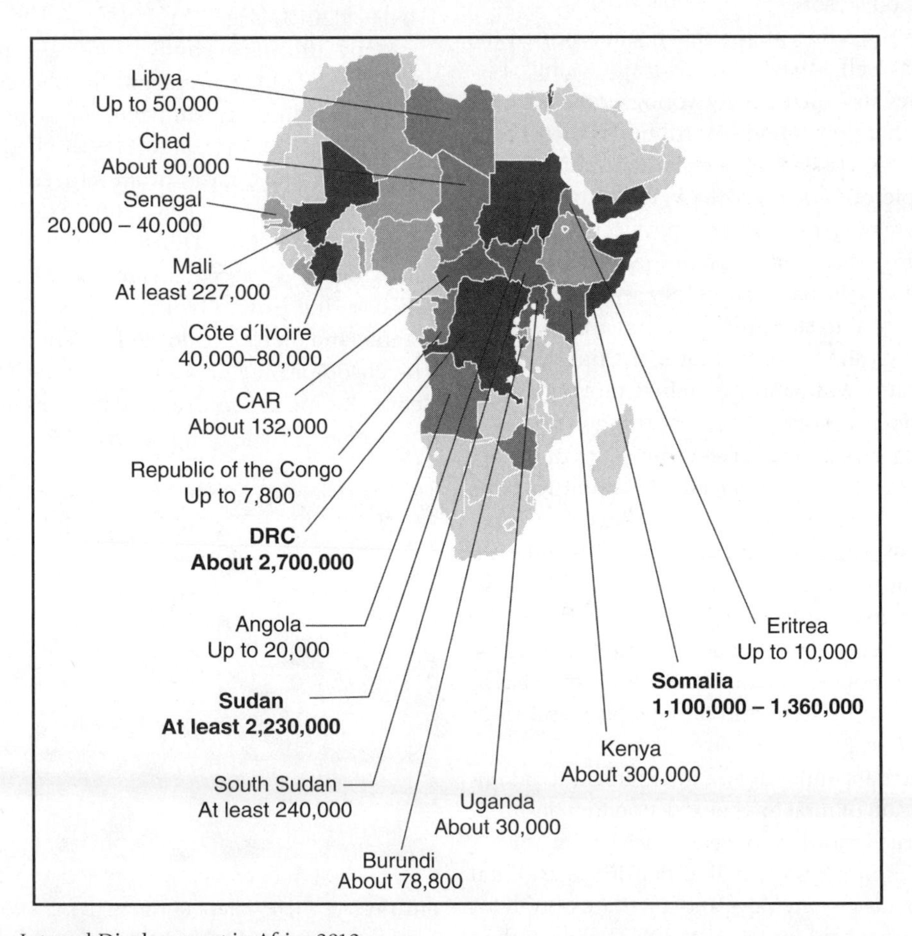


**FIGURE 7.3** Internal Displacement in Africa 2012.

ratio of refugees to local people are located in the region. Africa refugee figures for 2011 indicate a downward trend, reflecting a more peaceful region overall (even though there are some protracted conflicts), but there is a countervailing rising trend of IDPs. An inverse relationship exists between declining refugee figures and rising IDPs, indicating that barriers may be preventing people from moving across state borders.

Africa houses close to half of the world's IDPs (Forced Migration Online 2013) and nine of the 24 countries with the largest IDP populations in the world (Crisp 2010:2). Countries with large shares of their population displaced include Somalia (16%), Sudan (7%), Zimbabwe (7%), and DRC; South Sudan also records high levels, but given its recent emergence as a state, its IDP tally is impossible to calculate at this time. Official counts put the number of IDPs in Africa in 2011 at 9.7 million, and displacement is geographically widespread (Norwegian Refugee Council and Internal Displacement Monitoring Centre 2013). Combining the official figures of refugees and IDPs indicates that 12.4 million persons were displaced (the generic term covering both refugees and IDPs) in 2011 (UNHCR 2012).

Outstanding grievances over political power sharing and disputes over natural resources or land are some of the main drivers of contemporary displacement in Africa (Norwegian Refugee Council and International Displacement Monitoring Centre 2012). IDPs are often intentionally uprooted by governments on ethnic, religious, or political grounds, or as part of counterinsurgency campaigns. In some civil wars, the displaced are perceived as the enemy or the government objects to relief efforts in insurgent areas (e.g., Angola and Darfur); in other cases, the displaced are trapped between opposing sides in civil wars or are threatened by insurgents (e.g., DRC). IDPs frequently suffer the highest mortality rates. In Uganda, for example, the HIV/AIDS rate among IDPs is six times higher than in the general population (Crisp 2010:21).

Many people become internally displaced because physical barriers (distance and difficult terrains) impede their flight across borders; some flee to other parts of their own country because they prefer to remain in familiar surroundings. However, a large part of the explanation of rising IDPs is that hostile reactions to refugees have become more pervasive. Most displaced persons are not welcome across borders and face increasingly restrictive national admission policies and hostile receptions. African governments are more willing to reject refugees at frontiers and return them to their areas of origin, even when the conditions that motivated them to flee still exist. The plight of IDPs is further aggravated by the lack of an international agency with a mandate to aid them and offer them legal protection.

Official counts represent underestimates, as many displaced persons neither report to authorities in their countries of residence nor remain in official camps. There are wholly undetermined predicaments (e.g., the displacement of pastoralists throughout the Horn of Africa) that are neglected by both governments and donors. Moreover, many displaced persons are hard to track and blend into the local poor populations, an absorption often made easier by speaking the same language or belonging to the same cultural/ethnic groups as the local population. Indeed, UNHCR now recognizes that displaced persons are leaving rural camps in significant numbers (approximately half) and moving to urban locations (typically slums) where they believe they can earn a living, improve their livelihoods, and perhaps even send remittances to family members. Also, health care, education, and other services are generally better in urban centers. Urban locations enhance possibilities for communication with distant family members, and the existence of ethnic urban enclaves and other social networks supports the integration process. However, urban anonymity brings a different set of challenges: isolated, displaced persons are more easily denied access to government services, and individuals are prone to exploitation by employers, landlords, and other locals.

Refugees as well as IDPs are not adequately protected. The head of a UNHCR research unit, Jeff Crisp (2010:5), notes "that the industrialized States—rather than those in Africa—have taken the lead in eroding the right of asylum and undermining the principles of refugee protection. . . . introducing a vast array of measures specifically designed to prevent or dissuade the arrival of refugees." At the same time, the African tradition of hospitality and offers of asylum, prevalent in the two decades after independence, now operate in reverse. Indeed, negative attitudes toward refugees are growing.

The declining commitment to asylum in Africa is a result of pressing internal economic problems in every African state, concerns about the long-term cost of refugee assistance, inadequate international compensation to cope with refugee influxes, and misgivings that settled persons are not appreciative enough of their host's generosity. Refugees now exchange one form and degree of vulnerability for another when crossing into another state. Corralling refugees into "safe zones" is the norm.

Sources of insecurity in Africa's refugee camps and settlements are varied and numerous. Commonly exiled populations are viewed as posing threats to the local economy, local society, and political security. Refugee populations may be targeted by direct external military attacks. Despite the Organization of African Unity's Refugee Convention of 1974 (Article 2.6) and UNHCR's guidance of settling refugees "at reasonable distance from the frontier of their country of origin" (OAU 1974), refugees are often settled close to the border; geographical propinquity jeopardizes their safety and undermines their protection. Furthermore, refugees may be subject to numerous nonmilitary security threats involving violence, coercion, intimidation, and criminal activity. A long list of potential threats includes domestic and sexual violence, rape, armed robbery, abduction into militias, forced marriages, arbitrary arrests and punishment by refugee community leaders and members of local community security forces, fighting between different clans and subclans within the same refugee community, and armed confrontation between refugees of different nationalities (Crisp 2010).

The 2009 Kampala Convention (signed by most states) is a legally binding regional instrument that protects IDPs from arbitrary displacement and mandates protection and assistance during displacement until a durable solution can be implemented. However, in most instances, IDPs are weakly protected and the feeble rule of law further perpetuates their economic and political marginalization. While levels of violence and insecurity are difficult (if not impossible) to assess in Africa, many IDP camps are becoming increasingly dangerous places. Precarious and protracted situations are not helped by weak donor commitment to protecting IDPs and by the donors' focus on immediate "hot" situations. Donors have failed to designate IDPs as a special category for humanitarian assistance on the basis that singling out one group could lead to discrimination against others and generate more tension and conflicts. For example, the International Committee of the Red Cross provides assistance and protection to all civilian victims of armed conflict. Indeed, IDPs can face poorer treatment than refugees in the same location. For instance, in specific places, UNHCR provides returning refugees tools and seeds but does not extend the same support to IDPs.

Displacement has social, cultural, economic, and environmental consequences for both transient individuals and locals. Displaced persons lack economic and social capital. Their relocation to high-density camps adds significant stress on immediate food and water supplies and on health care, sanitation, and security systems. Establishing the rule of law in concentrated areas of displaced persons is difficult. Conflict can be anticipated when displaced and destitute peoples live alongside poor members of the local population; the potential for disorder and violent outbreaks is intensified when camp populations include persons responsible for terrible crimes in their country of origin. Although relief agencies provide considerable assistance, much of the burden falls upon national governments and already-overwhelmed municipalities (e.g., Johannesburg in the 2000s and Dar es Salaam in the 1990s).

It is increasingly difficult to distinguish among legal categories of people, and mixed migratory movements (encompassing refugees, asylum seekers, economic migrants, and other migrants) are occurring (Long and Crisp 2011). In many mixed movements, people travel alongside each other using the same routes and means of transportation but with different motivations and objectives. Many of those involved in these movements, while not having a claim to refugee status, can end up in situations where their human rights are violated; humanitarian needs may develop along their journey or occur upon their arrival. Mixed movements also include people who are recognized as refugees in a country of first asylum but who move on to another state in search of better protection, more economic opportunity, and/or reunion with their family.

For some refugees moving to a third country is the only way to secure permanent safety and basic human rights. Options and placements are limited: only 26 countries offered a resettlement base in 2011 (the United States and Canada admitted 80% of resettled refugees), and worldwide only 80,000 were resettled out of 800,000 applications (UNHCR 2012). In war-torn areas, some refugees eventually return home voluntarily, and processes of postconflict resolution can begin. Once disarmament and demobilization have taken place, the more difficult process of reintegrating ex-combatants and refugees proceeds. Besides mental trauma and other health challenges, creating new livelihoods or recreating old ones is difficult. For example, gaining access to land is fraught with difficulty as displaced persons often expropriate land during conflicts. Large numbers of refugees, however, never make the journey back home and instead blend in to their host countries.

There are more than 20 protracted refugee situations in Africa, and more than 2.3 million persons have been in exile for over five years (Crisp 2010:13). Wars in recent decades (e.g., Angola, DRC, Liberia, Rwanda, Somalia) have encompassed intense ethnic and communal antagonisms, high levels of organized violence, and deliberate targeting and displacement of civilian populations. The international community has failed to bring these predicaments to an end, which contrasts with refugee situations in northern Iraq, Bosnia, and Kosovo, where the United States and its allies directed speedy repatriation after the wars. Weaker geopolitical and economic stakes in Africa explain the international community's lack of urgency and commitment to repatriation. As a result, voluntary self-repatriation has become an African norm.

Some refugee populations become "warehoused" indefinitely in camps while the lucky ones escape to cities. Dadaab, in northern Kenya, is the largest refugee camp in the world, housing mainly 380,000 Somalis who have fled civil conflict and settled over the border (Al-Jazeera 2011:1). There have been mass refugee movements across Africa, and refugee camps have become more permanent settlements. For example, 500,000 Somalis have been in Kenya since 1991; 200,000 Eritreans relocated to Sudan and Ethiopia 40 years ago; and 70,000 Liberians have resided in Ghana and Côte d'Ivoire for 20 years. Warehousing is, basically, the practice of depriving millions of refugees the right to work, practice professions, operate businesses, own property, move about freely, and choose their place of residence. Spending years in refugee camps with enforced idleness deprives refugees of a decent life. Concentrating on long-term "care and maintenance" programs does little to promote self-reliance and/or integration into local communities. Skeptics suggest that relief organizations have a vested interest in perpetuating a relief model of refugee assistance, including the maintenance of large, highly visible camps.

More than 5 million refugees in Africa have been repatriated during the past decade, and an unknown number (perhaps even more) have voluntarily returned to their home communities. There is a well-established legal principle that voluntary repatriation would result in upholding the rights of refugees, but, in fact, substantial numbers return under conditions that do not meet standards of safety and dignity. In some instances, undue pressure is placed on refugees by governments, local communities, and militia forces to return home, as repatriation plays an important role in validating postconflict political order and in rebuilding community. Kenya and Somali signed an agreement in 2013 to initiate the voluntary repatriation of 500,000 Somalis from Kenya, but many Somalis who settled in Nairobi's Eastleigh suburb, known locally as "Little Mogadishu," have rebuilt their lives and have little incentive to take the risk and return home 20 years later.

Repatriation is far from being a smooth process; reintegration can be difficult. Returnees often experience several legal forms of insecurity. They may not have proof of their nationality and thus are not recognized as citizens of the country to which they return. Lack of official documentation such as identity cards and birth certificates can open the door to arbitrary arrest, inhibit in-country movement, and prevent returnees from finding a job and voting. Many going back to war-ravaged places experience severe material insecurity. Land-use patterns and the fabric of the local economy may have been dramatically altered. Depending still on emergency relief, they live hand to mouth, as they did in exile. This reality clashes with many unrealistic expectations about returning to their homeland.

## MIGRANT REMITTANCES TO AFRICA

"Migrant transfers" to developing countries have increased substantially since the 2000s, although the volume of remittance flows to Africa lags behind other world regions. Nevertheless, Nigeria has become the fifth largest remittance-receiving country in the world in 2011 (World Bank 2013), and repatriated earnings of Africans are increasing, showing a steady rebound after the 2007–08 world economic downturn and a temporary remittance decline. The substantial share of remittances to Africa is accounted for by several countries: Nigeria (US$21 billion) received half of remittances in 2011, in addition to Sudan (US$1.4 billion), Senegal (US$1.4 billion), Kenya (US$1.3 billion), South Africa (US$1.1 billion), and Uganda (US$953 million) (World Bank 2013).

In 2012, US$40 billion was remitted to Africa, up from US$9.1 million in 1990. Remittances have been growing in volume as well as relative to other sources of development finance. Remittances are at present one of the region's largest source of foreign inflows (Mohapatra and Ratha 2011). Remittances to Africa equaled 2.6% of gross domestic product (GDP) in 2009, higher than the average of 1.9% of GDP for all developing countries.

Indeed, the true size of remittances to Africa may be significantly larger because of the scale of undocumented internal migration, the prevalence of informal remittance channels, and weak data-collection efforts in the region (only half of African countries report remittance data, and countries such as DRC, Somalia, and Zimbabwe do not report any remittance data) (Mohapatra and Ratha 2011). Indeed, when African states improve their national data-collection efforts, they reveal significant information. For example, the Central Bank of Ghana reported US$1.6 billion in remittances in 2009, more than ten times the value reported by the International Monetary Fund's balance-of-payments statistics.

Intraregional, domestic, and many international remittances are sent through informal channels. Personally carried during visits home, sent through transport companies, or transmitted through informal hawala channels (Box 7.3), they reflect the fact that

### BOX 7.3 THE HAWALA SYSTEM OF TRANSFER IN THE HORN OF AFRICA

Some groups participate in the hawala system of transferring funds. Developed in ancient times and still active, this system operated in the medieval and colonial periods to facilitate trade across geographical, political, and cultural borders. Originally developed in the Middle East and South Asia for transferring money long distances and across borders, the system is used by many Islamic groups in African states with weak or absent formal financial institutions. It was established to transfer funds through clearing systems that minimize costly shipments of coins and bullion. Over time, it led to the development of networks of agents who could transfer funds, often without physical currency changing hands in propinquity, and later involved agents exchanging credits and debits with each other over long distances and balancing transactions. The hawala system operates like an informal Western Union transfer system but without government enforcement or legal recourse. Failure to deliver funds is rare; strong social bonds within the system seem to safeguard against fraud, and the system is self-enforcing (Schaeffer 2008). Moreover, the costs of using the hawala system are in the 2% to 5% range, much lower than formal financial institutions charge.

Somalis use the hawala system extensively, and it works by honor and outside of official purview (a major advantage to customers because of tax, immigration, currency control, and other concerns); it involves the transfer of money without money actually moving. It is estimated that US$1.6 billion per year is transferred by *hawaladars* (brokers), and the system is capable of delivering funds to the most remote parts of Africa. In some parts of the world (for example, Somalia and northern Mali since the 2012 Taureg rebellion), it is the only available option for legitimate funds transfer, and aid agencies make use of it. In Eastleigh, a neighborhood of Nairobi (locally known as "Mogadishu"), informal agents with radio or satellite phones are conduits in transferring money to Somalia. Not surprisingly, the hawala system become controversial in the United States after 9/11 after it was alleged that Al-Qaeda and other Islamic terrorist groups use the system to channel money for arms and military operations. (The National Commission on Terrorist Attacks Upon the United States later showed that SunTrust Bank in Venice, Florida, transferred the bulk of funds to finance the Twin Towers operations.) A vigorous debate has ensued about how to regulate the hawala system and about whether it is possible to strike a balance between protecting U.S. national security interests (there are proposals to get hawaladars to register in the United States) and ensuring that very poor Somalis have access to badly needed funds; in their war-torn country, alternative supply lines function poorly or not at all. Nevertheless, although many groups have strong codes of trust, Africans are wary of the "Nigerian" or "South African factor," whereby certain groups are distrusted and transferring via formal institutions is boosted.

trust in social networks plays an important role. Limited access to formal banking and cost are other incentives to use informal channels. Africa is the most expensive region to remit money to, with transfer costs averaging 12% in 2012. Abroad, certain offices/homes of Ghanaians, Ugandans, and others are known among migrants as providing back-office transfer services, but Africans also use formal transfer institutions (e.g., Western Union).

Data on remittance and regional migration flows within the region are very weak, and this is significant as intraregional migration is more important in Africa than in any other developing region. For many decades, the economies of South Africa's neighbors—Swaziland, Lesotho, Mozambique, Malawi, etc.—have depended heavily on financial transfers from migrants working in South Africa: US$1.4 billion was remitted from South Africa in 2011 (World Bank 2013). Lesotho is the most heavily remittance-dependent African country: these transfers accounted for 28.5% of GDP in 2010. However, intraregional remittances involve smaller sums than intercontinental remittances, largely because migrants to North America, Europe, and the Gulf states have higher incomes.

Africa's more constrained remittance inflows reveal that geography matters. There are large out-migrations from countries, but civil strife sends migrants across borders to other impoverished African countries rather than to the Global North. Not being geographically contiguous to wealthy countries is clearly a major impediment to large flows of migrants to wealthier countries (a sharp contrast to the geographical proximity of Mexico and the United States).

Major remittance outflow countries to Africa include the United States, France, Germany, Saudi Arabia, Israel, Kuwait, and Oman. Within Africa, there are remittance-sending countries, for example Nigeria, South Africa, Botswana, Angola, Côte d'Ivoire, and Gabon; Lesotho, Mali, Burkina Faso, and Nigeria are major recipient countries for African remittance flows. There are some prominent source–destinations dyads, including Mozambique–South Africa, Zimbabwe–South Africa, Ghana–Nigeria (1970s), and Burkina Faso–Côte d'Ivoire. Remittances are very important to poor countries, as mentioned, but they are also significant to more prosperous countries, such as Nigeria and Ghana, which have large, well-established expatriate communities in Europe and North America.

Remittances are a tangible link between migration and development. Migration and remittances are linked but do not always mirror each other. Distortions in remittance flows occur due to intercontinental flows surpassing continental flows, and refugee and remittance flows do not readily match as refugees quite often remit to or receive money in neighboring stable countries rather than in their home countries. Moreover, levels of remittances tend to differ by migrant group, migration intent, and duration (remittances generally increase in time). Nigerians, for example, are highly mobile: two thirds of all Nigerian households are estimated to have had emigrants, and 10% of Nigerians live outside their homeland (Sander and Maimbo 2003). West African immigrants to North America (e.g., Nigerians and Ghanaians) are far more likely to build their "dream house" in their home country rather than in the host country (Grant 2009) (Box 7.4). In addition to remittances, migrants spend, on average, US$1,000–$3,000 per visit back to their home country (Appiah-Yeboah, Bosomtwi, and Yeboah 2013). Skilled migrants also spend more time working abroad before returning home and in the process return with more to invest than lower-skilled workers.

As with other external resource flows, the effects of remittances are complex. Remittances are the most visible examples of how migration reshapes countries of origin (Yeboah 2008). Many economists and development experts (e.g., Easterly and Nyarko 2008) are highly positive about the contributions of remittances to national economic development. Remittances can be particularly important in augmenting private consumption and emergency spending, and in alleviating some transient poverty in receiving countries. Transfers can be spent on nutrition, education, and health care. Overseas communities have become an invaluable source of investment for business ventures and all sorts of bootstrap enterprises, and they have contributed to residential building booms in many African cities as migrants build houses from a distance.

Many geographers and anthropologists working at the household/individual level are less sanguine about the effects of remittances, however (Grant 2009). Remittances may unfairly advantage recipients and, in

## BOX 7.4 TRANSNATIONAL HOUSES

A dramatic feature of the residential geography of particular African cities (e.g., Accra, Kumasi, Lagos, Nairobi, Dakar) is the housing built by members of the diaspora or by intercontinental flows of remittances. A large part of the explanation of how individuals can pay between US$25,000 and US$300,000 for a home when per capita incomes are still low and mortgages are scarce involves the confluence of transnational forces.

Members of the diaspora are building transnational houses with greater frequency, so much so that they have become a salient feature in several types of neighborhoods.

Transnational houses are dwellings built with money earned abroad and often are of architectural designs and styling more reminiscent of housing in Europe and North America. Preferences are for single-family, villa-style dwellings (as opposed to extended-family traditional dwelling units such as compound housing) in low-density residential locations. Some of these houses are located in new secure gated communities that operate as residential enclaves with high proportions of absentee owners. Also, foreign builders with imported building materials construct many houses in gated communities. However, the majority of transnational houses are found in new suburban locations where land is available and where housing develops in a haphazard manner (Owusu-Ansah and O'Connor 2010). On the fringes of cities, houses built by monies sent from abroad are typically incrementally constructed based on cycles of remittance flows (local parlance now equates the height of a building to the length of time the owner has been abroad); as such, some transnational dwellings can take more than a decade to complete. Smaller numbers of transnational houses are built in traditional, indigenous neighborhoods, standing out as larger dwellings that are incongruent with traditional architectural styles in the vicinity.

The notion of a transnational house can also apply to transmigrants' split existence of working abroad and investing in a property in their native country in the hope of occupying the home on their return or retirement. In the interim, the transmigrant may occupy the home during vacations, and family caretakers are often allowed to live in it rent-free in compensation for house-sitting responsibilities. Increasingly, individuals who have lived and made money abroad reinforce the connection between migration and local urban development. A select few of them (with considerable net worth) have been able to obtain mortgages from lenders abroad. These wealthy individuals are far more attracted to gated communities, and, consequently, private developers employ international advertising and marketing techniques (websites, social media, brochures, in-flight magazines, international housing exhibitions) to pitch upper-income enclave communities to targeted buyers abroad. For example, ReconnectAfrica.com is a portal that has been developed to advertise properties to members of the Ghanaian, Nigerian, and South African diasporas. Usually, international housing exhibitions are coordinated with targeted communications to members of the diaspora. Real estate developers have also worked with banks with branches in Africa to organize payment transfers to pay for this housing.

The urban focus of transnational houses shows that expatriates have a reduced loyalty to their traditional villages and are more attracted to cities. Some expats have short-term investing horizons and build houses to rent to expats in Africa, earning rental incomes (often in dollars/euros, taking advantage of currency fluctuations). Other expats have longer-term horizons, opting to build in unserviced new areas where land is relatively cheap and where housing can potentially appreciate more in time when services are eventually delivered to the area. Others build immediately to take advantage of money earned but plan on full occupancy at a later date. About 30% of such homes are left vacant.

By investing in houses during their period abroad, migrants strengthen their membership rights in their community of origin for when they return. Many migrants care about their home communities and invest so as to contribute directly to their family members' housing needs. Housing investments result in indirect benefits, such as increasing family members' marriage prospects as transnational houses signal access to international connections and global funds. At the same time, these housing investments add to the development of more modern housing stock in native cities. In turn, hometown residents benefit from employment opportunities and increased demand for local construction materials. Significantly, housing investment may be a starting point for a broader investment relationship between migrants and their countries of origin.

turn, dislocate them from the dynamics of their immediate environments. Remittances exacerbate income inequalities, particularly in households receiving intercontinental remittances, which tend to be better off from the start. Large remittance flows into land and housing can cause "Dutch disease" (negative consequences due to large inflows of monies) effects, particularly in small open economies (e.g., Cape Verde and Swaziland), causing currency appreciation and rendering exports more expensive. Moreover, the benefits accruing to remittance-receiving families can further foster a culture of remittance dependency and even promote future migration streams. High levels of migration disrupt agricultural production and lead to the replacement of crops requiring high labor inputs with less labor-intensive varieties, which, in turn, has a negative effect on nutrition (e.g., substitution of cassava for yams is labor-saving, but cassava is nutritionally inferior).

Remittances can be a public moral hazard, whereby governments ignore some obligations, anticipating that inflows will fill voids in public expenditure. Other human costs associated with the cycle include the negative social effects on people left behind (hardship, stress, and missing parents and family members) as well as on those who leave. In the case of women who migrate without their families, raising children from abroad can entail tremendous pressures and sacrifices (Wong 2006).

Remittances do not burden taxpayers in rich countries but represent a substantial source of transfer to poor countries. Countries more open to immigration are the main source of remittance transfers: the United States, France, Italy, the UK, the Gulf states, Canada, and Australia. There are far fewer contributions from immigrant-resistant countries (e.g., Japan). However, it remains to be determined whether remittances will become the principal method of transfer between rich and poor countries. This would require a more liberal open-door policy to immigration. Perhaps, in the next round of global negotiations, "immigration not aid" could be the new mantra for development, replacing the older "trade not aid" slogan.

The flow of ideas embodied within the transfer of monies may actually have a more lasting impact on society. The communications revolution has facilitated an exponential growth in international phone calls, emails, and social media communication, and in international travel. It is no longer just the elites who are exposed to external flows of new ideas. Instead, the circulation of knowledge about what, where, and how to change things, for example, alters expectations about global and local societies (e.g., service quality, national regulations, and the behavior of politicians).

The diaspora has emerged as a major development actor in an increasingly interdependent, global world. Many of those involved have acquired skills, and some have accumulated capital and established contacts with business partners and potential investors in Africa. Moreover, a number of the diaspora are influential in their host societies and have risen to positions that can engender positive socioeconomic and political outcomes in their host countries in favor of their home countries.

Given the importance of remittances and their expanding 21st-century trajectory, data collection, research, and policy formulation are playing catch-up. Knowledge is especially thin compared to the large body of accumulated knowledge on official development assistance and foreign development assistance. A more active policy of engagement related to the African diaspora is now taking place. The governments of Senegal and Cape Verde are leading the way by their establishment of a ministerial department devoted to the diaspora and the management of migration (Ghana and Nigeria also tried but subsequently reversed course). In addition, the president of Senegal now appoints four Senegalese returnees to the senate to represent the interests of the diaspora. Ethiopia, Ghana, and others are attempting to engage the diaspora by issuing diaspora bonds to raise funds to boost national economies. Clearly, governments and the international community perceive the remittance arena as dynamic, and governments want to innovate. They have to be watchful that remittances do not become another destabilizing globalizing force and that they instead serve as another tool to work toward the MDGs. However, the role of the diaspora and its relationship to national governments in developing their own societies need considerable reflection.

## HUMAN TRAFFICKING AND SMUGGLING

International organizations (e.g., UNHCR, International Labor Organization, and International Organization for Migration), migration research institutes (e.g., Southern African Migration Project at Queens University in Canada, African Center for Migration & Society at Witwatersrand University in South Africa, and the International Migration Institute at Oxford University in the UK), NGOs, and individual researchers are increasingly drawing attention to the exploitation of vulnerable migrants by international human smugglers and human traffickers.

Human smuggling is when an individual crosses a country's international border without state authorization and with the assistance of paid smugglers. Costs of smuggling are typically paid up front and vary according to the length and difficulty of the journey. An average long journey from the Horn to South Africa costs approximately US$2,000, but many undertake the journey in stints, paying for each leg along the way, often working or waiting for sponsors to remit payments so their journey can be continued (Horwood 2009). Some long-distance migrants have no option but to enter into debt bondage agreements and involuntary service, often enforced by threats of violence. Many newly arrived migrants become part of their benefactor's workforce until smuggling debts are paid off, while others are allowed to obtain employment elsewhere and make periodic payments to their benefactors.

International trafficking occurs when unsuspecting people are lured and deceived by promises of a better life but are subsequently coerced into forced labor and sexual services. In the Palermo Protocol and other legal conventions, coercion (whether directly or indirectly via the threat of physical force against them or their relatives back home) is the key difference between human smuggling and trafficking. Drugs and witchcraft (*juju* can be used to get victims to take an oath of loyalty to traffickers) are used to control workers in the commercialized sex industry, and confiscation of identity documents is also common.

Some of the worst cases of human trafficking involve child migrants ending up in modern slave-like conditions (working as street hawkers, laborers, and sex workers) or, in others, having their organs removed. Forced labor typically takes place in the informal economy or on the margins of the formal economy, where workers are not legitimate entrants into the national labor market and so are severely deprived in terms of protection. Advocacy campaigns and documentary film (e.g., *Dark Side of Chocolate*) have shown trafficked child labor being deployed by various commercial interests—for example, cocoa plantations in West Africa—and as domestic help in homes, sometimes laboring for distant relatives (occasionally even in their own countries). Wealthier African families typically have several laborers working as house help, and some people remain in slave-like employment conditions that are hidden behind residential walls, without any labor protection or recourse. Modern-day slavery flourishes because of its profitability: it is the second largest illicit profit industry in the world (after narcotics), accounting for approximately US$30 billion per annum, and the fastest-growing criminal enterprise, expected to surpass drug trafficking in five years (Polaris Project 2013). Criminal syndicates are attracted to the profitability of sex trafficking because, unlike drugs, humans can be sold repeatedly.

The clandestine nature of the activities means that the full extent of trafficking cannot be ascertained. Hundreds of thousands of African men, women, and children at any given time are in the process of being recruited, entrapped, transported, and exploited by human traffickers (Fitzgibbon 2003). The global trafficking business is characterized by a high supply of desperate individuals and an equally large demand for inexpensive labor and commercial sex workers. Established cultural norms of child migration throughout many parts of Africa on the basis that "relatives" elsewhere are better positioned to take care of the children and the propensity to rely on child laborers make it very difficult to uncover evidence of child trafficking.

Human trafficking flourishes because of porous borders, the involvement of organized crime syndicates, the ease with which border officials look the other way, and the collusion between criminal gangs and the police (Horwood 2009). Only around half of African countries have enacted anti-trafficking legislation. Enforcement is an even greater challenge; it has concentrated on forced sex workers and has accomplished much less in terms of identifying, much less tackling, other forced labor or organ harvesting. Trafficking is conducted at low risk and high profits to the organizers, who establish the network at arm's length, making their detection and eventual conviction difficult. International organizations and NGOs have assembled many testimonies and photographic narratives of victims, but still surveillance and enforcement are weak. Victims rarely pursue legal justice and bear witness against traffickers, fearing retaliation or recrimination against their families and villages. Mistrust of police and their possible complicity further compel victims to remain silent. Fear and abuse are common, and trafficking survivors typically struggle with post-traumatic stress disorder.

Trafficking can span numerous states and link, directly or indirectly, specific tasks such as recruiting, transporting (via trucks, buses, bush guides, etc.), transferring, harboring, and receiving persons at their final destinations. There are cases where small syndicates organize and operate all of the stages, but more commonly the trafficker subcontracts tasks along the way and chooses from a variety of facilitators who compete and offer their specific trafficking services. These chains make prosecuting trafficking crimes especially difficult, as laws are written to target multiple criminal elements rather than isolated acts of kidnap, assault, rape, extortion, and illegal border crossing, for instance.

Most African states are affected by trafficking to various degrees, but impoverished and conflict zones are especially affected; they can serve as origin, transit, and even destination locales. There are general geographical patterns to origin, transit, and destination areas, but these are not discrete categories and often overlap. For example, some transit countries (e.g., Kenya and Mozambique) are also source and destination countries and also transit countries for smuggling people from the Horn to South Africa (e.g., DRC and Burundi). The Western Cape province in South Africa is a key trafficking point for people brought from Asia and the Middle East bound for North America and Europe (See Box 7.5 for a case study of migrant journeys to South Africa). Some smuggled migrants change their objective during their journey. For example, a refugee originally intending to reach South Africa may take up residence in one of the countries along the route or may decide to move outside the region. Others experience harsh forms of detention, relocation, and deportation along the route. Key safe houses may be used to hold and process smuggled persons or those to be channeled into the sex industry. However, most migrants are drawn to large and economic vibrant metropolitan areas such as Johannesburg, where the size of the immigrant population is hotly debated (See Box 7.6 on The Johannesburg Hub: How Many Foreigners Are Present?)

## BOX 7.5 MIGRANT JOURNEYS IN PURSUIT OF THE SOUTH AFRICAN DREAM

South Africa, with its strong economy and good infrastructural system, is a migrant hub. The country is the largest single recipient of asylum applications in the world: its refugee and migration policy offers all irregular migrants the chance to seek asylum once they are in the country. South Africa and Kenya have reputations as venues for solving documentation and visa issues. For example, after acquiring a Kenyan passport, irregular migrants headed south can acquire a visa for South Africa from facilitators in Nairobi for between US$400 and US$600 (Horwood 2009:13), and documents are transacted with similar ease in South Africa. Johannesburg is a major end destination as well as the most reliable departure destination to exit Africa by air.

South Africa has more than 210,000 asylum cases pending (half of the applications are from Zimbabwe) and 57,889 registered refugees (UNHCR 2013b). In addition, tens of thousands of people are trafficked to and through South African every year. Harwood (2009:7) estimates that between 17,000 and 20,000 males undertake a journey to South Africa every year from the Horn, but not all make it there. The number of women and children is estimated to be lower, possibly in the low thousands. South Africa is the top smuggling route in Africa for aspirant migrants, particularly for those originating in the Horn, Great Lakes, and Southern African regions. The journey to South Africa is perceived as less dangerous and difficult than crossing the Gulf of Aden into Yemen (a route with a reputation for violence and fatality) or traveling across the Sahara to Libya and on to Europe (a very difficult and expensive route).

On average, it is an eight-week journey for Ethiopian or Somali migrants to reach South Africa. The sizable Ethiopian and Somali diaspora (estimated to be larger than 2 million people) plays a critical role in financing many migrants' passage to South Africa (Horwood 2009). The key receiving payment locations appear to be in Nairobi, Addis Ababa, and various South African cities, but a range of alternative agents and intermediaries facilitate parts of deals, and as a result payments can be made at a variety of other locations as well.

Various modes of transportation are used to make the journey to South Africa. Choice of passage depends on the economic status of the migrant but also on the options presented at any given time. Kenyans have the greatest ease traveling to South Africa, and most obtain visas and subsequently travel by bus to South Africa or by long-haul truck. Kenyans rarely travel in groups except in buses and opt to pay bribes to ensure safe passage. In some instances, bus drivers assume smuggler roles by brokering deals with immigration officials on migrants' behalf. Some Kenyans are even able to make their way to South Africa without documentation.

Horn of Africa irregular migrants have a more difficult time in their passage, and many do not have the required funds to complete the journey, so working along the way and spending time in Nairobi is common. Typically, Somali and Ethiopian groups start off with 15 to 30 migrants, but the numbers expand to the low hundreds by the time they move into the countries bordering South Africa. Reportedly, grouping is a cost-saving measure to move large numbers of people at the same time, and it minimizes negotiations and bribes paid to officials. Usually, migrants are taken overland the entire way, but some routes combine boat and road travel or air and road, or other various combinations. A myriad of unofficial rural tracks (known as "*panya* routes" in East Africa) are used to crisscross countries. Migrants routinely have to walk some stretches along the journey, especially in border regions, sometimes for days at a time.

(*Continued*)

**BOX 7.5 (*Continued*)**

The overland route from Ethiopia and/or Somalia to South Africa with the fewest border crossings passes through only three countries (Kenya, Tanzania, and Mozambique), but few migrants are channeled along the most direct route (See figure 7.4). Most stop in Nairobi, which serves as a migrant interchange hub. Refugee camps and Nairobi neighborhoods with high concentration of Ethiopians and Somalis (e.g., Eastleigh and Kariobangi) are key staging points at which members of their clans/ethnic groups work out arrangements. In these locations, there are safe houses for smuggled persons as well as clandestine offices where travel documents can be obtained. Many irregular migrants reinvent themselves in Nairobi with various identification cards, birth certificates, or passports (Somalis are known to obtain Ethiopian forged identities because Somali passports are not accepted in South Africa).

Safe passage to South Africa is far from guaranteed. Migrants can face deportation along the way, some spend time in prisons (often until the smuggler can raise the funds for their release), and others are left waiting at safe houses, forests, or beaches for periods of time or become stranded indefinitely. Bribes, harassment, and exploitation are common hallmarks of the migrant experience. Irregular migrants' transit across countries is characterized by significant night travel, cramped traveling spaces (containers are among the most dangerous), hiding in woods and safe houses, robberies, beatings, arrests, and multiple shakedowns for bribes by various officials. As a result, the physical and mental conditions of migrants often deteriorate, and unknown numbers of people are killed along the way. Exposure to considerable exploitative situations and prejudice in South Africa is the price that many accept for new lives of "opportunity." For some migrants, the South African dream turns into a nightmare. Hundreds of Somalis now disillusioned about their lives in South Africa are returning every week to Somalia, where political conditions are stabilizing and now offer them an alternative. Nevertheless, others still arrive, and working for months and even years in South Africa is viewed as a necessary step in joining the diaspora and in raising enough funds for sponsoring the next generation of migrants.

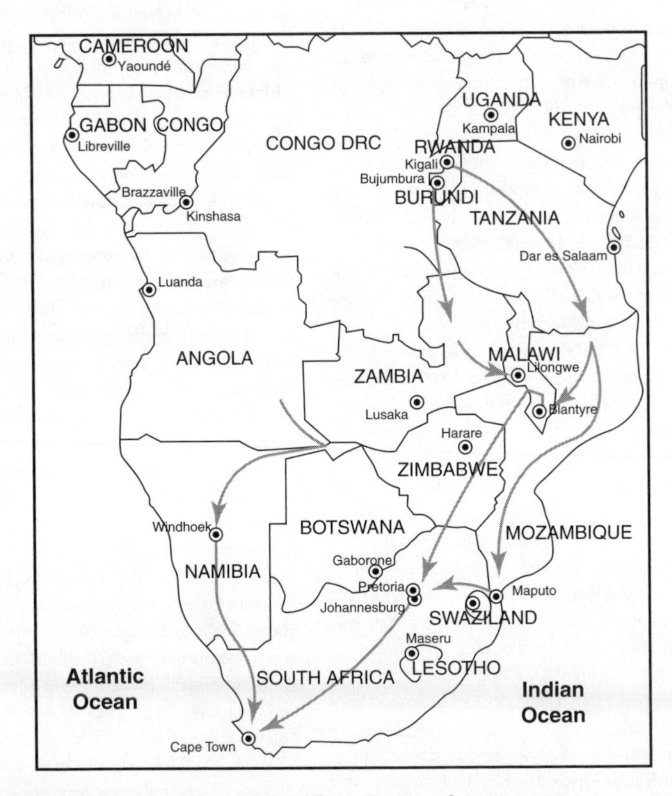


**FIGURE 7.4** Human Trafficking in South Africa. Adapted from Horwood 2009, p. 25.

### BOX 7.6 THE JOHANNESBURG HUB: HOW MANY FOREIGNERS ARE PRESENT?

In 1994, the South African police estimated that there were 2 million undocumented migrants in the country. In 1995, they came up with the figure of 8 million—20% of the South African population—an implausible figure that was highly questioned. A survey by the Center of Socio-Political Analysis of the Human Sciences Research Council (HSRC) revealed an estimate of 5 million to 6 million (the survey asked respondents how many foreigners they knew in their surrounding area). These figures confirmed popular suspicions, were quickly bandied about in official speeches, and became popular knowledge. The HSRC withdrew the figures in the early 2000s on the basis of a flawed methodology (Segatti and Landau 2011).

The effects of these counts on political rhetoric and popular discourse cannot be overstated. Their numbers shaped immigration policy and were used to shore up government failure to deliver jobs (the government alluded to a figure based on a flawed estimate that coincidentally corresponded to a significant proportion of the unemployed). They have fanned xenophobic fears and popular unexamined views that foreigners have "taken" South African jobs as well as South African women. Xenophobic attacks and police harassment of undocumented immigrants have been common. In 2008, violent attacks on immigrant entrepreneurs erupted in several locations of South Africa. The xenophobic riots, as they became known, resulted in 62 deaths and 670 wounded people. More than 150,000 were displaced and forced to leave South Africa. Tensions against international migrants also run high in townships: Somali, Bangladeshi, and Chinese shops are often the object of robberies, extortion, and attack (Fig. 7.5).

The Southern African Migration Project and the African Centre for Migration and Society generated more modest figures of about 3 million, a number that would need to be revised upward with the Zimbabwean crisis beginning in 2000. Surprisingly, these more reasoned estimates by leading migration academics were not incorporated into political or popular discourse.

**FIGURE 7.5** Somali Shop in Thokoza Township, Southeast of Johannesburg. Photograph by Dan Thompson.

Estimating the number of South Africans who are moving to urban areas is very difficult. Domestic migration is far more significant than international migration in terms of local government challenges. The importance of domestic migration is particularly evident in peri-urban areas around Johannesburg, Tshwane, and Cape Town, which the number of migrants increased by 20% to 35% over a seven-year period (Segatti and Landau 2011). By global standards, these are extraordinary figures.

## OUT OF AFRICA: U.S. AND EUROPEAN MOVEMENTS

A growing number of Africans are making their way to Europe and North America and recently to China and the Middle East. The number of self-identified members of the African diaspora living in the United States was 3.5 million in 2009 (Capps, McCabe, and Fix 2011). Although the number of black Africans as a proportion of the total U.S. foreign-born population is small (about 3%), they have been among the fastest-growing groups in U.S. immigration since 2000 (Capps, McCabe, and Fix 2011). Net African immigration to the United States is even larger, as white and biracial Africans are recorded in different U.S. census categories.

Large-scale voluntary migration from Africa to the United States is recent. Severe restrictions were placed on flows when the slave trade ended. It was not until the 1965 immigration reforms, when national origin quotas were lifted and caps on non-European migration were eliminated, that the door opened for African immigration. Small numbers of Africans have pursued higher educational opportunities for many decades, and some have stayed on, securing positions at universities, colleges, and hospitals. For example, Africans regularly taught Africa courses in universities. Still, distance, travel costs, and admission procedures proved to be significant obstacles for unskilled Africans.

President Barack Obama's two terms in office have bolstered awareness of the presence and sacrifices of

African immigrants and their families/descendants, and his success has gone some way to raising the profile of Africans in the U.S. and world imagination. The geography of African immigrant settlement in many ways mirrors some of the major patterns of African-American concentrations. African immigrants (as the U.S. black population) are concentrated in New York, Texas, Florida, California, and Illinois. Large metropolitan concentrations of African immigrants are found in Washington, D.C.–Arlington–Alexandria and New York–northern New Jersey–Long Island. Also, noteworthy is Minneapolis–St. Paul, where one in five of the immigrant population is African-born. However, unlike the black population, fewer Africans reside in rural Southeastern states (e.g., Louisiana, Mississippi, Tennessee, and South Carolina).

In general, in the United States, immigrants from Africa have a high educational attainment, higher than other immigrant groups and higher than the native-born population. The United States, along with Canada and Australia, disproportionately attracts better-educated African immigrants, while less-educated migrants move to the UK, France, and other European countries. There are, however, immigrants from refugee-originating countries in the United States (e.g., Somalia) who have attained much lower levels of education.

There are four main channels by which African immigrants can enter the United States legally. First, most are admitted through family reunification channels. Second, some groups have a high proportion of people accepted through employment (e.g., white South Africans and white Zimbabweans). Third, immigrants from countries such as Burundi, Somalia, DRC, Congo, and Rwanda tend to gain admission from refugee resettlement and asylum programs. Fourth, the diversity visa program has increased flows from underrepresented African countries by allowing immigration by individuals from countries such as Benin and Cameroon who would not qualify under the other schemes.

Multiple channels for immigration mean that the African-born population in the United States has never been so culturally, linguistically, and nationally diverse. Africa-to-United States migration is no longer dominated by flows from English-speaking countries, whereby Nigerians, Ghanaians, Kenyans, and Liberians heavily dominated in previous decades. African nationals from French-, Portuguese-, and Arab-speaking countries have moved beyond traditional migrant destinations in Europe and North Africa to the United States and elsewhere.

In Europe, labor migration from Africa has taken place for decades. Large flows of Senegalese and Malian males to France have occurred since the 1970s. Africa-wide flows to a greater number of European countries (the UK, Italy, Germany, Spain, Portugal, and Ireland) have occurred since the 1980s. There are an estimated 3.5 million to 8 million Africans in the European Union. North Africans, especially Moroccans, Algerians, and Tunisians, dominate the flows northward, but, in practice, many Europeans fail to differentiate between migrants from these countries and those from other African countries, leading to hype about an "African invasion of Europe" (de Haas 2008).

Many Africans enter Europe by traditional routes (e.g., ports and airports), but several nonconventional routes are well traveled (Fig. 7.6). Several well-traversed cross-border circuits are used prior to difficult sea crossings. Many migrants from East Africa/the Horn journey through transit circuits via Sudan–Egypt and/or Sudan–Egypt–Libya as well as routes from Nairobi (Kenya) to Bossasso (Somalia) and from Addis Ababa (Ethiopia) to Obock (Djibouti). Several West African routes are prominent. First, there is a land route drawing people from Gulf of Guinea countries (Ghana, Nigeria, and Côte d'Ivoire) that connects via the migration hub of Agadez (Niger) to Libya and Algeria. A second land route draws others from Gulf of Guinea countries and connects via Mali to the southern Algerian migration hub of Tamanrasset and/or via Senegal through Mauritania to Morocco. A third route runs from various West Africa ports to the Canary Islands (a Spanish island located off the Moroccan coast). A fourth route is via major airports in the West African region (e.g., Accra, Lagos, Dakar, Bamako, and Abuja). Many Africans ply these routes independently, but human smugglers and criminal syndicates guide other African migrants. Smugglers' help is typically solicited to make the illegal, perilous boat journey across the Mediterranean Sea to Europe. Thousands of Africans have perished in this dangerous crossing: 1,500 died in 2011, a record-high year.

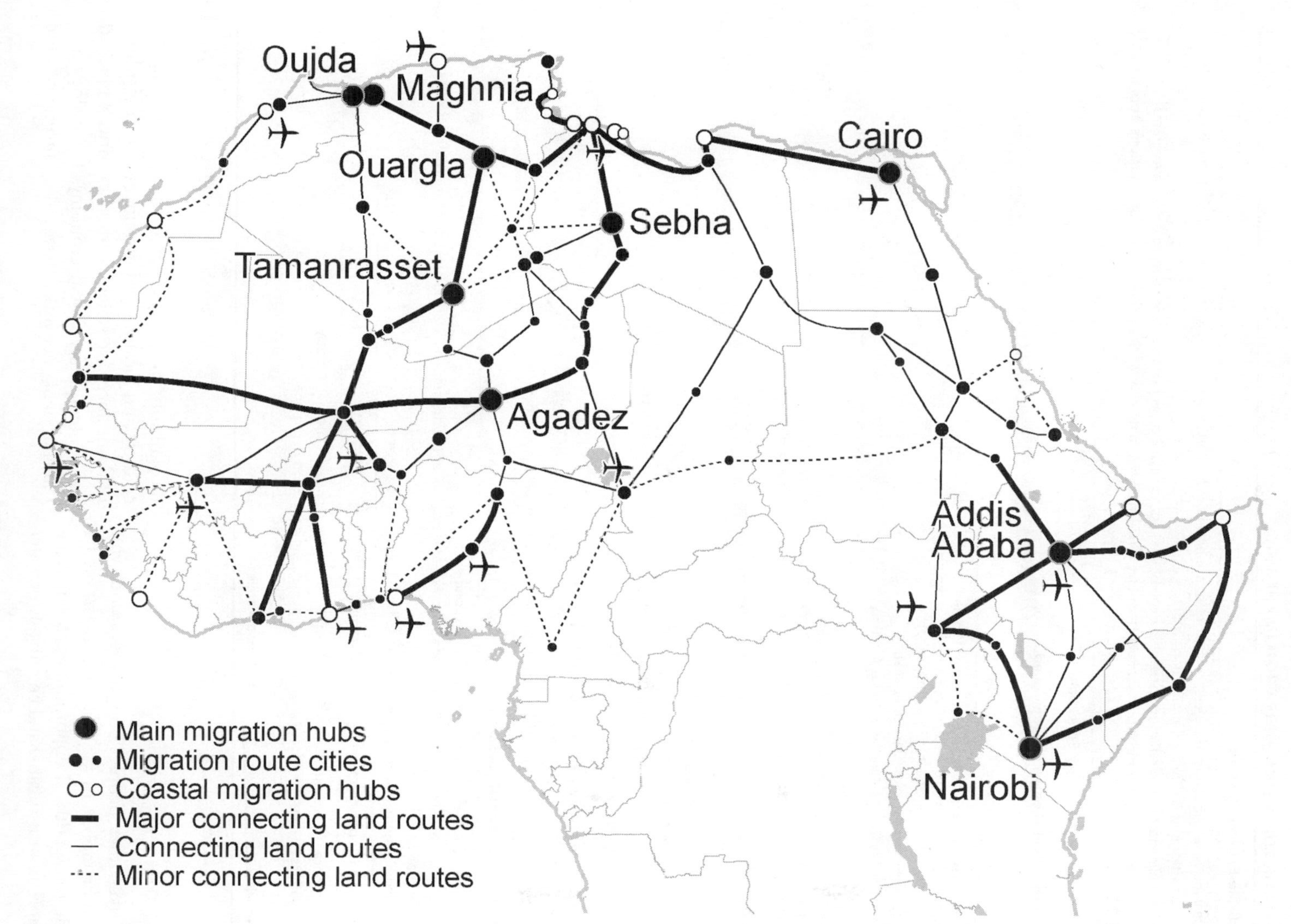


**FIGURE 7.6** Trans-Saharan Migration Routes. *Source:* Adapted from MTM iMap. 2012. Map on Irregular and Mixed Migration Routes.

**TABLE 7.2 MAINSTREAMING MIGRATION IN THE MDGs**

*MDG 1. Eradicate Extreme Poverty*

- Remittances: increase remittance transfers; create better opportunities for investment of remittances in productive and employment-generating activities and agricultural production
- Migrant rights: eliminate exploitation of vulnerable migrant workers; facilitate cross-border movements to enhance migrants' business opportunities; encourage urban agriculture among rural–urban migrants and facilitate the transfers of foodstuffs from rural to urban areas

*MDG 2. Achieve Universal Primary Education*

- Harness the contribution of the diaspora for education
- Promote remittances to be invested in education
- Remove barriers to migrants' accessing education

*MDG 3. Promote Gender Equality and Empower Women*

- Address gender discrimination in migrant households
- Identify obstacles to girls' migration for education
- Harness remittances for girls' and women's education

*MDG 4. Reduce Child Mortality*

- Address health needs of pregnant migrants and newborns
- Reduce out-migration of health care professionals from vulnerable states/localities
- Develop special programs for the children of migrants

*MDG 5. Improve Maternal Health*

- Ensure migrants' access to health care
- Eliminate gender-based violence against migrant women
- Promote exchange of health care professionals between over- and underserviced countries/locations

*MDG 6. Combat HIV/AIDS, Malaria, and Other Diseases*

- Understand the role of migration in the diffusion of diseases
- Improve access of migrants and other mobile people to health care and medicines
- Eliminate migration systems that forcibly separate migrants from spouses/partners

*MDG 7. Ensure Environmental Sustainability*

- Avoid resettlement in marginal or unsustainable environments
- Ensure migrants are not involved in unsustainable resource exploitation, and eradicate unsustainable practices triggering further population migration
- Eliminate discrimination against migrants' access to land, housing, and services

*MDG 8. Develop a Global Partnership for Development*

- Liberalize temporary freedom of movement of workers to supply services in another country
- Promote good governance of global migration and mainstream migration in regional, national, and urban and rural development plans
- Strengthen efforts to manage migration at all spatial scales

*Source:* Crush and Frayne 2010:11–13.

## CONCLUSIONS

Africans, like other people around the world, will continue to be mobile in the 21st century. The media have made the wealth of opportunities that may be possible abroad more visible and enticing. Far greater recognition of differential earning potentials by country and location and the emergence of a culture of remittances are salient features of Africa's place in the contemporary global economy. Technology and transnational networks have also prompted the movement of people across international borders. For the foreseeable future, the dynamics motivating people to leave their home country in search of work and opportunity look set to continue and even expand in the African region.

Opportunities to emigrate legally are severely limited, and this is unlikely to change. In a globalized world, irregular migrants remain highly vulnerable. There will always be a lucrative market for smugglers and traffickers and their foot soldiers, transnational criminals, corrupt officials, document forgers, and other facilitators of the illicit transport of people across borders. Migrants are generally unaware of their human rights, and in their calculus of alternatives and leaps of faith to fulfill their dreams, they come to accept (at some level) increased vulnerabilities, abuse, and violence as part of their migrant journey and often their everyday existence (Horwood 2009).

The movement of peoples across states borders in Africa reveals the frailty of African states' law-and-order regimes and their inability to control cross-border flows. If people can move undetected across borders, contraband, arms, narcotics, and terrorists can do so as well. The scope and magnitude of illegal movements reveal the power and successes of transnational criminals, and the profits earned from illegal activities further undermine governments' efforts to establish the rule of law. Terrorist groups are now trafficking players in the East Africa region and elsewhere. Clearly, greater political will and international support are needed to curtail international trafficking and human smuggling so that Africans are not deprived of their human rights and that "modern-day slavery" is abolished (Fitzgibbon 2003). As it stands, African states lack the capacity (financial, technical, legal, humanitarian) to exercise their protective responsibilities to their citizenry. Many people are forced to migrate because their states have failed them. Migration and development form a very complex dynamic, and no simple answers exist about whether underdevelopment leads to migration or whether development draws migrants; it is, however, high time that Africa national development plans and international development strategies, especially those related to achieving the MDGs, were better integrated (Table 7.2).

Unfortunately the MDGs were formulated before migration was recognized as a critical development cross-cutting issue. Migration researchers have elaborated on how strengthening migrants' rights, reducing migrant vulnerabilities, and harnessing the knowledge, skills, and resources of the diaspora can be used and incorporated into development policies (Crush and Frayne 2010). Going about this in sensible, systematic, and balanced ways will provide the greatest benefits to the greatest number of people.

## REFERENCES

Al-Jazeera. 2011. "Dadaab, the World's Biggest Refugee Camp." Available at http://www.aljazeera.com/indepth/features/2011/07/2011718284487 6473.html (accessed November 10, 2013).

Appiah-Yeboah, K., A. Bosomtwi, and M. Yeboah. 2013. "Factors Impacting Remittance by Skilled Ghanaians Abroad." *International Migration* 51(1):118–132.

Bakewell, O. 2007. "Keeping Them in Their Place: The Ambivalent Relationship between Development and Migration in Africa." Working Paper No. 8, International Migration Institute (IMI). James Martin 21st Century School, University of Oxford, UK.

Blatter, S. 2003. "Soccer's Greedy Neo-colonialists." *Financial Times,* December 17, p. 19.

Capps, R., K. McCabe, and M. Fix. 2011. "New Streams: Black Migration to the United States." Migration Policy Institute (MPI), Robert Schuman Centre for Advanced Studies, European University Institute. Available at http://www.migrationpolicy.org/pubs/africanmigrationus.pdf (accessed February 5, 2013).

Crisp, J. 2010. "Forced Displacement in Africa: Dimensions, Difficulties, and Policy Directions." *Refugee Survey Quarterly* 29(3):1–27.

Crush, J., and B. Frayne, eds. 2010. *Surviving on the Move: Migration, Poverty and Development in Southern Africa.* Cape Town: International Development Research Center (IDRC), the Development Bank of Southern Africa and South African Migration Programme (SAMP).

Darby, P. 2007. "Out of Africa: The Exodus of Elite African Football Talent to Europe." *Working USA: The Journal of Labor and Society* 10(12):443–456.

de Haas, H. 2008. "The Myth of Invasion: The Inconvenient Realities of African Migration to Europe." *Third World Quarterly* 29(7):1305–1322.

Dialogue on Mediterranean Transit Migration iMap. 2012. Map on Irregular and Mixed Migration Routes. Available at http://www.imap-migration.org/index.php?id=470 (accessed October 28, 2013).

Easterly, W., and Y. Nyarko. 2008. "Is the Brain Drain Good for Africa?" Brookings Global Economy and Development Working Paper No. 19. The Brooking Institute, Washington, D.C.

Fitzgibbon, K. 2003. "Modern-day Slavery? The Scope of Trafficking in Persons in Africa." *African Security Review* 12(1):81–89.

Forced Migration Online. 2013. "Africa." Available at http://www.forcedmigration.org/research-resources/regions/africa (accessed February 20, 2013).

Gill, N. 2010. "'Environmental Refugees': Key Debates and the Contributions of Geographers." *Geography Compass* 4(7):861–871.

Grant, M. 1995. "Movement Patterns and the Medium-Sized City: Tenants on the Move in Gwery, Zimbabwe." *Habitat International* 19(3):357–369.

Grant, R. 2009. *Globalizing City: The Urban and Economic Transformation of Accra, Ghana*. Syracuse, NY: Syracuse University Press.

Horwood, C. 2009. *In Pursuit of the Southern Dream: Victims of Necessity: Assessment of the Irregular Movement of Men from East Africa and the Horn to South Africa*. Geneva: International Organization for Migration.

Interactive Map on Migration (iMap). 2014. Map on Irregular and Mixed Migration Routes–West, North and East Africa, Europe, Mediterranean, and Middle East. Available at http://www.imap-migration.org/index.php?id=470 (accessed March 7, 2014).

Konseiga, A. 2005. "New Patterns in Human Migration in West Africa." *Vienna Journal of African Studies* 8:23–46.

Kress, B. 2006. "Burkina Faso: Testing the Tradition of Circular Migration." Available at http://www.migrationinformation.org/USFocus/display.cfm?ID=399 (accessed January 18, 2013).

Long, K., and J. Crisp. 2011. "In Harm's Way: The Irregular Movement of Migrants to Southern Africa from the Horn and the Great Lakes Regions." New Issues in Refugee Research, Paper No. 200, The United Nations High Commission for Refugees, Policy Development and Evaluation Service.

Mohapatra, S., and D. Ratha, eds. 2011. *Remittance Markets in Africa*. Washington, D.C.: The World Bank.

Norwegian Refugee Council and Internal Displacement Monitoring Centre. 2013. *Global Overview 2012: People Internally Displaced by Conflict and Violence*. Geneva.

Organization of African Unity (OAU). 1974. *OAU Convention Governing the Specific Aspects of Refugee Problems in Africa*. Addis Ababa.

Owusu-Ansah, J., and K. O'Connor. 2010. "Housing Demand in the Urban Fringe Around Kumasi, Ghana." *Journal of Housing and the Built Environment* 25:1–17.

Polaris Project. 2013. "International Trafficking." Available at http://www.polarisproject.org/human-trafficking/international-trafficking (accessed March 13, 2013).

Poli, R. 2007. "African's Status in the European Football Players' Labour Market." *Soccer & Society* 7(2–3):276–291.

Potts, D. 2011a. *Circular Migration in Zimbabwe and Contemporary Sub-Saharan Africa*. Woodbridge, UK: James Currey.

Potts, D. 2011b. "Making a Livelihood in (and Beyond) the African City: The Experience of Zimbabwe." *Africa* 81(4):588–605.

Prothero, R. 1965. *Migrants and Malaria*. London: Longman.

Reader, J. 1999. *Africa: A Biography of the Continent*. New York: Knopf.

Sander, C., and S. Maimbo. 2003. "Migrant Labor Remittances in Africa: Reducing Obstacles to Development Contributions." Africa Region Working Paper Series no. 64, The World Bank. http://www.worldbank.org/afr/wps/wp64.pdf (accessed January 19, 2013).

Schaeffer, E. 2008. "Remittances and Reputations in Hawala Money Transfer Systems: Self-enforcing Exchange on an International Scale." *The Journal of Private Enterprise* 24(1):1–17.

Scherrens, J. 2007. "The Muscle Drain of African Football Players to Europe: Trade or Trafficking?" MA thesis, Human Rights and Democratisation, University of Graz, Austria.

Segatti, A., and L. Landau, eds. 2011. *Contemporary Migration to South Africa: A Regional Development Issue.* Washington, D.C.: The World Bank.

Sports Business Daily. 2012. "Eto'o is the Best Paid Footballer; Ronaldo and Messi Round Out the Top 10." Available at http://www.sportsbusinessdaily.com/Global/Issues/2012/09/05/International-Football/Top-10.aspx (accessed February 4, 2013).

United Nations (UN). 2013. "Côte d'Ivoire." http://data.un.org/CountryProfile.aspx?crName=Côte%20d'Ivoire (accessed January 18, 2013).

United Nations High Commission for Refugees (UNHCR). 2012. *A Year of Crises: UNHCR Global Report 2011*. Geneva: UNHCR.

United Nations High Commission for Refugees (UNHCR). 2013a. "UNHCR Country Operations Profile: Burkina Faso." Available at http://www.unhcr.org/pages/49e483de6.html (accessed February 24, 2013).

United Nations High Commission for Refugees (UNHCR). 2013b. "UNHCR Country Operations Profile: South Africa." Available at http://www.unhcr.org/pages/49e485aa6.html (accessed February 24, 2013).

Wong, M. 2006. "The Gendered Politics of Remittances in Ghanaian Transnational Families." *Economic Geography* 82(4):355–381.

World Bank. 2013. Annual Remittance Data. Available at http://econ.worldbank.org/WBSITE/EXTERNAL/EXTDEC/EXTDECPROSPECTS/0,,contentMDK:22759429~pagePK:64165401~piPK:64165026~theSitePK:476883,00.html (accessed January 20, 2013).

Wouterse, F. 2011. "Continental vs. Intercontinental Migration: An Empirical Analysis of the Immigration Reforms on Burkina Faso." Organization of Economic Cooperation and Development (OECD), Development Center, Working Paper No. 299. Paris: OECD.

Yeboah, I. 2008. *Black African Neo-Diaspora: Ghanaian Immigrant Experiences in the Greater Cincinnati, Ohio Area*. Lanham, MD: Lexington Books.

Zoomers, A., and G. Nijenhuis. 2012. "Does Migration Lead to Development? Or Is It Contributing to a Global Divide?" *Societies* 2:122–138.

## WEBSITES

The Guardian 2012. *Burkina Faso: Human Trafficking in West Africa*. http://www.guardian.co.uk/global-development/gallery/2012/jun/28/burkina-faso-human-trafficking-africa-in-pictures#/?picture=391321486&index=5. Human trafficking in Burkina Faso in pictures.

Mac Bryla. "Football player transfers in 60 seconds." Available at http://vimeo.com/77775892). Mapping African soccer player export flows, 1900–2013.

MikiMistrati 2010. *Dark Side of Chocolate.* Documentary film available at http://www.youtube.com/watch?v=7Vfbv6hNeng (accessed December 5, 2013).

Reconnect Africa. Available at http://www.reconnectafrica.com (accessed December 8, 2013). Online magazine for African professionals in the diaspora.

UNICEF 2010. *Not My Life*. Available at http://www.youtube.com/watch?v=P9UMqH2v1is (accessed December 5, 2013). Glenn Close narrates a film about the human trafficking of children.

CHAPTER 8

# WATER

## INTRODUCTION

Water is a vital substance for human survival. Safe water is required for drinking, hygiene, and food preparation; adequate water is needed to produce energy and to support economic activities in both urban and rural settlements. An essential resource for economic development, water is critical not only to food production but to most production processes, from microchips to potato chips. Historical patterns of human settlement have reflected the distribution of adequate water resources; for example, the emergence of early civilizations along the Nile River was closely tied to the availability of water.

A now-salient theme in books, at international conferences, and in media reports is that global water demand is spiraling out of control and a water crisis is looming (Jones 2010; Oakland Institute 2011a). The strain on water resources is greater now than it has been at any time in human history, and the current global water trajectory looks unsustainable. Water demands have increased on many fronts: rising population and the urban revolution require more food and more from agriculture, and industrial and domestic needs have increased significantly. Industrial and domestic water demand quadrupled in the second half of the 20th century. In Africa, increased water scarcity, the shrinking of some water bodies (e.g., Lake Chad, Lake Turkana), desertification, and deforestation have been documented. Water is one of the most local resources, and its movement is expensive and difficult (e.g., 1 $m^3$ weighs .35 tons). Considerable research is devoted to quantifying how much water is currently available, assessing areas of "water poverty," and projecting future water withdrawals (Jones 2010). Some innovative techniques for mapping water resources at the national and regional levels are emerging (MacDonald et al. 2012).

Water is moving up on the 21st-century political agenda. It is becoming one of the most important security challenges, and future water scarcity may threaten economic and social gains, undermine the global system, and quite possibly trigger widespread conflict and even war (although violent water conflicts have been rare to date). Sporadic conflict has occurred, for example, among Kenyan and Ethiopian pastoralists over access to a declining Lake Turkana, and ongoing projects to build a number of dams (for hydropower generation) on the Omo River (Ethiopia) that feeds the lake will escalate tensions (Fig. 8.1). Broad international frameworks and transboundary agreements are urgently needed to avoid conflict.

Water and sanitation goals were incorporated into the United Nations (UN) Millennium Development Goals (MDGs) in 2000. MDG 7c sets a target of halving the number of people without access to safe water and sanitation by 2015. The UN now provides a global water audit via its World Water Assessment Program and convenes regular World Water Forums to inform political leaders, policymakers, and the public about

**FIGURE 8.1** Lake Turkana, Kenya: Water Amidst a Barren Landscape. *Source:* © Toby Adamson/Design Pics/Corbis.

water management and its development dimensions. The World Economic Forum, the annual meeting of influential corporate leaders at Davos, Switzerland, also recognizes water as a key global issue. At the 2009 meeting, water rights were hotly debated. Many stakeholders favored a basic right to water, but international financial institutions (e.g., the World Bank) and corporate interests advocated water privatization. An uneasy compromise was reached at Davos: governments committed to provide citizens with up to 25 liters of water per day for drinking and hygiene but acknowledged that water use above this level would incur charges (Jones 2010:10).

The tilt toward privatization has sparked a global water justice movement. Since 2003, grassroots activists have held an annual concurrent Alternative Water Forum focusing on their vision of water as a resource that should be held in common, opposing the water privatization movement. African nongovernmental organizations have formed around such a "water commons" vision and have played an integral part in developing a world water vision. In July 2010, the UN endorsed the grassroots vision, passing a landmark resolution that access to water and sanitation is a "human right that is essential to the full enjoyment of life" (Resolution 64/292) (UN 2010). The vast majority of countries supported the resolution (no state opposed it), but Canada, the United States, the UK, Australia, and Botswana abstained from voting, weakening the global agreement. The politics surrounding the resolution ensures that the private-versus-commons debate remains unresolved and is sure to become even more contentious.

## AFRICA'S WATER

At an aggregate level, Africa is the world's second driest region, surpassed only by Australia. However, overlooking temporal and spatial climate variability (for the moment), Africa has abundant rainfall and relatively low levels of water withdrawal for the three major uses of water—agriculture, community use, and industry; combined, these three account for approximately 3.8% of total annual water resources (UNEP 2009). Water resources are unevenly distributed throughout the region, plentiful in some counties and scare in others. Central Africa has abundant water, with 50% of Africa's water resources, whereas West Africa contains approximately 23% of the continent's water (UNEP 2010). Existing (known) water resources are limited in other African regions, but aquifers may represent an untapped resource (UNEP 2010).

In Africa, access to water is a critical issue. Distribution and reliability problems impede development throughout the region; the result is that two thirds of Africans lack proper sanitation facilities and 40% lack reliable access to safe water (WHO/UNICEF 2008). Lack of access to safe drinking water and adequate water for hygiene is a major cause of health and disease throughout the region. Moreover, large portions of the available water supplies are undermined by human activities such as rapid urbanization without proper water and sanitation infrastructure (e.g., discharges of untreated wastes, solid waste thrown into storm drains, and liquid leaching from refuge dumps), poor land use and agriculture practices (fertilizer and pesticide runoff), and mining and industrial activities (petroleum refining), many of which contaminate the region's water supplies.

Africa's climate is a fundamental source of its water challenges. Straddling the Equator and extending beyond both tropics, the climates of Africa are dominated by the Hadley cells. These circulation cells create both the excessive rains on the rising limb along the intertropical convergence zone around the equator and the desiccation of the adiabatically heated air on the sinking limb around the subtropics, suppressing convective activity and rain formation. They contribute to a highly uneven distribution of water, ranging from desert to equatorial rainforest. Climatic conditions also result in high evaporation losses of water, reflected in higher unusable and nonrenewal resources compared to other world regions (UN-Water Africa 2001). Fluctuations in climatic variability in recent decades have led to an increasing frequency of droughts, particularly in the Sahel and the Horn of Africa. At the same time, elsewhere in Africa, seasonal flooding and water overabundance appear to have increased in frequency (e.g., floods in the Zambezi Valley in 2000 resulted in a major flood in Mozambique).

Africa's water geography is composed of big rivers, large lakes, vast wetlands, and limited but widespread groundwater. Large rivers and lakes represent much of Africa's surface and groundwater resources. The region is home to 63 shared river basins. Rivers such as the Congo, Nile, Zambezi, and Niger hold vast amounts of water and flow through multiple states. Some of the world's largest groundwater lakes are located in the region. For example, Lake Victoria or Victoria Nyanza is the world's second largest freshwater lake, and Lake Tanganyika is one of the world's deepest and largest lakes. Africa's inland lakes support vital fisheries, particularly important to populations in Uganda, Tanzania, Kenya, Congo and the Democratic Republic of Congo (DRC). Wetlands are found in almost all countries and cover about 1% of the region's total surface. Examples of significant wetlands are the marshland along the White Nile in Central Sudan and the Okavango Delta, which straddles large areas of Botswana and Namibia and includes smaller areas of Angola and Zimbabwe. Wetlands are important for water storage and filtration as well as for food and wildlife habitats.

Countries with abundant water are Congo, Gabon, Liberia, Equatorial Guinea and Central African Republic, DRC, and Gabon; the latter two countries are among the world's top 10 natural water-available states (Table 8.1).

**TABLE 8.1 WATER AVAILABILITY IN AFRICA**

| Item | Country | International Rank | Water Resources [Total Resources: Total Renewable Per Capita (m³/capita/year)] |
|---|---|---|---|
| 1. | Congo | 7 | 275.679 |
| 2. | Gabon | 9 | 133.333 |
| 3. | Liberla | 15 | 79.643 |
| 4. | Equatorial Guinea | 21 | 56.893 |
| 5. | Central African Republic | 28 | 38.849 |
| 6. | Sierra Leone | 31 | 36.322 |
| 7. | Guinea | 36 | 27.716 |
| 8. | Guinea Bissau | 39 | 25.855 |
| 9. | Congo (DR) | 41 | 25.183 |
| 10. | Madagascar | 46 | 21.102 |
| 11. | Cameroon | 49 | 19.192 |
| 12. | São Tomé and Príncipe | 51 | 15.797 |
| 13. | Angola | 55 | 14.009 |
| 14. | Mozambique | 61 | 11.814 |
| 15. | Namibia | 65 | 10.211 |
| 16. | Zambia | 66 | 10.095 |

(Continued)

**TABLE 8.1 WATER AVAILABILITY IN AFRICA** *(Continued)*

| Item | Country | International Rank | Water Resources [Total Resources: Total Renewable Per Capita (m³/capita/year)] |
|---|---|---|---|
| 17. | Botswana | 71 | 9345 |
| 18. | Mali | 75 | 8810 |
| 19. | Gambia | 87 | 6332 |
| 20. | Chad | 90 | 5453 |
| 21. | Côte d'Ivoire | 92 | 5058 |
| 22. | Swaziland | 93 | 4876 |
| 23. | Mauritania | 95 | 4278 |
| 24. | Senegal | 96 | 4182 |
| 25. | Benin | 99 | 3954 |
| 26. | Togo | 109 | 3247 |
| 27. | Niger | 111 | 3107 |
| 28. | Uganda | 115 | 2833 |
| 29. | Ghana | 119 | 2756 |
| 30. | Tanzania | 124 | 2514 |
| 31. | Nigeria | 125 | 2514 |
| 32. | Sudan | 129 | 2074 |
| 33. | Ethiopia | 137 | 1749 |
| 34. | Eritrea | 139 | 1722 |
| 35. | Comoros | 140 | 1700 |
| 36. | Zimbabwe | 143 | 1584 |
| 37. | Somalia | 144 | 1538 |
| 38. | Malawi | 145 | 1528 |
| 39. | Lesotho | 147 | 1485 |
| 40. | South Africa | 150 | 1154 |
| 41. | Burkina Faso | 152 | 1084 |
| 42. | Kenya | 154 | 985 |
| 43. | Morocco | 155 | 971 |
| 44. | Egypt | 156 | 859 |
| 45. | Cape Verde | 158 | 703 |
| 46. | Rwanda | 159 | 683 |
| 47. | Burundi | 161 | 566 |
| 48. | Tunisia | 162 | 482 |
| 49. | Algeria | 163 | 478 |
| 50. | Djibouti | 164 | 475 |
| 51. | Libya | 174 | 113 |

*Source:* Njoh and Akiwumi. 2011. Water availability in Africa, p. 455.

DRC, in particular, contains enormous water resources (275,670 m$^3$ per capita). In DRC, the population size and the water availability are incongruent; it holds vast amounts of water for its population size (Njoh and Akiwumi 2011). Central Africa has huge potential to export water, as do parts of southwest and southeast Africa and West Africa. In particular, the Congo and Zambezi rivers have much untapped potential for fulfilling human needs.

Water scarcity is a critical development challenge in other areas. At a regional level, water capacity pressures are most salient in East Africa, the Sahel, and the Horn and in the south, especially in the Kalahari Desert. Many African states fall within the bottom third of the world's worst water-resourced states (Jones 2010:13). For example, Burundi, Rwanda, Burkina Faso, and South Africa have very low levels of natural water resources. Other within-country water-scarce pockets are evident; for example, Windhoek (Namibia), Nouakchott (Mauritania), and Arusha (Tanzania) are water-stressed cities in their respective countries.

Water availability is not always correlated with water accessibility. For instance, South Africa's water availability ranking is 150 out of 180 countries, but a large majority of Johannesburg's urban residents have pipe-borne water access, and some of this water is provided by the neighboring country of Lesotho. At the other end of the spectrum, South Sudan and Sudan have low levels of water stress but rank lowest in access to municipal water: less than 15% of Sudanese have access to utility water in its most basic form (i.e., stand-posts). Decades of war, the accompanying lack of maintenance, and policy neglect have created extreme water vulnerability in both countries. Again, factors beyond nature play important roles in producing contemporary water and sanitation gaps in Africa.

Water availability is an obvious determinant of agriculture productivity. In many African countries, rain-fed agriculture is the most important driver of economic growth. Agriculture accounts for 20% of the African gross domestic product and 60% of total employment; farming is the main livelihood in rural areas (UN-Water Africa 2006). Agriculture accounts for approximately 85% of water withdrawal (UN-Water Africa 2001); however, agricultural productivity in Africa has traditionally been constrained by the haphazard nature of water availability. Constructing dams and irrigated enclosures can enhance the productivity of land, but both are expensive to build and operate. Irrigation is poorly developed, and 70% of water used for irrigation is lost and not used by plants (UN-Water Africa 2001:15). Given the economic situation in the 2010s and projected future water stresses, Africa cannot afford to spend its constrained resources on channeling water that is wasted.

For the most part, Africa lags behind other regions of the world in terms of harnessing water resources and creating large-scale water infrastructure (e.g., dams to supply hydropower and irrigation), and African dams are concentrated in Southern and West Africa and absent in the Central Africa and Sahel regions. On average, there is one dam to every 683,000 persons, compared to one dam per 168,000 persons in the rest of the world (UNEP 2010). Large dams in Africa have resulted in some of the largest human-created lakes in the world (e.g., Lake Volta [Ghana], Lake Kariba [on the Zambia/Zimbabwe border], and Cahora Bassa [Mozambique]), but large portions of the interior are not dammed and rely solely on rainfall and natural resources.

Government commitments to increase water coverage have failed to keep pace with population growth. Only 12% of rural populations have been provided with home water connections, and less than 50% have benefited from any other form of improved water source (WHO/UNICEF 2012). Urban coverage rates show more improvement: 50% of urban residents have access to piped-water connections, and 74% have benefited from some other form of improved water coverage (WHO/UNICEF 2012). Significant intra-urban disparities, however, exist in water consumption. Informal settlements have the lowest levels of piped-water access (19% of dwellings), and 40% of slum dwellers have access within 200 meters (UN-HABITAT 2007). Average daily rates of urban water consumption show high variability between formal housing areas and informal or slum areas. Slum dwellers consume, on average, 25 liters per day while other urbanites consume 50 liters per day, and elites may consume as much as 2,000 liters per day (UN-HABITAT 2012).

Knowledge about Africa's renewable water stores is still at an early stage of development. Many contend that increasing reliable water supplies throughout Africa depends on the development of groundwater (MacDonald et al. 2012). Groundwater is used extensively in agriculture, and many cities depend on it for municipal water (e.g., Lusaka, Windhoek, Kampala, and Addis Ababa are heavily dependent; Lagos, Abidjan, Cape Town, and Pretoria are partly dependent). Groundwater is present in most African environments and generally does not require treatment because it is naturally protected from contamination, although in some environments elevated levels of iron, fluoride, or arsenic can be a problem. Groundwater mapping is an emerging research frontier; MacDonald et al. (2012), for example, have mapped aquifer potential to indicate the areas where the greatest borehole potential exists (Fig. 8.2).

Their mapping shows that some areas identified as being water-insecure (Kenya, Chad, Western Sahara, Mali, and especially the North African region) actually have ample underground reservoir supplies. Groundwater availability is estimated to be 0.66 million $km^3$, more than 100 times the annual renewable freshwater resources and 20 times the freshwater stored in African lakes (Mac Donald et al. 2012). As the largest and most widely distributed store of freshwater in Africa, groundwater appears to provide an important buffer to climate variability and change. However, considerable work needs to be undertaken on developing water management and groundwater conservation practices, and pricing mechanisms remain controversial (Strand 2010).

Water resources are being discovered in Africa's desert environment. For example, in 2013, the Kenyan government announced the discovery of two massive stores of underground water by a French exploration company, Radar Technologies, that was partially funded by UNESCO; it employed cutting-edge satellite technology to explore the Turkana Desert for water (Plaut 2013). Until this discovery this barren area, inhabited by 700,000 pastoralists, had been severely challenged by widespread poverty and intermittent famines. The largest aquifer of the find is estimated to hold 250 billion $m^3$ of water, equivalent to the volume of water in Lake Turkana.

Water is the 21st century's great challenge in Africa. Water scarcity is not simply contingent on geography and climate: population growth, rapid urbanization, poor planning, increased poverty, higher standards of living, lack of resources, and competition between sectors (industry, agriculture, municipal services) for available freshwater are also significant factors. They contribute to water stress or water scarcity in regions where the quantity and quality may not be enough to provide adequate safe drinking water, food, and hygiene, and this can severely constrain economic development and overburden health care systems. Water availability has wide implications for economic and development performance, health, food security, transboundary cooperation, and environment change. Water is also attracting international investors and proposed projects will add new and different pressures on water availability. In the context of projected global climate change, the prudent management of water is a critical development challenge.

## WATER AND SANITATION: COLONIALISM TO THE PRESENT

Inadequate African water and sanitation systems have colonial origins. Under colonial rule, infrastructure developed by the colonial powers was oriented toward cities and the elites and was highly uneven, reflecting concerns from Europe (about spatial planning and protecting the health of Europeans) rather than those from African. Rural areas were generally excluded from colonial water infrastructure, except for irrigation projects.

Development of infrastructure was driven by the colonial financial logic that projects were to be generated from Africa's own natural resources. Labor for many projects was mainly provided by Africans' forced labor. For example, British colonial authorities employed forced labor to develop the Mombasa water supply project (1911–17) (Njoh and Akiwumi 2011). The French, under the auspices of the Office du Niger, attempted to irrigate the Sahara in an area downstream from Bamako for cotton production in the 1930s. The project failed for a number of reasons, among them the inability to persuade enough Africans to relocate to the irrigated area, the lack of appropriate seed varieties, technical difficulties, and higher-than-expected costs.

Public health in the colonies was driven by a very different rationale. Whereas public health initiatives in Europe were designed to benefit the masses, the same

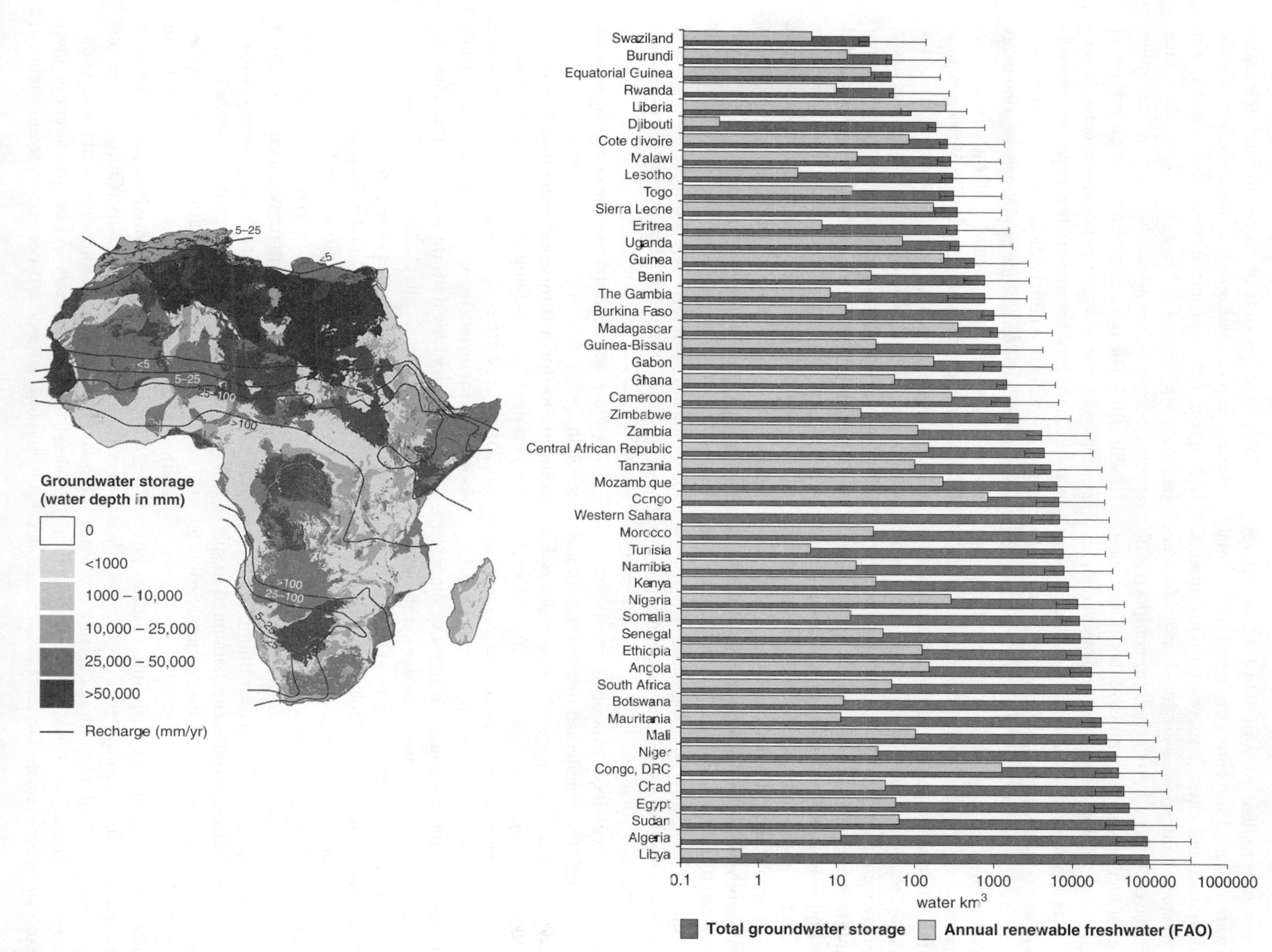


**FIGURE 8.2** African Stores of Underground Water. *Source:* Courtesy of IOP Science. Map 2 from MacDonald et al. 2012.

logic was not applied to Africa. Instead, initiatives there were designed to protect the health of colonial administrators and elites. This was partially attained through spatial and social segregation and infrastructure enclaves. Colonial enclaves were equipped with sanitation facilities such as piped water, flush toilets, and electricity. Colonial business and residential enclaves were separated by a building-free zone or *cordon sanitaire* from indigenous settlements that lacked such services. In Accra, a graveyard was planned as a buffer to separate the colonial district from the indigenous district or bazaar area. In practice, colonial urban infrastructure ignored indigenous communities except to contain the residents. Colonialism turned Africa into a nightmare for Africans and a paradise for settlers.

Another difference between the Western universal water and sewerage systems and the Africa partial system is that the Western model assumes that urban space is both relatively homogenous and spatially coherent (Gandy 2008). The African reality could not be further from this Western model: it is characterized by fragmentation and extreme polarization. Even if the colonialists wanted to extend the colonial infrastructure (and this was considered only at a later stage in urban development), it could never be transposed with the same effect. The fragmented basis of the colonial African infrastructure may explain why progress has been so slow on improving access to water and sanitation. Environmental control was limited and restricted in colonial urban Africa. After the colonial period, independent governments sought to improve the water and sanitation infrastructure by expansion, based on the principle that water is a social good to be provided free of charge or subsidized. However, poor governments could not afford to keep expanding or even to maintain the existing infrastructure. Poor economic performances resulted in growing fiscal constraints and fierce competition for ever-dwindling public resources. Nevertheless, governments continued to subsidize water costs heavily until the economic crises of the 1980s.

Inadequate financing for water infrastructure, poor governance, and corruption in many Africa states led international financial institutions to support water privatization in Africa and other regions. The privatization rationale is based on the fact that private companies can raise large amounts of money to fund new infrastructure for areas with sparse or no coverage and areas with aging, inefficient infrastructure. For instance, it is estimated that 50% of the water in African urban water systems is lost before it can be used by consumers (UNEP 2010). The privatization tilt has enabled several powerful water services corporations to enter African markets. Many private water utility companies receive financial backing from international private equity firms. With an emphasis on making profits, privatization treats water as a commodity that can be allocated by the government and sold to the highest bidder. Water privatization remains controversial in Africa and in much of the developing world. In general, because farmers are generally poor and because rainwater, rivers, and lakes are considered a gift from God/nature, charging farmers the full cost for water is problematic. Governments to date have failed to come up with mechanisms to manage water in a socially sustainable way. For example, the government of Tanzania in 2005 reclaimed control of the water sector after a two-year experiment with privatization led by Biwater (UK); there was a public outcry after the escalation of water tariffs and delays in extending the water supply system.

There are several important postcolonial legacies involving water. First, African urban areas were poorly equipped to deal with the rapid urbanization of the last two decades. A striking characteristic of rapid urbanization in Africa is an enlargement of urban water footprints. Inefficiencies and wastage of water means that many cities now draw on water both near and far (e.g., Nouakchott, Addis Ababa, Dakar) and several cities draw from sources far away (e.g., Dar es Salaam, Cape Town, Johannesburg) (Table 8.2).

For example, a pipeline from Thiès supplied Dakar, and the Akaki wells contributed to Addis Ababa's needs. Many cities extract water from rivers that flow more than 15.5 miles (25 km) from their urban perimeters, and several now depend on international transboundary rivers for water supply. The city of Johannesburg in 1970 relied on the Vaal River and a dam to meet its water needs, but by 1990 its expanded water needs required capturing water from eight additional rivers (Crocodile, Olifants, Tugela, Pongola, Komati, Limpopo, Usutu, and Malibamatso). These nine rivers will be insufficient to meet Johannesburg's water and population requirements in

**TABLE 8.2 AFRICAN URBAN WATER SOURCES IN THE 1990s**

| Country (Total: 29) | Urban Area (Total: 67) | Ground Water | | Surface | River | | | | |
|---|---|---|---|---|---|---|---|---|---|
| | | Near | Far | Collect | Use | National | International | IBT | Def. |
| Algeria | Algiers | no | no | reservoir | no | no | no | no | yes |
| Botswana | Gaborone | no | no | dam | n + f | Ngotwane, Metsimotlhabe | Marico | yes | yes |
| | Serowe | yes | no | no | no | no | no | no | yes |
| | Selibe Phikwe | no | no | dam | far | Motloutse | Shashe | yes | no |
| | Mahalapye | (n.d.) | no | dam | far | Motloutse | no | yes | no |
| | Francistown | no | no | dam | far | no | Shashe | yes | no |
| | Lobatse | no | no | barrage, dam | n + f | Nnywane, Ngotwane | no | yes | yes |
| | Molepolole | wf | no | no | no | no | no | no | (n.d.) |
| | Kanye | wf | no | no | no | no | no | no | (n.d.) |
| Burkina Faso | Ouagadougou | no | no | dam | far | no | Nazinon | (n.d.) | yes |
| Cape Verde | Praia | w | no | desal.sea | no | no | no | no | yes |
| | Sao Vicente | w | no | desal.sea | no | no | no | no | yes |
| | Sal | w | no | desal.sea | no | no | no | no | yes |
| Côte d'Ivoire | Abidjan | yes | no | no | no | no | no | no | no |
| Democratic Republic of Congo | Kinshasa | (n.d) | no | no | near | Ndjili | no | no | no |
| Djibouti | Djibouti | w | no | no | no | no | no | no | no |
| | Tadjourah | w | no | no | no | no | no | no | no |
| Egypt | Cairo | w | no | reuse-s,1 | near | no | Nile | no | no |
| | Alexandria | no | no | no | near | no | Nile | no | no |
| Ethiopia | Addis Ababa | yes | wf | dam, reuse-s | far | Legedadi, unspecified | no | gw, yes | yes |
| Ghana | Accra | no | no | dam | n + f | Densu | Volta | yes | yes |
| | Tema | no | no | dam | n + f | Densu | Volta | yes | yes |
| | Kumasi | no | no | dam | n + f | Owabi, Ofin | no | yes | no |
| | Winneba | no | no | (n.d) | near | Ayensu | no | no | (n.d.) |

(Continued)

**TABLE 8.2 AFRICAN URBAN WATER SOURCES IN THE 1990s** *(Continued)*

| Country (Total: 29) | Urban Area (Total: 67) | Ground Water | | Surface | River | | | IBT | Def. |
|---|---|---|---|---|---|---|---|---|---|
| | | Near | Far | Collect | Use | National | International | | |
| Lesotho | Maseru | no | no | no | near | no | Caledon | no | no |
| | Morija | yes | no | no | no | no | no | no | (n.d.) |
| | Mapoteng | yes | no | no | no | no | no | no | (n.d.) |
| Liberia | Monrovia* | w | no | dam | near | no | St. Paul | no | (n.d.) |
| Libya | Tripoli | (n.d.) | fgw | (n.d) | no | no | no | fgw | (n.d.) |
| | Benghazi | (n.d.) | fgw | (n.d) | no | no | no | fgw | (n.d.) |
| | Sirt | (n.d.) | fgw | (n.d) | no | no | no | fgw | (n.d.) |
| | Misratah | (n.d.) | fgw | (n.d) | no | no | no | fgw | (n.d.) |
| Malawi | Blantyre | no | no | (n.d.) | near | no | Shire | no | (n.d.) |
| Mali | Timbuctu | yes | no | (n.d) | no | no | no | no | no |
| | Gao | yes | no | (n.d) | no | no | no | no | no |
| Mauritania | Nouakchott | bh | wf | (n.d) | no | no | no | gw | yes |
| | Nouadhibou | w | w | reservoir | no | no | no | gw | no |
| | Akjoujt | w | no | (n.d) | no | no | no | no | no |
| Mozambique | Maputo | no | no | dam | near | no | Umbeluzi | no | (n.d) |
| Namibia | Windhoek | bh | no | dam, reuse-1 | far | Swacop | no | no | no |
| | Tsumeb | w | no | no | no | no | no | yes | no |
| | Walvis Bay | no | no | dam | near | Kuiseb (rb) | no | no | yes |
| | Swakopmund | bh | no | (n.d) | far | Kuiseb (rb) | no | no | yes |
| Niger | Niamey | no | no | (n.d) | near | no | Niger | no | (n.d) |
| Nigeria | Lagos | no | no | (n.d) | near | Ogun | no | no | no |
| | Maiduguri | yes | no | dam | near | Ngadda | no | no | no |

| | | | | | | | | | |
|---|---|---|---|---|---|---|---|---|---|
| Senegal | Dakar | bh | wf | no | no | no | no | gw | yes |
| | Thiés | bh | no | no | no | no | no | no | no |
| Somalia | Mogadishu | yes | no | (n.d) | no | no | no | no | (n.d) |
| South Africa | Cape Town | no | no | reservoir | far | Erste. Palmiet, Steenbras, Berg, Riviersonderend | no | no | yes |
| | Johannesburg | no | no | treated wáter to Vaal Dam | far | Vaal, Crocodile, Olifants, Tugela, Pongola | Limpoo, Komati, Usutu, Malibamats'o | yes | no |
| | Durban | no | no | no | n + f | Mzimkulu, Mooi, Mkomaas, Mgeni, Iilovo | no | yes | yes |
| | Port Elizabeth | no | no | no | n + f | Great Fish, Sundays | Orange | yes | yes |
| Swaziland | Mbabane | no | no | dam | near | no | Mbuluzi | no | no |
| | Manzini | no | no | storage | near | Mzimnene | no | no | no |
| Tanzania | Dar es Salaam | no | no | storage reservoir | far | Ruvu | no | yes | yes |
| Tunisia | Jerid | dw | (n.d.) | (n.d.) | (n.d.) | (n.d.) | (n.d.) | (n.d.) | (n.d.) |
| | Nefzaoua | dw | (n.d.) | (n.d.) | (n.d.) | (n.d.) | (n.d.) | (n.d.) | (n.d.) |
| Tunisia | Gabès | dw | (n.d.) | (n.d.) | (n.d.) | (n.d.) | (n.d.) | (n.d.) | (n.d.) |
| | Medenine | dw | (n.d.) | (n.d.) | (n.d.) | (n.d.) | (n.d.) | (n.d.) | (n.d.) |
| Zambia | Lusaka | w | no | no | far | Kafue | no | yes | no |
| Zimbabwe | Harare | no | no | dam | near | Manyame | no | no | yes |
| | Bulawayo | no | no | dam, reuse-1 | far | Ncema, Insiza, Unzingwane, Inyankuni | no | yes | yes |
| | Mutare | no | no | dam | n + f | Odzani | Pungwe | yes | yes |
| | Beitbridge | no | no | off-river storage dam | near | no | Limpopo | no | no |
| | Masvingo | no | no | dam | far | Mutirikwe | no | yes | no |

*Source:* Showers (2002).

Key: Near: <25 km; Far: ≥25 km; n + f: near & far; IBT: interbasin transfer; w: wells; dw: deep wells; bh: borehole; dbh: deep borehole; sp: springs; ships: brought by ship; no: do not use; div.dam: diversion dam; (rb): bore hole in dry river bed; desal.sea: desalinate sea water; desal.brac: desalinate brackish water; gw: ground water that is an IBT; (n.d.): no data.

2030. The projected water deficit could be catastrophic unless planners devise alternative solutions and better manage urban water supplies.

Second, women and girls have been most adversely affected by restrictions on access to water resources. In general, women perform more than two thirds of water-collection duties, and some African women expend as much as 40% of their daily nutritional intake traveling to collect water (UNEP 2010). An average African household consumes about 40 to 60 liters of water daily for drinking, cooking, cleaning, personal hygiene, etc. Extensive water-collection duties undertaken by women in rural areas can mean that they travel as much as 3.72 miles (6 km) a day. On average, women spend more than half an hour per round trip for water; this time could be spent on other livelihood duties, and it reinforces a gender division of household labor. Women have a higher risk of exposure to water-borne diseases: washing clothes, bathing children, drawing water from surface sources, and, in some regions, working in flooded rice fields increase their health risks. Women can even be placed in harm's way when collecting water in conflict or postconflict areas where land mines are still present (Fig. 8.3).

**FIGURE 8.3** Women Collecting Water in Angola.
*Source:* © Lynn Johnson/National Geographic Society/Corbis.

Third, the colonial and modernization approaches to water management overlooked indigenous water knowledge, indigenous watershed management, indigenous water values (water bodies having spiritual and living attributes), and indigenous water justice issues (how people can have cultural rights and rights of nature). Indigenous knowledge is acquired by local people through informal experiences and intensive understanding and learning from their environment. It can contribute to water sustainability and should be incorporated into the decision-making process and development activities, but integration of indigenous knowledge into planning and management is weak (Butt, Shortridge, and WinklePrins 2009). UN-Water Africa (2006) started compiling some best practices in indigenous water conservation and water knowledge (Box 8.1).

## BOX 8.1 INDIGENOUS KNOWLEDGE OF WATER

The Fulani of Mauritania practice a detailed art of detecting groundwater. Their indicators are based on knowledge of topography (e.g., shallow aquifers can be found near natural ponds or in depressions of mountains), plant species (especially tap-rooted trees such as *Bauhinia rufescens*, *Tamaris senegalensis*, *Capparis decidua*, and *Acacia albida*, but also perennial grasses, such as *Vetivera nigritana* and *Panicum anabaptistum*), and the health or vigor of the plants, such as the greenness of leaves during the year. Other indicators are based on fauna (e.g., wild boars only live where they can dig and find moist soil; other animals that prefer to stay around moist places are caimans, amphibious lizards, tortoises, bands of butterflies, some bird species, and many termite hills). The Fulani also are familiar with the geological strata in their area and know that they must dig through the whole layer of red or gray clayey soil and arrive at the sandy layer before finding groundwater. A good-quality groundwater that is clear, sweet, and has a good mineral content is indicated by the presence of *Guiera senegalensis*, *B. rufescens*, termite hills, and the depth of wells (the deeper, the better quality). The best-quality natural ponds are indicated by the presence of water lilies, followed by *Acacia nilotica* and *Mitragyna inermis*. Bad, diseased water is indicated by the presence of the grass *Echinochloa pyramidalis*. Water quality is also tested by immersing a leather container in it. The best water leaves the leather intact; as water quality deteriorates, the leather's color will change to white, black, red, or finally yellow/orange. Water quality is also evaluated by its effect on livestock, especially their behavior after drinking (whether they are content or not) and the yield of milk.

*Source:* Excerpted from UN-Water Africa 2006:306.

Fourth, African governments have continued to embrace Western models to develop water infrastructure systems. Increasingly, there is a privatization and large private investment tilt in agriculture, with accompanying plans to build large irrigation projects throughout the region in an effort to bring African agriculture into the 21st century. Agriculture has become a fashionable investment frontier, especially since the global financial crisis. Since 2005 large land investments have been under way in Africa. Increasingly claims are made that the investors are not only after land but also water (See Box 8.2).

Fifth, African informal entrepreneurs are filling the gap and providing affordable water supplies for the urban poor in the form of "water sachets" (Fig. 8.4). Water sachets are 500-mL (approximately 2 cups) sealed plastic sleeves of purified drinking water that is of generally high quality (but not always) and low cost (prices were US$0.05–$0.07 in Accra in 2013), compared to expensive bottled water (Stoler 2014:182). Now ubiquitous in urban West Africa, water sachet consumption is related to deficiencies in piped water coverage and failure to build new water infrastructure and/or properly manage existing

## BOX 8.2 WATER GRABBING

Private-sector expectations of higher world food and commodity prices, mainly linked to projected demographic growth, have created a thirst for land that is rain-fed and for land with high irrigation potential. Researchers now draw attention to "the hidden agenda of land acquisition in Africa" (Woodhouse and Ganho 2011), where murky land deals conceal even murkier water deals.

The term "peak water" is gaining currency as the state of peak water is being approached in many regions of the world. It is based on an assumption that population pressures on water will lead to a peaking in major withdrawals followed by a subsequent decline as production of water exceeds natural recharge rates. Some even forecast that by the middle of the 21st century water will be the single most valuable commodity, more valuable than precious metals and oil. Private equity firms (e.g., Emergent Asset Management [UK], the largest agricultural fund manager in the region) market land investments on the basis that African land is undervalued. Other private equity firms (e.g., Chayton Capital [UK]) claim that the real value is in African water surrounding some lands (GRAIN 2012). According to J. Minaya, Global Fund Manager at TIAA-CREF, investment in farmland is "probably the most efficient way for us to get exposure to water . . . When you really look into buying water, at the end of the day it is a water play" (quoted in GRAIN 2012:16). Major U.S. universities (e.g., Harvard and Vanderbilt) are allegedly investing in hedge funds that participate in African land and water deals (Vidal and Provost 2011). Both Africa's land and water resources are seen as underutilized and therefore highly undervalued.

The implications of international commercial withdrawals on local, regional, and national water resources have been overlooked until recently. Some label the trend of commercial withdrawals as "water grabbing": when powerful actors (multinational corporations [MNCs], private equity firms, and individuals) take control of or reallocate water resources (from local communities) for their own benefit, without considering local access, rights, and livelihoods (fishing, farming, herding) that depend on the resource (Mehta, Vedwisch, and Franco 2012). Water grabbing can span a continuum from agroindustrial farming to the capturing of water for hydropower plant transmission channels. Large land deals are shrouded in secrecy, making it impossible to determine which water rights are bundled in the deals and which are not (if any). According to GRAIN (2012), contracts typically do not mention water rights, opening the door for international companies to build dams and irrigation canals at their discretion. In other words, land investment is water investment, but it is not articulated as such in land acquisition contracts.

Importantly, land acquisition involves a corporate commercial transfer and legal capture of local people's previously established (customary) rights to water. Powerful actors use all means at their disposal (legal, bureaucratic, and technical) to divert water and profits away from local communities. Privatization and commodification processes remove water control from local communities. No longer the purview of state or traditional control, new entanglements among domestic and international players assume wide-ranging control of water resources. The BRIC (Brazil, Russia, India, and China) countries, Europe, the Gulf states, the UK, the United States, Egypt, and MNCs are the major international players in the land and water deals. Ironically, many of these investors come from regions facing water shortages (e.g., China, India, the Gulf states).

Land/water grabbing may be unleashing "hydrocolonialism" around major African rivers (e.g., Nile, Niger) and freshwater lakes (e.g., Turkana and Victoria) and on fertile and fragile wetlands (e.g., the Niger and White Nile deltas) (GRAIN 2012). Land and water investors have rushed into Africa's newest state, South Sudan, and acquired 8% of the land in a short time. Investors charged in after a 2005 peace accord to acquire land around the fertile Nile in both South Sudan and Sudan. In total, 12.1 million acres (4.9 million hectares) has been acquired, an area greater than the Netherlands. Virtually all of this land will require irrigation, and water-thirsty crops (e.g., sugar, rice, jatropha) are

*(Continued)*

**BOX 8.2 (*Continued*)**

scheduled to be introduced. According to GRAIN (2012), so much land has been acquired along the Nile Basin (24.7 million acres [10 million hectares]) from Uganda to Egypt (including South Sudan, Sudan, and Ethiopia) that irrigating it would require more water than is available in the entire Nile Basin; this could result in environmental disaster, with commercial farms promoting several harvests per year without adequate consideration of seasonal flows. Water requirements for large investor projects have not been quantified. Large withdrawals will affect the volume of water in transboundary basins; consequently, intrastate and interstate water disputes are likely to occur because transboundary water management has been neglected in the land sell-off (Jägerskog et al. 2012). Hydrological suicide is even possible.

According to the Oakland Institute (http://www.oaklandinstitute.org), an independent think tank and advocacy organization, twice the volume of water used in African agriculture in 2005 would be needed if 40 million of the 125 million acres of acquired land came under cultivation, putting considerable stress on Africa's freshwater supply (Oakland Institute 2011b). Additional pressure on water resources would adversely affect small farmers, pastoralists, and fishermen, people whose livelihoods rely on water resources and access. Negative impacts would be felt not only by those in the immediate areas but also by users downstream, possibly even across international borders. Irrigation projects (dams, canals, drains) will divert water from rivers and lakes and restrict or interrupt the flow of water downstream; for example, the Niger River originates in Guinea, flows through Mali, Niger, and Benin, and enters the Atlantic Ocean in Nigeria (a 4,180-km journey), and Lake Turkana stretches from northern Kenya into Ethiopia.

Water is critical to land deals, especially in the semiarid and arid regions of the world and especially when investors plan to grow thirsty agrofuel crops (e.g., sugar, corn, jatropha). Investors typically want to secure water rights as part of the deal; accessing abundant water resources is often explicit in investors' business plans. Motivated by potential revenues from water fees and the prospect of improving agricultural productivity, many African governments are giving up water rights to investors without considering the impacts on customary users or on future water management. Agreements are sometimes accompanied by African governments' commitment to build supporting infrastructure such as dams. For example, Mali's Office du Niger is building the Talo Dam (irrigation) and planning another at Djenne (irrigation) and the Fomi Dam (irrigation and power) in the upper reaches of the Niger River; these dams will affect the annual flooding of the wetland downstream. Some experts are concerned that the levels of water withdrawals from the Niger River will result in changes in water flow and availability, water contamination through new intensive agricultural uses, and changes in fish stocks and wildlife, all of which threaten the existence of the river.

Indigenous water access can be affected after the agreements. For example, with the Sun Biofuel (UK) acquisition in Kisarawe, Tanzania, locals are having to sneak back onto the newly acquired land to access their old water sources and to "steal the water" or buy it at inflated prices elsewhere (Oakland Institute 2011b:39). Sun Biofuel's land parcel (2,029 acres [8,211 hectares]) used to belong to 11 villages before it was leased for 49 years to grow jatropha (Oakland Institute 2011b:17). Customary law is unclear about whether groundwater is part of the land and therefore under the private owners' possession, because customary law evolved before contemporary repurposing of water for agroindustrial use.

Water grabbing is part of the state-sponsored taking of resources from citizens, and the grabber in the first instance is typically the state, bolstered by a bewildering array of foreign investors (Hall 2011). The water dimension of large-scale acquisition means that the impacts are likely to be far more geographically extensive than might be anticipated from the initial land deals. Many questions remain unanswered about the impact of land/water grabs. What is the "market price" for customary land with water rights cleared of its inhabitants and leased by governments? Does agroindustrial land and water use signal the end of indigenous water-management systems that deploy techniques ranging from water conservation and harvesting to stream diversion and irrigation in line with intraseasonal interruptions and annual variation in rainfall in Africa? Should water rights be linked to land? What rights do locals have to water? When most water resources are allocated for irrigated agriculture, how will governments deal with competition and scarcity in droughts or even seasonality of water availability? Ultimately, we do not know the extent to which water demand from these projects will displace existing water use and result in increased agricultural risk and impoverishment (Woodhouse and Ganho 2011).

Water grabbing raises the crucial issue of who owns water in Africa. Throughout the continent, water tends to be vested in and managed by national governments. In most places, locals hold customary uses (human consumption, agriculture, transportation, fishing) but lack formal rights. Clearly, transparency is missing in land and water negotiations. Existing land and water rights need to be incorporated into land agreements. Adequate compensation is essential so the people have the capital to diversify into an alternative livelihood (Zoomers 2010). Local communities should share in the benefits, rather than lose out, from foreign investments in land/water resources. Environmental sustainability needs to be preserved so that industrial agriculture does not deplete or divert larger amounts of water from human and other environmental uses, cause soil erosion, and eliminate biodiversity. The matrix of land deals involving strategic waterways needs to be made transparent. Discussion among river basin organizations and intergovernmental entities is needed to assess how land deals will affect the political relations between stakeholders in the shared basins (Jägerskog et al. 2012). Finally, the ethics of deploying water for export agriculture (biofuels, food, other commodities) needs to be considered, especially in countries where the food supply is insecure. When food security is at risk (e.g., during droughts), domestic supplies need to be given priority, and foreigners should not have an overriding right to export at all times (Zoomers 2010).

**FIGURE 8.4** Water Sachets in Accra, Ghana. Photograph by Justin Stoler.

resources. Sachet water improves water access in water-stressed neighborhoods, particularly low-income and slum communities (Stoler 2014). However, discarded plastic sleeves are a sanitation and environmental menace, causing litter, clogging drains and gutters, and thus intensifying the likelihood of urban flooding in the rainy season. This in turn poses health risks by increasing the public's exposure to untreated waste (Stoler 2014). Producing an immense plastic waste burden, sachets remain an unsustainable water-delivery vehicle. Governments have considered banning sachet water, taxing its street trade, and taxing the plastic used in production, but government regulation of the activity has been slow (Stoler 2014). Nevertheless, the Nigerian government is considering a national ban on sachet water along with plastic bags in 2014; Benin City, Nigeria, banned sachet water in 2013. An alternative solution is to develop a comprehensive recycling process for discarded sleeves. A Ghanaian NGO, Trashy Bags, is making an effort to collect used sachets and recycle them into stylish carrier bags, but an urban or national program is needed to address the larger environmental issue.

## WATER AND POVERTY

Africa is widely acknowledged as the poorest and least-developed world region. Multiple linkages exist among poverty, water, and the environment (Fig. 8.5).

Poverty is a significant indicator for low levels of access to safe water and sanitation and for lack of water for other uses, such as irrigation. It is often the reason why people have to consume unhealthy water. Poverty is widespread in rural areas, where the majority of the population depend on rain-fed agriculture. Water poverty is being exacerbated by population increases and accompanying shifts of the population into cities, manifested in the phenomenal growth of overcrowded slums. Whereas the average rural inhabitant in Africa consumes approximately 25 liters of water per day, urbanities consume 10 to 20 times more (Jones 2010). Poor residents are unequipped to deal with municipal water rationing as water valves are turned on and off to accommodate high/low water demand (this is common in Accra, Ghana), and often water is not stored safely and poses a health risk. In West Africa, informal water sachet vendors compensate for some of the shortfall, and some of this water is of good quality but much of it is not (Stoler et al. 2012).

Urban slums cause problems for the water supply and contribute to other water related problems. Lack of sanitation infrastructure in urban slums results in open defecation and is manifested in the "flying toilet" phenomenon (plastic bags containing urine and fecal matter are thrown from buildings into gutters). Evidence from low-income residential areas in Ghana shows that even the presence of a working toilet can give a misleading impression that water and sanitation service exists. The local reality is that many residents have to pay informal user fees to heads of households for the privilege of using these in-house facilities; those who can't afford it may be deprived of these services (Melara et al. 2013). Another factor contributing to "flying toilets" is the lack of public toilets and their uneven distribution. Even when public toilets are provided, they involve a fee and are poorly maintained. Their inconvenient locations often pose security risks at night. Children are especially vulnerable because they may not have the money to pay to use a community toilet. Water contamination caused by lack of centralized sewage treatment and waste disposal affects the broader community. Typically, every liter of polluted water can pollute up to 10 liters of freshwater.

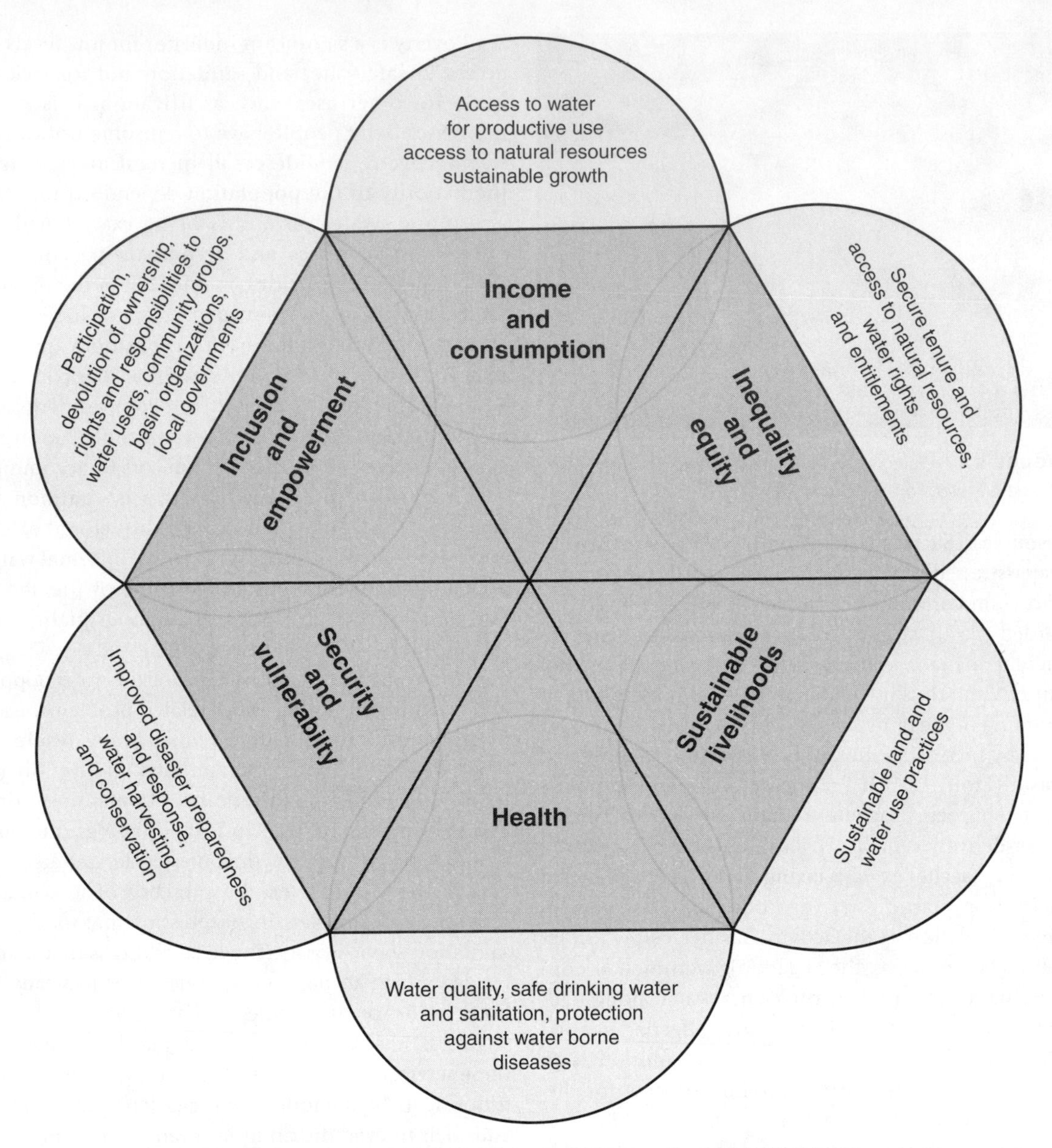


**FIGURE 8.5** Linkages Between Poverty, Water, and the Environment. *Source:* From UNEP 2010:30.

Water plays a central role in the ecology of diseases. Water poverty enhances exposure to water-related diseases. Several categories of water-related diseases are present; among them are water-borne disease (e.g., diarrhea, hepatitis A, cholera, malaria, typhoid), water-based disease (guinea worm, intestinal worms), and water-related vector diseases (malaria, dengue, yellow fever). There are also diseases related to the lack of water, called water-washed diseases (dysentery, lice, scabies, trachoma). More than half of Africans suffer from one of the main water-related diseases (WHO 2008). The only way to break the continued transmission cycle is to meet people's basic needs—drinking water, washing and bathing facilities, and sanitation—and to improve their hygienic behavior.

On the one hand, poverty is manifested in the lack of access to improved water resources. On the other hand, wealth is often linked to the overconsumption of water. The contrast can be sharp. For example, a family of eight living in a squatter camp in Cape Town (South Africa) uses about 120 liters of water a day, collected from a tap a few hundred meters away. In contrast, a couple in a nearby rich neighborhood with a large garden can use 2,000 liters per day (UNEP 2010). Access to water and the use of water are some of the clearest indicators of inequality throughout Africa.

## TRANSBOUNDARY WATER ISSUES

Worldwide, there are 263 transboundary river basins, defined as basins shared by two or more riparian states (UNEP 2010). River basins support complex ecosystems and human settlements. Approximately 60% of the world's population depends on river basins. Africa is well endowed with 63 river basins that contain 93% of the continent's total surface water resources, and 77% of the region's population reside around them (UNEP 2010:38). Surface water flows across basins, and sub-basins provide common water sources, aquatic habitats, and transport networks. In Africa, 24 watersheds and 15 lakes cut across the national boundaries of two or more countries. Water interdependency is high: a high percentage of total flows in downstream countries originate outside their borders. For example, 94% of total flow in Botswana and Mauritania originates outside of these countries' borders.

Complex human–environment interactions make equitable transboundary basin management challenging. Different geographical distributions of resources and populations within large basins mean there are significant intraregional and interregional differences. In general, upstream areas are more advantageous locations than downstream ones. The type and degree of dependence on river resources vary among states. For example, Uganda is heavily dependent on the Nile for hydroelectric power generation, whereas South Sudan primarily uses the river for agriculture and transportation. Furthermore, expanding populations in river basins contribute to deforestation, land conversion, agriculture, livestock, industrialization, waste disposal, and fishing pressures (UNEP 2010).

The Nile Basin provides a good illustration of the complex interplay of uses and dependencies. The basin covers a 4,225-mile (6,800 km) area from the Burundian highlands to the Egyptian coast in the Mediterranean. Almost one quarter of Africa's population resides in the Nile River Basin, which straddles 11 countries and very different climate zones. This means that some countries are net users of water (e.g., Sudan and Egypt), while others are net contributors (e.g., Kenya and Ethiopia; the latter alone contributes 50% of the total water budget of the Nile due to higher rainfall, lower highland temperatures, and lower levels of evapotranspiration). Population is concentrated in three Nile Basin areas: (1) that surrounding Lake Victoria in Kenya and Uganda, (2) the Ethiopian Highlands, and (3) that along the banks of the river in Egypt. Since the formation of the Nile Basin Initiative in 1997, countries have attempted to manage the Nile's resources across national boundaries; the 21 development projects that are under way (dams, irrigation, and other water-diversion projects) in several of the basin's countries will have major impacts on resource use, irrespective of regional and national boundaries.

The Jonglei Canal project in South Sudan is the most controversial Nile Basin initiative. Completion of this canal is expected to save significant water from evaporation and transpiration and could be of immense value to communities downstream in Sudan and Egypt, but it will have serious repercussions on the Sudd wetland. The project commenced in 1977 but was put on hold in 1983 (two thirds of the total length had been dug) because of military conflict in the area. The giant excavator shown in Figure 8.6 became a rusting monument to a project that was one of the triggers of the renewal of civil war in Sudan. The Sudd is a vast wetland in South Sudan where the Nile meanders for 400 miles or 644 km on its journey through the country. The Sudd overflows to 80,000 $m^2$ in the wet season but contracts to 8,300 $km^2$ in the dry season (UNEP 2010:81). Annual patterns of flooding are crucial to

**FIGURE 8.6** The Jonglei Canal, South Sudan.
*Source:* Photograph by George Steinmetz.

flora, fauna, birdlife, wild animal migrations, and local livelihoods. Many fishermen depend on the area, and pastoralists depend on the seasonal flooding to access grass and water for livestock. It is estimated that canal completion will result in the diversion of 55 million $m^3$ of water daily, resulting in a loss of 36% of pasture and 20,000 metric tons of fish (UNEP 2010). Environmentalists further claim that the project could affect climate, groundwater recharge, and water quality (Howell, Lock, and Cobb 1989; UNEP 2010). Proponents claim the benefits will outweigh the negative impacts on the wetlands: the project will enhance downstream irrigation and allow more food crops to be grown (e.g., rice), and a road on the bank of the canal will reduce the travel distance from Khartoum to Juba by 186 miles (300 km).

Even when agreements can be reached to construct these large international projects, the outcomes are far from straightforward. For example, the Lesotho Highland Water Project is the world's largest interbasin transfer water exchange. The project enabled Lesotho (a landlocked mountain kingdom completely surrounded by South Africa) to become a major water exporter to South Africa by transferring water north to meet the water needs of Johannesburg's expanding population. An engineering triumph, this massive project won international acclaim, but it also exacerbated rural impoverishment in Lesotho. The project's dams submerged some of Lesotho's most fertile land, which previously had supported several thousand agricultural families. Lesotho receives much-needed income from the project (earning US$40 million per year in water exports plus $400,000 for hydroelectric power exports), and it has enabled hydroelectric power sufficiency in the country. Moreover, the project enables water-poor South Africa to share the benefits of a shared water resource, making an invaluable contribution to Johannesburg's water supply, without which the economy of the extended regional area (Gauteng province) would suffer. However, the future sustainability of water exports to its wealthier neighbor is now in question: 30% of Lesotho's water points (boreholes, wells, and springs) are drying up.

Transboundary aquifers are another important water source, but research is lacking. Just as there are internationally shared river basins, there are underground transboundary aquifers in many parts of Africa. Some contain huge freshwater resources of excellent-quality water that can provide safe drinking water for current and future populations. Africa is endowed with large aquifers in the large shared subregional sedimentary structures of the Sahara, Central and Southern Africa, and West and East Africa. Large shared aquifer resources often represent the only source of drought security in semiarid regions. However, the watersheds in many aquifers' recharge zones are threatened by accelerated land degradation and desertification. Thus, management issues and transboundary implications extend beyond water balance and control of hydraulic systems to include land-use protection in recharge and discharge areas.

## VIRTUAL WATER

Africa's water is traded out of the region in occluded ways. "Virtual water" is a concept that involves the complex interactions among water, commodities, and long-distance trade. It was developed by Tony Allan, who was awarded the 2008 Stockholm Water Prize for this contribution to water scholarship. The virtual water concept is used to examine the volume of freshwater needed to produce a product measured at the production location and considers the increasingly international travels of embedded water in commodities (Hoekstra and Chapagain 2007). Virtual water is incorporated into exports and represents a subtraction from an area's indigenous water. Therefore, it represents a water addition or a water import in the receiving country.

A plethora of organizations (the UN's Food and Agriculture Organization [FAO], the National Geographic Society, the UN Organization for Education, Science and Culture [UNESCO]) employ the concept to measure commodity/country water footprints (the direct and indirect quantities of freshwater used in a product). Water Footprint Network, a Dutch-based nonprofit organization, coordinates the main portal for information about water footprints and virtual water. Consideration of water footprints sheds light on the wide differences in global consumption patterns and the extent to which consumption in one area affects water resources in another area.

Increasingly, more virtual water is being incorporated into calculations related to commodities traded from Africa to urban centers in Europe, China, the Middle East, and elsewhere. The per capita water footprint (measured by usage from washing, drinking, and the production of food and other consumer goods) in Southern, Central, and Eastern Africa is among the lowest of the world (Fig. 8.7). On a daily basis, North Americans and Europeans on average consume at least 3,000 liters of virtual water, compared to 1,100 liters consumed in Africa, reflecting unequal consumption power as well as more elaborate consumer tastes and demands for imported foodstuffs, and especially meat-rich Western diets. Most of the water footprint is accounted for by products purchased in supermarkets, particularly meat, and very little is related to water used in homes. This is strikingly at odds with public water conservation campaigns in the United States that emphasize water-saving strategies (e.g., more efficient toilets, advice on lawn-watering practices). However, African diets appear to be moving in the same direction as a middle class emerges, as societies continue to urbanize, and as large supermarket chains gain more of a foothold in urban

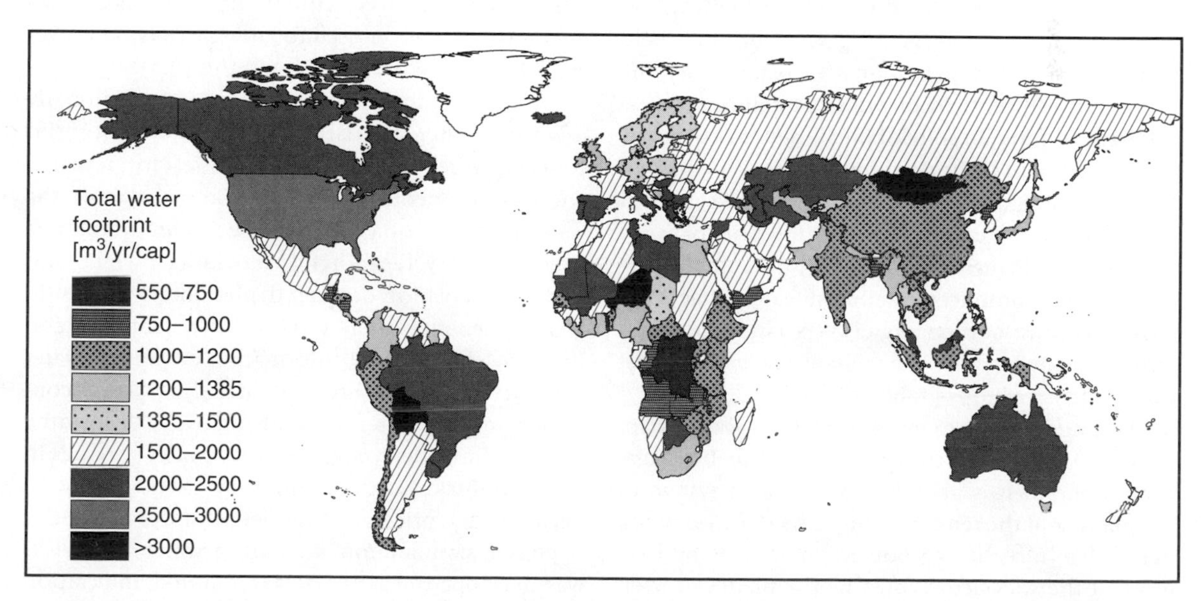


**FIGURE 8.7** Average Water Consumption Footprint Per Person. *Source:* From Mekonnen and Hoekstra (2011).

markets. Virtual water transfers from Africa to other regions will grow, especially in the context of land and water grabs. For example, European firms are acquiring 3.9 million hectares of African land to meet their 10% biofuel target by 2015. The amount of water required for biofuel plantations is high: 1 liter of ethanol from sugarcane requires 18.4 liters of water and 1.52 $m^2$ of land (UN-Water Africa 2012).

Virtual water has implications for the alteration of the natural distribution of water resources and their sustainability. It often involves inequitable transfers from poor and water-scarce rural regions to wealthy urban centers (and sometimes abundant water centers), and it has longer-term national security implications for the water-originating regions, especially when the population is expanding. In general, higher-value crops such as sugar and vegetables are more water-intensive than cereals, and meat and dairy are also more water-intensive. Product water footprints are complex, and water use varies by method of production, local soil types, variety of product, time of year, and, above all, location and how much water has to be applied to make things grow (e.g., in dry vs. wet climates). The exact water footprint can be calculated for any commodity, but a tomato produced in an irrigated Ghanaian field will require different amounts of water than one produced in drier field in the same country or in a different country. In spite of measurement obstacles, it is important to understand the movement and transfer configurations within the international trade system. For example, the water footprint of maize is three times higher in Nigeria than in Germany (the most efficient producer in terms of water usage). The water footprint of bioenergy is very large and varies tremendously by crop—jatropha is among the most water-intensive crops (Table 8.3).

During the growth cycle of a plant, water is applied to the field, whether by rainfall or by irrigation. Some of it is absorbed by the root system of the plant, and the remainder finds its way into rivers, percolates into the soil, or evaporates. At harvest, most of the water consumed by the plant has been lost to the atmosphere through evapotranspiration. A small portion of the water is locked inside the plant itself. Water is "virtual" because once crops are grown, the water used to grow them is no longer available.

**TABLE 8.3 VIRTUAL WATER NEEDED TO PRODUCE COMMODITIES**

| Product and Measure | Liters of Water |
|---|---|
| *Foods* | |
| Beef, 1 kg | 15,500 |
| Rice, 1 kg | 3,400 |
| Sugar cane, 1 kg | 1,500 |
| Wheat, 1 kg | 1,350 |
| Tomatoes, 1 kg | 180 |
| Potatoes, 1 kg | 160–250 |
| Coffee, 250-mL cup | 280 |
| *Fuels* | |
| Biodiesel: | |
| Jatropha, 1 liter | 20,000 |
| Soybean, 1 liter | 10,000 |
| Bioethanol: | |
| Cassava, 1 liter | 2,926 |

*Sources:* Jones (2010:166, Table 8.1) and Gerbens-Leenes, Hoekstra, and van der Meerb (2009:10222).

About one fifth of the water consumed globally is virtual water. Virtual water and water footprint concepts offer exciting opportunities for research and development. For instance, knowledge of the water footprint of a river basin or of an urban area would be very meaningful, providing insight into water conservation or water depletion, which would greatly benefit water sustainability management. Product labeling with information on virtual water content and source area would help educate consumers about water stewardship (e.g., informing them on how their purchases affect water supplies in other countries). At present, water use is hardly reflected in the price of consumer products (Box 8.3). A global standard on measuring water footprints was developed in 2011. However, unlike the carbon footprint, where it matters less where it happens, geography matters in the water footprint.

## BOX 8.3 WATER FOOTPRINT OF CUT FLOWERS IN KENYA: WHAT IS THE TRUE COST?

Kenya's cut-flower industry has been hailed as an economic success, and flower exports now rank as Kenya's third largest foreign exchange earner after tea and tourism. The industry provides employment to 25,000 people directly (plus another 25,000 indirect employees) and income and infrastructure such as schools and hospitals for the community around Lake Naivasha. Most farms pay workers more than the legal minimum wage and provide housing, free medical services, schools for children of farmworkers, and modest sports facilities.

Lake Naivasha, situated 50 miles (80 km) northwest of Nairobi, lies in the fertile Rift Valley of Kenya. Lake Naivasha is also the second largest freshwater lake in the country. The total irrigated commercial farm area around Lake Naivasha is approximately 10,996 acres (4,450 hectares), and cut flowers are grown on about half of the irrigated area. Production in the Lake Naivasha area accounts for 95% of Kenya's cut-flower exports (Mekonnen and Hoekstra 2011:12). Growers concentrate on roses, but other flower varieties (e.g., carnations) are also produced. Both roses and carnations require high rates of fertilizer application (especially compared to vegetable crops such as baby corn and cabbage, also grown in the area), and leaching runoff has become an issue. Most of this area's cut flowers are exported to the European Union; urban markets in the Netherlands (69%), the UK (18%), and Germany (7%) are the main export destinations (Mekonnen and Hoekstra 2011:12). The total virtual water export of fresh flowers was 21 million m$^3$/year in 2005. Vegetables grown in the region account for another 8.5 million m$^3$/year (the main export designations are the United Arab Emirates, France, and the UK).

The average total water footprint of a rose is 9 liters per stem. According to Severino Maitima of the Ewaso Ngiro Water Authority in Kenya, "a flower is 90% water. We are on the driest countries in the world and we are exporting water to one of the wettest" (quoted in Vidal 2006:1). The capturing of water from rivers such as Ngiro by UK- and other European-owned flower companies (usually between 10 p.m. and 2 a.m.) means that there is much less water available for small-scale farmers. Local activists claim the river's course has been shortened by 60 miles in some areas (Vidal 2006).

There are major controversies surrounding the flower industry at Lake Naivasha. Local environmental groups contend that companies exporting flowers use pesticides at levels that are unacceptable to the flower industry in Europe and are poisoning the lake and killing the fish stock. Beginning in the 1990s, increased attention has been paid globally to the ethical sourcing of commodities and to the social and environmental conditions in the cut-flower market, and conditions are now better monitored. The burgeoning flower industry has also attracted migrants to the region in search of jobs, but in the process, social problems have arisen and many workers live in poor conditions, mainly seasonal workers and, particularly, single women. There are major differences between farms that have adopted codes that comply with fair trade and European retailer codes and those that do not comply (about 25% of the farms). Conditions of employment and living conditions are generally much better on compliant farms than on noncompliant farms. For example, half of the workers on noncompliant farms have no access to tap water on their house plot (Omosa et al. 2005).

## CONCLUSIONS

Providing safe water for Africa's growing population is a mounting challenge, especially for large numbers of new immigrants moving into urban slums, where municipal services are virtually nonexistent. Most states in Africa will not achieve the MDG safe water target (halve the portion of the population without sustainable access to safe drinking water) and the MDG sanitation target (halve the portion of the population without sustainable access to basic sanitation). At present, Africa has the lowest provision of piped water in the world. Incidences of water-related and water-borne diseases and diarrhea (1.29 cases per person annually [UNEP 2010]) are the highest in the world. To meet MDG targets, 300 million people would need to gain access to an improved drinking water source and 400 million would need to secure access to an improved sanitation facility. In the meantime, unsafe drinking water and inadequate sanitation structures impede human health and drain community assets.

Water scarcity is not entirely a natural phenomenon. Some of the problems have colonial origins; municipal and regional water utility budgets fail to keep pace with population growth; and existing infrastructure is not adequately maintained. In addition, water is wasted by leakages in all major water-use sectors: agriculture, industry, and municipalities. UNEP (2010:153) estimates that over half of the water introduced into distribution systems cannot be accounted for. For example, Kibera (Nairobi) receives about 20,000 $m^3$ of water per day and 40% is lost through leakage, illegal taps, and dilapidated infrastructure (UNEP 2010). As a consequence, great inequalities in water provision and costs borne by households are evident in Africa. In urban areas, high-income households derive most of the benefits from municipal water provision, where water is typically provided at prices below the amount required to cover operations and maintenance costs. Kenyan slum dwellers pay five to 10 times more for water per unit than Kenyans residing in high-income areas and even more than consumers in world cities such as New York and London. In rural areas, women shoulder the largest burden in fetching water; they make multiple trips each day to collect household water.

There is also a growing rural–urban disparity in access to improved drinking water and sanitation: rural access is greatly lagging (drinking water coverage is 51% and sanitation coverage is 29%) (UNEP 2010). Although water and sanitation are inextricably linked, sanitation investment has lagged behind water investment. In Africa today, there are more working phones than working toilets. However, the extent of mobile phone penetration provides an example of how innovation and entrepreneurship could yield economic benefits and improve well-being if water and sanitation technologies were introduced. One novel approach is being put forth by the World Toilet Organization (WTO) to accelerate progress in sanitation provision: the promotion of toilet ownership as a status symbol and an object of desire (just like cell phones) but with significant health benefits.

Globally, more water will be needed to provide for future food security. Africans need to secure more water now, a time when water resources are under mounting pressures. As Africa's population rises and becomes more urbanized, increasingly more water will be needed to meet food requirements. Africa's water infrastructure is underinvested and underdeveloped to meet this challenge; for example, the continent's irrigation capacity is so underdeveloped that only 7% of cultivated land is under modern irrigation (UNEP 2010). Lack of irrigation investment has contributed to an expansion of rain-fed agriculture onto marginal lands with uncertain rainfall. Moreover, the water that is bundled into international land and water investment deals (without clarifying specific water uses/costs) may well subtract from the water available for Africans. Without investment in irrigation, dependence on food imports will increase, and both water and food security will remain major burdens.

Moreover, it remains to be determined how countries will navigate the conflicts that may emerge within Africa's 63 water basins, especially because some of the massive irrigation plans will affect the volume of water flows and the seasonality of water downstream. The impact of climate change remains the big unknown, and Africa is already experiencing extreme rainfall vulnerability.

Effective institutional and technical capacities for managing water are lacking. There is an insufficient water knowledge base about Africa: lack of data, dearth of skilled staff, poor dissemination of water data, weak cross-national collaboration (except in Southern Africa), and a lack of legal frameworks for ownership, allocation, and management of water. Governments need to ensure that water is an integral part of the development and planning processes, and they urgently need to change the perspective that water is a stand-alone public utility.

Sanitation, water, and irrigation infrastructures require considerable upfront investments beyond what poor countries can afford. It is also not feasible to pass the upfront costs on to users, so unless there is a major global initiative to provide soft loans, privatization and more international commercial involvement seem the most likely ways that infrastructures will be provided. However, privatization support should only be solicited within a framework that is sensitive to safeguarding access to water as a human right. UN-Water Africa has outlined an Africa Water Vision for 2025, recommending an accelerated increase in water resource investment by 10% by 2015 and 25% by 2025 to meet all increased demands. UN-Water Africa (2009:17) outlined 10 key pillars to guide Africa's water management and policy:

1. Provide safe and adequate water and sanitation for all, urgently.
2. Make equitable and sustainable use of Africa's water resources.
3. Ensure sustainable development and management of water resources for all.

4. Use water resources wisely to promote agricultural development and food security.
5. Develop water resources to stimulate socioeconomic development.
6. Treat water as a natural asset for all in Africa.
7. Share management of international water basins to stimulate efficient mutual regional economic development.
8. Ensure adequate water for life-supporting ecosystems.
9. Manage watersheds and floodplains to safeguard lives, land, and water resources.
10. Price water to promote equity, efficiency, and sustainability.

## REFERENCES


Butt, B., A. Shortridge, and A. WinklePrins. 2009. "Pastoral Herd Management, Drought Coping Strategies, and Cattle Mobility in Southern Kenya." *Annals of the Association of American Geographers* 99(2):309–334.

Gandy, M. 2008. "Landscape of Disaster: Water, Modernity, and Urban Fragmentation in Mumbai." *Environment and Planning A* 40:108–130.

Hall, R. 2011. "Land Grabbing in Southern Africa: The Many Faces of the Investor Rush." *Review of African Political Economy* 38 (138): 193–214.

Howell, P., M. Lock, and S. Cobb, eds. 1989. *The Jonglei Canal: Impact and Opportunity*. Cambridge, UK: Cambridge University Press.

Gerbens-Leenes, P. W., A. Y. Hoekstra, and Th. van der Meer. 2009. Water footprint of bioenergy. *Proceedings of the National Academy of Sciences* 106(25): 10219–10233.

GRAIN. 2012. "Squeezing Africa Dry: Behind Every Land Grab Is a Water Grab." Available at http://www.grain.org/article/categories/14-reports (accessed July 18, 2012).

Hoekstra, A. Y., and A. K. Chapagain. 2007. "Water Footprints of Nations: Water Use by People as a Function of Their Consumption Pattern." *Water Resources Management* 21(1):35–46.

Jägerskog, A., A. Cascão, M. Hårsmar, and K. Kim. 2012. "Land Acquisitions: How Will They Impact Transboundary Waters?" Report no. 30, Stockholm Water Institute, Stockholm.

Jones, J. 2010. *Water Sustainability. A Global Perspective*. London: Hodder Education.

MacDonald, A. M., H. C. Bonsor, B. É. Ó. Dochartaigh, and R. G. Taylor. 2012. "Quantitative Maps of Groundwater Resources in Africa." *Environmental Research Letters* 7(2):1–7. Available at http://iopscience.iop.org/1748-9326/7/2/024009/pdf/1748-9326_7_2_024009.pdf.

Mehta, L., G. J. A. Vedwisch, and J. Franco. 2012. "Introduction to the Special Issue: Water Grabbing? Focus on the (Re)Appropriation of Finite Water Resources." *Water Alternatives* 5(2):193–207.

Mekonnen, M. M., and A. Y. Hoekstra. 2011. "National Water Footprint Accounts: The Green, Blue and Grey Water Footprint of Production and Consumption." Value of Water Research Report Series No.50, UNESCO-IHE.

Melara, J., R. Grant, M. Oteng-Ababio, and B. Ayele. 2013. "Downgrading—An Overlooked Reality in African Cities: Reflections from an Indigenous Neighborhood of Accra, Ghana." *Applied Geography* 36 (1):23–30.

Njoh, A., and F. Akiwumi. 2011. "The Impact of Colonialization on Access to Improved Water and Sanitation Facilities." *Cities* 28:452–460.

Oakland Institute. 2011a. "Understanding Land Investment Deals in Africa: Land Grabs Leave Africa Thirsty." December. Available at http://polarisinstitute.org/files/OI_brief_land_grabs_leave_africa_thirsty_1.pdf (accessed June 4, 2012).

Oakland Institute. 2011b. *Understanding Land Investments Deal in Africa. Country Report: Tanzania.* Available at http://www.oaklandinstitute.org/sites/oaklandinstitute.org/files/OI_country_report_tanzania.pdf (accessed June 5, 2012).

Omosa, M., M. Morris, A. Martin, B. Mwarania, and V. Nelson. 2005. "Assessing the Social Impact of Codes of Practice in the Kenyan Cut Flower Industry." National Resources Institute, University of Greenwich. Available at http://www.nri.org/projects/nret/final_kenya_briefing_paper.pdf. (accessed November 7, 2012).

Plaut, M. 2013. "Kenya Water Discovery Brings Hope for the Drought Relief in Rural North." Available at http://www.theguardian.com/global-development/2013/sep/11/kenya-water-discovery-drought-relief (accessed September 16, 2013).

Showers, K. 2002. "Water Scarcity in Urban Africa: An Overview of Urban-Rural Linkages." *World Development* 30 (4): 621–648.

Stoler, J. 2014. "The Sachet Water Phenomenon in Accra: Socioeconomic, Environmental, and Public Health Implications for Water Security." In *Spatial Inequalities, Health, Poverty and Place in Accra, Ghana*, eds. J. Weeks, A. Hill, and J. Stoler, pp. 181–190. New York: Springer.

Stoler, J., G. Fink, J. Weeks, R. Appiah Otoo, J. Ampofo, and A. Hill. 2012. "When Urban Taps Run Dry: Sachet Water Consumption and Health Effects in Low-Income Neighborhoods of Accra, Ghana." *Health and Place* 18:250–262.

Strand, J. 2010. "The Full Economic Costs of Groundwater Extraction." World Bank Policy Research Working Paper 4494. The World Bank Development Research Group Environment and Energy Team, Washington, D.C.

United Nations. 2010. General Assembly Adopts Resolution Recognizing Access to Clean Water, Sanitation as Human Right. Available at http://www.un.org/News/Press/docs/2010/ga10967.doc.htm

United Nations Environment Programme (UNEP). 2009. *Africa's Water Vision*. UNEP: Nairobi.

United Nations Environment Programme (UNEP). 2010. *Africa. Water Atlas.* UNEP: Nairobi.

UN-HABITAT. 2007. Global Urban Indicators. Database. Available at http://www.cityindicators.org/deliverables/global%20urban%20ind icators%20databa_12-4-2007-1028705.pdf.

UN-HABITAT. 2012. *State of World Cities 2012/13*. Nairobi: UN-Habitat.

UN-Water Africa. 2006. *African Water Development Report*. New York. United Nations Economic Commission for Africa. Available at http://www.uneca.org/awich/AWDR_2006.htm (accessed August 5, 2012).

UN-Water Africa. 2009. *Africa Water Vision 2025*. Addis Ababa: Economic Commission for Africa.

UN-Water Africa. 2012. *Managing Water Under Uncertainty and Risk*. Available at: http://unesdoc.unesco.org/images/0021/002156/215644e.pdf (accessed November 7, 2012).

Vidal, J. 2006. "How Your Supermarket Flowers Empty Kenya's Rivers." *Guardian*, October 20. Available at http://www.guardian.co.uk/uk/2006/oct/21/kenya.world (accessed August 6, 2012).

Vidal, J., and C. Provost. 2011. "US Universities in Africa 'Land Grab.'" *Guardian*, June 8. Available at http://www.guardian.co.uk/world/2011/jun/08/us-universities-africa-land-grab (accessed August 6, 2012).

WHO. 2008. *The Global Burden of Disease: 2004 Update.* WHO: Geneva.

WHO/UNICEF. 2008. *Progress on Drinking Water and Sanitation. Special Focus on Sanitation.* Joint Monitoring Program. Available at http://www.who.int/water_sanitation_health/monitoring/jmp2008/en/index.html (accessed May 29, 2012).

WHO/UNICEF. 2012. "A Snapshot of Drinking Water and Sanitation in Africa." Available at http://www.wssinfo.org/fileadmin/user_upload/resources/1251454622A_Snapshot_of_Drinking_Water_in_Africa_Eng.pdf

Woodhouse, P., and A. Ganho 2011. "Is Water the Hidden Agenda of Agriculture Land Acquisition in Sub-Saharan Africa?" Available at http://www.iss.nl/fileadmin/ASSETS/iss/Documents/Conference_papers/LDPI/12_P_Woodhouse_and_A_S_Ganho.pdf (accessed June 4, 2012).

Zoomers, A. 2010. "Globalisation and the Foreignisation of Space: Seven Processes Driving the Current Global Land Grab. *Journal of Peasant Studies* 37(2): 429–447.

## WEBSITES

Trashy Bags Organization. Available at http://www.trashybags.org (accessed December 5, 2013).

Water Footprint Network. Available at http://www.waterfootprint.org (accessed December 5, 2013).

CHAPTER 9

# HEALTH

## INTRODUCTION

Africa is a region with enormous public health challenges. Broadly defined, public health is the art and science of preventing disease, promoting population health and well-being, and extending life through organized local and global efforts (McMichael and Beaglehole 2009:2). Constituting 13% of the world's population, Africa carries 24% of the global disease burden without an adequate corresponding health care infrastructure. For instance, the continent accounts for less than 1% of global health expenditures, its share of the global supply of doctors is 2%, and its share of health workers is 1.3% (Mills et al. 2009). Moreover, Africa's immense disease burden and frail health care systems are embedded in the broader contexts of rapid urbanization, poverty, underdevelopment, and weak governmental institutions. Some countries face additional health burdens from prolonged conflict and intermittent humanitarian emergencies (e.g., large-scale migrations, famine, and economic crises).

Throughout the region, medical facilities, clinics, and laboratories are in short supply Those that exist are routinely in poor condition with outmoded, nonworking equipment and erratic or no electricity supplies (in some rural areas). Africa's health care infrastructure is in a very unhealthy condition and there is a major shortage of physicians (Fig. 9.1).

Loss of African medical professionals to the "brain drain" in the Global North is disturbingly high. Nine of the top 20 countries with the highest emigration rates of health care professionals are in Africa (Sanders et al. 2009:172). The World Health Organization (WHO 2006) acknowledged the severity of the problem and determined that 36 of 57 countries with critical shortages of health care professionals were located in the region. The hemorrhaging of health care professionals reflects active recruitment by institutions in the Global North, underinvestment in African medical systems and health care facilities, lack of established posts, low pay, and substandard working conditions.

The physician outmigration stream from Ethiopia, Kenya, Malawi, Nigeria, South Africa, Tanzania, Uganda, Zambia, and Zimbabwe represents a US$2.17 billion loss of return from public investment in medical education (based on underwriting the entire education cost of physicians and their practical training prior to emigrating with proper qualifications) (Mills et al. 2009:1). The benefit of Africa-educated nurses and physicians to medical systems is US$2.7 billion in the United Kingdom and US$846 million in the United States, representing a 50% net loss of investments for Africa (even after accounting for remittances) (Mills et al. 2009:4). As a result, the World Health Assembly (the decision-making body of the WHO) adopted the first code of practice for the international recruitment of African health care personnel in 2010, recognizing the global shortage of health care professionals but

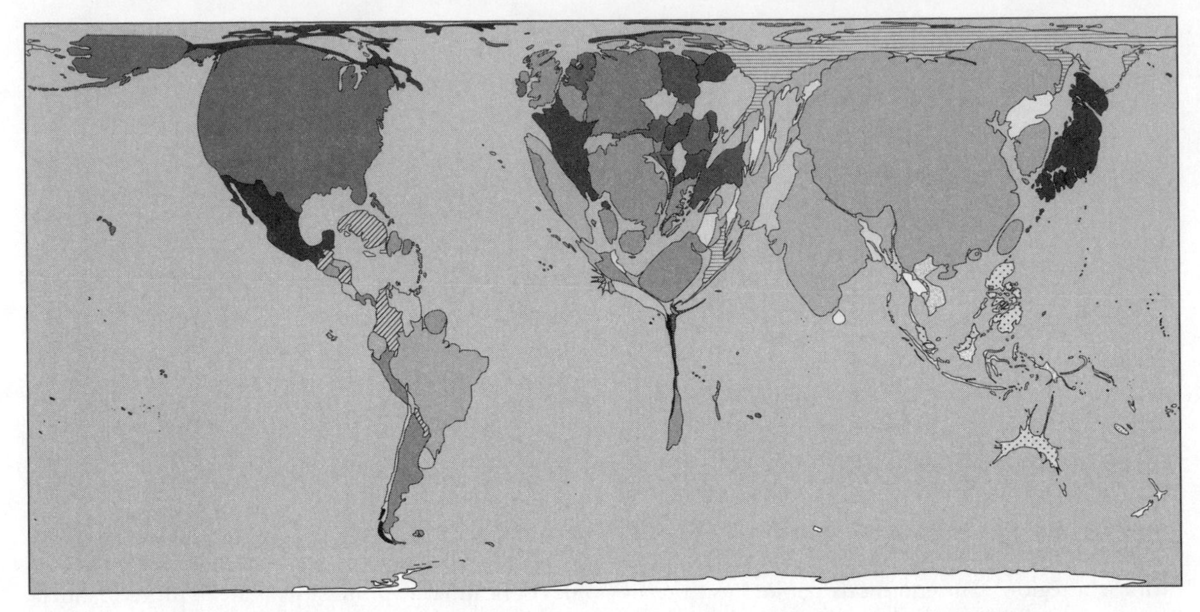

**FIGURE 9.1** Global Distribution of Physicians Working. *Source:* © Copyright Sasi Group (University of Sheffield) and Mark Newman (University of Michigan).

also the negative effects on home country's health systems of "poaching" African professionals.

Africa's health care worker ratio is 2.3 per 10,000 population compared to 24.2 in the United States. Some of the lowest physician ratios per population in the world occur in Niger (0.2:10,000) and Ethiopia (0.3:10,000) (Henry J. Kaiser Family Foundation 2012). This critical shortage of health care workers translates into a deficit of doctors and nurses of 2.4 million professionals (Naicker et al. 2009). Statistics mask the real human tragedy of these personnel shortages. The majority of Africans will never visit a physician's office in their lifetime, and because of the vastly inadequate number of trained and employed health care workers, people cannot enjoy good health that would enable them to flourish. According to WHO (2006:xvi), "the lack of health workers is a major factor in the deaths of large numbers of Africans who would survive if they had access to healthcare." This means that the Africa health care infrastructure underperforms in preventing and treat injuries and diseases and in relieving pain and suffering.

Making matters worse, the region has a chronic shortage of medicines, and counterfeit pharmaceuticals are rampant. With a few notable exceptions (e.g., the South African pharmaceutical manufacturer and supplier Aspen is the leading generic drug manufacturer in the Southern Hemisphere), there is a dearth of domestic pharmaceutical production in Africa, which means that countries are heavily dependent on imports. Indian and Chinese generic companies have emerged as prominent players. The global import business has many loopholes and blind spots that enable counterfeit medicines to enter the African marketplace. The main dangers to health produced by fake medicine are (1) failure to provide effective treatment, (2) direct harm, and (3) drug resistance. Various failures in manufacturing, importation, handling, and regulation prevail. Some medicines are accidently or deliberately falsified, producing dire results (they are sometimes a direct cause of death; e.g., 80 children died in Nigeria in 2009 after ingesting a teething mixture named My Pikin). Drugs are not dispensed properly in many locations. For example, imported Chinese and Indian medicines are routinely sold in the informal economy, sometimes in open-air markets alongside fruit and vegetables. Estimates are highly unreliable, but approximately 30% of pharmaceuticals sold in Africa could be counterfeit and/or substandard. In a recent study of tuberculosis (TB) medicines in Africa, 16.6% of them failed and 7% were outright fakes, containing no active ingredients (Binagwaho et al. 2013). The effects of counterfeit medicine are difficult to quantify, but they may have disastrous effects on human health. For example, malaria parasites can build up resistance if diluted medicines are used (see Dan Rather's 2010 report on counterfeit antimalarial medicines).

Despite human and physical infrastructural constraints, the global HIV/AIDS pandemic has spurred a historic and unprecedented mobilization of attention and resources for Africa. HIV/AIDS has elevated the profile of African public health among development and foreign policy practitioners, generating new global institutions (e.g., UNAIDS and Global Fund to Fight AIDS, Tuberculosis and Malaria) and mobilizing new constituencies, including religious organizations and private foundations (e.g., Bill and Melinda Gates Foundation, Ford Foundation, Aga Khan Foundation), corporations (e.g., Pfizer, Bristol-Myers Squibb & Gilead), and nongovernmental organizations (NGOs; e.g., Save the Children, Oxfam, Christian Aid). The focus on African health has been reinforced by a burgeoning public health research community. NGOs play a strong global advocacy role, appealing for increased resources from donor countries and organizing specific campaigns aimed at accelerating access to badly need medicines. Public policy experts now position international health as a global security issue, and 39 new pathogens (that we know of) have emerged since 1967. Some (e.g., HIV/AIDS and Ebola) have African origins.

The key challenge in Africa is to break out of the vicious cycle of ill health and poverty, which impedes development. Africa is, and will remain for the foreseeable future, a preoccupation of global public health policies and interventions. There is an obvious globalization and public health paradox: the sanitary revolution (the introduction of clean water and sewage disposal on a mass scale) that was very effective in improving public health in the Global North centuries earlier has yet to be transferred to Africa, even though it had been acknowledged as the greatest medical advance since 1840 in the Global North (Ferriman 2007). Indeed, international public health policies involving Africa have focused more on treating symptoms and sickness than on the causes of ill health but there are health technology exceptions (see Box 9.1). Nevertheless, the vast majority of people living in Africa have yet to benefit from the advances in medical and public health research that other regions have enjoyed. Without any doubt, Africa is the continent where suffering related to poor health is concentrated and where illness needs to be tackled.

## AFRICA'S DISEASE BURDEN

WHO has taken the lead in compiling regional profiles of disease burdens (including mental health disorders), commencing in 2004 and updating them in 2008, in an attempt to shape health policies by using regionally accumulated bodies of evidence rather than just well-intentioned advocacy. WHO's (2008a) statistics for Africa are alarming: 1.7 million die each year from HIV/AIDS, 900,000 from entirely preventable diarrheal diseases, 700,000 from malaria, and 400,000 from TB (Table 9.1). The latter two diseases are very treatable, but treatment is sporadic and drug-resistant TB is a new public health concern in the region.

WHO's global disease burden data show a large number of annual deaths in Africa due to diseases, specifically HIV/AIDS, parasitic infections, and maternal and nutritional causes. Centuries-old disease threats (e.g., malaria and TB) continue to pose high risks because of mutation, rising resistance to antimicrobial medicines, and weak health care systems. Largely attributable to the HIV/AIDS pandemic, adult life expectancy remains low in Africa, hovering between 40 and 60 years in most countries but with considerable spatial variation (Fig. 9.2).

Child mortality rates on the continent are the worst in the world. Among the 10.4 million deaths in children under five worldwide, 4.7 million (45%) occur in Africa. High child mortality rates occur even though cost-effective interventions are available to prevent major causes of death (WHO 2008a:15). A child born in Africa today is more likely to lose his or her mother from childbirth complications or by HIV/AIDS and will have a life expectancy of 47 years, during

## BOX 9.1 TECHNOLOGY BRIGHT SPOTS IN HEALTH ARENAS

There are several bright spots where technology is being deployed to improve health care services in Africa. Rwanda has implemented a national system with legal and technical oversight to manage entire pharmaceutical supply chains and to combat substandard and falsified medicines (Binagwaho et al. 2013). The Rwandan government purchases high-risk drugs, such as those for treating TB, exclusively from manufacturers certified by WHO, and the Rwandan Ministry of Health oversees their distribution to hospitals and clinics. The Ministry of Health performs quality control by testing each imported drug shipment as well as by sampling systematically anti-TB and antimalarial drugs on a quarterly and annual basis. In addition, the Rwandan government has trained health care workers who handle the medicines to spot and report substandard and falsified products. Rwanda's comingling of policing and public health is now being suggested as the antidote to the counterfeit drugs problem. The country has not recorded a single case of drug-resistant TB (Binagwaho et al. 2013), but Rwandan pharmacovigilance (the science and activities related to detection, assessment, understanding, and prevention of adverse effects or any other drug-related problem) needs strengthening by international protocols. A global treaty and leadership by WHO are essential to address the manufacturing and trade of substandard and falsified medicines and to make pharmaceutical falsification an international crime.

Nigeria and Ghana are leading the effort to deploy technology to intensify pharmaceutical surveillance. In Nigeria, the National Agency for Food Drug Administration Control (NAFDAC) uses hand-held spectrometers (laser light passes through pills to assess their chemical composition) to verify the authenticity of key pharmaceuticals. NAFDAC's TrueScan monitoring has resulted in a progressive decline in counterfeiting: from 40% in 2001 to 6.4% in 2012 in medicines sampled. In Ghana, MPedigree (http://mpedigree.net) developed a technology platform that connects mobile phone networks to a central registry: callers use toll-free numbers to verify pedigree information on pharmaceutical products from participating manufacturers. Consumers can get drug authentication information at the point of purchase. If the medicine is authenticated, an approval text message is sent, confirming that the medicine is certified in its current jurisdiction and has not been blacklisted since entering the supply chain. The platform was rolled out in Ghana in 2008 and subsequently was introduced into Kenya and Nigeria (all malaria medicines in Nigeria are integrated into the mobile phone-based consumer verification system). The technology facilitates trust between an African consumer and a pharmaceutical manufacturer that is typically some distance away. MPedigree verification codes had appeared on 10 million medicines by February 2013. MPedigree is being tested in South Asia, and this African innovation is being imitated aggressively worldwide.

Other technological advances are enabling a rise of telemedicine, which offers enormous potential to extend access to health care throughout the region, even to remote locations. Project Masiluleke (Project M) (in Zulu the word means "giving warm and wise counsel") is a South African mobile health initiative that commenced in 2008 and promotes HIV/AIDS awareness, education, and treatment. Project M sends over 1 million messages daily and connects with rural South Africans in peripheral locations with sparse access to health information and services. It is now working on self-testing for HIV (with mobile support during the testing) and follow-up mobile support for HIV-positive individuals. Self-testing for HIV/AIDS is controversial and there are access problems and other barriers to overcome, but it may be successful in getting a population tested that otherwise avoids clinics. Considerable interest and efforts are under way to develop "lab-on-a-chip" medical testing using mobile phones that could potentially diagnose HIV, TB, malaria, hepatitis, etc.

Mobile health (mHealth) is rapidly evolving in many spheres. For example, remote health care workers can now confer with specialists in tertiary medical facilities to diagnose and treat illness. Chinese and Indian companies are using video-related health care technology in the region that allows India-based doctors to treat African patients remotely in three regional hospitals (Nigeria, Republic of Congo, and Mauritius) by linking these hospitals to specialist Indian facilities through the pan-African e-Network Project, a joint venture between the Indian government and the African Union.

which he or she will have to grapple with poverty, poor health care, droughts, floods, and civil conflict and may be compelled to migrate.

### DOUBLE BURDEN OF DISEASE

As societies pass through various stages of development, they show significant changes in patterns of health and disease. Abdel Omran's (1971) model of epidemiological transition is widely applied in health geography and public health. It specifies three processes of the epidemiological transition: (1) replacement of common infectious diseases (e.g., malaria and TB) by noncommunicable diseases (NCDs) or degenerative disease (e.g., hypertension, stroke, and diabetes), with injuries as the primary cause of death; (2) a shift in peak morbidity and mortality from the young to the elderly; and (3) a situation in which mortality predominates rather than one in which morbidity is dominant.

**TABLE 9.1 LEADING CAUSES OF DEATH IN AFRICA, 2004**

| | Cause of Death | Deaths (millions) | % of Total Deaths |
|---|---|---|---|
| 1 | HIV/AIDS | 1.74 | 14.2 |
| 2 | Cardiovascular diseases | 1.7 | 13.9 |
| 3 | Lower respiratory infections | 1.4 | 11.1 |
| 4 | Perinatal conditions | 1.1 | 8.6 |
| 5 | Diarrheal diseases | 0.9 | 7.2 |
| 6 | Malaria | 0.7 | 5.9 |
| 7 | Malignant neoplasms | 0.7 | 5.3 |
| 8 | Tuberculosis | 0.4 | 3.3 |
| 9 | Respiratory diseases | 0.4 | 3.2 |
| 10 | Unintentional injuries | 0.35 | 2.9 |
| 11 | Digestive diseases | 0.3 | 2.4 |
| 12 | Road traffic accidents | 0.3 | 2.4 |
| 15 | Diabetes | 0.14 | 1.8 |

*Source:* World Health Organization, *The Global Burden of Disease: 2004 Update*.

The quality of the health care system greatly affects the stages. North America and Europe have passed through the transition, and countries of the Global South are at different transition points. Latin America's path is not unidirectional: some diseases are near eradication by immunization (e.g., polio) but others are reemerging (e.g., TB, malaria, and dengue fever). Furthermore, in contexts of extreme income inequality, epidemiological profiles are polarized, with higher morbidity and mortality rates among the poor, who have high rates of infectious and nutrition-related diseases, while NCDs predominate among the wealthier classes.

The African region is experiencing a "double burden of disease" based on the coexistence of communicable diseases and NCDs (Agyei-Mensah and de-Graft Aikins 2010). Indeed, African slum dwellers may be carrying a quadruple burden of disease: in addition to the double burden, they experience very high rates of personal injuries, accidental deaths, and nutritional deficiencies. Wealthy Africans, especially urbanites, have a higher risk of chronic diseases—diabetes, hypertension, cancer, and chronic respiratory conditions. The rapid

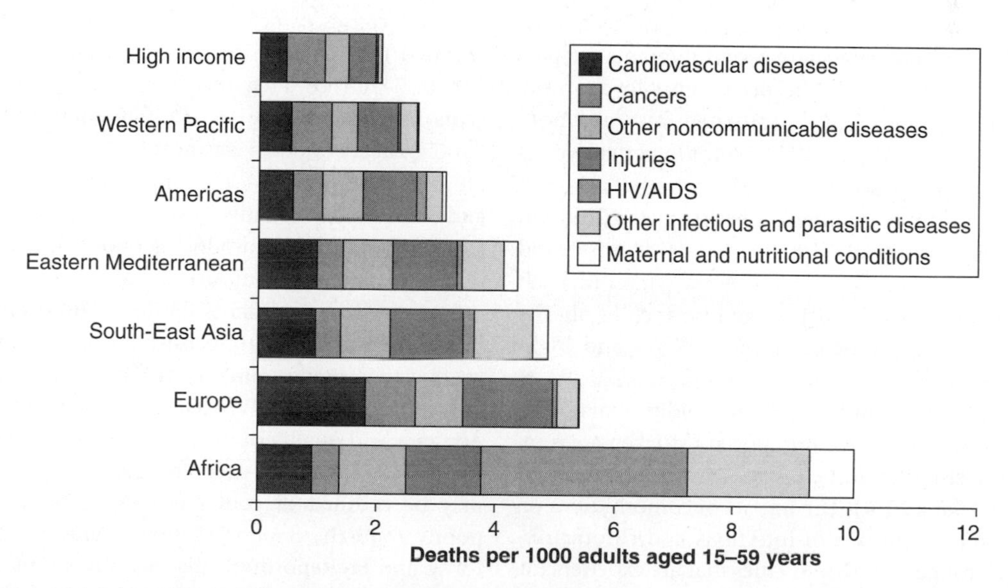


**FIGURE 9.2** Adult Mortality in Africa Compared to Other World Regions.

expansion of chronic disease is driven by more sedentary lifestyles (increased inactivity and reliance on technology), dietary changes, and increasing alcohol and tobacco consumption. Alarmingly, available data suggest that age-specific mortality rates for chronic disease in Africa are as high as the worst-affected developed regions.

Community studies also reveal the urban poor to be experiencing increased rates of chronic diseases. Moreover, there is also mounting evidence of adverse interactions between some chronic diseases and infectious diseases, especially among the poor. For example, diabetes intensifies the risk of developing TB, and antiretroviral (ARV) therapy enhances the risk of diabetes and cardiovascular risk factors (de-Graft Aikins et al. 2010). Africa recorded 10 million diabetes cases in 2006, making it the 15th highest leading cause of death, a ranking that is expected to climb as the number of metabolic disease cases doubles by 2025 (WHO 2008a). Indeed, the incidence of diabetes may be much higher in Africa as NCD data are scant. The rate of undiagnosed diabetes in Africa is acute: the International Diabetes Federation (2012) estimates that 78% of cases are undiagnosed. Accounting for only 20% of health spending targets, chronic diseases are marginalized within African public health systems.

The relationship between poverty and Africa's rising burden of chronic disease needs to be underscored. Urban poverty is directly associated with diets, and slum dwellers and recent migrants are particularly at risk. For example, poor diets—low in fruits and vegetables and high in saturated fats—are more prevalent. Salt intake, a risk factor for hypertension, is also much higher. Current obesity rates in Africa are high; for example, women's obesity rates tripled in Ghana, rising from 10% in 1993 to 30% in 2008, and men's obesity rate was 46% in Accra (Agyei-Mensah and de-Graft Aikins 2010). In addition, child obesity is an emerging problem among middle- and high-income groups.

A key factor is the role of comorbidities: major infectious diseases prevalent in poverty-stricken areas increase the risk of chronic diseases (Agyei-Mensah and de-Graft Aikins 2010). The impact of comorbidities on the cumulative burden of infectious and chronic diseases is strongest in African cities that are experiencing high levels of rural–urban migration, urbanization, and urban poverty. As such, the epidemiological transition cannot be separated from the urban revolution that is unfolding in Africa.

Africa's health transition represents an enormous challenge because of limited resources, limited preparedness, and widespread poverty. Whereas previous epidemics, including HIV/AIDS, caught Africa unprepared, this time around there is little learning about preparedness. Lessons from the epidemiological transition model are not been heeded in Africa.

## AFRICA'S HEALTH DATA LACUNA

WHO's (2006) *Global Burden of Disease* is the best available region-wide source of data on many disease indicators, but overall health data are sparse. Medical recordkeeping is abysmal in many countries, especially in rural areas. Only a few African countries maintain vital registration systems that are more than 50% complete (South Africa, 90% complete; Kenya, 60%; and Mauritius, 90%) (Bradshaw et al. 2010). Far more common are scant registration systems (e.g., Mozambique, 5% complete; Zambia, 16%), and many countries collect only urban data (Rao, Bradshaw, and Mathers 2004:83). Fifty-five percent of births are not registered, and half of adult causes of death are not recorded (Bradshaw et al. 2010). Data on child mortality are even sparser.

Most causes of death are based on national data reported to WHO, which, along with various ministries of health, uses region-wide projections based on econometric modeling. WHO is attempting to standardize causes of death certification, but reported data are highly variable. Lack of information technology and skilled administrative personnel and inconsistencies in monitoring are producing poor health information systems throughout Africa.

South Africa maintains the most uniform records in the region. Still, imprecise data on deaths could produce inaccurate numbers as high as 20%, particularly with regard to HIV/AIDS (Bradshaw et al. 2010). Accuracy and quality of cause-of-death data are highly variable: inaccuracies occur because causes of death may be certified or coded incorrectly, especially by poorly trained personnel (often relying on oral autopsy and lay reporting); distinctions are not always drawn between the principal underlying cause of death and contributory causes of death; and poor handwriting in non-electronic record systems also may lead to improper coding. In Africa, death often occurs outside of health facilities, and information is often obtained from the family and village leaders in rural contexts, individuals with limited knowledge and no training in cause-of-death attribution. The major concern about cause-of-death information is that it is used to inform health policy and allocations of health spending for various programs. Lessons from the past are troubling. For example, inaccurate records of cause of death during the denial days of HIV/AIDS in South Africa led to ill-informed national policies from the mid-1980s to 2000, and it was only after efforts were made to improve the quality of data collection that real changes in policy and programs were implemented.

Most people in Africa are born and die without a trace in any legal record and official statistic. The absence of reliable data about births, deaths, and causes of death is at the root of health data inaccuracies, and the poor are disproportionately affected. For example, most official health statistics use hospital-based data and incorporate community reporting, but these latter data miss some of the urban poor who remain outside of these systems. Existing data rely heavily on major cities, giving little or no coverage to small and medium-sized urban centers. Vital statistics about births, deaths, and causes of death provide crucial information for policy, planning, and evaluation of all sectors of development.

## SICKNESS AND HEALTH: THE COLONIAL LEGACIES OF INHERITED HEALTH SYSTEMS

Africa has long been portrayed as "the land of disease" and "the white man's grave," representations derived from the Europeans' inability to survive in the region before the modern medical age. The colonial era was accompanied by vast improvements in hygiene and health and medical care for some members of colonial society. The development of tropical medicine, the implementation of modern medical infrastructure, and improvements in urban infrastructure contributed to a decrease in mortality and facilitated urban population growth.

Colonial health systems produced enduring legacies, and the policy behind those systems targeted the health of Europeans. Hospitals in colonial outposts were exclusively for the colonial elite, and urban and residential planning was used to protect European populations from the threat supposedly posed by "unhealthy Africans." Outside of official colonial health apparatuses, there were missionary efforts to provide medical care that linked a health mandate to a proselytizing mission. Protestant and Roman Catholic missions pioneered Western medicine and public health, building hospitals throughout rural Africa (Good 1991). For example, Swiss missions built hospitals in rural South Africa, Tanzania, and Ghana. Church-based hospitals and health care programs had enduring legacies: in many African countries a century later they were providing 25% to 50% of available services (Good 1991:1).

For the most part, rural Africans had no access to Western colonial health care. Instead, rural (and urban poor) populations relied on traditional medicine and indigenous healers. Colonial government campaigns against specific diseases targeted colonial capitals and ports. There was very little interest on the part of colonial powers in promoting health care in rural areas except in crisis situations when disease could undermine stability in the colony, particularly by diminishing the productivity of the labor force. When wider campaigns were launched against sleeping sickness, yellow fever, and plague outbreaks, they were crafted in response to particular crises and represented isolated rather than sustained efforts. As a result, the health care system handed over at the end of the colonial period was oriented toward cities, served the elite, was fragmented in its spatial organization, and was geared toward curative instead of preventive medicine. The inherited professional cadres and structures fashioned for Western health systems were inappropriate for African health needs. The patchy, selective, and intermittent nature of colonial interventions made many Africans skeptical about the benefits of public health programs.

Colonial health systems in Africa were far from homogenous and exhibited considerable variation inside countries and from country to country. Core parts of the system in the Belgian Congo were so neglected that on the eve of independence in 1960 there was not a single indigenous graduate in medicine in the country. By contrast, health systems in Ghana, Kenya, and South Africa were more promising, comprising a good number of trained doctors, including medical specialists.

Ironically, in the colonial era, European medical knowledge benefited from pioneering field research in the region. The Liverpool School of Tropical Medicine, for example, operated its first overseas field laboratory in Sierra Leone, and research conducted in Freetown helped identify the vector for malaria. British doctor Ronald Ross was awarded the Nobel Prize for Physiology and Medicine in 1902 largely on the basis of field research in Africa and of other research in India.

In the decade after independence, substantial progress was made in improving the reach of health care services in many African countries. Most African governments increased spending in the health sector to extend primary health care and to develop national public health systems that redressed some of the inequalities of the colonial era. Despite these increased expenditures, however, African governments failed to build adequate health care structures. Deteriorating macroeconomic conditions in the early 1970s resulted in African governments pulling back from investment in public health. Governments were forced to reduce their per capita expenditures on health (Senegal and Côte d'Ivoire cut health expenditures by half [UNICEF 1990]): many facilities had to close their doors, medical personnel were reduced, and funding for needed repairs, equipment, and pharmaceuticals dried up. Severe cutbacks compelled health care workers to moonlight as a survival strategy, and absenteeism became more frequent, adding further stress on already fragile health systems. These factors led to even poorer service delivery and a hemorrhaging of health professionals from the national systems. In the worst cases (e.g., the Democratic Republic of Congo [DRC] and Somali), per capita government expenditures were reduced to less than US$1 by 1985 (Vogel 1993).

## THE NEXUS OF INTERNATIONAL HEALTH POLICY AND NATIONAL HEALTH POLICIES, 1980–PRESENT

With few exceptions, health conditions worsened in Africa until the 2000s. The economic crises of the 1970s and 1980s led to declining real incomes and cutbacks in public health expenditures and accelerated the brain drain (e.g., the number of physicians in Ghana decreased from 1,700 in 1981 to 800 in 1984) (Oppong 1997:13). Government cutbacks in health expenditures were mandated by the International Monetary Fund because African governments were insolvent (strapped with rising debts and declining terms of trade), meaning that health care spending was halved in the 1980s. Cutbacks disproportionally affected poor rural populations; more powerful urban constituencies ensured that reductions in national capital expenditures were the last resort. Meanwhile, governments confronted new sources of health care demand and costs as the spread of HIV/AIDS meant health systems had to do more with fewer resources. Already-compromised national health systems almost collapsed, health care provision became even less inadequate and in many cases nonexistent, and people were left to resort to what private or traditional medicine had to offer. Those who could afford health treatment (political and business elites) went abroad for treatment; the rest were left to suffer (Chabal 2009).

The reality on the ground in Africa became dissociated from the international health policy consensus advice. African governments signed the Declaration of Alma-Ata (renamed as Almaty, Kazakhstan) in 1978, vowing to deliver "health care for all by 2000" by providing community-based, affordable, and accessible health care (as had been done in China). However, financing such a comprehensive health system became an insurmountable obstacle for African governments. More and more governments moved further from the development of primary universal health systems recommended by prominent international health organizations at this time.

In time, the Alma-Ata Declaration was criticized as being too broad, too idealistic, and too unrealistic in its implementation timetable. Instead, international donors, agencies, and scholars recommended and promoted a more cost-effective strategy that could deliver measurable goals as an interim strategy—later labeled the selective primary health care (SPHC) for disease control. UNICEF and other major international health organizations got on board with this targeted approach.

SPHC was promoted by neoliberal thinking and fatigue about spending money on programs that might never attain a utopian goal and a "children's health revolution." Instead, SPHC involved the reprioritization of programs around four interventions: growth monitoring, oral rehydration techniques, breastfeeding, and immunization. The neoliberal political context forced more international agencies to concentrate on short-term technical programs with clear budgets (prioritized programs to address a few targeted diseases, the leading causes of mortality) rather than on broadly defined health programs. Not everyone agreed with the SPHC approach: three major criticisms are (1) its failure to address the root causes of ill health in Africa (i.e., poverty); (2) its emphasis on top-down, externally driven, technological approaches that failed to empower communities to tackle their own health problems; and (3) its focus on only targeted diseases.

New international funds became available in the 2000s with the Millennium Declaration that acknowledged the fundamental importance of health in the development rubric. In the 2000s, the emphasis switched back to an integrated approach: six of the eight Millennium Development Goals (MDGs) aimed to improve an aspect of health to accelerate development: MDG 1 aimed to reduce extreme poverty and hunger; MDG 4, to reduce child mortality; MDG 5, to improve maternal health; MDG 6, to combat HIV/AIDS, malaria, and other diseases; MDG 7, to ensure environmental sustainability; and MDG 8, to provide access to affordable essential drugs. The emphasis on addressing diverse factors (e.g., clean water, universal education, and food production) to improve well-being was widely welcomed by development experts, but the levels of international and national commitments needed to achieve the MDGs were lacking. At the same time, African governments recognized more and more that national health systems need to be strengthened and, above all, that there was a dire need for much greater financial and capacity building in health care delivery systems.

In 2001, African Union heads of state signed the Abuja Declaration, establishing national health care expenditure targets of 15% of the national budget. Since that time, some countries (e.g., Botswana, Burkina Faso, Rwanda, and Tanzania) have risen to the challenge and reversed decades of underinvestment in health services. Tanzania embarked on health service reform by decentralizing services and making them more accessible in remote rural areas (child mortality was reduced by 40% in two districts within five years) (Uthman 2012). Most African countries, however, have not met the Abuja commitment.

The implications of modest national expenditures are that, in most African countries, governments and private national sources (e.g., employers and private insurances) pay only half of the total average national health expenditure. The other half is met by out-of-pocket spending, a ratio that rises to 70% in Nigeria and 90% in DRC. These costs represent a huge burden on many who can ill afford to shoulder the burden of health costs. Ghana and South Africa have taken steps toward universal coverage, but even in these countries many medicines and services are not included and require out-of-pocket payments. The result of fragmented coverage across Africa has meant that private health financing fills some of the void (e.g., for-profit hospitals and clinics and nonprofit providers such as aid organizations and missionary hospitals). The Economist Intelligence Unit (EIU) (2012) notes that in Ethiopia, Nigeria, Kenya, and Uganda more than 40% of people in the bottom 20% income bracket receive their health care from private, for-profit providers. The remainder and the poorest Africans have little or no access to public health care coverage, and they frequently also lack access to the prerequisites of health: clean water, sanitation, and adequate nutrition. The global financial crisis of 2008 to the present added even more pressure on external financing for African health systems, and the immediate horizon looks shaky and somewhat bleak. African governments are going to need to make up for shortfalls and/or turn toward more private sector provision.

More recently, African governments have reengaged and reaffirmed their commitments to provide universal primary health care. The Ouagadougou Declaration in 2008 on primary health care and health care systems in Africa is the most comprehensive statement of intent. It acknowledges that the sources of many African health issues are products of weak national health systems (without denying other societal, behavioral, and environmental determinants of health). According to the Ouagadougou Declaration, major challenges that beset health systems include lack of financing, poor health infrastructure, and geographical and socioeconomic inequalities (e.g., the spatial mismatch in clinics and medicines and restrictive access for those most in need but unable to pay). The declaration calls for action from Africa's governments and the international community on primary health care development, and a framework for implementation was outlined with nine cross-cutting priorities for African countries (Fig. 9.3).

The main challenge is to develop a comprehensive health care system using a coordinated, cross-sectoral "diagonal" approach (Ooms et al. 2008). "Vertical" health programs aiming for disease-specific results (e.g., AIDS and malaria), popular among international donors and NGOs, are too focused. The ultimate goal of horizontal integration of health programs is to create environments where stakeholders can cooperate within a framework of shared policies and strategies, and where health policies can be aligned and mutually reinforcing. Of course, much more global funding and commitment from national governments is necessary to ensure that health allocations are transparent and accountable and that proper integration takes place within national health care systems. It is interesting to consider whether the Global Fund could be converted into a larger global health fund

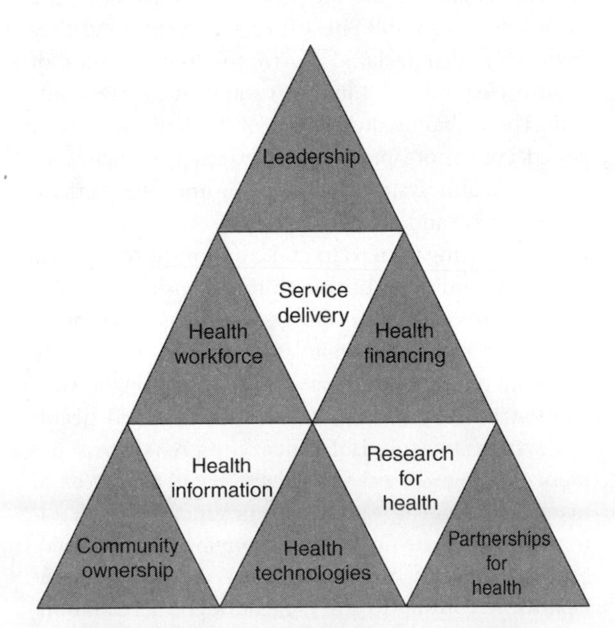


**FIGURE 9.3** Nine Health Policy Priorities for African Countries.

that could champion the transition toward comprehensive horizontal integration.

There is considerable debate about what to emphasize post-2015 (i.e., after the MDG deadline) to improve and strengthen African health systems. EIU (2012) makes five recommendations: (1) primary and preventive care that can educate people about healthy lifestyles and that can tackle the double disease burden; (2) empower communities as health care providers (stop the brain drain and augment the professional cadre by training lay health care workers; (3) implement universal coverage; (4) make telemedicine ubiquitous; and (5) encourage local pharmaceutical manufacturing.

## HIV/AIDS: AN EXTRAORDINARY AFRICAN STRUGGLE

Acquired immunodeficiency syndrome (AIDS) is a fatal disease caused by the slow-acting human immunodeficiency virus (HIV). The virus multiplies in the body until it causes immune system damage, leading to AIDS. HIV emerged in Africa in the 1930s in the vicinity of Cameroon and Congo, and it traveled to the United States and Europe in subsequent decades. The virus diffused spatially in the early independence years of Zaire (formerly the Belgian Congo) as Belgians returned to Brussels and as 10,000 Haitian contract workers returned to Port-au-Prince (Haitians were hired on short-term contracts as French-speaking teachers and professionals for government positions). Around this time (during the 1970s), there are also reports of medical missionaries and some Africans in several European countries seeking treatment of what may have been AIDS (Chin 2007). During the 1970s and 1980s, increasing "sparks" of HIV-infected persons traveled throughout Africa and other world regions. These "sparks," however, were limited and remained silent and unrecognized. According to Chin (2007:37), it was not until "such 'sparks' were introduced into gay bathhouses and/or IDU [intravenous drug users, mainly heroin users] shooting galleries during the 1970s, explosive HIV epidemics ensued."

HIV diffused across the globe and became a pandemic in the 1980s. HIV/AIDS is the greatest disease problem the world has ever faced. The first strain (HIV-1) was identified in laboratories in 1983 (largely on the basis of screening homosexual men), and labs identified a second strain in 1985 (HIV-2) (among heterosexual West Africans, many of them of Guinea-Bissau origin). Since then, the virus has shown a proneness to mutations: more than four different groups and nine different subtypes have been identified. Worldwide, HIV-1 predominates. As recently as 2009, a new strain was discovered in a Cameroonian woman and designated as HIV-1, group P (AVERT 2012). The virus is transmitted mainly by sexual fluids but also by blood (from mother to child in the womb, by birthing, by breastfeeding, and by blood transfusions) and by IDUs sharing needles.

Unsafe medical and dental practices and carelessness may be responsible for a smaller, but largely unknown, portion of transmission in Africa (Oppong and Kalipeni 2003). While IDU needle sharing is no more of an issue in Africa than in other world regions (IDU needle sharing can spike in some refugee camps and among some sex workers), a much greater problem is the repeated use of unsterilized needles and syringes for injections to treat various ailments such as malaria. A WHO study revealed that in 2000 approximately 19% of injections were administered unsafely (quoted in Reid 2009:2). Unsafe needle practices especially affected populations relying on bush doctors, roadside dentists, and itinerant pharmaceutical vendors. Indeed, in some contexts blood tests for malaria and sexually transmitted diseases constitute a potential source of HIV infection. Moreover, public awareness of HIV risk from various other skin-piercing procedures (e.g., tattooing, unsterile dental care, shaving with an unsterilized razor) remains weak in the region.

Africa is worst-afflicted region in terms of HIV/AIDS transmission: 16 million people have already died from AIDS there, some 23.5 million people are living with HIV/AIDS, and 1.8 million new infections occurred in 2011 (UNAIDS 2012). The region remains the most heavily burdened by the HIV/AIDS pandemic (accounting for 69% of the global HIV burden), although the number of new infections has fallen by 30% in the past six years (UNAIDS 2012).

The geography of AIDS is a complex mosaic, but there are some broad national patterns. Ten countries in southern Africa (Angola, Botswana, Lesotho, Malawi, Mozambique, Namibia, South Africa, Swaziland, Zambia, and Zimbabwe) bear a disproportionate share of the global AIDS burden—in total, they account for 34% of people living with AIDS worldwide. Swaziland has the highest adult prevalence in the world: in 2011, approximately 26% of its population were living with AIDS (UNAIDS 2012). In Swaziland, AIDS kills 50 people a day and HIV infects another 55, and this occurs in a country that has only two physicians for every 10,000 people. South Africa continues to be home to the world's largest population of HIV-positive people: 5.6 million in 2011. HIV prevalence in West and Central Africa is relatively lower, ranging from 2% in most countries (Benin, Burkina Faso, DRC, Gambia, Ghana, Guinea, Liberia, Mali, Mauritania, Niger, Senegal and Sierra Leone) to 5.3% in Cameroon (UNAIDS 2012). HIV prevalence in East Africa shows a considerable range, from approximately 2% to 4% in Ethiopia, South Sudan, and Rwanda to 6% to 7% in Kenya and Uganda (Fig. 9.4) (see Box 9.2 for the sharp reduction in HIV incidence rates in Uganda).

Within regions and within countries, there is considerable geographical variability in the burden of HIV (Kalipeni and Zulu 2012). There can be wide variations among neighboring countries (e.g., Côte d'Ivoire

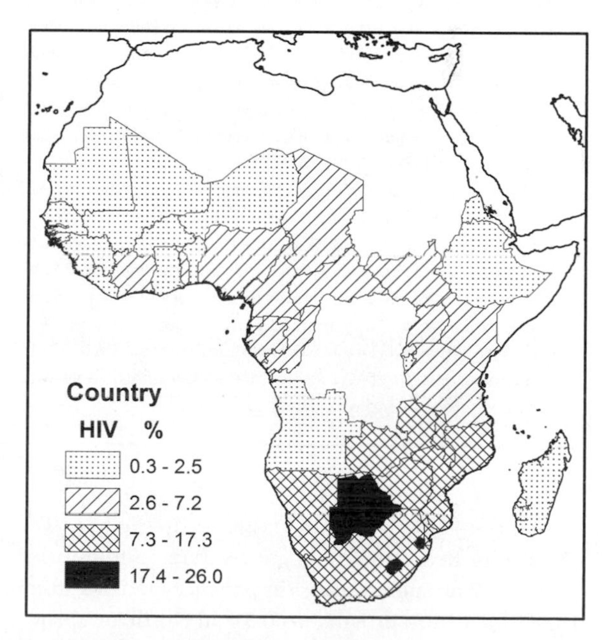


**FIGURE 9.4** HIV Prevalence in 2011.

## BOX 9.2 AIDS IN UGANDA: AN UNFINISHED SUCCESS STORY

Uganda's sharp reduction of its AIDS incidence has won international acclaim, and its experience inspired a wave of aid programs and public health strategies to combat the disease in Africa and other world regions. Uganda was the first country to devise a national response based on strong government commitment, vigorous civil society participation, donor support, and multiple public messaging campaigns. Locally known as "slim," the first case of HIV was identified in 1982, and the disease proliferated in the late 1980s (Allen 2006). Over 1 million Ugandans have died of HIV/AIDS, and 1.2 million children have been orphaned by the devastating epidemic.

The Ugandan AIDS prevention program began in 1987 with extensive public health messages promoting family values, faithfulness, and "zero grazing." In the 1990s, the government implemented the "ABC approach" (abstain, be faithful, and use a condom), and Kampala instituted national safe-sex education programs and made condoms widely available. Other key elements in the government's approach were decentralization, community mobilization, and the empowerment of the NGO community. The AIDS Support Organization (TASO) emerged to play a pivotal in providing family assistance, testing, counseling, AIDS support services (dispensing ARVs), training, and advocacy. TASO has become the largest NGO in Uganda and the largest HIV/AIDS support organization in Africa. Grassroots campaigns using community mobilization—local chiefs, churches, community, care groups, and village meetings (which even President Museveni would hold)—combined with free ARV medicines from the 2000s onward (supply and access are still not perfect) funded by international donors allowed the country to achieve noteworthy, measurable successes. Social communication led to behavior changes, a major factor behind the reductions in HIV (e.g., a 65% reduction in casual sex occurred from 1989 to 1995) (Ministry of Health 2012).

The Ugandan case is widely celebrated (but contested) as evidence of moving from worst to best practice. Prevalence rates at antenatal clinics fell from 30% in 2000 to 10% in 2005 (Parkhurst 2002:79). Overall prevalence rates in Uganda declined from 15% in 1992 to 6.4% in 2005. Nevertheless, recent reports from Uganda indicate that infection rates have begun to increase, especially in urban areas (rising from 6.4% in 2005 to 7.3% in 2011) (Ministry of Health 2012:25). Uganda is now one of only two Africa countries (along with Chad) where AIDS is on the rise.

Some researchers and grassroots organizations have questioned the accuracy of previous official figures, believing that real figures were higher and that the success stories involved selective use of information (Parkhurst 2002). For example, surveillance data from antenatal clinics are generally biased and exaggerate recorded declines in prevalence. The extrapolation of data from urban clinics to the entire population (when 87% of the population is rural) is problematic. Undoubtedly, there was pressure to show results and to maintain the narrative of success to keep international funds flowing into Uganda rather than to contextualize the evidence. At the same time, the international community was under pressure to present successful examples of HIV-1 prevention, given the groundswell of media and public attention to the profound effects of AIDS in Africa. Parkhurst (2002:80) emphasizes that "the standard of proof for policy recommendations seems to have been lowered to provide the international community with the African success story it wants, or even needs."

Others consider that the plateauing of AIDS incidence levels reflects normalizing effects of having waged war on high-risk populations (e.g., sex workers, truck drivers, the army, migrant men) and the dulling of people's attention after public campaigns become routine. Harder-to-target general populations naturally exhibit slower containment rates. "AIDS fatigue" may be creeping in, and more risky sexual behavior may be on the increase. Multiple partners are still common among wealthy men and among people in poverty who travel for employment and trade. The Ministry of Health's (2012:23) latest survey showed that 25% of married men had multiple partners. At the same time, national prevention efforts are discriminatory and overlook one particular high-risk group: homosexuals. Uganda maintains a hardline approach toward homosexuality: the Anti-Homosexuality Bill (2009) criminalizes same-sex relations, meaning that gays are not included within Kampala's boundaries of prevention.

Other experts (e.g., Allen 2006) contend there is an uncomfortable truth about the Ugandan HIV/AIDS success story: it is a story of mythmaking. Allen (2006) maintains that behavior change does not result from advocacy campaigns alone; social norms are stubborn to dislodge. For example, women have more limited agency when it comes to sex: they face pressure to behave according to accepted social and moral norms so they rarely are in positions to abstain from sex and/or negotiate condom use (Allen 2006). Moreover, the cornerstone of AIDS prevention—ABC—can be a viable prevention option for women only if it is implemented as one component of a package of interventions aimed at redressing deep-rooted gender imbalances.

compared to Liberia and Ghana). According to UNAIDS (2011a), in Kenya there is a greater than 15-fold variation in HIV prevalence across its provinces, ranging from 13.9% in Nyanza province to 0.9% in North East province. Nyanza province is home to 25% of HIV-infected Kenyans. Spatial clustering of HIV incidence occurs in the two Kenyan provinces bordering Uganda (Western and Nyanza), where the HIV incidence is almost four times the rate in the adjoining Rift Valley province (UNAIDS 2011a). Even in small African countries (e.g., Benin), a wide variation in HIV prevalence is documented.

HIV infections peaked in Africa in 1998 (Fig. 9.5). There is mounting evidence that the region has turned AIDS around, and sustained progress in halting new infections has occurred: since 2001, the rate of new infections dropped by 73% in Malawi, 66% in Ghana, and 41% in South Africa (UNAIDS 2012). Successes have been registered in AIDS prevention: 12 West and Central African countries have attained prevalence rates under 2% (UNAIDS 2012). One of the brightest signs of progress against the epidemic is the growing number of babies born free of HIV: Botswana has virtually eliminated mother-to-child transmission of HIV. More than 7 million Africans have access to life-saving ARV treatments (UNAIDS 2012). A focus on AIDS treatment combined with reductions in the price of medicines (which have come down to US$10 per person per year from US$10,000 in the mid-1990s) has resulted in more than half of diagnosed HIV-positive Africans receiving ART (UNAIDS 2012). Domestic spending on HIV/AIDS almost doubled between 2008 and 2011, and some governments have been steadily decreasing their reliance on donor AIDS support. Even though progress in stopping AIDS has been significant, there is still some distance to go before UNAIDS' (2011b) three zero targets can be met: (1) zero new HIV infections; (2) zero AIDS-related deaths; and (3) zero discrimination.

Through another lens, AIDS can be as much a social problem as it is a medical problem. Epstein (2007:xvii) underscores that "it is not that Africans have more sexual partners, over a lifetime, than people in Western countries do—in fact, they generally have fewer." Many Africans, however, are more likely to have more than one overlapping or concurrent long-term partnership ("concurrency" means sleeping with more than one partner at closely spaced intervals for months or years) at a time (compared to the general pattern of serial monogamy in the Global North). Still, the theory of concurrency is highly controversial and very difficult to measure (men overstate the number of sexual partners and women understate them). Concurrent relationships propagate interlocking networks, but there are differences of opinion on whether they link people into giant networks that create a virtual superhighway for HIV transmission (Epstein 2007) or whether this claim propagates a polysexual and promiscuous myth about Africans (Lurie and Rosenthal 2010). Of course, concurrency is not only cultural but has roots in urban poverty as a driver of transactional sex. The tendency for some members of the urban poor to form transactional sexual relationships is intensified by high levels of poverty in the context of the rising consumption lifestyles that create desires for phones, makeup, and fashions. Studies of young women in slums show some woman are often involved in transactional relationships with men with other partners (even though these same women have only one partner and are not prostitutes) (Epstein 2007). However, other studies in rural areas find that women who engage in paid sex do so out of desperation (Oppong and Kalipeni 2003). Therefore, sexual activity with multiple partners may be better understood as a survival response dictated by the vicious economic system engendered by colonization and globalization (Oppong and Kalipeni 2003).

There is a salient gender dimension to HIV/AIDS in Africa. Three quarters of all women with HIV/AIDS live in Africa. African women are disproportionately affected: they represent 60% of the population living with HIV/AIDS (UNAIDS 2012). Generally, women are at a greater risk of heterosexual transmission of HIV, and the majority of HIV transmission in Africa occurs during heterosexual encounters. Women's particular vulnerability is a product of persistent gender inequalities embedded in African social relations and economic realities. Adolescent girls are especially vulnerable, biologically and socially, to sexual infection. In many places, male identity is very much linked to sexual performance: men feel pressure to have many sexual conquests to prove their masculinity. At the same time, girls are often socialized to be subservient to men. Biologically, women are twice as likely to acquire HIV from an infected partner during unprotected sex than men are (AVERT 2012). Moreover, women's childbearing role means that women also have to contend with issues such as mother-to-child transmission of HIV and the responsibility of caring for AIDS patients and AIDS-affected children. AIDS widows experience the most negative scenarios: for example, losing out on inheritances (e.g., property, land, or rights to use land), and

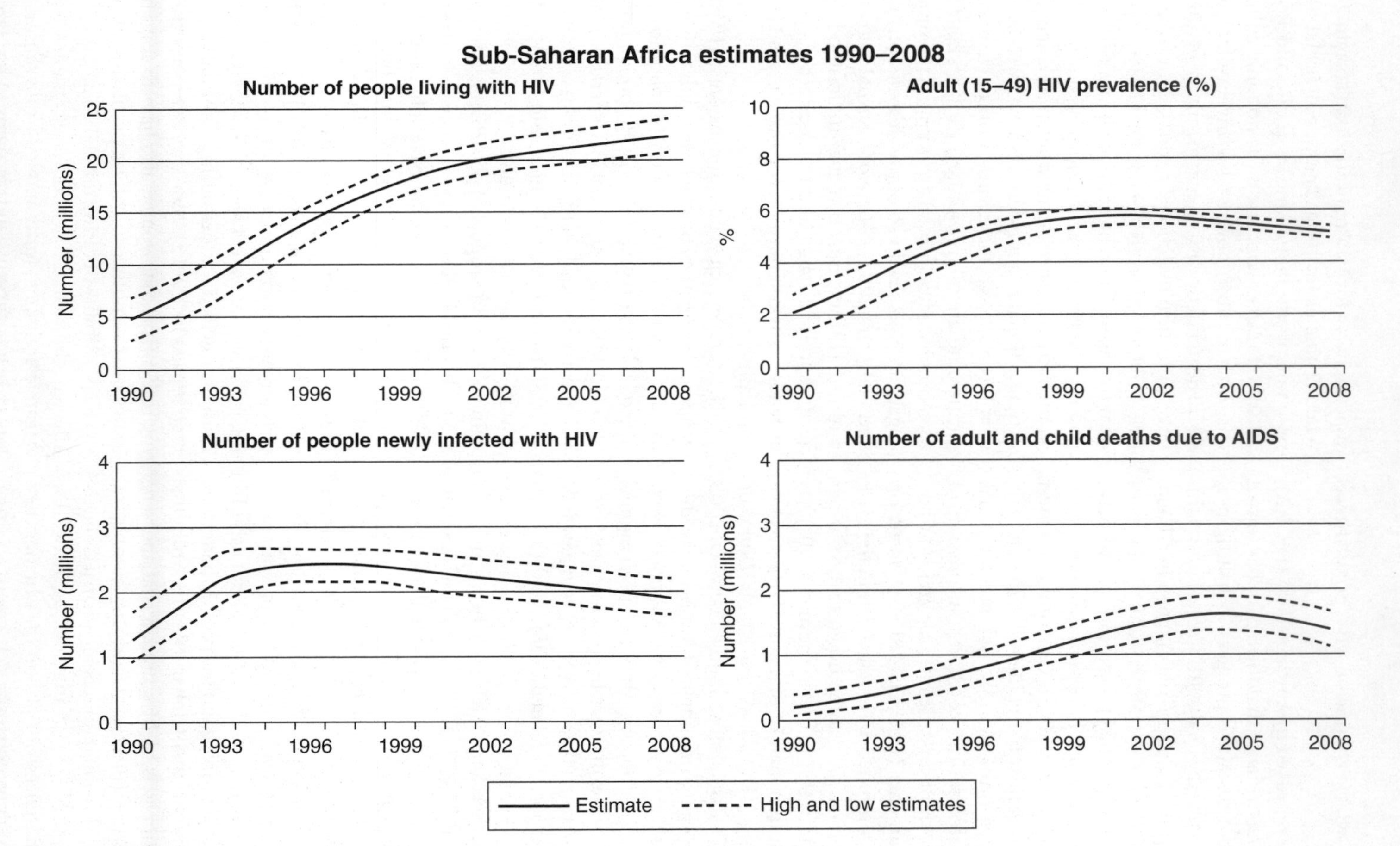


**FIGURE 9.5** African AIDS Epidemic 1990–2008.

being brought into relationships (not of their own choosing) with their deceased husband's male relatives and/or forced to return to their maternal homes. Loss of inheritance income, increased workload, and loss of support structures are usual (Economic Commission for Africa 2012).

AIDS is responsible for leaving large numbers of children across Africa without one or both parents, and it has dealt a massive blow to the African family. Globally, 90% of all AIDS orphans (16.6 million children) live in the region. Countries with the largest number of AIDS orphans are Nigeria (2.5 million), South Africa (1.9 million), Kenya (1.2 million), and Uganda (1.2 million) (UNICEF 2010) (Fig. 9.6). One fifth of children in Botswana, Lesotho, and Swaziland are AIDS orphans.

The scale of the orphan crisis is magnified by interim periods of suffering between the time parents become ill and when they die from AIDS. The sickness period and the eventual death of one and/or both parents take a huge emotional toll (see Young Carers' *Through Our Eyes* 2012 movie). Typically, access to basic necessities (e.g., shelter, food, clothing, and education) becomes more difficult, especially if the person infected with HIV/AIDS is the main breadwinner or if other household members have to forgo working to stay at home and become a caregiver (Wangui 2009). With less money available for food based on loss of income and medical costs, nutritional status can decline for all household members. Rising debts and funeral bills make the predicament even worse. Some children are forced to fend for themselves; others become

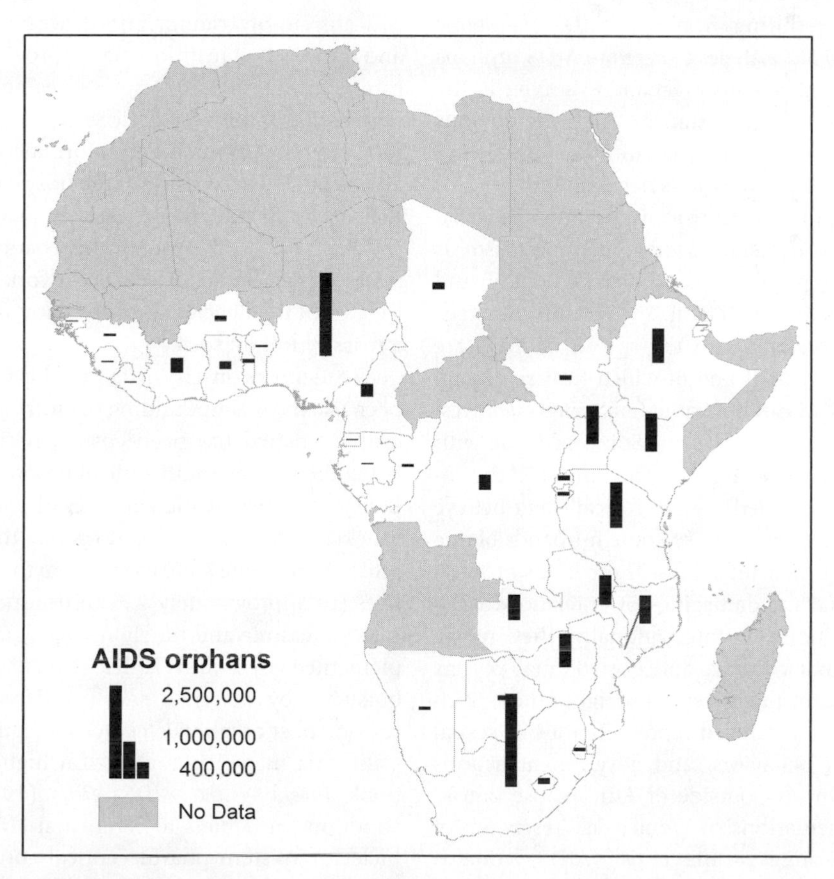


**FIGURE 9.6** AIDS Orphans in Africa, 2009.

child heads of households; and others are raised by relatives (often in female-headed houses). It is not uncommon for grandparents to be thrust into the often-unwanted roles as substitutes for fathers and mothers. Caregiver households are likely to become poorer and struggle to care for an additional person.

After the death of parents, various poverty contexts put pressure on AIDS orphans to provide for or contribute to their households. Begging and/or working typically take precedence over school attendance. AIDS contributes to the number of street children and dropouts. Compared to their peers, orphans are three to four times more likely to contract HIV in their teens, making them a very vulnerable population. Children are often stigmatized by society through their association with HIV, which in turn can be a trigger for social isolation. Overall, the orphan burden is a window into the potential massive social breakdown and dislocation in Africa resulting from the AIDS pandemic (UNICEF 2010). Nevertheless, there are AIDS orphans who manage to make ends meet and to survive in the face of very difficult circumstances. There are obvious long-term implications of generational rupture and traumatized children on Africa's development.

The AIDS epidemic in Africa is essentially a social process whereby the spread of a microorganism is shaped by social, political, economic, ideological, and cultural contexts (Kalipeni et al. 2003). Unfortunately, colloquial interpretations of the spread of AIDS are shrouded with myths, some of which further stigmatize sufferers. AIDS can be blamed on moral transgressions, "bad spirits," and sorcery. Some of those with extreme religious views blame it on sins (i.e., the innocent remain protected). Some African men believe AIDS is a disease of women, so some husbands blame their spouses for their infections. These kinds of views reinforce unequal and damaging power relations and a profound mystification. Internationally, they reproduce the way that colonial and postcolonial powers represented African practices and social politics. Such gender-based and racialized representations of sexual practices, social behaviors, and government actions generated within and outside of Africa cause untold damage. Representations of women as "reservoirs of infection," "Africans as promiscuous," "AIDS victims as depraved," and "African governments as incompetent" are insidious stereotypes (Oppong and Kalipeni 2003). Looking at the macro pattern, Craddock (2003:5) emphasizes that "it is clear that AIDS has been exacerbated by deepening poverty experienced by the majority of African countries over the past 20 years; that it has spread in the aftermath of war, civil unrest and refugee movements; that migration patterns necessitated by underemployment in chronically underfinanced economies ensure both an increase in the rate of transmission and spread from urban to rural areas; and that governments shackled by poor terms of trade and crippling debts have neither the finances nor the personnel to address the problem adequately."

The way that patients face AIDS involves other social processes. Paying for treatment requires pooling collective resources. Family and kin are the first line of defense, and patients must draw from extensive social and solidarity networks. Without solidarity, ordinary Africans simply cannot afford access to health care, and the family is instrumental in providing in-hospital care. Chabal (2009:267) notes that "hospitals in Africa require the family feed, clean, provide bedding and look after the patient. In addition, medicine must usually be purchased outside of the hospital, which means that a crucial part of treatments also hinges on the social network." Therefore, when national health care systems are lacking, the social network has to make up for part of the deficiency. In so many ways, AIDS takes a massive toll on society.

Leadership on HIV/AIDS in the region has often been lacking. Compounding the infrastructure deficits, outright denial has been costly, the most notorious being the former South African government of Mbeki. Researchers (e.g., Chigwidere et al. 2008) claim that Mbeki's stance (which went against the medical consensus) in the late 1990s resulted in the loss of 330,000 lives (or approximately 2.2 million person-years) because a feasible and timely ARV program was not implemented in South Africa. Mbeki's position was bolstered by maverick scientist deniers outside the region, most notably Peter Duesberg, the University of California, Berkeley author of a highly controversial book *Inventing the AIDS Virus* (1998). It is still common for deniers to claim that HIV/AIDS is a vehicle for Western pharmaceutical companies to reap economic rewards in "pharmacolonialism."

New infection rates in Africa appear to have peaked in 1998, although the struggle is far from over. Africa's HIV/AIDS crisis continues to be an enormous challenge at all levels because of lack of infrastructure, finance, ignorance, misunderstanding, and absence of a cure. HIV prevention strategies focus heavily on encouraging individuals to alter their sexual behavior, avoid having sexual relations at an early age, reduce the number of sexual relationships, use condoms, and use voluntary testing. Specific donor programs vary in their emphases. Under U.S. President George W. Bush, one third of all available funding under the President's Emergency Plan for AIDS Relief (PEPFAR) in 2006 was reserved for abstinence-only programs, and in 2008, 50% of PEPFAR was allocated for abstinence and fidelity programs, excluding comprehensive programs. Ideological and religious views rather than evidence-based interventions drive many policies. Although the Obama administration has reversed some of the Bush-era restrictions, the broad thrust remains.

The assumption that sexually active adults will have only one uninfected lifetime partner has been questioned. Societies and individuals do not respond so easily to advice requiring changes in human behavior, particularly those involving complex issues such as sexual relations. Prevention demands more than instruction and information; it requires alleviation of the problems of people most at risk. Empowering African women on many levels is now regarded as a vital step in allowing them to take more control in sexual relationships and to reduce their vulnerability to AIDS (Ouma and Kalipeni 2009). Limiting sexual transmission will necessitate empowering large numbers of people so that they can change their behaviors and stop engaging in activities that are now central to some poor people. Alternatively, a focus on "condoms, needles, and negotiation" (CNN) is proposed as a better alternative than abstinence and fidelity.

There are many critics of the Global North's approach and strong arguments that the wrong emphases are being pursued in Africa. Epstein (2007) believes that "our greatest mistake may be to overlook the fact that, in spite of everything, African people often know how to solve their own problems." Kalipeni et al. (2003) contend that biomedical understandings are privileged at the expense of integrating biomedical and social scientific knowledge.

Certainly a vaccine would be a massive boost, but most experts believe it will take at least another decade. Too often policymakers have opted for simple solutions and limited interventions (e.g., ABC) that offer a glimmer of hope of interrupting the epidemic without threatening underlying power dynamics and altering gross inequalities. Many leading medical researchers claim a truly effective AIDS vaccine may be a biological impossibility. A promising development is the U.S. Food and Drug Administration's approval of the first HIV prevention drug in 2012, Truvada; however, its annual cost of US$13,200 per person will be an insurmountable barrier to most people, so access is likely to be very limited, at least in the immediate future. Controversy surrounds some of Truvada's drug trials, largely those conducted in Africa, over the difficulty of getting African women to participate fully in clinical trials (many women stopped taking the medicine on a daily basis), meaning that the trials produced more conclusive results for men. The availability of Truvada and other medicines coming into the pipeline raises the question of whether their introduction could lead to a decline in safe-sex practices, a core focus of the past two and half decades of HIV/AIDS prevention.

The history of AIDS in Africa is, therefore, far from complete, and the impacts of the AIDS epidemic will continue to affect future generations, households, communities, businesses, public services, and national economies in the region. The epidemic has drastically reduced the workforce in many countries, and the costs of caring for a growing generation of AIDS orphans could trim gross domestic product growth rates by 1% to 1.5% in the worst-affected countries. According to the UN Food and Agriculture Organization (FAO, 2005), 8 million agricultural workers had died of AIDS in the 25 worst-affected countries by 2003, and 18 million more AIDS-related deaths are projected by 2020, although this forecast needs to be reduced because of the rolling out of ARVs since 2005 and sharply reduced AIDS mortality rates. Nevertheless, the loss of a significant portion of the agricultural workforce decreases food production and increases food insecurity as human and material resources diverted toward covering the costs of AIDS care intensify other vulnerabilities. There is no precedent for understanding the effects that AIDS deaths will have on small countries with high infection rates such as Botswana and Swaziland; the entire population structure of these countries is being radically altered.

## GLOBAL HEALTH FUNDING PRIORITIES, EMPHASES, AND OMISSIONS: NEGLECTED TROPICAL DISEASES AND MENTAL HEALTH

The global archipelago of international, governmental, and NGO entities that has emerged to channel funds, consultants, and medicines to African programs has major impacts on health emphases in the region. Major international organizations focus intensively on infectious diseases. For example, WHO allocates 87% of its total budget to infectious diseases, 12% to NCDs, and less than 1% to injuries and violence. The Global Fund, the most prominent health donor, allocates all of its funding to HIV/AIDS, TB, and malaria. Between 2004 and 2011, the Global Fund invested US$5.5 billion to dedicated support in Africa: approximately half of the funds were directed to HIV/AIDS, one third targeted malaria, and the remainder focused on TB eradication. Concentration on the "big three" infectious diseases is justified because Africa accounts for 90% of all malaria deaths, an inordinate share of all people living with HIV/AIDS, and nearly one third of all TB cases.

There are serious debates over whether the amount of HIV/AIDS funding in Africa is disproportionate. HIV is the leading killer, contributing 12.5% to the disease burden, but it receives 40% of all health funding (England 2007:344). General funding allocations do not correspond with causes of death or other measures of the prevalence of illness in Africa. An interesting question is how and why some global funding came to target HIV/AIDS exclusively and not broaden to tackle, for example, diabetes as well. No doubt the fact that HIV/AIDS travels and diabetes is stationary is a key part of the explanation. Moreover, because the North has generally conquered the scourge of infections, there was a sense that the same could be achieved in the South—but this is turning out to be much more difficult than anticipated.

The central argument for extraordinary support for HIV/AIDS is that a pandemic requires an unprecedented response. The devastating impacts on individuals, families, and communities pose major obstacles to development and threaten to reverse development gains. Health care systems will remain dysfunctional throughout Africa until AIDS is controlled and reversed. HIV/AIDS is such a sensitive issue—at its core is sex, gender inequality, sex work, homosexuality, drug use, stigma, and discrimination—that only an aggressive and well-financed approach is going to make inroads. Few dispute the humanitarian rationale for fighting HIV/AIDS, but there are questions about the disproportionate level of spending and about its crowding out other health concerns. Public health experts emphasize that the costs per disability-adjusted life-years averted are higher for HIV/AIDS than immunizations, malaria, traffic injuries, childhood illnesses, and TB. Thus, concentrating on an alternate set of interventions might produce wider benefits; for example, bed nets, immunization against pneumonia, family planning, etc. What if more money were spent on research and drug and vaccine development for other health challenges? This as an intriguing and provocative question.

Prioritizing funding for infectious diseases has led to distortions by increasing the profile of certain diseases and by influencing policy priorities and recipient health care systems. Inevitably, resources are drawn away from other important and looming health challenges. For example, maternal health, bacterial and parasite diseases, and chronic diseases receive less funding and scant international attention.

A major omission is a focus on tropical diseases, which affect the poorest 500 million Africans. Neglected tropical diseases (NTDs) are a group of chronic, disabling, and disfiguring conditions that most commonly occur in extreme poverty contexts. Combined, NTDs produce a burden of disease that may be equivalent to half of Africa's malaria disease burden and more than double that of TB (Hotez and Kamath 2009:412). The leading NTDs in Africa are helminth infections (parasitic worm infections such as hookworm, schistosomiasis, and ascariasis) (Table 9.2). Hookworm infections affected 29% of Africans in 2003 and accounts for 192 million cases per year (Hotez and Kamath 2009). Nigeria registered the largest number of cases (38 million)

**TABLE 9.2 GEOGRAPHICAL DISTRIBUTION AND ESTIMATED BURDEN OF THE MAJOR HELMINTH NTDs IN AFRICA**

| Disease (Number of Cases in SSA) | Country with Highest Prevalence | Country with Second Highest Prevalence | Country with Third Highest Prevalence | Country with Fourth Highest Prevalence |
|---|---|---|---|---|
| Hookworm infection (198 million) | Nigeria 38 million | DR Congo 31 million | Angola and Ethiopia 11 million cases each | Côtes d'Ivoire 10 million |
| Schistosomiasis (192 million) | Nigeria 29 million | Tanzania 19 million | DR Congo and Ghana 15 million cases each | Mozambique 13 million |
| Ascariasis (173 million) | Nigeria 55 million | Ethiopia 26 million | DR Congo 23 million | South Africa 12 million |
| Trichuriasis (162 million) | Nigeria 34 million | DR Congo 26 million | South Africa 22 million | Ethiopia 21 million |
| Lymphatic filariasis (382–394 million at risk) | Nigeria 106 million at risk | DR Congo 49 million at risk | Tanzania 31 million at risk | Ethiopia 30 million at risk, Kenya 29 million at risk |
| Trachoma (30 million) | Ethiopia 10.3 million | Sudan 3.6 million | Tanzania 2.1 million | Kenya and Niger 2.0 million cases each |
| Yellow fever (180,000) | Côtes d'Ivoire 16 reported cases in 2006 | Mali 5 reported cases in 2006 | Cameroon, CAR, Ghana, and Guinea 1 case each in 2006 | |
| Human African trypanosomiasis (50,000–70,000) | DR Congo 10,369 | Angola 2,280 | Sudan 1,766 | Congo 839 |
| Leprosy (30,055) | DR Congo 6,502 | Nigeria 5,381 | Ethiopia 4,611 | Mozambique 1,830 |
| Leishmaniasis (visceral) (19,000–24,000 new cases) | Sudan 15,000–20,000 new cases | Ethiopia 4,000 new cases | Kenya and Uganda not determined | |
| Dracunculiasis (9,585) | Sudan 5,815 | Ghana 3,358 | Mali 313 | Nigeria and Niger <100 cases each |
| Buruli ulcer (>4,000) | Côtes d'Ivoire 2,000 | Benin and Ghana 1,000 each | | |

*Source:* Hotez, P. J., Kamath, A. 2009. "Neglected Tropical Diseases in Sub-Saharan Africa: Review of Their Prevalence, Distribution, and Disease Burden." *PLoS Negl Trop Dis* 3(8): e412. doi:10.1371/journal.pntd.0000412.

(Hotez and Kamath 2009). Hookworms affect children disproportionately: illness results in school absenteeism and hurts educational performance. Schistosomiasis cases are largely concentrated around contaminated water bodies, rivers, and reservoir dams, affecting 25% of the African population (countries with high prevalence include Tanzania, with 19 million cases, and DRC and Ghana, with 15 million cases each) (Hotez and Kamath 2009). Data on other NTDs such as typhoid and salmonella are simply not available, and stepped-up surveillance is urgently needed to present a more adequate assessment of Africa's disease burden. It is important to note that the treatment of many NTDs is inexpensive and simple.

Many other health challenges—for example, mental health and traffic accident deaths—are almost completely ignored (see Box 9.3). Vehicular deaths (currently ranked 12th among causes of death in Africa) are projected to increase in the future as vehicles increase with urbanization and the emergence of more middle-class drivers.

The strongest criticism for abandoning dedicated disease support is that more aid should be allocated to

## BOX 9.3 MENTAL HEALTH IN AFRICA

Mental health issues rank at the bottom of the list of African policymakers' health priorities. Most African countries have no mental health policies and very weak legal protections for the mentally ill. African policymakers focus on communicable diseases and fail to grapple with the impact of mental disorders, even where there is a direct link between the two. For example, HIV/AIDS has mental health consequences. Those with the disease who tend to be ostracized and rejected by their communities. In many instances, rural AIDS patients are forced to relocate from their home communities to urban locations where they can be cared for and/or concealed by more tolerant family members. There is also the immediate psychological impact of explaining the illness and imputing causality, which often puts into sharp relief modern versus traditional explanations of illness. Traditional explanations of weakness and susceptibility to influences such as witchcraft are common for those who do not participate in modern clinical treatments (e.g., ARV therapies, counseling). Effects on bereaved families include dissolution of households, loss of income, depletion of savings, and changes in families' spending patterns; a direct casualty is often spending on girls' education. AIDS-affected children are often sent to live elsewhere, are twice as likely to work compared to their peers (and in the most exploitative labor arrangements and vulnerable jobs), and shoulder larger proportions of household chores (Economic Commission for Africa 2012). Even more worrisome is the fact that female AIDS orphans are more likely to suffer from sexual abuse and exploitation and to contract HIV/AIDS later in their teen years because they are more prone to engage in sexual relationships with "sugar daddies" to obtain physical and material security.

Poverty contexts increase the mental health risks for children in particular. Poverty does not harm all children, but many are part of the extremely vulnerable population with greater developmental risks. Several mechanisms contribute to children's vulnerability. Some are direct, such as exposure to infectious diseases and environmental toxins, inadequate health care and nutrition, and overcrowded and economically fragile households. Other mechanisms are indirect, such as the impact of discrimination and marginalization on self-esteem and the lack of successful role models to prompt and support children's aspirations.

In most parts of Africa, people's attitudes toward mental illness are outdated and strongly influenced by traditional beliefs in supernatural causes and remedies. Policymakers tend to believe that mental illness is largely incurable and/or unresponsive to medical treatments. Besides these attitudes, a major problem is the lack of trained specialists in psychiatry, neurology, and other mental health fields. WHO's (2011b) *Mental Health Atlas* reports that median ratios of mental health human resources per 100,000 population in Africa are the lowest of any world region (the ratio is 0.05 per 100,000 for psychiatrists and 0.04 for psychologists). Africa is off the global mental health map, and human resource and policy environments in Africa even show evidence of deterioration (WHO 2011b). In many African countries the ratio of psychiatrists to population is as low as 1 per 5 million, compared to 13 per 100,000 in the European Union (WHO 2013).

As a consequence, traditional healers and religious groups carry a large care burden for the mentally ill. Many traditional healers are strongly against any medication intake, and in some cases this can be detrimental to those who would be better served by a different treatment regimen. For example, in Ethiopia, one of the most progressive Africa states, 85% of emotionally disturbed people follow traditional healers because of the dearth of psychiatrists (10 psychiatrists serve the entire population of 61 million) (Gureje and Alem 2000:476). Only South Africa and Ethiopia provide state benefits to people with neurological disorders. WHO estimates that mental diseases will become one of the leading global health challenges by 2030, but African countries have yet to face their current, not to mention their future, mental health issues. Social environments where disease, poverty, and migration are common have disruptive effects on good mental health. Moreover, in many African environments, dislocation, war, and civil unrest add elements of trauma to the mental health matrix. Africa lacks a broad view of mental illness as a major cause of morbidity and as a burden on those afflicted, their families, and society.

strengthening health care systems within countries and in urban centers; funding should be focused on evidence-based health priorities rather than health issues that draw the attention of global funders and particular international funding constituencies (e.g., Christian groups, celebrity advocates, social justice advocates). Health policies and programs should be driven less by global advocacy and more by grounded research, local contexts, and African agency.

### URBAN HEALTH

African urban environments expose individuals to a range of debilitating diseases and conditions. Urban health inequalities begin at birth (with undernutrition) and are reproduced and intensified over a lifetime by high exposure to other problems. Long-held assumptions among development specialists that urban dwellers (compared to rural dwellers) benefit from an urban bias in terms of health care are being reconsidered. An emerging consensus is that African urbanites' health care advantage is eroding and that a new "urban penalty" is actually emerging.

Researchers initially chronicled the urban penalty in 19th-century Europe during an intensified and rapid period of industrialization and urbanization. Chronic conditions of urban poverty—overcrowding, dirty and unsanitary conditions, high levels of disease transmission—along with concentrations of hospitals, orphanages, and prisons contributed to higher urban mortality levels (that was particularly a consequence of a spike in TB). Edwin Chadwick's (1842) report on the *Sanitary Conditions of the Laboring Population in the Great Britain* (Chadwick is considered the father of public health reform in the UK) documented severe unsanitary urban conditions. He wrote about the poor state of urban drainage, noting that "many of the streets are unpaved and almost covered with stagnant water, which lodges in numerous large holes which exist upon their surface, and into which the inhabitants throw all kinds of rejected animal and vegetable matters, which then undergo decay and emit the most poisonous exhalations. These matters are often allowed, from the filthy habits of the inhabitants of these districts, many of whom, especially the poor Irish, are utterly regardless both of personal and domestic cleanliness, to accumulate to an immense extent, and thus become prolific sources of malaria, rendering the atmosphere an active poison" (1842:93). Eventually, urban mortality rates declined in 19th-century Europe and the United States because of a sanitary revolution that encompassed improved housing and sanitation infrastructure and immunization and nutrition campaigns.

A corollary can be drawn with contemporary African cities that face enormous and escalating challenges—unplanned urban growth, extensive migration, a phenomenal growth of slums (with accompanying deficiencies in housing, infrastructure, and environmental conditions, few decent jobs, and insecurity of tenure), urban poverty, and increasing urban disparities—manifested in differential access to health care and social services. With inadequate and often unstable income and low asset bases, urban slum-dwellers are drawn into unhealthy work in informal economies and have unhealthy diets (higher salt and saturated fat intakes, which increase NCD risks). Well-being is further compromised by long workdays, lengthy work commutes, and limited or no access to public health systems. Many of the starkest contrasts in health outcomes are now observed in neighborhoods within cities rather than in rural areas (Weeks, Hill, and Stoler 2014). Indeed, health conditions may deteriorate to the extent that established urban poor residents can have worse health than new, poorer international migrants. For example, in Johannesburg's inner city, "a healthy migrant phenomenon" has been identified whereby Zimbabwean immigrants have better health on a number of indicators than poor South Africans (Vearey et al. 2010).

Sixty-two percent of urban Africans live in slums (UN-HABITAT 2011), concentrated in very unhealthy urban zones. Confronted with the worst urban conditions and very heavy health burdens, slum-dwellers encounter extra burdens of government neglect and animosity (e.g., threats of evictions and exclusion from national health initiatives). Growing and expanding populations are channeled into slums; this causes urban environments to deteriorate even further and intensifies residents' vulnerability to, for example, communicable disease transmission (e.g., TB, acute respiratory infections, and meningitis) (Sclar, Garau, and Carolini 2005). Inadequate water and sanitation result in high incidences of diarrheal diseases and worm infections. Vaccine-preventable diseases (measles, diphtheria, and whooping cough) spread more rapidly in overcrowded urban areas and among nonimmunized slum-dwellers (Sclar, Garau, and Carolini 2005). Inadequate drainage and sanitation increase the risk of diseases such as malaria, West Nile virus, and dengue fever, as the respective mosquito vectors may breed in flooded areas, open drains, or rainwater trapped in piles of refuse. Cooking with solid fuels in overcrowded and inadequately ventilated shacks creates indoor air pollution, another health hazard, and the risks of fires are higher (Weeks, Hill, and Stoler 2014). Moreover, the standard practice of strapping infants to their mothers' backs while cooking increases the infants' proneness to respiratory illnesses. Furthermore, many slums are at high risk from extreme weather events, which are becoming more frequent and intense because of climate change.

Africa's new urban penalty is manifested in the urban poor dying from infectious and chronic diseases in disproportionate numbers. The penalty can take different forms within the same poor household. For example, child malnutrition and maternal obesity coexist within the same low-income Johannesburg household (Vearey et al. 2010). Many diseases show much higher rates of prevalence in slums than in other areas of the same city. For example, HIV prevalence rates in South African slums are more than double that of non-slums, and the more common mental disorders (depression and anxiety) exhibit a similar pattern (Rehle et al. 2007). The urban poor (particularly adolescents) also have greater risks of injuries and violence. Occupational injury can quickly diminish earnings and hasten poverty. Urban poverty typically goes hand in hand with increased food insecurity. For example, in Nairobi, Kenya, nearly half of slum households have insecure food supplies (Faye et al. 2010). Harpham (2010:2) shows that over the past decade the share of urban preschoolers in the group of underweight children has increased in urban Africa. This finding is concerning on many levels: malnutrition impedes children's motor, sensory, cognitive, and social development and weakens their immune systems. This extensive cumulative evidence underscores that urban poverty is both a cause and a result of ill health.

Relative deprivation can be more harmful for health than absolute deprivation (Harpham 2010). Some poor people are overwhelmed by stressful life experiences (unlike the theme of the creativity and resilience of poor people in urban livelihoods research). Making social comparisons (i.e., knowledge of how others live) can have a powerful negative effect on well-being, intensifying anxiety and undermining psychological good health (Harpham 2010). With economic growth and the emergence of an epidemiological transition in Africa, some groups attain higher living standards while others continue to encounter severe deprivation. As inequalities become more apparent, the relative aspects of poverty affect the poor person's subjective experience, with potentially damaging effects on health. Wilkinson (1996:215) argues that "to feel depressed, cheated, bitter, desperate, vulnerable, frightened, angry, worried about debts or job and housing security; to feel devalued, useless, helpless, uncared for, hopeless, isolated, anxious and a failure: these feelings can dominate people's whole experience of life, coloring their experience of everything else. It is the chronic stress arising from feelings like these, which does the damage." Wilkinson (1996) emphasizes that the material environment becomes a constant reminder of the oppression of self-failure, social exclusion, and devaluation as a human being. Bad social feelings create toxic environments.

Neighborhood health indicators show extreme variation within cities. A city of Cape Town study (Bradshaw et al. 2006) shows that income inequality is manifested in different health and well-being experiences at the neighborhood level. Large differences were recorded in the premature mortality rates among the poor townships (e.g., Nyanga, Mitchell's Plain, and Khayelitsha) and the wealthier "white" outer suburbs (e.g., South Peninsula and Blaauwberg) (Bradshaw et al. 2006) (Figs. 9.7 and 9.8). All diseases were more prevalent in the townships, and HIV/AIDS featured most prominently as cause of death. However, NCDs accounted for the majority of diseases in more affluent areas. Other studies confirm evidence of geographical and racial differences in children's injuries. For example, fatal burns and pedestrian road injuries occur much more frequently among black and "colored" township children than among whites and Asians who live outside of the townships (Burrows, van Niekerk, and Laflamme 2010).

Urban health burdens vary significant within places and between different places in the same city. The Nairobi Urban Health and Demographic Surveillance System (NUHDSS), the first urban-based demographic surveillance system in Africa, studied 70,000 people in two slums over an initial four-year period and found significant variation among different stages of life cohorts and slum subareas, and between families, even though the community was affected by similar environmental and poverty conditions. Women-headed households, the elderly, and new arrivals had the lowest incomes and exhibited the poorest health (Zulu et al. 2011). High mortality levels were observed among children under five years old, a cohort with exceptionally low levels of vaccinations, high levels of malnutrition,

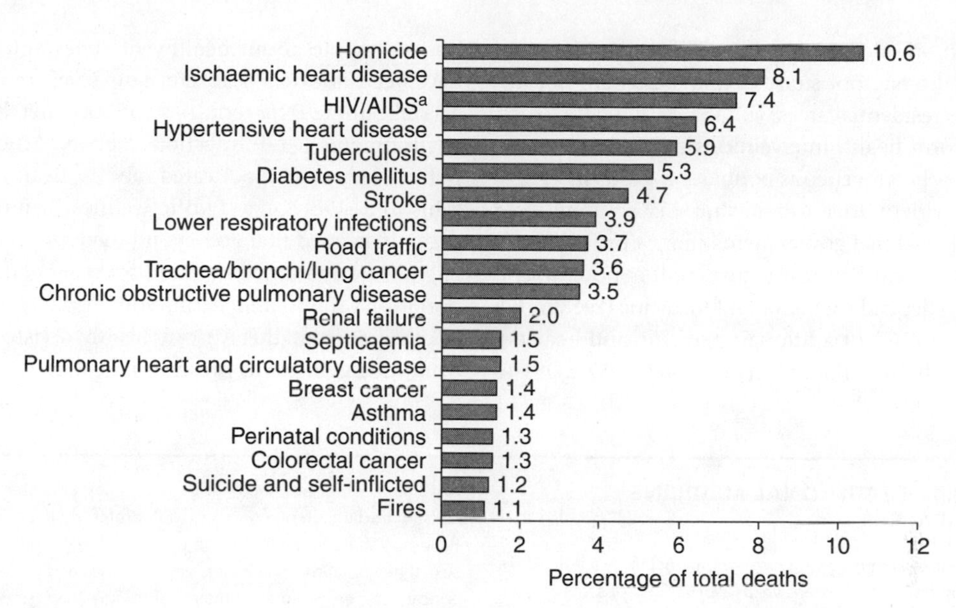


**FIGURE 9.7** Leading Causes of Death, Cape Town, South Africa, 2001. *Source:* From Bradshaw et al. (2010).

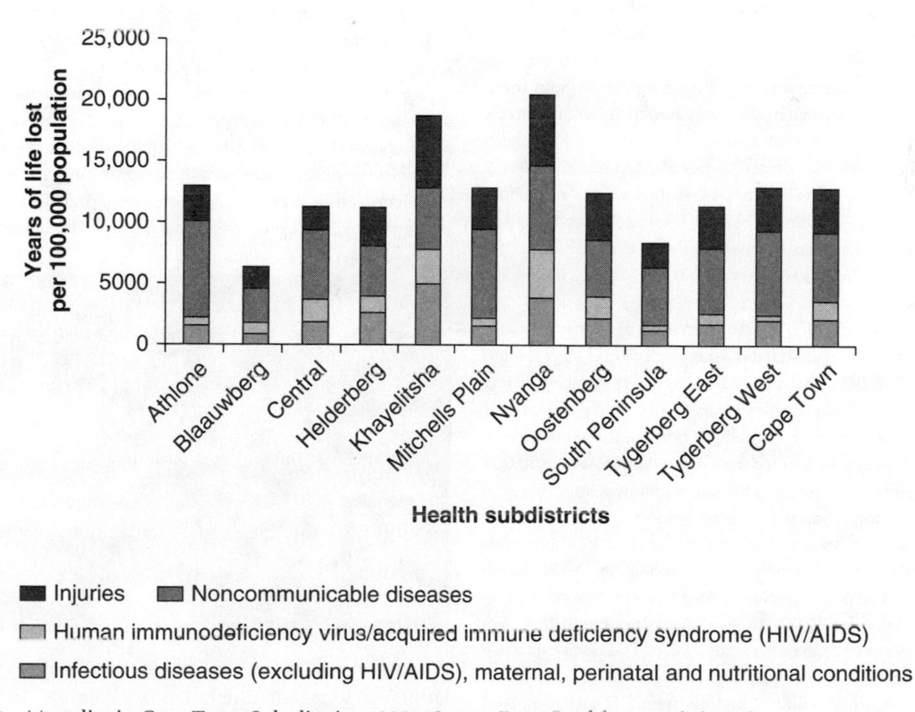


**FIGURE 9.8** Mortality in Cape Town Sub-districts, 2001. *Source:* From Bradshaw et al. (2010).

and a high prevalence of infectious diseases. The main lesson of the Nairobi study is that slums are not homogenous residential areas with identical health issues; thus, uniform health interventions may not always be effective for heterogeneous population groups.

It is evident that urbanization is so rapid and recent in Africa that governments cannot provide basic physical and health infrastructures so that poor urbanities have to depend on traditional medicine (see Box 9.4 for a discussion of traditional medicine and healers). Public health education efforts (beyond HIV/AIDS) to inform people about healthy lifestyles and nutrition have been noticeably absent. Low levels of awareness exist about the long-term risks of poor diet, hazardous environments, and infectious agents. Much of the public remains uneducated about health warning signs and symptoms. Public wellness campaigns are urgently needed that go beyond medical professionals to embrace a range of expertise from agriculture, transportation, water and sanitation, food security, and housing so that the African health debate can shift from curing illness to preserving health.

## BOX 9.4 TRADITIONAL MEDICINE AND HEALERS

Traditional African medicine, or ethnomedicine, is a holistic discipline involving the extensive use of indigenous herbs combined with aspects of African spirituality. The term "traditional" acknowledges the longevity and genealogy of healing knowledge of Africa (which is unwritten for the most part). Contemporary healers' knowledge is derived from the medical systems practiced by Africans before the arrival of Europeans and the implementation of colonial public health regimens.

Colonial authorities cast doubt on the integrity of traditional healing practices, referring to them as unscientific and inferior because they used knowledge derived from magical/religious theories of causation. Traditional healing was associated with illiteracy, irrationality, and the chaos of traditional societies, and a sharp contrast was drawn with "official," scientific biomedicine. Many colonies banned traditional healers, labeling them as "witch doctors" and branding traditional healing knowledge as fetishistic and "primitive." Biomedical knowledge was established as the authoritative knowledge, and biomedical technology continued its hegemony.

The contemporary reality is that up to 80% of Africans consult traditional healers for health-related issues, including life-threatening diseases (WHO 2008b), and many Africans collect medicinal plants for themselves or their families (King 2012). There are several healing traditions, namely herbalism, bone-setting, midwifery, and spiritual healing. In South Africa, for example, over 350,000 traditional healers practice, and the government is attempting to regulate the sector. Given the network among traditional healers in referring patients to one another and also to local hospitals and clinics, the system requires regulation (King 2012). Some 4,000 of the 6,400 plant species in Africa are employed for medicinal purposes, and 20,000 tons of plants are harvested from nature each year rather than from cultivation. The decision to use indigenous healing systems or biomedical health systems (or both) is influenced by culture, history, personal attitudes, and philosophy; preference varies by country, location (urban versus rural), and, especially, income. Indeed, many educated Africans consult traditional practitioners but prefer to keep it secret. Patients visit healers for a range of illnesses (e.g., mental health, sexual health, common viruses, various aches and pains, vitamin deficiencies, etc.) and even malaria (Fig. 9.9).

There are several strengths to traditional African medicine. It preserves part of African cultural traditions. Medical services, herbal remedies, and individual and group psychotherapies are delivered in affordable ways, and large swaths of the population have good access. Recent times have coincided with an exponential increase in the prevalence of locally produced herbal mixed medicine preparations (some reportedly prepared by African healers) that are widely available at informal markets, supermarkets, and pharmacies. Traditional medicine is suitable for treating a broad spectrum of diseases, including conditions not diagnosed by biomedical practitioners (e.g., social and psychological illnesses). However, at the same time, traditional healers recognize that some diseases (e.g., HIV/AIDS) are better treated by biomedicine. Traditional medicine is considered to be

**FIGURE 9.9** Traditional Practitioners: Malaria Sufferer Drinking Liquid Mixture in Lusaka, Zambia.
*Source:* © John Stanmeyer/VII/Corbis.

superior in treating psychic and psychosomatic disease because of the knowledge that healers possess about the social and ethnic backgrounds of their patients (Shizha and Charema 2011).

Community-based studies show that traditional healers have considerable success in treating some mental illnesses. Persons with psychiatric problems are considered to benefit more from traditional treatments: because of the traditional belief that outside forces (e.g., past wrongdoing, lack of balance between patient and social environment and/or the spiritual world) are the cause of mental disorders, acceptance and inclusion in their communities are more likely. In contrast, Western medicine and pharmaceutical companies offer biochemical explanations for mental illness that may stigmatize the patient. Concentrating on the deeper causes allows traditional healers to focus on prevention (as opposed to curative practices, which prevail in biomedicine).

Some components of African traditional medical knowledge have been transferred to Western medicine. For example, a traditional African medicinal cure that has developed a wide following is the use of an extract from the bark of *Prunus africana* (a small evergreen tree) to treat moderate prostate problems. Traditional healers make a tea from the extract for treating other illnesses, such as fevers, malaria, and psychiatric disorders, but these remedies have not been transferred to Western medicine. A lucrative export trade of the prunus bark has developed, mainly from harvests from Cameroon, DRC, Kenya, and Madagascar. The trade has expanded so much that unsustainable harvesting now threatens the tree's extinction. It takes 12 to 15 years to produce the bark, and 40- to 50-year-old trees produce the best medicinal yield, so tree replenishment cannot keep pace with harvesting.

There are major limitations to traditional African medicine. There is a lack of standardized training and much variation in the quality of services provided. Many traditional healers consider their medical knowledge to be personal property so they act to protect and keep it secret, which means that best practices cannot be diffused. Disagreements about which herb is effective in treating a medical problem and/or what dosage is most effective have no way of being resolved or assessed, and the variances may result in serious side effects and have life-or-death implications. Such lack of consensus makes it difficult for biomedical practitioners and traditional healers to cooperate and trust each other. The accumulated traditional knowledge base is affected by limited resources to produce and distribute pharmacopoeia monographs (scientists who specialize in the toxicology of medicinal plants and botanists could benefit and help in verification). Patients are provided with no safeguards and instead are expected to totally surrender to the healer (although some contend that licensing and regulations in conventional medical systems provide weak safeguards). There is limited documentation and insufficient data on safety and efficacy, and the quality of traditional medicine varies dramatically throughout the region.

The role of traditional medicine in national health systems has been boosted by the declaration of an annual African Traditional Medicine Day on August 31. WHO has long recognized the central role that traditional medicine can play in providing primary health care, and there are growing examples of integration. Many states in Africa are developing mechanisms to involve traditional medical practitioners within the activities of ministries of health: the Ministry of Health, Botswana, is at an advanced stage of integration. In KwaMhlanga, South Africa, a 48-bed hospital combining traditional African medicine with homeopathy and Western healing methods, as well as traditional Asian medicine, was established. Founded by a traditional African healer, the hospital is the first of its kind in Africa (Shizha and Charema 2011). A few training institutions, such as the Kwame Nkrumah University of Science and Technology (Ghana), have established a Department of Herbal Medicine for the training and continuing education of traditional health practitioners. The West African Health Organization in collaboration with WHO is developing a West African pharmacopoeia and producing a monograph on medicinal plants in the subregion.

The deficiencies and shortfalls that prevail in African health systems are prompting considerations of a rapprochement between biomedicine and traditional medicine. The latter has survived for centuries despite Western prejudices. If traditional healing methods were not effective they would not have endured for so long. The schism that exists between biomedicine and indigenous healing practices disadvantages patients who could benefit from both approaches. It is clear that the arsenal of biomedicine cannot cope with the many different health problems and that its medicines are less available and/or unaffordable in African countries. A sound argument can be made for depending on local resources whenever possible rather than furthering the hegemony of external, expensive medical expertise, infrastructure, and pharmaceuticals. Traditional healers are an available local resource that can fit within the reorientation back to primary health care systems. Moreover, modern urban lifestyles may require even greater attention to the psychosocial aspects of healing, which fits within the scope of traditional medicine.

## CONCLUSIONS

There is obviously much more to Africa's unfinished public health agenda than the control of HIV/AIDS, TB, and malaria. The continuing urban revolution is bringing looming public health challenges. The double or quadruple disease burden makes the region's public health challenge even more daunting than that of 19th-century Europe, and an African sanitary revolution has yet to be launched. Basic infrastructure is lacking. WHO (2011a:399) estimates that 50% of Africans have access to a working toilet but that 68% of Africans own a mobile phone.

A shift in focus from selective disease-specific intervention to a more comprehensive health systems approach is urgently required. This will necessitate a seismic shift in policy, time horizon, and scale of investment. Much more global, national, urban, and rural funding is needed to mount a large-scale response. To convince donors that resources are going to be effectively used, African governments must develop a health care infrastructure with enhanced human capacity as well as integrate and regulate traditional health care systems. Improved and new data (e.g., for better projecting trends in chronic disease and mental health challenges) and better recordkeeping are needed so that policy effectiveness can be assessed at all spatial scales. Africa's current and future health challenge is immense: the region carries almost one quarter of the global disease burden; diseases such as polio and cholera are reappearing in certain places; the epidemiological transition is bringing new challenges; global environmental changes are unfolding and expanding; and possibilities of new diseases remain a threat.

African governments need to live up to their Abuja Declaration commitment to allocate 15% of their national budgets to fortifying health care systems. They need to augment national health care policies and develop urban and rural health plans. Many scholars and practitioners argue that urban and rural governments have been "let off the hook" by ceding health concerns to national debates instead of accounting for their health (non)achievements and (non)provision of basic services to all. It is unjust how the poor are expected to meet local, regional, and national governments halfway by investing their time, labor, and money in developing community-based services. In an era when water privatization is being ushered in, NGOs (e.g., CHF International) are expanding their role in urban water and sanitation delivery via community-driven approaches that enlist the support and participation of the community to improve access and delivery services. In sharp contrast, middle- and upper-income groups are not expected to participate to have services such as safe water, sanitation, or other municipal services.

Urban health disparities are a salient feature of the health situation in contemporary Africa. Urban and rural health plans need to be designed to fit the particular community rather than using a one-size-fits-all template. Moreover, urban and regional governments must take into account the diverse health needs of diverse communities to deliver spatially targeted, multilevel, and multisectoral health responses.

Africa is accumulating a large "health debt," much of which is urban-centered (Sclar, Garau, and Carolini 2005:903). This debt will be far more expensive to pay off (if it even could be) three decades from now through conventional curative methods than it would be to prevent the problem now by improving slums, water supplies, sanitation, and waste management and providing other public health interventions. Slums, urban poverty, and poor health are all linked and should be addressed together.

Health issues are not accorded the attention they deserve in development policy and in rural and urban planning. So many measures that relate to health and well-being fall to organizations that are not health agencies and that do not understand their roles. For example, slum upgrading is seen as a housing rather than a health intervention. An integrated approach is needed in the post-MDG development agenda that enlarges the scope of health to incorporate healthy lifestyles, better nutrition, NCDs, gender equality, and mental health.

## REFERENCES

Agyei-Mensah, S., and A. de-Graft Aikins. 2010. "Epidemiological Transition and the Double Burden of Disease." *Journal of Urban Health* 87(5):879–897.

Allen, T. 2006. "AIDS and Evidence: Interrogating Some Ugandan Myths." *Journal of Biosocial Sciences* 38(1):7–28.

Averting HIV and AIDS (AVERT). 2012. "HIV Types, Subtypes, Groups and Strains." Available at http://www.who.int/hiv/topics/treatment/en/index.html (accessed May 3, 2012).

Binagwaho, A., R. Bate, R. Gasana, M. Karema, C. Mucyo, J. Mwesigye, F. Biziyaremye, C. Nutt, C. Wagner, P. Jensen, and A. Attaran. 2013. "Combatting Substandard and Falsified Medicines: A View from Rwanda." *PLOS Medicine* 10(7):1–3.

Bradshaw, D., P. Groenewald, D. Bourne, H. Mahomed, B. Nojilana, J. Daniel, and J. Nixon. 2006. "Making COD Statistics Useful for Public Health at Local

Level in the City of Cape Town." *Bulletin of Health Statistics* 84(3):211–217.

Bradshaw D., V. Wyk Van-Pillay, R. Laubscher, B. Nojilana, P. Groenewald, N. Nannan, and C. Metcalf. 2010. "Cause of Death Statistics for South Africa: Challenges and Possibilities for Improvement." *South African MRC Burden of Disease Research Unit*. Available at http://www.mrc.ac.za/bod/cause_death_statsSA.pdf (accessed July 4, 2013).

Burrows, S., A. van Niekerk, and L. Laflamme. 2010. "Fatal Injuries among Urban Children in South Africa: Risk Distribution and Potential for Reduction." *Bulletin of the World Health Organization* 88(4):267–272.

Chabal, P. 2009. *Africa. The Politics of Suffering and Smiling*. New York: Zed Books.

Chadwick, E. 1842. *Report on the Sanitary Conditions of the Laboring Population in Great Britain: A Supplemental Report on the Results of the Special Inquiry into the Practice of Internment in Towns*. London: Clowes & Sons.

Chigwedere, P., G. Seage, S. Gruskin, T-H. Lee, and M. Essex. 2008. "Estimating the Lost Benefits of Antiretroviral Drug Use in South Africa." *Journal of Acquired Immune Deficiency Syndrome* 49(4):410–415.

Chin, J. 2007. *The AIDs Pandemic: The Collision of Epidemiology with Political Correctness.* Oxon, UK: Radcliffe Publishing.

Craddock, S. 2003. "Beyond Epidemiology: Locating AIDS in Africa." In *HIV & AIDS in Africa: Beyond Epidemiology*, eds. E. Kalipeni, S. Craddock, J. Oppong, and J. Ghosh, pp. 1–12. New York: Wiley.

de-Graft Aikins, A., N. Unwin, C. Agyemang, P. Allotey, C. Campbell, and D. Arhinful. 2010. "Tackling Africa's Chronic Disease Burden from the Local to the Global." *Globalization and Health* 6(5):1–7.

Duesberg, P. 1998. *Inventing the AIDS Virus.* Washington, D.C: Regnery Publishing.

Economic Commission for Africa. 2012. *The Impact of HIV/AIDS on Families and Communities in Africa*. Addis Ababa: UN Economic Commission for Africa.

Economist Intelligence Unit (EIU). 2012. *The Future of Healthcare in Africa*. London: EIU.

England, R. 2007. "Are We Spending Too Much on HIV?" *British Medical Journal* 17(334):345.

Epstein, H. 2007. *The Invisible Cure: Africa, the West, and the Fight against AIDS*. New York: Macmillan.

Faye, O., A. Baschieri, J. Falkingham, and K. Muindi. 2010. "Hunger and Food Insecurity in Nairobi's Slums: An Assessment Using IRT Models." CEPS/INSTEAD Working Paper no. 2010–33. Available at http://www.ceps.lu/pdf/11/art1572.pdf (accessed July 15, 2013).

Ferriman, A. 2007. "*BMJ* Readers Choose the "Sanitary Revolution" as Greatest Medical Advance since 1840." *British Medical Journal* 334(7585):111.

Food and Agriculture Organization (FAO). 2005. *Agricultural Workers and Their Contribution to Sustainable Agriculture and Rural Development*. Rome: FAO.

Good, C. 1991. "Pioneer Medical Missions in Colonial Africa." *Social Science Medicine* 32(1):1–10.

Gureje, O. and A. Alem. 2010. "Mental Health Policy Development in Africa." *Bulletin of the World Health Organization* 78(4): 475–482.

Harpham, T. 2010. *Urban Health in Africa: What Do We Know and Where Do We Go?* Berlin: Lit Verlag.

Henry J. Kaiser Family Foundation. 2012. "Global Health Facts. Physicians (per 10,000 Population)." Available at http://kff.org/global-indicator/physicians/ (accessed July 4, 2013).

Hotez, P., and A. Kamath. 2009. "Neglected Tropical Diseases in Sub-Saharan Africa: Review of Their Prevalence, Distribution, and Disease Burden." *PLoS Neglected Tropical Diseases* 3:1–10. Available at http://www.plosntds.org/article/info%3Adoi%2F10.1371%2Fjournal.pntd.0000412) (accessed May 2, 2012).

International Diabetes Federation. 2012. *IDF Diabetes Atlas*, 5th ed. Available at http://www.idf.org/diabetesatlas/5e/Update2012 (accessed July 5, 2013).

Kalipeni, E., S. Craddock. J. Oppong, and J. Ghosh. 2003. *HIV & AIDS in Africa. Beyond Epidemiology*. New York: Wiley.

Kalipeni, E., and L. Zulu. 2012. "HIV and AIDS in Africa: A Geographical Analysis at Multiple Scales." *GeoJournal* 77:505–523.

King, B. 2012. "We Pray at the Church in the Day and Visit the Sangomas at Night": Health Discourses and Traditional Medicine in Rural South Africa." *Annals of the Association of American Geographers* 102(5):1173–1181.

Lurie, M., and S. Rosenthal. 2010. "Concurrent Partnerships as a Driver of the HIV Epidemic in Sub-Saharan

Africa? The Evidence Is Limited." *AIDS Behavior* 14(1):17–24.

McMichael, A., and R. Beaglehole. 2009. "The Global Context of Public Health." In *Global Public Health. A New Era*, 2nd ed., eds. A. Michael and R. Beaglehole, pp. 1–22. New York: Oxford University Press.

Mills, E., S. Kanters, A. Hagopian, N. Bansback, J. Nachega, M. Albertsin, C. Au-Jeung, C. Mathers, T. Boerma, and D. Ma Fat. 2009. "Global and Regional Causes of Death." *British Medical Bulletin* 92(1):7–32.

Ministry of Health. 2012. *Uganda AIDS Indicator Survey 2011*. Kampala: Ministry of Health.

Naicker, S., J. Plange-Rhulle, R. Tutt, and J. Eastwood. 2009. "Shortage of Healthcare Workers in Developing Countries—Africa." *Ethnicity & Disease* 19(2):1–60.

Omran, A. 1971. "The Epidemiological Transition Theory: A Theory of the Epidemiology of Population Change." *Milbank Memorial Fund Quarterly* 49(1):509–538.

Ooms, G., W. Van Damme, B. Baker, P. Zeitz, and T. Schrecker. 2008. "The 'Diagonal' Approach to Global Fund Financing: A Cue for the Broader Malaise of Health Systems?" *Globalization and Health* 4(6):1–7.

Oppong, J. 1997. "Obstacles to Acceptance of Location-Allocation Models in Health Care Planning in Sub-Saharan Africa." *East African Geographical Review* 19(2):12–22.

Oppong, J., and E. Kalipeni. 2003. "Perceptions and Misperceptions of AIDS in Africa." In *HIV & AIDS in Africa: Beyond Epidemiology*, eds. E. Kalipeni, S. Craddock. J. Oppong, and J. Ghosh, pp. 47–57. New York: Wiley.

Ouma, V., and E. Kalipeni. 2009. "Socio-cultural Predictions of HIV/AIDS-Related Health and Preventive Behaviors in Kisumu District, Kenya." In *Strong Women, Dangerous Times: Gender and HIV/AIDS in Africa*, eds E. Kalipeni, K. Flynn, and C. Pope, pp. 109–131. New York: Nova Science.

Parkhurst, J. 2002. "The Ugandan Success Story? Evidence and Claims of HIV-1 Prevention." *Lancet* 360(7):78–80.

Rao, C., D. Bradshaw, and C. Mathers. 2004. "Improving Death Registration and Statistics in Developing Countries: Lessons from Sub-Saharan Africa." *Southern African Journal of Demography* 9(2):81–99.

Rehle, T., O. Shisana, V. Pillay, K. Zuma, and A. Puren. 2007. "National HIV Incidence Measures—New Insights into the South African Epidemic." *South Africa Medical Journal* 97(2):194–199.

Reid, S. 2009. "Injection Drug Use, Unsafe Medical Injections, and HIV in Africa: A Systematic Review." *Harm Reduction Journal* 6(24):1–14.

Sanders, D., E. Igumbor, U. Lehmann, W. Meeus, and D. Dovlo. 2009. "Public Health in Africa." In *Global Public Health. A New Era*, 2nd ed., eds. A. Michael and R. Beaglehole, pp. 161–183. New York: Oxford University Press.

Sclar, E., P. Garau, and G. Carolini. 2005. "The 21st Century Health Challenge of Slums and Cities." *Lancet* 365(9462):901–903.

Shizha, E., and J. Charema. 2011. "Health and Wellness in Southern Africa: Incorporating Indigenous and Western Healing Practices." *International Journal of Psychology and Counselling* 39(9):167–175.

UNAIDS. 2011a. *The Kenyan AIDS Epidemic 2011 Update*. New York: UNAIDS.

UNAIDS. 2011b. Global HIV/AIDS Response Progress Report. Available at http://www.who.int/hiv/topics/treatment/en/index.html (accessed May 3, 2012).

UNAIDS. 2012. "Africa Fact Sheet." Available at http://www.unaids.org/globalreport/Global_report.htm (accessed July 5, 2013).

UN-HABITAT. 2011. *State of the World's Cities 2010/2011: Bridging the Urban Divide*. Sterling: Earthscan.

UNICEF (United Nations Children's Fund). 1990. *Economic Crisis, Adjustment, and the Bamako Initiative: Health Care Financing in the Economic Context of Sub-Saharan Africa*. New York: UNICEF.

UNICEF. 2010. *Children and AIDS. Fifth Stocktaking Report, 2010*. New York: UNICEF.

Uthman, O. 2012. "Health and Medical Issues in Sub-Saharan Africa." In *Africa South of the Sahara 2012*, 41st ed., ed. I. Frame, pp. 54–61. New York: Routledge.

Vearey, J., I. Palmary, L. Thomas, L. Nunez, and S. Drime. 2010. "Urban Health in Johannesburg: The Importance of Place in Understanding Intra-Urban Inequalities in a Context of Migration and HIV." *Health & Place* 16(4):694–702.

Vogel, R. 1993. *Financing Health Care in Sub-Saharan Africa*. Westport: Greenwood Press.

Wangui, E. E. 2009. "Livelihood Strategies and Nutritional Status of Grandparent Caregivers of AIDS Orphans in Nyando District." *Qualitative Health Research* 19(12):1702–1715.

Weeks, J., A. Hill, and J. Stoler. 2014. *Spatial Inequalities. Health, Poverty and Place in Accra, Ghana*. New York: Springer.

WHO (World Health Organization). 2006. *World Health Report 2006—Working Together for Health*. Geneva: WHO.

WHO (World Health Organization). 2008a. *The Global Burden of Disease, 2004 Update*. Geneva: WHO.

WHO (World Health Organization). 2008b. "Traditional Medicine." Available at http://www.who.int/mediacentre/factsheets/fs134/en/ (accessed May 16, 2012).

WHO (World Health Organization). 2011a. "Public Health Round Up." *Bulletin of the World Health Organization* 89, 396–497.

WHO (World Health Organization). 2011b. *Mental Health Atlas 2011*. Geneva: WHO.

WHO (World Health Organization), Regional Office for Europe. 2013. "Data and Statistics: Prevalence of Mental Disorders." Available at http://www.euro.who.int/en/what-we-do/health-topics/noncommunicable-diseases/mental-health/data-and-statistics (accessed September 4, 2013).

Wilkinson, R. 1996. *Unhealthy Societies: The Afflictions of Inequality*. London: Routledge.

Zulu, E., D. Beguy, A. Ezeh, P. Bocquirt, N. Madis, J. Cleand, and J. Falkingham. 2011. "Overview of Migration, Poverty and Health Dynamics in Nairobi City's Slum Settlements." *Journal of Urban Health: Bulletin of the New York Academy of Medicine* 88(2):185–199.

## WEBSITES

Caroline Ruport. *Young Carers Through Our Eyes* (2012). Documentary film available at http://www.youtube.com/watch?v=MOKdJ49ZB6s (accessed December 5, 2013). Life as seen through the eyes of children living in HIV/AIDS-afflicted families.

Dan Rather. 2010. *Fake Drugs in Africa*. Available at http://www.youtube.com/watch?v=Tooxeb3Byrg (accessed December 5, 2013). Documentary about importing counterfeit pharmaceuticals in Africa.

CHAPTER 10

# LAND AND FOOD

## INTRODUCTION

Africa is a distinct agricultural and food region. The colonial legacy that enforced primary commodity and regional cash crop specializations left an indelible imprint on the region's agricultural and food geographies. Colonial incorporation of African agriculture into the world economy emphasized export commodities such as bananas, coffee, cotton, cocoa, groundnuts, oil palm, and tea. Cash crops were produced on both smallholder plots and large-scale plantations to provide inexpensive supplies for emerging food and manufacturing industries in Europe. The colonial legacy and smallholder traditional African agricultural systems have remained embedded with the rural fabric. Despite the passage of time and various agricultural reform efforts, the region's agriculture and food systems have, for the most part, remained intact. The green revolution that catalyzed agricultural production in Asia and Latin America did not happen in Africa.

Agriculture remains the mainstay of African economies, employing approximately 60–70% of the workforce and contributing 30% of gross domestic product (GDP) on average (Thornton et al. 2011). The region accounts for the highest proportion of agricultural land (44%) relative to all land (FAOSTAT 2011). Agricultural and development research has registered only modest successes in improving agricultural efficiency, yields, and incomes, and even these successes are further constrained by their geographical unevenness and limited scalability. Overall, African economies have experienced declines in agricultural productivity and food production since the mid-1970s, which accentuated food vulnerabilities. Paradoxically, Africa has become a net importer of food and agricultural products since the 1980s, despite its vast agricultural potential.

Agriculture development in Africa faces daunting challenges. Climate change and increasing climate variability are acute. Projected rises in temperature of 2 to 4 degrees Celsius will affect patterns of crop and livestock production and result in diminished agriculture options in many subregions. Rapid population growth rates (Africa's population is growing, on average, at 2.4% per annum, nearly double the world's population growth rate) mean that the region will have to accelerate the amount of food produced within the region or increase food imports to satisfy demand and gaps (Rakotoarisoa, Iafrate, and Paschali 2012). Acute food gaps in areas where food production is low add another spatial dimension to this challenge. Macro trade policy environments in Africa are characterized by a growing liberalization of trade, whereby states have been dismantling food import tariffs, and consumers, for better and (mostly) worse, rely more on imported foods, which are highly sensitive to fluctuations in global commodity prices (Moseley, Carney, and Becker 2010). Most experts forecast worsening food insecurity throughout Africa in coming decades.

With the low food prices and global food stability of the 20th century ending, food scarcity will likely become a 21st-century norm. The FAO Food Price Index in 2011 reached an all-time high, eclipsing the 2007–08 food price hikes that resulted in various urban protests in particular African countries (e.g., Burkina Faso, Côte d'Ivoire, Guinea, Mauritania, Mozambique, Senegal). Some of the worst rioting took place in Maputo, Mozambique, in 2010: following demonstrations against government bread price increases in excess of 25%, a three-day riot left a dozen people dead and more than 400 injured. Hunger protests are becoming more frequent and widespread, and food price hikes disproportionately affect the poor. Food expenditures as a percentage of total expenditures for the poor are high, ranging from 28% (e.g., Swaziland and Mauritius) to 73% (e.g., Tanzania and Nigeria). By comparison, food accounts for approximately 10% of total expenditures in the United States (Rakotoarisoa, Iafrate, and Paschali 2012:24). Not surprisingly, governments consider maintaining affordable and sufficient food supplies to be a national security imperative. Food policy is emerging as a critical issue in urban, national, and global policy arenas, creating a new geopolitics of food, and contemporary inflows of foreign development assistance are altering the control of, access to, and use of African agricultural land.

Global players have turned their attention to food and farmland in Africa. National governments in the Gulf states and South Korea, London and Wall Street institutional investors, agroindustrial corporations (e.g., Lonrho), and private individuals from around the world are heavily targeting Africa agriculture. South African farmers are also acquiring land outside of their home state (See Box 10.1 for a discussion of South African farmers entry into the African farmland arena) African governments are opening up their agricultural land to new investments as a vehicle to increase agriculture spending, spur investment, provide jobs, and enhance food security. Cotula (2013:35) notes that "the worlds of high finance and big corporations might seem light years away from the everyday life of villagers who grow crops and graze livestock in rural Africa," but the two worlds are increasingly intersecting.

There is much debate on how to enhance food security in Africa (Moseley, Carney, and Becker 2010). Going beyond "sufficient food," FAO (2006) has outlined a multidimensional definition of food security: (1) that food is available at all times; (2) that everyone has a means of access to it; (3) that it is nutritionally adequate in terms of quantity, quality, and variety; and (4) that it is stable for the preceding three dimensions over time, which acknowledges that weather conditions, political instability, and/or economic factors (e.g., unemployment, rising food prices) influence food security. Major food security challenges confronting the region are chronic undernutrition and famine. With one quarter of Africa's population undernourished (many of them children), Africa's food security challenge is urgent (Rakotoarisoa, Iafrate, and Paschali 2012:8).

The food crisis has produced different responses. One camp proposes boosting supplies by intensifying modern farming methods, crop research, and food processing in African countries. For those in this group (food companies, crop breeders, and international

### BOX 10.1 SOUTH AFRICANS' NEXT "GREAT TREK" INTO AFRICA

AgriSA's participation in large land investments in the region is an intriguing development. Settler land grabbing took place in South Africa during the colonial and apartheid eras, and now again white (African) investors are eyeing land across Africa. The South African media refer to the current development as the "new Boer trek north" (*Mail and Guardian* 2011:1). This trend is bolstered by an alliance between South African agroindustrial and government complexes that desire a food-secure future (or for different reasons), given climate change scenarios and their anticipated negative effects that will constrain food production in the Southern Africa region. Some 1,000 farmers and agricultural entrepreneurs now engage in agrobusiness beyond South Africa's borders (*Business Day* 2011). South African farmers are pushed by anticipated national land reform: one third of white-owned land is in the process of being transferred to black owners by 2014 (even though the record of transfers is slow and considerable inertia prevails). White farmers are propelled to seize opportunities farther north because of South African changes in labor rights, increases in minimum farmworker wages, extensions of tenure rights to farmworkers and their families, and unresolved land claims by former black land occupiers (Chamberlain and Rogerson 2012). Buoyant at the prospect of reaping large rewards from an untapped agricultural potential, these South African farmers are at the crest of the investment wave (Hall 2011). Spearheaded by AgriSA, Pretoria provides support by engaging in bilateral investment talks with a number of governments in the region.

South African expansion into other areas of Africa not only centers on farmland but also incorporates a tourism investment focus (to develop gaming, hunting, and safari operations within enclosures). Many tourism initiatives center on coastal Mozambique (Maputo to Beria, and Nacala and Quirimbas archipelagoes) as well as on areas in Tanzania. This entire expansion is alluded to as the "South Africanization of the region": not in the sense of South Africa becoming a colonizer of the region (but this is an element) but more in the sense that the changes under way, such as control over land, labor, and commodity chains, are altering agrarian structures to that of a settler territory (i.e., South Africa), marking a departure from historical colonial powers and arrangements (Hall 2011). This process may narrow the contrast that exists between South Africa, Kenya, and other former settler colonies that engage in agricultural food production for global markets and former nonsettler African colonies—since the colonial period, the latter have been unable to participate in supplying the global food market.

Prior to the current surge, only sporadic movements of South African farmers northward occurred (e.g., to Zambia, Mozambique, and Nigeria), and they mainly involved individual or small-group relocation for individual reasons. Current migration is quantitatively and qualitatively different: it is more centrally organized and coordinated and larger in scale and scope (Hall 2011). AgriSA, an association representing 70,000 South African farmers, is enabling large concessions for newly formed consortia of farmers and agribusinesses. South African investors are taking a comprehensive agricultural investment perspective, viewing farmers and land as elements within an agroindustrial infrastructure with construction, engineering, and financial services that can be developed to support investments. By 2010, AgriSA was engaged in discussions with 22 African countries about land acquisitions. South African farmers, through AgriSA, participate in several of the largest land deals (e.g., 2.5 million acres [1 million hectares] in Mozambique and a Republic of the Congo deal that could involve 20.5 million acres [10 million hectares] in southwest Congo, where in the initial phase 217 million acres [88,000 hectares] were transferred to 70 South African farmers. Major deals are also under negotiation for sugarcane production in Mozambique and Sudan's Nile Delta (Hall 2011). It is a matter of conjecture whether the relocation of South African farmers to other African countries, when coupled with land reform in their home states, might result in more food insecurity in South Africa (as transpired in Zimbabwe, but under extreme political pressure) or whether South African transnational farmers could produce for their home market.

In tandem with South African farmer migration, a geographical expansion of South African agribusiness, especially the sugar multinational corporations (e.g., Illovo and Tongaat-Hulett), is occurring. The South African-based sugarcane industry is expanding production with ethanol and sugar-refining plant investments in Mozambique, Zambia, Tanzania, and Malawi.

South African investment has recently been controversial. For example, Tanzania in the early 2000s received a surge of South African capital flows and commodities following the lifting of the apartheid ban. Stunned by the rapid influx of several thousand South Africans and their buying power, Tanzanians either welcomed (officially) or rebuked (unofficially) the surge (Schroeder 2008, 2013). South Africans were portrayed by some as "economic invaders," exporting their expertise and flooding and dumping goods on Tanzania. Commentators attributed the South African incursion as contributing to a rapid reconfiguration of the Tanzanian national economy. South Africans' economic foothold raised fears that the influx would be accompanied by the import of apartheid values (Schroeder 2013). Many observers perceived that South Africans were taking over everything of value in Tanzania (mining, retail, finance, and tourism—especially wildlife safaris), resulting in "the United States of South Africa" (Schroeder 2008). The heralded and visible presence of South African enclaves, with concentrations in particular residential neighborhoods, and the emergence of de facto all-white bars, restaurants, rugby clubs, and nightclubs catering to South African clientele led to a spike in race consciousness. For South Africans, Tanzania offered a fresh start and an opportunity for social mobility in another African country. However, it was not long before all South Africans were stereotyped and shouldered with the baggage of the former minority-ruling Pretoria government and scapegoated for an underdelivering Tanzanian economy.

South African investment in Tanzania and the presence of South African enclaves were not interpreted so much as representing a regional power extending its economic muscle but rather as a white invasion. This raises the intriguing question of what is means to be white but also African in the land investment landscape across the entire region, which is opening in one sense (by market reform) but closing in another (by the revival of economic nationalism). For good and bad, it still matters greatly where the investor is from, the purpose of the investment, how many good-paying jobs are created in the process, and who loses out in the process. Ultimately, it matters a great deal which investor face represents the new South Africa: the black-majority ANC political party or white corporate business. Lines of political solidarity, once very strong and clear, between Tanzanians and South Africans have become contested in the contemporary round of international land investment.

development agencies), the green revolution of the 1960s was a stunning success and needs to be followed by a 21st-century African green revolution. International donors such as AGRA view increased productivity by farmers and the use of genetically modified organism (GMO) technologies as key to enhancing food security. The alternative view is skeptical of, or even hostile to, the global agrofood industry. This group, influential among nongovernmental organizations (NGOs) and some consumer associations, asserts that modern agriculture produces food that is tasteless, nutritionally inadequate, and environmentally disastrous; worst of all, modern agriculture locks Africa into dependent relations with global agribusinesses, creating multiple dependencies (e.g., seeds, technology, and external expertise). Different solutions to increase food supplies are proposed: land reform, enhancing land tenure security, and promoting traditional smallholder and urban agriculture. Proponents of this policy contend that building local capacity to produce, distribute, and control local supplies is the only viable option to enhance food security.

Food is very connected to domestic agriculture and land use. Food security depends on the availability of land, labor, favorable farming conditions (soil, water, climate), and agricultural infrastructure (storage, markets, logistics). At the macro level, government policy plays a major role in influencing what is grown and where, how output enters the market, and the extent to which and the types of foods that are exported and imported. International development agencies (e.g., World Bank and FAO) play steering roles in shaping governments' agricultural policies and practices.

A debate rages over whether Africa should continue to produce cash crops (which now also encompass biofuels—fuels produced from renewable resources, especially plant biomass, and vegetable oils), refocus on food crops, do both or move in a fairtrade and/or sustainable trade pathway (See Box 10.2 and 10.3 for an elaboration of fairtrade and sustainable trade). Many governments are attempting to do both, swayed by the perspective that land is underused in Africa and has a high potential for expansion. A criticism of the past was that agricultural and food production followed a one-size-fits-all approach, which overlooked the diversity among countries, regions, and places. Governments, international agencies, farmers, and entrepreneurs are assessing Africa's land and foodscapes with fresh eyes. Rapid changes are afoot with increased foreign development investment in farmland, an international humanitarian focus in improving African food security, and a movement to kick-start the African green revolution.

## BOX 10.2 THE FAIR TRADE DEBATE

Fair trade products can be traced back to 1988, when the first fair trade coffee of Mexican origin was retailed in Dutch supermarkets (FLO 2012). Once a small grassroots movement, fair trade has become mainstream. Since 1997, Fairtrade Labelling Organizations (FLO), also known as Fairtrade International, has set international fair trade standards and regulates the use of the fair trade labels on products. The fair trade logo is currently the most widely recognized ethical logo and is proudly displayed at Starbucks and numerous other retail establishments (Fig. 10.1). Fair trade sales worldwide climbed to US$6.6 billion (€5 billion euros) in 2011; the U.S. market accounts for 20% of total sales (FLO 2012). African farmers' revenue share was US$175 million in 2011 (FLO 2012).

Fair trade in Africa involves the participation of 378 certified producer organizations from 29 countries (FLO 2012). FLO is also working with additional African producers (e.g., in DRC and Zimbabwe) to help farmers' output meet certification standards so that they can become full-fledged fair trade members. Bananas, cocoa, coffee, cotton, sugar, and tea are the largest product categories, and the fair trade product list is continually expanded (e.g., wine and gold are recent additions). South Africa and Kenya retail Africa-produced fair trade products, and fair trade product launches are being planned for other African countries.

Fair trade cocoa producers in Ghana, the Kuapa Kokoo farmers, have even been successful in launching a joint venture chocolate company (holding a 45% stake)—Divine Chocolate. In 2013, Divine captured 20% of chocolate sales in the United Kingdom, and U.S. sales are increasing. Celebrities have expressed their love for this product. Chris Martin, of Coldplay, claims, "Fair Trade chocolate tastes better. . . . But also it's amazing to go to a rich, green area like this and know that for every bar of Divine chocolate you're eating, you're helping out the people who grew it for you more than if you eat Nestlé" (quoted in Ryan 2011:98). However, the "Divine miracle" has an occluded economic geography: Divine Chocolate is manufactured

**FIGURE 10.1** Fair Trade Logo. *Source:* Owned and licensed by Fairtrade International.

in Germany rather than in Ghana, maintaining the latter's entrenched role as a raw material provider. Moreover, Kuapa Kokoo's large membership dilutes farmers' individual returns. In 2007, for example, the organization received a dividend check for UK£47,309, which resulted in a US$1.80 payment for each of the 40,000 farmers (Ryan 2011).

Fair trade strengthens the position of farmers in the value chain. Fair trade pricing incorporates buyers' payment of a "social premium" to the producer organization, a portion of which (US$18 million in 2011) is earmarked for investment in community-based development projects, mainly schools. Consumers purchasing fair trade products of African origin are, in turn, assured that they are contributing to improving the livelihoods of African farmers and workers. African farmers are guaranteed stable prices even if world market prices fall, enhancing security. Most scholars agree that small-scale producers linked to fair trade are better off than producers without this connection.

Despite its huge achievements and growing popularity, the fair trade movement has many critics. Free market proponents contend the fair trade model is not a long-term strategy for development and reflects the demands of Northern consumers rather than the needs of African farmers. Paradoxically, a rise in consumer consciousness may have paved the way for retailers with an eye on the bottom line to develop politically savvy marketing strategies (Freidberg 2003). The operational costs of the fair trade system are high (inspection, certification, campaigns for products, new market development, organizational development) and swallow a good portion of the price premium, reducing the amount paid to farmers. For example, about 25% of expenditures were allocated to producer support in the 2011 FLO operating budget.

Critics on the left contend that fair trade has evolved along a philanthropic rather than a rights-based path, a trajectory that inadvertently obscures the power dynamics in the global trade system. Moreover, the way fair trade is promoted has put the responsibility for ensuring that farmers and workers are fairly paid on the shoulders of moral shoppers. As such, it lets global companies, importers, manufacturers, and retail chains "off the hook" and free to make substantial profits (profit margins are identical for fair trade and non-fair trade goods). Fair trade initiatives have not accomplished much in altering the general conditions of unfairness of the North–South trade system: unfortunately, contemporary reality is still in sync with unfairness. As a result, it is argued that fair trade operates more as an addendum to market relations than as an alternative. Furthermore, it makes the export orientation of African economies marginally fairer but at the same time serves as a distraction from self-sufficiency and sustainable development. Indeed, airfreighting African fruits and vegetables to distant markets contributes to global carbon emissions. Despite its visibility and consolidation of market shares in many countries, fair trade volume is still miniscule, accounting for less than 1% within product categories (Barratt Brown 2007:273).

Fair trade is well entrenched in African agriculture, involving 700,000 producers from the region. Fair trade has made good progress in improving the livelihoods of the poorest farmers by creating a more humane niche within the free market economy, but it needs to be scaled up. However, it is important not to airbrush the complex dynamics of the international political economy and the role that governments play in establishing higher prices for commodities such as cocoa, elements affecting the livelihoods of farmers. In many ways, 25 years of fair trade is not long given the history of agriculture and trade systems, and it may be fairer to assess the project after a longer operating time. Nevertheless, fair trade is a useful movement for African producer organizations to ally with, to strengthen their collective influence, and to be supported by Northern social justice activists and conscious consumers. Its functioning builds an unusual alliance among small farmers, NGOs, Northern consumers, and global business leaders who occupy different ideological space but find common ground in fair trade campaigns. The concept of "fair" is quite esoteric and difficult to quantify (unlike price), especially as the producers are faceless and far away; the basic question of how much is "fair" to each part of the value chain is never addressed. Overall, the fair trade movement has not yet made a significant contribution to international trade and development policy.

## BOX 10.3 ETHICAL WILDFLOWER TRADE AND SUSTAINABILITY: FLOWER VALLEY, SOUTH AFRICA

Closely related to fair trade is the concept of ethical trade, which gained currency from the mid-1990s onward. Ethical trade campaigns focus on cleaning up international trade by ensuring that agricultural practices conform to international standards in relation to working conditions, child labor, wage levels, good agricultural practices, resource conservation, and sustainable practices. Projects have recorded mixed successes. For example, ethical trade of micro-vegetables from Zambia rose and then fell (Freidberg 2003), but a South African niche wildflower trade is maintaining an upward trajectory (Bek, Binns, and Nel 2010; McEwan, Hughes, and Bek 2012).

A best-practices example is Flower Valley, a 1,433-acre (580 hectares) farm 137 miles (220 km) from Cape Town and situated between Stanford and Gansbaai on the western coastal edge of the Agulhas Plain. The area is characterized by considerable spatial and social inequality. Land ownership is based on historical patterns of ownership: white farmers own most of the land, whereas the black population experiences high unemployment (exceeding 50%) and few decent job opportunities. The area is very biologically diverse, located within the Cape Floral Kingdom, the smallest and richest of the six floral kingdoms of the world. Flower Valley epitomizes biological diversity and is home to 2,000 native fynbos species ("fine-leaved bush"). Perhaps the best-known variety is the king protea flower, the national flower of South Africa (Fig. 10.2). Cultivated and wild fynbos are major components in flower farming in this region, although globally they are not well known (accounting for 0.001% of the global flower market share), but the flowers are popular because of their long vase life and various colors.

Pristinely located, the farm came under threat due to unsustainable harvesting, commercial viticulture, the development of a nearby nuclear plant, and the presence of alien species. Many of its neighboring farms practice intensive horticulture cultivation, and adjacent lands have been converted to more lucrative viticulture as the boutique pinot noir growing extends further into the area. Flower Valley has resisted these lures and has evolved into a best-practices farm with sustainable local development that brings wildflowers to consumers, incorporating sustainable harvesting and cultivation practices.

The Flower Valley Conservation Trust (FVCT) was established in 1999 to develop export markets for sustainably harvested wild fynbos. The practice of "wild harvesting" entails picking plant stems and flowers from plants occurring naturally in the landscape (as opposed to using plantations for commercial horticulture). A core objective has been to establish flower-picking schedules and rates that allow for natural rejuvenation within a plant productivity cycle. The project gained momentum in 2002 when the trust received a Shell Foundation Legacy Project award. Funding allowed the farm to become linked to a local packaging enterprise, Fynsa, then located in the town of Stanford. Fynsa aimed to produce the first ethically branded fynbos bouquet in South Africa, and this initiative took off after the Shell Foundation and UK retailer Marks & Spencer joined forces in a small-scale supplier endeavor to develop the producer end of supply chains.

**FIGURE 10.2** Flower Valley Farm, South Africa. Photograph by Flower Valley Conservation Trust.

Flower Valley is registering successes and profits from domestic and international bouquet sales. For example, Marks & Spencer's UK bouquet sales climbed from 330,000 in 2006 to 470,000 in 2010 (Shell Foundation 2012). The sustainable harvesting story was included on bouquet packaging. Fair trade was considered but rejected based on cost. Instead, FVCT decided to secure accreditation from the Wine and Agri-Industry Ethical Trade Association, whose Ethical Trade Initiative provides a guarantee of respect for workers' housing and tenure rights, freedom of association, prohibition of harsh and inhumane treatment, and prohibition of child labor (Bek, Binns, and Nel 2010). FVCT is now working to secure buy-ins from the Protea Producers of South Africa (the industry representative body), which in turn works with large retailers (Walmart, Tesco, Marks & Spencer, and Pick n Pay), and the association is working on creating a single ethical label and base code for horticultural products.

Flower Valley provides employment for 13 full-time workers, offering them a decent standard of living, in contrast to the local norm of part-time employment in the seasonal fynbos industry. Farmworker compensation at Flower Valley also includes provided on-farm housing and schooling for children. In addition, 75 more jobs directly related to sustainable fynbos harvesting have been created (Shell Foundation 2012). The farm has experimented with various initiatives to provide additional income streams such as making handmade stationery that incorporated dried wild fynbos, which proved essential after Flower Valley Farm was badly burned in 2006, but this effort was discontinued to concentrate on the core activities of sustainable harvesting and opening the farm to visitors to learn about sustainability and the environment. To strengthen the educational mission of the farm, links have been developed with a local horticultural college (Green Futures), and internships and training on the farm began in 2013. Plans are under way to create a learning and education hub on the farm that will bring hundreds of students and eventually tourists to learn about sustainable horticulture and the environment. Such nature-based educational tourism effort is badly needed in an area not known for responsible tourism (e.g., a large flow of adventure tourists bypass the lowland fynbos en route to shark-cage diving at Gansbaai). The planned tourism enterprise is already offering hiking trails linking Flower Valley farm with neighboring farms to showcase the fynbos and milkwood forest unique to the area, and its accommodations will serve as a research station for students and researchers. Year-round employment has been created by training workers in alien species harvesting. Failure to clear alien species can result in losses of R1,000 per hectare and of many fynbos species, and Flower Valley's workers are now in demand for invasive alien species clearing on neighboring farms.

Flower Valley initiatives show that ethical trade and sustainable harvesting can be commercially viable. According to the United Nations Development Programme, "Flower Valley is a world leader in understanding ecological sustainability of flower harvesting and translating this into certification."(cited by Flower Valley Conservation Trust 2011:1) Flower Valley is a small project, but its accomplishments illustrate an alternative and sustainable land future.

There is still much work to be done in educating consumers, who are price-conscious and demand flower colors in accordance with seasons (e.g., red, green, and silver bouquets for the December holiday season and red for Valentine's Day) (personal communication, Roger Bailey, Conservation Manager, Flower Valley, June 2012). Nature does not always produce flowers in accordance with the Hallmark calendar, and sameness rather than diversity punctuates mass consumer tastes. Nevertheless, Flower Valley demonstrates that local pro-activism can be used to harness global resources, inform corporate players, and educate global consumers about sustainable land and growing practices in distant places.

This case shows that conservation can take place on private land and does not have to be limited to state land, the norm in Africa. It also illustrates that highly scalable global supply chains can be developed; that unconventional alliances among NGOs, large corporations, foundations, farmworkers, and biodiverse farms (the strangest of bedfellows) are possible; and that partnerships can be employed to provide opportunities to marginalized communities and to develop fair trade supply chains (Kilgallen 2008). The ultimate measure of its success, however, will be whether its core principles can be disseminated and successfully adopted in other settings (Bek, Binns, and Nel 2010).

## WHY HAS AFRICA BECOME A NET FOOD-IMPORTING REGION?

Since 1980 agriculture imports have grown consistently faster than agriculture exports, producing a deficit that climbed to US$22 billion in 2007 (Rakotoarisoa, Iafrate, and Paschali 2012:1). The composition of Africa's agricultural exports has not changed much since the 1960s despite efforts to accelerate nontraditional exports (e.g., flowers, semiprocessed fruits and vegetables). Coffee, cocoa, tea, and spices are the major food exports, accounting for 35% of agricultural exports. There are individual success stories in export diversification, such as Kenya's development of cut flowers, fruits, and vegetable export niches, but most nontraditional export efforts barely make a dent in altering the composition of agricultural exports.

Food import dependency is so high for some countries (e.g., Burundi, Gambia, and Somalia) that total export revenues cannot cover the costs of food imports. Food dependency is aggravated by the fact that 88% of Africa's imports originate outside of the region (mainly from Europe and Asia). There has been a failure to foster more intraregional food trade: only 20% of African food trade is conducted within the region (Rakotoarisoa, Iafrate, and Paschali 2012:7).

Africa's food imports dramatically increased after the 1980s. Increasing population created more demand, and domestic production did not rise to meet this demand. Some debate continues about whether per capita food consumption increased due to changes in income, dietary patterns, and urbanization. At the national level, per capita consumption remained fairly stable for staples, meats, and dairy. Roots (such as cassava and taro) produced and consumed within the region attenuated some food imports. However, fundamental changes were recorded at the urban and regional scales, where a dietary transformation occurred.

Taste changes are evident. The proliferation of supermarkets, fast-food outlets, and corporate marketing of processed foods (e.g., tomato paste, cheese, cured/salted meat, margarine) encourages urban and rural households to abandon locally produced foodstuffs. Growing tourism in many countries (Mauritius, Tanzania, Ghana) also has an impact on food marketing and product availability, and tourist consumption has demonstration effects. The increasing occurrence of noncommunicable diseases among Africans is evidence of poorer diets (higher in salt and oil), reflecting a switch from a plant-based diet to meat and processed food consumption.

New consumer demand for imported food commodities has occurred in many places—for instance, wheat in Nigeria. Although wheat is not part of the traditional diet in most of Nigeria, by the 1980s bread was among the most important food items consumed by large swaths of the population, so much so that Nigeria was snared into a "wheat trap" (Andrea and Beckman 1985). High demand for bread is based on convenience, low cost (compared to other staples), compatibility with local diets, and consumer preferences. The Nigerian government allowed wheat imports to increase throughout the 1970s to ensure that poor urban populations had access to inexpensive foods. Faced with financial troubles in the 1980s, the government moved to curb wheat imports and banned its importation (1987–90), trying to unsettle the entrenched position of wheat products by encouraging domestic consumption of indigenous alterative staples. It also introduced import-substitution policies to accelerate domestic wheat production, but returns were disappointing and farmers found their land to be poorly suited for wheat cultivation. When the wheat ban was lifted in 1990, high levels of wheat imports resumed. In 2013, Nigeria imported 4 billion metric tons of wheat, a staggering increase from the 385 million metric tons imported in 1970. Nigeria's "wheat trap" illustrates the entrenchment of the commodity into the fabric of Nigerian culture, a new dependence on an imported foodstuff, and a cultural taste transfer. The government has since implemented a policy requirement that a certain percentage of cassava flour be incorporated into bread baking (10% in 2002 and increasing to 40% in 2013) as a novel attempt to find a substitute for wheat.

Much of the explanation of why the region became a net food importer is due to the poorly developed agricultural sector. This started in the colonial period when the colonial occupiers concentrated on minerals, forests, and tropical stimulants such as tea and coffee, and saw food production as a local activity to feed colonial cities (Freidberg 2009, 2010). At that time, logistics and refrigeration systems were not sufficiently developed to deliver fresh products from Africa to Europe. Freidberg (2010:277) states that "the fruits of empire (like vegetables, meats and other fresh goods) simply could not survive the long distances to metropolitan markets."

Efforts to diversity food production and to develop new nontraditional food exports have been both costly and risky. For example, in the 1990s, new green bean production zones were established in Burkina Faso to serve the French export market in the winter season, although the product had been introduced in colonial times (Freidberg 2004). The scheme brought early euphoria about an invigorated African effort to reap rewards from serving European markets, and the possibilities for African replication seemed boundless. However, this new trade placed Burkinabè smallholders at the mercy of strict international food safety and agricultural practice standards, fickle Western consumers (and French chefs), and delivery disruptions that could ruin the product. Freidberg (2010:195) emphasizes, "if a truck breaks down or a plane arrives a few hours late, the beans wither. At that point, they're worth less than the cardboard cartons they travel in." Typically, farmers bore the brunt of these disruptions. In addition, African producers have to contend with price fluctuations, changing international certification standards, European media food scares that cast aspersions on commodities originating from Africa, and occasional food gluts; the at-times-oversaturated Burkina marketplace meant some product was left to rot (Freidberg 2004). To date, food production for export has not resulted in an equivalent drive for developing domestic food production.

Contemporary African agriculture is beset by numerous other challenges: overreliance on primary agriculture, low-fertility soils, minimal use of external inputs, environmental degradation, and minimal value addition and product differentiation. Ninety-five percent of the crops in Africa are rain-fed, so food production is vulnerable to adverse weather conditions.

Low yields and low productivity reduce the amount of food available and undermine domestic agricultural competitiveness. Fertilizer use remains the lowest among world regions (10% of the world average). Some interpret low levels of fertilizer application as a hindrance to preserving soil fertility and controlling pests and diseases. Commitments to agricultural R&D are exceptionally low by international standards (except in Botswana and South Africa)—on average less than 0.5% of agricultural budgets.

Agricultural infrastructure is lacking (e.g., irrigation dams, storage facilities, cold chains [transportation of temperature-sensitive products in refrigerated packaging through a supply chain], and modern slaughterhouses), which means that the proportion of land under irrigation is low and farmers often struggle to tap into available water. Moreover, poorly developed transport networks make it difficult and costly to distribute food beyond local markets. Road systems are so deficient that in some countries food production surpluses can occur in one area while another area in the same country suffers severe shortages. Compounding the poor road situation, transport cartels operate throughout the region, and their market dominance is a disincentive for other transport companies to invest in modern trucks and logistics.

Astonishingly, 23% of African food production is wasted or lost (Lipinski et al. 2013:9). Food loss/waste refers to the edible parts of plants and animal products produced for human consumption that are prevented from entering the human food chain. Spills, spoils, bruising, and wilting cause product losses, and products are discarded routinely because they fail to meet exporters' size, length, and aesthetic criteria. For example, French beans that are too short or not straight enough are discarded. Most product loss occurs at the site of production (e.g., crops damaged in harvesting/picking, crops left behind in fields) and from poor and improper handling or storage (food eaten by pests, degraded by fungi and diseases, and damaged in storage and transportation).

National food and trade policies are outmoded. Trade policy for food staples is ad hoc, discretionary, and unpredictable. Export/import bans are widely applied: for example, countries ban imports during good harvest years to ensure domestic production is consumed first and also limit exports during periods of low yields. Trade distortions hamper the development of stable food markets and intra-African trade and push cross-border food trade into informal channels. Furthermore, liberal market environments open economies for food imports from outside the region.

African food policy needs revamping at all levels, from implementation to point of sale. Typically, elites within government ministries dictate food policy. There is always a willingness to accept food aid: on humanitarian grounds this cannot be disputed, but food aid causes market distortions. Alternative agriculture is poorly developed: most fair trade schemes and certified organic production are geared toward export markets. Traditional food markets, where most Africans obtain their food, are in an unhealthy state (overcrowded and lacking refrigeration, clean water, and sanitary facilities), and their deterioration is driving consumers into retail supermarket chains and fast-food outlets and toward processed imported foods. Government investment and donor support is urgently needed to improve traditional markets and to bring them into the 21st century.

## CONTEMPORARY AGRICULTURE IN AFRICA: "THE GREAT AFRICAN LAND GRAB"?

Only a short time ago in Africa, agricultural land was of little interest to outside investors, but times have changed. Many investors are now motivated by rising returns in agricultural and land values. The food and financial crises have turned agricultural land into a strategic asset that is deemed a sound investment. By many measures, agriculture is underinvested. For example, agricultural investment accounts for less than 0.2% of equity market capitalization, and farming's share of the global agricultural value chain is only 22% (less than packing, distribution, and supplying) (Deininger and Byerlee 2012:706). Abundant land (or even free land) is supposedly available in Africa based on calculations that only 20% of the region's potential production is realized; for example, only 2% of land in the DRC is farmed (with basic tools), leaving 3 million acres (10–12 million hectares) of potential farmland underused (Deininger and Byerlee 2012:709). Private equity firms market African farmland investment funds to prospective investors by emphasizing that land is "undervalued" and therefore an excellent investment opportunity. Notions such as "available," "idle," or "empty" land, however, are heavily contested. Typically, such lands are owned collectively under systems of customary and traditional landholding arrangements, whereby communities make use of these lands for grazing, hunting, fishing, and forest farming (Anseeuw, Cotula, and Taylor 2012).

From the perspective of institutional investors, most African smallholder agriculture achieves only a fraction of its potential productivity. Countries such as DRC, Mozambique, Sudan, South Sudan, Tanzania, and Zambia offer high levels of available land and the highest yield gaps. Such knowledge is herding investors toward African land, and international investors are partnering with a large array of local actors in securing land investment deals. According to Chamberlain and Rogerson (2012:6492), by mid-2011, nearly 9.6 million acres (39 million hectares) of African land was under negotiation or already managed by foreign investors. To put the scale of these land acquisitions in perspective, the total African land under negotiation is greater than the combined agricultural land of Belgium, Denmark, France, Germany, the Netherlands, and Switzerland (Deininger and Byerlee 2012:705).

Labeled "land grabs" as well as the "foreignization of African lands" by the media and NGOs, land investments are controversial. Carmody (2012) argues that a "second scramble for Africa" is under way, with resource-rich but land-poor countries competing to appropriate large areas of fertile land for production and financial return. Cotula (2013) emphasizes that these agricultural investments will insert African production into the global food system. Although it is difficult, if not impossible, to ascertain the full investment intent, inconsistencies exist between what is specified in contracts and what transpires and when. The majority of capital flows target land for food production, about one third is geared for biofuel plantations, and the remaining land is procured for gaming, marine fisheries, and unspecified uses (Hall 2011:194). Regional geographical concentrations are emerging: Southern Africa is labeled "the new Middle East of biofuels," and the Nile Valley is heralded as the world's rice and wheat basket (Future Agricultures 2011:2).

Combinations of global processes, availability, and affordability of land drive Africa-based investments. Although average land prices conceal considerable within-country variation, foreign investors have acquired farmland in Ghana, Tanzania, and Zambia for between US$100 and US$800 per hectare (very inexpensive in international market price terms). Myriad investors are hedging against the risks of global food price increases and future food deficits and windfalls from biofuels. There are, however, some African investors who are propelled by a different thought process. For example, AgriSA (the large South African agricultural trade association that represents agrobusiness and agroproducers), one the largest investors in Africa land, is motivated primarily by its own domestic political context, where "white farmers" are coming under increasing political pressure and a major round of land reform is on the horizon.

The turn to Africa land investment has accelerated since the mid-2000s, following earlier Latin American and Eastern European commercial successes with large-scale farming. Since then, large international and multiplayer land acquisition deals have gained traction and show a recent acceleration, despite some well-publicized setbacks. Two examples of negative results are (1) Daewoo's (South Korea) investment in Madagascar engendered a public outcry that led to project abandonment, and (2) Sun Biofuel (UK) shut down its large project in Tanzania after four years of operation due to changes in the global financial climate. (Watch a video documentary from *The Guardian* 2011). Lease contracts typically last 30 to 99 years are heavily sought on higher-valued land (those with irrigation potential and/or proximity to markets). African land appears to be a lucrative investment when bundled with water rights (projected to be the most important physical commodity and asset in the near future, surpassing oil) and/or access to marine resources, wildlife, and forests. According to Citigroup's chief economist, Willem Buiter, "no surprise, then, that so many corporations are rushing to sign land deals that give them wide-ranging control over African water" (quoted in GRAIN 2012b:14).

## BENEFITS AND RISKS OF LAND INVESTMENTS

Land investment deals are announced as "win–win solutions" for all parties involved (Deininger et al. 2010). Investors are viewed as the way to overcome the dearth of investment in rural economies since the 1970s. Land investments are promoted to revitalize agricultural economies, to enhance food security, and to develop agricultural infrastructure. Land investment stakes are high: they encompass various and complex processes of agrarian change and may affect the ability to feed Africa's growing populations as well as rural sustainability. Governments and other land authorities are hoping that their "leasing" of natural resources (land, soil, water) to foreign commercial interests can spur national development.

Land investments are running far ahead of our knowledge on their impact (1) on local people's livelihoods and dispossession, (2) on the environment, (3) on national sovereignty, and (4) on food and water security (GRAIN 2012b). Tensions run highest when domestic food production is converted to food and/or biofuels for export (Hall 2011). Clearing forests for food production is less controversial as it is typically assumed that some of this food will end up in domestic markets (Hall 2011).

Land grabbing, or "the farms race," in Africa is described as a neocolonial push by foreign investors and governments to annex key natural resources. Critics contend that rich countries are "buying poor countries' soil fertility, water and sun to ship food and fuel back home, in a kind of neo-colonial dynamic" (Hall 2011:194). The situation resembles settler colonialism and anticolonial land struggles. This time around, however, new global drivers, refracted through the contemporary configurations of land relations and political economies, are producing more geographically diverse patterns of involvement. The complexity of current deals and the wide array of international actors involved make the contemporary land investment situation murkier and faster evolving than in preceding periods. Colonialists appropriated considerably more land, but it took time to move settlers, to establish funding mechanisms, and to develop plantations that were horizontally articulated with the mother colony. In contrast, contemporary land investment projects appear to be more sophisticated, aiming for horizontal and vertical integration into global commodity chains, with projects spanning the entire continuum of financing, production, processing, and distribution. For example, Lonrho, one of the largest foreign agribusiness companies in Africa, focused traditionally on processing and distribution but is expanding its operations to engage in production in Angola, Mali, and Malawi (Cotula et al. 2011:s102).

Information on the scale, location, players, and project emphases is emerging via a burgeoning scholarly literature on the initial effects on host communities (e.g., Allan et al. [2012] is a comprehensive handbook on land and water investments in Africa), but because of the newness of the phenomenon, researchers rely heavily on NGO reporting. NGOs are at the forefront of compiling and producing land investment data and providing timely on-the-ground assessments. However, high levels of uncertainty surround the reported figures, and some deals come to light only after local resistance and exposés. In many of the land acquisitions, details pertaining to the identities of investors, terms, distributions of rents, and commitments made on behalf of national and regional authorities are kept secret.

A leading reporting NGO is GRAIN, a Barcelona-based international nonprofit organization that works to support small farmers and social movements in their quest to secure community-controlled and biodiverse food systems. It has compiled the most comprehensive data on large acquisitions in Africa (e.g., on more than 400 land deals) (GRAIN 2012a). A second leading nonprofit on land grab information is the Oakland Institute, a California-based policy think tank dedicated to advancing public participation in and fair debate on critical social, economic, and environmental issues. The Oakland Institute has complied detailed reports on land investment deals in the most active investor African host states (e.g., Cameroon, Ethiopia, Mali, Mozambique, Sierra Leone, South Sudan [e.g., Oakland Institute 2011], Tanzania, and Zambia). A third useful source is the Land Matrix, led by the International Land Coalition, an civil society organization headquartered in Italy with a regional branch in Kigali (Rwanda), that seeks to monitor and map land investments in real time online and to provide data free to users (see Land Matrix 2013).

Figures on land deals, however, need to be treated with caution: some deals are never recorded by investment agencies, others are rescinded, and the media focus their reporting on the total land associated with various deals rather than on what is currently used (e.g., an investment deal may pertain to a large parcel of land, but only a small part of the parcel may be used in the early stages of production). For example, a land audit in Mozambique found that one third of acquired land was unused, and 15% was used in ways that did not comply with the agreed investment plans (Deininger and Byerlee 2012:705). The largest land acquisitions in Africa (in terms of land size) are South African acquisitions in the Republic of Congo and Mozambique. AgriSA, for example, has signed deals for 494,211 acres (200,000 hectares) in the Republic of Congo with an option to expand to 2.5 million acres [10 million hectares], so on-the-ground research verification is vital to assess the status of projects (Chamberlain and Rogerson 2012). One the one hand, many of the specifics waver between agreement and implementation: deals may be poorly conceived, without technical and financial rigor, and reality always sets in during project implementation. On the other hand, researchers (e.g., Carmody 2011; Cotula et al. 2011) argue that the current data on allocations are conservative estimates and may be underestimations: some land deals are concluded quickly, and many are missing from data registers.

## GEOGRAPHIES OF FOREIGN INVESTMENT IN AFRICAN AGRICULTURAL LANDS

There are many faces to the investor rush, but some entities prefer not to reveal themselves. The geography of foreign investment in African agricultural lands is also diverse, illustrating the rise of regional powers, private equity firms, and other joint-venture foreign companies that complicates the old North–South dynamic and extend far beyond the periphery–metropole dynamic of the colonial era. There is a new South–South dynamic, with most of the BRIC countries (Brazil, India, China, and South Africa, but not Russia) prominently engaged; South Korea, Egypt, and the Gulf states also participate at the vanguard. The media have overly focused on the role of Chinese investors; they are significant but are not the largest player. Claims are made (but not corroborated) that Beijing brings all of its own inputs, including prison labor (Hall 2011), but these reports are misleading in terms of what is actually transpiring on the ground.

Gulf and BRIC states are motivated to acquire land to produce food beyond their borders to sustain their growing populations: offshore farming is a way for countries that are highly dependent on imported food (e.g., India and the Gulf states) to increase their national resource base. Governments of food-insecure countries are seeking to outsource domestic food production by incorporating large foreign farms, thereby extending their farming hinterlands and domestic food supplies. For instance, the Qatari government has gone as far as to establish, in 2008, an investment-related company—Hassad Food—to bolster its national food security program.

Land investments in Africa reflect shifts in policy of particular states and a new geopolitics of food. For example, Saudi Arabia is phasing out wheat production by 2016, and the government has turned toward supporting agricultural investments (rice, wheat, sugar, animal products, etc.) by Saudi companies in countries with a high agricultural potential to enhance national food security. Throughout the Gulf region, cereal agriculture is in an irreversible decline, and heavy dependence on food imports forced a strategic change in agricultural policy. With a population predicted to double from 30 million in 2000 to almost 60 million by 2030, the Gulf states face a dire situation of food insecurity. Governments are being proactive and establishing funds that provide financial services and backing (subsidies, loan guarantees, insurance) to companies engaged in land-based international investments (e.g., the Abu Dhabi Fund for Development and King Abdullah's Initiative for Saudi Agricultural Investment Abroad). The geographical proximity of Sudan and Ethiopia and the abundance of affordable land have made these host countries attractive to Gulf investors (Cotula 2013).

Governments are taking direct roles in land acquisitions, and government-to-government deals are evident. The Khartoum government (Sudan) is signing land deals with other governments. Their willingness to engage with external governments has encouraged the Egyptian and South Korea national governments to become active players in acquiring land. For example, the government of Egypt fronted an acquisition of 1.97 million acres (800,000 hectares) in Uganda in 2008 and then allocated the land among seven Egyptian companies with experience in cereal production.

Some land investments combine players from different countries; for example, AgriSol's Tanzanian land acquisition is via a joint venture between Pharos Financial (Qatar) and Summit Group (United States). Often land deals are concluded via intricate arrangements whereby a new subsidiary company represents powerful local interests as well as international backers. Powerful local interests may be invited to join the advisory board of international management companies, and international companies often take a controlling stake in a local company that, in turn, becomes the local driver in land deals. For example, Gabriel Paulino Matip is a major player in the South Sudan land scene and a member of the advisory board of Jarch Management (US). He comes from a well-connected military family and is the surviving eldest son of Paulino Matip Nhial, former deputy commander in chief of the Sudan People's Liberation Army, widely regarded as a founding father of South Sudan. This management company is one of the largest land investors in South Sudan and has acquired 988,422 acres (400,000 hectares) for cereal, flower, fruit, and vegetable production. Another widespread practice is for leases acquired by domestic investors to be sold to foreign, highly speculative interests, obscuring domestic and international bases of investment. As such, the eventual lessees may not be the original investor but rather other parties. Indeed, some international firms purposely operate from tax havens in the United Kingdom, Singapore, and Mauritius (for tax and reputation reasons), concealing their identities. Overall, an occluded geography of who owns what and of who controls the land results.

Renewable fuel targets set outside of Africa are drivers of acquisitions of large plantations. Biofuels contribute 4% of total energy used worldwide in 2011 (Cleveland and Morris 2013:401), but their share is expected to increase. Significant research funding for biofuel research and global powers' establishment of national/regional targets to increase the contribution of biofuels in total energy consumption drive a biofuel tilt. For example, the European Union has established a 10% renewable energy target for 2020 and expects to import 60% of its biofuels to meet that target. The United States has established annual biofuel targets that will rise to 36 billion gallons in 2022. Southern African countries have most embraced the biofuel trend in energy self-sufficiency drives.

Besides the growing world demand for palm oil, projected future opportunities in carbon credits encourage oil palm plantations. In the burgeoning carbon trade market, oil palm is eligible for carbon credits related to reforestation. It is native to West Africa, and smallholders dominated its production until new seed varieties and investment in large-scale plantations began to transform production in the region. West Africa has reemerged as a new frontier for foreign investment in oil palm, and mega-acquisitions by Singaporean, Malaysian, and Indian interests have taken place in Cameroon, Gabon, and Liberia (Table 10.1). Significant investment is also occurring in an axis from Gambia to Angola and in Tanzania, Burundi, and Madagascar.

The land rush also involves domestic players (local and national elites and dynamic farmers), who have been acquiring areas close to urban centers as well as other fertile areas (Cotula 2013). Africa's growing middle class in sprawling cities is expanding its food requirements; serving its demands is another driver of land commercialization. Local players have been acquiring land to store value, run new agricultural ventures, and position themselves as intermediaries with international players and/or as attractive ventures for external buyouts. As such, the land rush intermingles diverse players in different arenas in local-global processes.

Ethiopia, Mozambique, Madagascar, the Republic of the Congo, Sudan, South Sudan, Uganda, and Tanzania attract most investor interest (Table 10.2). Some countries (e.g., Namibia) have not attracted foreign land investors. The percentage of agricultural land earmarked for investors can be very high. For example, 95.1% of land in the Republic of the Congo is presented to foreign investors. This implies that land currently being used for permanent grazing (mostly

**TABLE 10.1 TWENTY LARGEST LAND GRABS IN AFRICA (BY AREA)**

| Host | Land Grabber/Base | Size (hectares) | Production |
|---|---|---|---|
| Benin | Green Waves (Italy) | 250,000 | sunflowers |
| Cameroon | Biopalm Energy (India) | 200,000 | oil palm/energy |
| Rep. Congo | Atama Plantation (Malaysia) | 470,000 | oil palm |
| Gambia | Karuturi (India) | 311,000 | maize, oil palm, rice, sugar |
| Gabon | Olam International (Singapore) | 300,000 | oil palm |
| Liberia | Sime Darby (Malaysia) | 220,000 | oil palm |
| | Golden Agri-Resources (Singapore) | 220,000 | oil palm |
| | Equatorial Palm Oil (UK) | 169,000 | oil palm |
| Madagascar | Madabeff (UK) | 200,000 | beef |
| Mozambique | AgriSA (South Africa) | 1,000,000 | unspecified |
| Nigeria | T4M (UK/Vietnam) | 300,000 | rice |
| Sierra Leone | Long Vab 28 Company (Vietnam) | 200,000 | rice |
| South Sudan | Jarch Management (US) | 400,000 | cereals, flowers, fruit/veg., oil seeds |
| | Nile Trading (USA) | 600,000 | unspecified |
| Sudan | Government of Egypt | 400,000 | maize, sugar, wheat |
| | Government of South Korea | 690,000 | wheat |
| Sudan | Sayegh Group (UAE) | 1,500,000 | unspecified |
| Tanzania | Karuturi (India) | 311,700 | wheat, rice, palm, oil, sugar |
| | AgriSol (US/UAE) | 325,000 | beef, biofuels, poultry |
| Uganda | Government of Egypt | 800,000 | maize, wheat |

*Source:* GRAIN (2012a).

communal grazing grounds) and possibly forests will need to be converted into cropland (Chamberlain and Rogerson 2012). Most hosts have participation from a range of foreign investors (e.g., there are 28 different projects in Ethiopia and Madagascar), but in some host countries, such as the Republic of Congo and Uganda, only a few investors dominate the scene.

African states are motivated to seize investment opportunities. Development aid deficits are driving countries to make large tracts of land available for foreign investment. Development aid for the agricultural sector has been in decline (falling from 18% of total development spending in 1979 to 4.6% by 2007), and governments are acutely aware that they need to increase agricultural productivity to feed growing urban populations. Investors compensate those relinquishing their rights to occupy or use land during the lease period via various mechanisms. Where land is owned by the state (most typical, as states acquire land prior to advertising the properties), formal lease payments and royalties are captured by the state, but these payments for the most part tend to be low, based on the government's anticipation that major benefits will accrue from employment generation and infrastructure development (e.g., irrigation systems), capital, technology, know-how, and market access. Heavy investment, the promise of new infrastructure (production, processing, and transport facilities such as roads and

**TABLE 10.2 MAIN AFRICAN HOST COUNTRIES OF LARGE-SCALE FOREIGN LAND INVESTMENT**

| Host | Area (1,000 hectares) | Number of Projects | Land Agriculture (%) |
|---|---|---|---|
| Mozambique | 11,066 | 22 | 22.4 |
| Republic of Congo | 10,040 | 3 | 95.1 |
| Madagascar | 3,719 | 28 | 8.3 |
| DR Congo | 3,048 | 3 | 13.6 |
| Zambia | 2,677 | 9 | 11.5 |
| Sudan | 2,151 | 12 | 1.6 |
| Ethiopia | 1,456 | 28 | 4.2 |
| Uganda | 1,024 | 4 | 7.3 |

*Source:* Chamberlain and Rogerson (2012:6492).

ports), and social infrastructure (e.g., schools and clinics) are most desired by host governments. As such, land investments are meant to catalyze economic development and alleviate the prevailing agricultural investment gap. Higher farm productivity is expected to improve national food security for hosts. A few African countries have provisions for land rental fees to be shared at the local level. These can involve fees payable at the local level (e.g., Madagascar) or rent-sharing agreements with traditional land custodians (chiefs and family heads) (e.g., Ghana). More frequently, local direct payments at the community level are limited to recompense for loss of harvests; loss of access to land/water is not considered in compensation (Cotula et al. 2011). In the Office du Niger area of Mali, compensation is paid in kind in the form of irrigated land (five hectares per household) (Cotula et al. 2011). Importantly, most deals involve leasing or other concessions rather than sale.

Foreign investors are pampered because foreign capital is a necessary condition for economic growth. Besides tax concessions, the repatriation of capital is guaranteed, and very attractive compensation is offered in the event of expropriation (Zoomers 2010:433). Embassies play a key role in the support of investors: they welcome potential investors and also provide a sounding board about risks and pitfalls.

Large farms in Africa have a checkered past. Plantation crops such as sugarcane in Southern Africa and oil palm registered some successes, but commercial agricultural has concentrated on traditional export crops (cotton, cocoa, tea, and coffee) and has largely been based on smallholder production integrated into global commodity circuits. For the most part, attempts at large-scale farming (even with subsidized credit, machinery, and inexpensive land) have been a failure in the postcolonial era. Attempts in the 1970s and 1980s involved large investments into developing large-scale farming by mechanization and extensive irrigation (e.g., sorghum and sesame [Sudan] and wheat [Tanzania and Nigeria]), but the projects were unprofitable and subsequently abandoned. Encroachment on traditional users of land engendered tension and conflict in Sudan, and large-scale farming did not appear to be more successful than the smallholders it replaced.

For a considerable time, the perspective held sway that smallholder farm productivity was the centerpiece in rural African development. Growth in smallholder agriculture was shown to have a disproportionately higher impact on poverty reduction than growth in other sectors. However, disillusionment with smallholder-based efforts to improve productivity in Africa has set in. The apparent success of Brazil in establishing a vibrant large-scale mechanized farming system caught the attention of institutional investors and led some countries to view this development as a faster path to modernizing agriculture and increasing productivity. The tilt was further reinforced by the apparent export competitiveness of large farms in Latin America and Eastern Europe.

Importantly, the World Bank appears to have reversed its long-held position that favored small farms, expressed in an "inverse size-productivity relationship." At present, the World Bank contends that African land is underused, with low productivity rates, and requires intensification (World Bank 2010). It also maintains that land deals, if properly regulated, could facilitate the transfer of land rights from less to more efficient producers. Agricultural intensification is further advocated on the basis of Africa's low rural population density and low mobility. While not completely abandoning small farms, the World Bank is skeptical that small farm growth is the fastest path toward poverty reduction.

## ARGUMENTS FOR AND AGAINST LARGE-SCALE FOREIGN INVESTMENT IN AFRICAN AGRICULTURAL LAND

Investors focus on the higher-value land with greater rainfall or irrigation potential, better soil, and access to infrastructure and markets: they are drawn to the most fertile land. Removing the best land from local production can have disproportionate (negative) impacts on access to resources, on food and water security, and, in particular, on livelihood options for women. Inadequate attention has been given to likely scenarios where entrepreneurial and skilled local people will benefit from job creation from land investment while vulnerable groups like women lose access to their means of livelihood without adequate compensation. Local people suffer asset losses but seldom receive adequate or even the promised compensation. Positive effects might materialize at some time in the distant future, but poor locals may end up subsidizing rich local and international investors: in effect, governments are trading away local livelihoods too readily and too cheaply. Hall (2011:205–206) argues that land is priced too low "because the property rights of those with uses and claims on the land are not recognized either in law or in practice."

Land grabbing forces local communities to endure enclosure and often triggers their relocation to more marginal areas (Zoomers 2010). Experience suggests that if land investments fail to materialize, land transfers are difficult, if not impossible, to reverse after displacements have occurred. It is wrong that Africa, the world's hungriest region, cannot feed its own population but is an emerging food supplier to foreign markets. As McMichael (2009:288) puts it, "this new international division of labor constitutes an asymmetrical form of corporate 'food security,' based on a dialectic of Northern 'overconsumption' and Southern 'underconsumption,' as the South exports high-value goods at the expense of its own food supplies, in turn addressed through the imports of cheap staple food, which destabilizes local food producers of the South." Carmody (2012) claims the commodification of African lands and their incorporation into global agroinvestment portfolios are ways for the global agroindustry to tighten its control on the food chain from the place of production to the place of consumption.

In cases where biofuel crops are planted, there is direct competition with resources for food crops (Chamberlain and Rogerson 2012). Biofuel plantations have introduced new crops into the region; for example, *Jatropha curas* (oil extracted from the seeds of this Latin American shrub is refined to produce biodiesel) is being widely planted. The uptake of jatropha and sugarcane (for ethanol) occurs throughout the region (e.g., Angola, Republic of the Congo, Madagascar, South Africa, Tanzania, Zimbabwe, Zambia). There are worries that large-scale biofuel monocropping will have harmful (and unknown) environmental effects.

Rather than improving the food security of the host country, land deals may actually be hurting it (Chamberlain and Rogerson 2012). Deals made are not only about land and agricultural potential; they also incorporate water and mineral concessions and even extend to labor. Governments seem reluctant to come to grips with investors' plans that are inconsistent with local visions and national plans for development. They have no way of enforcing vague commitments from investors pertaining to job creation and infrastructure spending. Large land investments appear to operate according to different logic (compared to other foreign investment) in that a country's probability of being targeted is associated with weak governance and a disregard for the rights of the majority of citizens.

Land investments entail risks, and returns on investments are not guaranteed. Commercial risks are significant: operating a large plantation business in Africa is challenging even for a major agrocorporation with extensive African experience, never mind a newcomer with little or no regional experience. Investors can become hostage to the host state. Politics can play a factor and land investments can be exploited by opposition parties in electoral campaigns. The Daewoo deal in Madagascar illustrates how public opposition to a deal contributed to riots that culminated in a change of government. Extensive media coverage of the 3.2-million-acre (1.3 million hectares) deal between Daewoo Logistics (South Korea) and the government of Madagascar to grow maize and palm for oil, mainly for the Korean export market, came to a head. According to the plans, the land would have been prepared by laborers from South Africa. The arrangement became politically charged, and a new government put an end to the deal.

Another potential risk exists if host governments decide to impose export restrictions to protect local food security and political stability in times of food crises. Foreign investors usually secure commitments from host states not to impose such restrictions, but agreements may be disregarded under intense domestic pressure. The various 2008 food protests showed that governments cannot afford to ignore citizens' demands for access to food, especially when it is being produced domestically. Moreover, investors cannot underestimate the international-level risks to their reputations for being tied to corrupt political regimes and for poor business practices and initiatives that drive farmers off the land and into desperation. A growing number of investment deals collapse or never become fully operational. These failures have inflicted considerable hurt, leaving behind a trail of broken promises and devastated livelihoods.

The Ethiopian registry of lands available for international investors classifies them as "wastelands" (Future Agricultures 2011). However, evidence indicates that a good portion of the land allocated to investors in long-term leases is not "wasteland." Many national land classifications are open to question, and land registration systems are incomplete in most of Africa. Moreover, shifting cultivation and dry-season grazing practiced by pastoralists in some areas is not fully acknowledged by the government authorities who are negotiating international land deals. This is not entirely surprising, as government land authorities charged with promoting foreign investment in agricultural land have incentive to conclude land deals, and they compete with other areas to attract funds. There is even evidence of land deals being made for refugee settlement areas and national parks. For example, AgriSol's land acquisitions in Tanzania involve land parcels in Katumba and Mishamo, areas occupied by Burundian refugees in the aftermath of the 1972 war (GRAIN 2012b).

Africa land investment winners and losers are evident. Winners include complex constellations of international financial capitalists, agrocorporations, national and local politicians (who benefit from opportunities in rent-seeking, patronage, etc.), rural and urban elites partaking in land investments, and individuals and groups playing intermediary roles and possibly consuming the fruits/products of the land, particularly if the deals result in less expensive fuel and food—from motorists in the United States to pension-holders whose retirement portfolios include investments in African land, to the growing population of Africa's sprawling cities and other cities around the world (Cotula 2013).

The losers, on the other hand, are rural Africans whose land and livelihoods have been turned upside down by various land investment arrangements without having much say in the process (Cotula 2013). Twenty-five of the deals in Africa resulted in dispossession of more than 25,000 people (Cotula 2013:128). Most experts believe that the publicly available information on dispossession underestimates the aggregate impact. Adverse effects extend to groups within any given project's vicinity, for example herders, fishermen, traditional healers, wood and water fetchers, and farmers affected by enclosures. The building of support infrastructure such as dams and irrigation canals also affects agriculture downstream. Small-scale producers risk being marginalized by the transition to large-scale agriculture, unable to compete with large farmers and/or corporations who are in control of the land and labor and who have government support (Cotula 2013). The livelihoods and culture of millions of Africans are at risk (Cotula 2013).

## URBAN AGRICULTURE

Traditionally, agriculture has been seen and assessed as a rural activity. This perspective is so entrenched that food security policies focus on rural areas, with a strong bias toward production-based solutions. By contrast, urban agriculture and urban food insecurity are neglected topics.

Urban agriculture refers to crops (grains, vegetables, and fruit), herbs, and livestock production within the built-up areas of cities, towns, and peri-urban areas (Thornton 2008). Encompassing a range of activities, from small vegetable backyard gardens on family plots in compounds to farming activities on community lands by associations or neighborhood groups, they vary in size, intensity, and contribution across social and spatial contexts. Most commonly practiced on vacant plots, urban agriculture can be found on fallow public spaces and to a lesser extent private spaces, wetlands, and underdeveloped areas. Urban agriculture rarely takes place on lands specifically designated for agriculture.

Given foreign investment in land and the accompanying transformation of rural lands, urban agriculture is becoming more important. If properly promoted and supported, urban agriculture can play a critical role in Africa's quest for food security (Table 10.3). Concerns about climate change and calls to reduce the carbon footprint associated with transporting food over long distances are making urban agriculture very relevant in the 21st century.

Mougeot (2005) estimates that urban agriculture contributes 15% of all food consumed in African cities. Approximately 30% to 40% of households participate in urban agriculture and related sectors (Arku et al. 2012; FAO 2010). In cities such as Libreville (Gabon), Kumasi (Ghana), and Lusaka (Zambia) more than half of urban dwellers engage in agriculture, and after petty trade, urban agriculture is the next largest informal employment activity, even though much is done on a part-time basis (Arku et al. 2012:9). The real value, however, may be not in monetary rewards but rather in building social cohesion and community social capital, especially among the poorest households, which are the most active participants. Where women dominate in urban agriculture (e.g., South Africa and in peri-urban environments), their work outside the home is critical to building women's social networks (Shackleton, Pasquini, and Drescher 2009).

Urban farmers produce a wide variety of staples (e.g., maize, cassava, potato, plantains, and sorghum) as well as other vegetables and fruits. Urban agriculture is geared mostly toward consumption within the household, but feeding needy neighbors and selling "over the fence" are regular occurrences. For most farmers, urban agriculture contributes extra income but is not the primary income source. In regions where urban agriculture is well developed (e.g., West and East Africa), it accounts for 20% to 50% of the urban poor's household income and savings (FAO 2010). The more sophisticated growers sell produce to informal traders and contribute to the livelihoods of market sellers. Urban agriculture also benefits auxiliary businesses such as fencing, composting, transporting, processing (milling, packing), and storage.

Urban agriculture is being reconsidered after the sharp increases in food prices and the global economic downturn that exacerbated unemployment. Despite policy elevation in the last decade and a half, urban agriculture is only weakly incorporated into municipal urban plans. It suffered from various negative historical antecedents. Colonial cities banned agriculture within city limits and promoted peri-urban agriculture as a spatial planning strategy to create a buffer separating elite residential housing areas from traditional

**TABLE 10.3 THE BENEFITS OF URBAN AGRICULTURE FOR FOOD SECURITY**

- Reduces urban food deficits in the face of Africa's rapid population growth and changing consumption patterns
- Provides household food security for those unable to realize food security under prevailing market conditions
- Enables nutritional diversification through crop diversity
- Provides nutritional security for HIV/AIDS-affected families, thereby augmenting antiretroviral therapies
- Eases the reliance on food imports and long-distance transport of food from rural areas
- Contributes to employment and household finances
- Allows for the emergence of niche markets (African indigenous vegetables/ wild foods)
- Helps stabilize food market prices
- Plays a role in linking poor and more affluent households in food provisioning

*Source:* Arku et al. (2012:10).

villages. Moreover, urban developers and industrialists with considerable political clout and economic muscle shaped urban policy arenas and were successful in pressuring for the conversion of untidy, vacant, and green spaces as well as wild lands into nonagricultural uses. Only in more recent times is urban agriculture being pro-actively incorporated in urban planning: Addis Ababa (Ethiopia), Bulawayo (Zimbabwe), Cape Town (South Africa), Dar es Salaam (Tanzania), Durban (South Africa), Kampala (Uganda), and Kumasi (Ghana) represent the urban agricultural policy and planning vanguard. However, despite its potential, urban planning regulations constrain the contribution that urban agriculture could make to the food supply and food security in the region. A neglect of the potential of urban agriculture to feed the poor means that food insecurity can loom large even in cities (e.g., Cape Town) that are proximate to vibrant agricultural areas where food is produced for wealthy domestic and international consumers (See Box 10.4 for a discussion of the paradox of food insecurity in the Cape Town).

There are various efforts under way to increase urban farming efficiency and to better integrate the activity into urban economies. A self-help group in Kibera (Nairobi) (consisting of 1,000 urban farm women) has developed a "vertical garden" system (with assistance and training from the French NGO Solidarités International) for producing wild spinach and other indigenous vegetables. Vertical gardening involves growing crops on the vertical surfaces of structures that are either freestanding or part of a building. This method allows more crops to be grown on the same area. Their vertical farming effort uses recycled sacks or biodegradable cement bags filled with soil and punctured with holes for growing seedlings. More organized community groups are engaging in commercial production, selling to informal markets and selling weekly or biweekly organic produce to subscriber clients. Generally referred to as community-supported agriculture initiatives, "fresh organic produce boxes," as they are colloquially known, are being sold to affluent, environmentally conscious urban consumers. Profits are invested back into projects, and in the process, important bonds are

### BOX 10.4 CAPE TOWN: THE PARADOX OF URBAN FOOD INSECURITY IN A FOOD-SECURE COUNTRY

Cape Town is one of wealthiest cities in Africa as well as one of the most unequal (McDonald 2007). Inequality is expressed in food consumption patterns and access to food. The city boasts a vibrant food scene with trendy restaurants, well-stocked supermarkets, and diverse specialty "foodie" establishments. There is an array of food markets, from gourmet markets to community fruit and vegetable markets to informal food markets. The city is also home to several major horticultural areas. The largest and best known is the Philippi horticulture area, which contributes half of the city's soft vegetables (Battersby 2011b). Moreover, the city is close to an agriculture and viticulture heartland that produces for domestic and, especially, international markets. Grapes, pears, apples, olives, and, of course, wine are produced for wealthy consumers.

Paradoxically, poor Capetonians experience food insecurity despite their proximity to an agricultural area that produces 21% of the national agricultural output (Battersby 2011a). Almost half of Capetonians have insecure food supplies, and their diets are deficient in nutrients. It is a surprise to many that urban food insecurity is so evident in South Africa, which is considered to have a secure food supply: it is the richest country in Africa, and access to food is a constitutional right. Cape Town's burgeoning food-insecurity problem remained largely invisible to policymakers until recently (Crush and Frayne 2011a). Evidence from Cape Town and other African cities overturns the view that food insecurity is largely a rural phenomenon.

Many Capetonians experience recurring shortages of food. Not everyone is food insecure every day; the majority of residents experience food insecurity once a week or once a month. In Cape Town, 11% of the population experience food shortages every day, 35% more than once a week, and 35% once a month (Battersby 2011b:20). Hunger forces people to beg in the streets, a common scene in the Cape Town central business district and more affluent suburbs. Food insecurity is concentrated in townships and informal settlements. For example, 89% of Khayelitsha's residents lack secure food supplies (Battersby 2011b); by contrast, food security is concentrated in "white" suburbs.

Food is the largest single expense for many urban residents, accounting for 39% of monthly expenses. Though food may be widely available, access is a critical issue, and constrained access affects the composition of diets. Half of the population rarely or never consumes eggs or meat, 47% of diets do not contain fresh fruit (even though fruit is grown throughout the region), and 34% of the population rarely consumes vegetables (Crush and Frayne 2011a:535). The diet of those in poverty shows movement away from traditional high-protein sources, such as samp

*(Continued)*

**BOX 10.4 (*Continued*)**

(cornmeal) and beans, and toward nonnutritive food groups such as fats, sugar/honey, and coffee/tea, with a reliance on maize meal products (e.g., sorghum porridge and white bread) (Battersby 2011b). Only small amounts of protein food are consumed (e.g., chicken, meat stews, and soybeans) and fresh fish is not consumed despite proximity to the sea (Battersby 2011b). Poverty is manifested in skipping meals, limiting portion sizes, borrowing food, purchasing food on credit, consuming less expensive and less nutritious meals, and peddling assets to pay for food.

NGOs, faith-based organizations, and schools are stepping in alleviate some of the food deficits. Indeed, rural networks participate in transferring food to migrant family members in Cape Town (Frayne 2007). Cape Town's escalation of food insecurity challenges the myth that people in urban areas are healthier and wealthier and fare better than people in rural South Africa.

Rapid urbanization and suburban sprawl in Cape Town are engulfing agricultural land, and the land with the highest agricultural potential is disappearing more quickly than the land with medium or lower potential (Geyer et al. 2011). For example, Somerset West, a suburban development, now threatens Philippi and other urban agricultural areas. Urban agricultural land use cannot compete with the more lucrative land use of residential development.

A large disconnect exists between the local urban food economy, in which fresh fruits and vegetables go directly to consumers, and the modern, integrated, large-scale farm/food economy, which centers on supermarket chains. Despite the newly recognized importance of sustaining urban agriculture, the activity is undermined by various negative perceptions, and smallholder production is not integrated into agroindustrial commodity chains. Cape Town initiated an Urban Agriculture Policy in 2007 to raise the importance of urban farming and to improve the health of poor urban residents.

The "supermarket revolution" is a countertrend, transforming the foodscape dramatically (Crush and Frayne 2011b; Weatherspoon and Reardon 2003). Supermarkets are at the end of a large-scale agroindustrial complex that has taken over food production, processing, marketing, and retailing. The leading players (e.g., Pick n Pay, Shoprite, Woolworths, and Spar) have aggressively targeted all consumers, including the poor, and this effort has been accompanied by a drive to locate outlets in township shopping centers, often near highways and away from population concentrations. Physical access is difficult, and the cost of food and transportation to the store is likely prohibitive for poor Capetonians. Trade liberalization allows supermarkets to pack their shelves with products, where the originating area is occluded, excluding local smallholder production. This has ramifications because supermarket shopping is becoming more common in the everyday lives of the urban poor, peaking at payday and declining thereafter. Supermarkets are encroaching on the market share of informal retailers, street traders, and small local producers. Supermarkets and fast-food outlets are gaining market share and having a major impact on the nutrition of the poor.

Cape Town may be the African city furthest along the supermarket food source path. Consumption patterns of food among the poor are drifting toward poorer-quality and energy-dense foods. Supermarket penetration is still in the early stages in most African countries, but some are hurrying along this path. Shoprite, for example, is active across the southern African region and is making inroads in West Africa (Ghana and Nigeria), and Massmart (the African affiliate of Walmart) had established operations in 12 African countries by 2013.

cemented between informal urban farmers and well-off households within a sustainable urban agriculture system—an important urban agriculture coalition in the making. For example, Harvest of Hope, a nonprofit community organization of informal organic producers based largely in the Cape Flats (Cape Town), organizes urban farmers (mostly women in small groups of three to eight producers) to grow vegetables for their own households, township schools, and community subscriber projects.

There are several benefits of urban agriculture. It makes significant contributions toward ensuring food security and providing employment opportunities for the urban poor. It allows access to additional and more nutritious food. Urban households engaging in farming activities consume greater quantities of food (up to 30% more), meaning greater dietary diversity (FAO 2010). Consumption of greater quantities of vegetables, fruits, and meats translates into higher energy/caloric intake and better diet composition. Local urban agricultural production for local markets helps reduce the transport costs involved in shipping goods over long distances to markets. Local production also hedges against supply disruptions due to market failures and bottlenecks. Urban farming can preserve local biodiversity, especially wild crops and seedlings that are at risk of extinction due to uncontrolled urban expansion and by the introduction of alien species into agricultural production. Indigenous wild crops (e.g., sweet potatoes [Central Africa], yams, pumpkin, and wild spinach [Southern and Eastern Africa]) could be preserved through sustainable wild urban farming (Shackleton, Pasquini, and Drescher 2009). Indigenous varietals often contain higher levels of nutrients and minerals and require less water and fertilizer (Shackleton, Pasquini and Drescher 2009).

There are some negative aspects to urban agriculture, however. Health and environmental issues predominate (exposure to diseases and pesticides and increases in mosquitoes, flies, pests, and odors from cattle rearing), there are administrative concerns (noncompliance with zoning and city ordinances), and perceived social concerns surface from time to time (criminals using the cultivation areas as well as squatters being drawn to the area) (Arku et al. 2012). Many urbanites perceive urban cultivation and urban livestock husbandry as holdovers of rural habits rather than as 21st-century activities. Other skeptics maintain that urban agriculture is an unproductive use of land (compared to high-rent uses) and makes only a modest contribution to the city's financial base.

The overall negative health-related effects of urban agricultural practices are inconclusive, but there have been food scares and anxieties. Crops can be potential pathways for biological and chemical contaminants. The use of chemicals by urban farmers can be dangerous. Pathogens have been found on vegetables irrigated by wastewater, but typically levels are below those posing a threat to human health (Shackleton, Pasquini, and Drescher 2009). Some urban farmers have skin problems associated with handling wastewater (Shackleton, Pasquini, and Drescher 2009:23). There is no way to assess the chemical residue on food grown and distributed in informal economic circuits, where there is no monitoring for pesticides and food and hygiene regulations are lax.

Of critical importance is the availability of reliable and clean water for urban farming. Unfortunately, urban agriculture operates within environments where other industries (formal and informal) have been known to discharge pollutants and contaminate local water supplies. For example, e-waste dumping and burning contaminates the Korle Lagoon, where cattle drink in the city of Accra (Ghana). It is well documented that desperate and hungry people grow vegetables near sewers and that there is no oversight in place. When cholera outbreaks have occurred (e.g., Dar es Salaam in 2007), urban farming has been banned temporarily, even though a direct connection between standing water/wastewater and this disease was never substantiated.

Institutional intransigence needs to be overcome before urban agriculture can be elevated within policy arenas. Officials need to rethink urban planning policy, provide stewardship informed by science and social science, and coordinate multiple levels of responsibility (i.e., health, agriculture, water, environment). Integration of urban agriculture into African city planning needs to be accompanied by further development of water supply infrastructure. Also, inclusive zoning codes are needed so that urban agriculture can be incorporated into land-use plans. Current insecure tenure and eviction fears undermine the entire activity. Clearly, urban agriculture can reduce vulnerability to food price hikes, contribute to local food movements, and provide a more sustainable use of the natural resource base that also assists in "greening" the city. It is high time for African urban planners to deal with the food and land use realities of cities and to move away from ideas inherited from colonial systems. However, the soil and water used in urban agriculture are often contaminated, and urbanites know it. Therefore, the entire urban agriculture infrastructure needs attention if local food is to be a trusted alternative and a critical component of food security.

## CONCLUSIONS

Population growth, urbanization pressures, and increases in meat and dairy consumption are making food a critical global issue. Power shifts and the larger roles of China, India, and the Gulf states in Africa are adding more pressures on the region's land and food systems. Middle Eastern and Asian entities are engaging in offshore production, and South African farmers are commencing onshore farming beyond their home state. Foreign investors and their African business partners are acquiring potentially high-producing food lands and removing some from the food economy by using them for biofuel and oil palm plantations. The consequences of contemporary developments are far-reaching: plantation agriculture is squeezing local landholders, and once land is under foreign control it may be very difficult and even catastrophic to reverse the process by canceling international agreements.

It appears that most new food and biofuel production is being geared for export markets, and new production systems lack local integration (e.g., machinery and seeds are imported). Until recently, the majority of Africans had been heavily reliant on agriculture for their livelihood, but if the contemporary land investment trend continues, the poorest members of agrarian societies risk losing their major assets: land, water, and homes.

These developments are all part of the global financial shakeup that has resulted in a shift in focus toward land and food: Africa is targeted heavily as an underinvested agricultural region. Consequently, fresh food from Africa will be added to the list of crops controlled by agroindustrial complexes, and Africa-produced food will travel outside the region. The conversion of large parcels of land to biofuel development (particularly jatropha) is another significant development. Many of the biofuel projects are controlled by British, Italian, and Chinese entities. That so much land is being offered to foreign investors or sovereign wealth funds is evidence that many African governments do not have a clear land and/or food policy in place (Chamberlain and Rogerson 2012). They appear to be gambling on foreign investors to provide the infrastructure to catalyze rural development. Furthermore, it appears that governments believe that other sectors can provide the growth, employment, and incomes that will enable food security that is not contingent only on what is produced locally. This raises the critical question: where will Africa's food come from, and at what costs?

In addition, governments of investor counties have an ethical obligation to consider the effects of policies that directly or indirectly increase pressures on land in Africa (Cotula 2013). They have a responsibility to regulate and consider the African operations of their home-based companies on the basis of long-term sustainability. Developing a more balanced international investment legal framework would greatly facilitate the process, but public opinion can put more pressure on governments to persuade their companies to act more responsibly.

In the second half of the 20th century, a consensus was built around promoting small-scale African agriculture because of higher productivity rates and an assumption that small-scale farming was the best way to alleviate rural poverty and to provide local food. This consensus has broken down: there are still efforts to promote small-scale agriculture, but the tilt toward large farms is gaining ground. If all of the land deals under negotiation come into full operation, large farms will dominate the African agricultural landscape. This trend in Africa's agricultural and food systems will lead to a different (and perhaps a global) norm. Clearly, this is the time for critical reflection. A more sensible approach is to acknowledge and plan for a variety of diversified agricultural systems and to recognize that each plays a unique role. Agriculture does not have to be viewed as involving only large modernized farms that are part of global commodity chains.

The land investment story currently unfolding reflects deep global economic and social transformations and will have profound repercussions for the region. African agriculture is moving toward more capital-intensive, mechanized, and labor-saving production systems and away from smallholder labor-intensive production. Decisions and agreements being put in place will have major impacts on many people's livelihoods and food and water security; both rural and urban societies will be affected. The processes under way may well develop into the corporatization of African agriculture, and negotiated deals may turn out to be irreversible (even at times of local food and water shortages).

Food and land are multidimensional arenas that affect other policy domains such as health, economy, and development. Clearly, a future challenge for Africa's land and food systems lies in greater transparency as well as greater coordination of local, national, and international interventions and of the various stakeholders, including small farmers, communities, governments, and NGOs. Alternative agricultural visions and sustainable models need to be promoted. In particular, urban agriculture should receive more attention in policy arenas so that the practice is better incorporated into urban planning and food and health policies. To date, governments have been reluctant to consider land reform to bolster food security, so land reform and food security remain poorly integrated (Kepe and Tessaro 2012). Investing in farmers rather than farmland seems a better and safer route to provide long-term food security. It is time for rural and urban Africans to become central to the decision-making processes shaping the future of African agricultural and food systems (Cotula 2013).

## REFERENCES

Allan, T., M. Keulertz, S. Sojamo, and J. Warner, eds. 2012. *Handbook of Land and Water Grabs in Africa: Foreign Direct Investment and Food and Water Security.* New York: Routledge.

Andrea, G., and B. Beckman. 1985. *The Wheat Trap.* London: Zed Books.

Anseeuw, W., L. Cotula, and M. Taylor. 2012. "Expectations and Implications of the Rush of Land: Understanding the Opportunities and Risks at Stake in Africa." In *Handbook of Land and Water Grabs in Africa: Foreign Direct Investment and Food and Water Security*, eds. T. Allan, M. Keulertz, S. Sojamo, and J. Warner, pp. 421–435. New York: Routledge.

Arku, G., P. Mkandawire, N. Aguda, and V. Kuuire. 2012. "Africa's Quest for Food Security: What Is the Role of Urban Agriculture?" African Capacity Building Foundation Occasional Paper no. 19. Harare: ACBF.

Barratt Brown, M. 2007. "'Fair Trade' with Africa." *Review of African Political Economy* 34(112):267–277.

Battersby, J. 2011a. "Urban Food Insecurity in Cape Town, South Africa: An Alternative Approach to Food Access." *Development Southern Africa* 28(4):545–561.

Battersby, J. 2011b. "The State of Urban Food Insecurity in Cape Town." Urban Food Security Series no. 11. Kingston and Cape Town: Queens University and African Food Security Network.

Bek, D., T. Binns, and E. Nel. 2010. "Wildflower Harvesting on Cape Aghulus Plain: A Mechanism for Achieving Local Economic Development?" *Sustainable Development.* Online, doi: 10.1002/sd.499.

*Business Day*. 2011. "Growing Demand for SA's Agricultural Skills." Available at http://farmlandgrab.org/post/view/18672 (accessed June 22, 2012).

Carmody, P. 2011. *The New Scramble for Africa.* Cambridge: Polity Press.

Carmody, P. 2012. "A Global Enclosure. The Geo-Logics of Indian Agro-Investment in Africa." In *Handbook of Land and Water Grabs in Africa: Foreign Direct Investment and Food and Water Security*, eds. T. Allan, M. Keulertz, S. Sojamo, and J. Warner, pp. 120–132. New York: Routledge.

Chamberlain, W., and C. Rogerson. 2012. "Agricultural Land Grabs in Africa: Scope, Patterns and Investors." *African Journal of Agricultural Research* 7(48):6488–6501.

Cleveland, C., and C. Morris. 2013. *Handbook of Energy. Volume 1: Diagrams, Charts and Tables.* New York: Elsevier.

Cotula, L. 2013. *The Great African Land Grab? Agricultural Investments in the Global Food Systems.* New York: Zed Books.

Cotula, L., S. Vermeulen, P. Mathieu, and C. Toulmin. 2011. "Agricultural Investment and International Land Deals: Evidence from a Multi-Country Study in Africa." *Food Security* 3(1):s99–s113.

Crush, J., and G. Frayne. 2011a. "Urban Food Insecurity and the New International Food Security Agenda." *Development Southern Africa* 28(4):527–535.

Crush, J., and G. Frayne. 2011b. "Supermarket Expansion and the Informal Economy in Southern African Cities: Implications for Urban Food Security." *Journal of Southern African Studies* 37(4): 781–807.

Deininger, K., D. Byerlee, J. Lindsay, A. Norton, H. Selod and M. Strickler. 2010. *Rising Global Interest in Farmland: Can It Yield Sustainable and Equitable Benefits?* Washington, D.C.: World Bank.

Deininger, K., and D. Byerlee. 2012. "The Rise of Large Farms in Land Abundant Countries: Do They Have a Future?" *World Development* 40(4):701–714.

Fairtrade International (FLO). 2012. *Annual Report, 2011–2012: For Producers, with Producers.* Bonn: FLO.

Food and Agricultural Organization (FAO). 2006. "An Introduction to the Basic Concept of Food Security." Available at http://www.fao.org/docrep/013/al936e/al936e00.pdf (accessed July 19, 2013).

Food and Agricultural Organization (FAO). 2010. "Fighting Poverty and Hunger: What Role for Urban Agriculture?" FAO Policy Brief no. 10. Available at http://www.fao.org/docrep/012/al377e/al377e00.pdf (accessed July 9, 2012).

FAOSTAT. 2011. "Arable Land." Available at http://faostat.fao.org/site/377/default.aspx#ancor (accessed July 19, 2013).

Flower Valley Conservation Trust. 2011. *Flower Valley Conservation Trust Annual Report.* Bredasdorp, SA: FVCT.

Frayne, B. 2007. "Migration and the Changing Social Economy of Windhoek, Namibia." *Development Southern Africa* 24(1):91–108.

Freidberg, S. 2003. "Cleaning Up Down South: Supermarkets, Ethical Trade and African Horticulture." *Social & Cultural Geography* 4(1):27–43.

Freidberg, S. 2004. *French Beans and Food Scares: Culture and Commerce in an Anxious Age.* New York: Oxford University Press.

Freidberg, S. 2009. *Fresh: A Perishable History.* Cambridge, MA: Belknap Press of Harvard University.

Freidberg, S. 2010. "Freshness from Afar: The Colonial Roots of Contemporary Fresh Food." *Food & History* 8(1):257–278.

Future Agricultures. 2011. "Land Grabbing in Africa and the New Politics of Food." Policy Brief 41, June, pp. 1–8.

Geyer, H., B. Schloms, D. du Plesiss, and A. van Eden. 2011. "Land Quality, Urban Development and Urban Agriculture within the Cape Town Urban Edge." *Town and Regional Planning* 59(1):41–52.

GRAIN. 2012a. "Data Set with over 400 Global Land-Grabs." Available at http://www.grain.org/article/entries/4479-grain-releases-data-set-with-over-400-global-land-grabs (accessed June 25, 2012).

GRAIN. 2012b. *Squeezing Africa Dry: Behind Every Land Grab Is a Water Grab.* Barcelona: GRAIN.

Hall, R. 2011. "Land Grabbing in Southern Africa: The Many Faces of the Investor Rush." *Review of African Political Economy* 38(128):193–214.

Kepe, T., and D. Tessaro. 2012. "Integrating Food Security with Land Reform: A More Effective Policy for South Africa." CIGI-African Initiative Brief No. 4. Available at http://www.cigionline.org/publications/2012/8/integrating-food-security-land-reform-more-effective-policy-south-africa (accessed July 19, 2013).

Kilgallen, K. 2008. "Corporate Social Responsibility: Ground Control." *Retail Week* (July 11), p. 27.

Lipinski, B., C. Hanson, J. Lomax, L. Kitinoja, R. Waite, and T. Searchinger. 2013. "Reducing Food Loss and Waste." Working Paper, Installment 2 of Creating a Sustainable Food Future. Washington, D.C.: World Resources Institute.

*Mail and Guardian.* 2011. "Boers Are Moving North." Available at http://mg.co.za/article/2011-05-03-boers-are-moving-north/ (accessed June 22, 2012).

McDonald, D. 2007. *World City Syndrome: Neoliberalism and Inequality in Cape Town.* New York: Routledge.

McEwan, C., A. Hughes, and D. Bek. 2012. "Futures, Ethics and the Politics of Expectation in Biodiversity Conservation: A Study of South African Sustainable Wildflower Harvesting." *Geoforum* 37(6):1021–1034.

McMichael, P. 2009. "A Food Regime Analysis of the 'World Food Crisis.'" *Agriculture and Human Values* 26(4):281–295.

Moseley, W., J. Carney, and L. Becker. 2010. "Neoliberal Policy, Rural Livelihoods and Food Security: A Comparative Study of Gambia, Côte d'Ivoire and Mali." *Proceedings of the National Academy of Sciences USA* 107(13):5774–5779.

Mougeot, L., ed. 2005. *Agropolis: The Social, Political and Environmental Dimension of Urban Agriculture.* London: Earthscan.

Oakland Institute. 2011. *Understanding Land Investment Deals in Africa. County Report: Sudan.* Oakland, CA: Oakland Institute.

Rakotoarisoa, M., M. Iafrate, and M. Paschali. 2012. *Why Has Africa Become a Net Food Importer? Explaining Africa Agricultural and Food Trade Deficits.* Rome: FAO.

Ryan, O. 2011. *Chocolate Nation: Living and Dying for Cocoa in West Africa.* New York: Zed Books.

Schroeder, R. 2008. "South African Capital in the Land of Ujamaa: Contested Terrain in Tanzania." *African Sociological Review* 12:20–34.

Schroeder, R. 2013. *Africa after Apartheid: South Africa, Race and Nation in Tanzania.* Bloomington: Indiana University Press.

Shackleton, C., M. Pasquini, and A. Drescher. 2009. *African Indigenous Vegetables in Urban Agriculture.* London: Earthscan.

Shell Foundation. 2012. "Trading up: M&S Partnership—Flowers from South Africa." http://www.shellfoundation.org/pages/core_lines.php?p=corelines_inside_content&page=trading&newsID=108 (accessed July 1, 2012).

Thornton, A. 2008. "Beyond the Metropolis: Small Town Case Studies of Urban and Peri-Urban

Agriculture in South Africa." *Urban Forum* 19(3): 243–262.

Thornton, P., P. Jones, P. Ericksen, and A. Challinor. 2011. "Agriculture and Food Systems in Sub-Saharan Africa in a 4°C+ World." *Philosophical Transactions of the Royal Society A* 13(369):117–136.

Weatherspoon, D., and T. Reardon. 2003. "The Rise of Supermarkets in Africa and Implications for Agrifood Systems and the Rural Poor." *Development Policy Review* 21(3):333–355.

Zoomers, A. 2010. "Globalisation and Foreignisation of Space: Seven Processes Driving the Current Global Land Grab." *Journal of Peasant Studies* 37(2):429–447.

## WEBSITES

GRAIN. http://www.grain.org (accessed December 6, 2013).

*The Guardian*. 2011. "Sun Biofuels have left us in a helpless situation. They have taken our land." Video available at http://www.theguardian.com/environment/video/2011/nov/09/biofuel-tanzania-video (accessed December 6, 2013).

Harvest of Hope. http://harvestofhope.co.za (accessed December 6, 2013).

Land Matrix Organization. 2013. http://www.landmatrix.org (accessed December 6, 2013).

Oakland Institute. www.oaklandinstitue.org (accessed December 6, 2013).

Solidarités. "Kibera sack garden" http://www.youtube.com/watch?v=1neYXUKbdKg (accessed December 6, 2013).

World Bank. 2010. *Rising Global Interest in Farmland: Can it Yield Sustainable and Equitable Benefits*. Washington D.C: The World Bank.

CHAPTER 11

# CLIMATE CHANGE

## INTRODUCTION

Climate has always been featured prominently in African development research (Toulmin 2009). The region's population has lived and adapted to a high degree of climate variability and its associated risks, moving in time and space from hunter-gatherers to pastoralists to more modern agricultural and urban livelihood systems (Yanda and Mubaya 2011). As in other world regions, the ever-changing climate has dramatically altered Africa's environments. For example, several thousand years ago, a much moister Sahara contained lakes and forests (Toulmin 2009). However, changes in the Earth's atmosphere are occurring at an historically unprecedented speed and scale. Accelerated changes in climate over the past century have led to a new consensus on the rate of climate change in the current global warming trend.

Two processes drive climate change in Africa (Parnell and Walawege 2011). First are slow-onset changes, such as increasing atmospheric temperature and humidity levels, changing precipitation patterns, sea-level rise, and desertification. These changes occur over several decades, and the environmental impacts are regarded as semi-permanent. Second, there is an increasing frequency and severity of extreme events like droughts and floods that amplify the effects of the first set of changes and extend their cumulative effect. It is now widely acknowledged that Africa is the most vulnerable world region and the least equipped to cope with these changes (Boko et al. 2007; Parnell and Walawege 2011). The region's vulnerability is partly driven by geography but also by its low adaptive capacity resulting from dysfunctions in national economies and health, education, infrastructure, and governance systems. There is also mounting evidence that African environments are changing in profound ways—and not just in climate terms: environmental health is deteriorating in both urban and rural contexts.

The scientific basis of global climate change and its African impacts has gathered momentum in recent decades. Almost 100 years ago, Swedish scientists calculated that human activities could warm the earth by adding more carbon dioxide. A warming trend was mentioned at the first United Nations (UN) Conference on the Human Environment in 1972 in Stockholm, but environmental experts were more preoccupied with other natural resource issues (e.g., oil and pollution). In the 1980s, policymakers viewed an emerging global warming challenge (if they did not ignore it altogether) as an environmental issue of peripheral concern that could be handled by environmental ministries. It was not until 1988 that the Intergovernmental Panel on Climate Change (IPCC) was established to assess and synthesize the latest scientific, technical, and social sciences literature on global warming. By the 1990s, climate modeling had become more sophisticated, actual patterns of change in regional climates were being observed, and a consensus developed around reducing greenhouse gas (GHG) emissions. By the early years of the 21st century, the climate change conundrum was seen not only as an emerging threat to international security and peace but also as a grave threat to humanity.

The IPCC is the most influential and authoritative agency on climate change science. In 2007, the panel and former U.S. Vice President Al Gore were jointly awarded the Nobel Prize for their efforts to integrate and disseminate greater knowledge about anthropogenic (human-induced) climate change, and for laying the foundations for the development of measures to counteract these changes. Thousands of the world's leading scientists and experts have worked under the auspices of the IPCC to produce major updated assessments in 1990, 1995, 2001, 2007, and 2014. IPCC reviews and assessments are based on the most recent scientific, technical, and socioeconomic information produced worldwide for comprehending climate change. Importantly, the IPCC publishes syntheses of the most recent accumulated knowledge, but it does not conduct research or monitor climate data itself. IPCC reports are based on published materials, two to three years old, and some worry that climate change is occurring faster than the scientists are able to study, record, and publish the changes.

Governments have tended to regard the IPCC as the authoritative source of information and analysis on climate change. IPCC assessments of climate change, for the most part, are based on the runs of models as well as a range of data and scenarios. Not surprisingly, IPCC reports and protocols are criticized heavily for privileging an understanding of climate change at the global scale produced by Western scientists and elite institutions, and for most heavily relying on quantitative data analysis and computer modeling. The diversity of data required and the need for consistency among different scenarios pose substantial challenges for climate researchers. This is especially the case in African contexts, where limited data from the past and present are available.

The climate change debate is very complex, and many expert knowledge communities promote different agendas. Climate change is an international research priority area, a hot topic in publishing, and a subject that receives increasing media coverage in Africa and elsewhere. The range of climate perspectives is expanding all the time (Table 11.1).

*The Carbon Map* is an excellent primer for visualizing and making sense of climate change responsibility and vulnerabilities (KILN 2013). Climate change has a large following on social media, and trends are always changing. *The Guardian* (2011) newspaper, which represents a vanguard of climate change coverage, has developed a list of the top 50 climate Twitter accounts to follow.

Nevertheless, climate skeptics and deniers abound (see climate skeptic website). Skeptics are generally distrustful of global climatic modeling and reason that the weather cannot be predicted five days in advance, never mind 50 years or more into the future. Indeed, climate modeling does involve huge data gaps. For example, the surface temperature record is woefully inadequate: climate models are based on a surface temperature database that covers only 40% of the landmasses and misses huge swathes of the world's oceans. Skeptics attack the climate change consensus on different fronts: some argue that global warming is caused by physical rather than human-induced changes; some remain unconvinced that we have

**TABLE 11.1 COMMUNITIES IN THE DEBATE ON CLIMATE CHANGE**

- The conspiracy community
- The denial community
- The catastrophe community
- The funding community (we're interested but we need to steer climate research according to our mandates)
- The scientific community (use my data but I am not self-aware)
- The social science community (we understand people but which model should we use?)
- The African government community (wait to see what global experts tell us)
- Adaptation community (make adjustments and hope for the best)
- Mitigation community (try to steer it according to climate modeling scenarios)

identified the causes of warming global temperatures; and others see it as part of a conspiracy theory perpetuated by opportunists (seeking political, financial, ideological, or institutional gain). A leading mainstream climate skeptic, Richard Lindzen, Professor of Atmospheric Sciences at the Massachusetts Institute of Technology (MIT), argues that the atmospheric system is more robust than given credit insofar as the system has fluctuated considerably throughout human history. Lindzen and other skeptics also believe that carbon dioxide is overemphasized as a climate engine in global warming science. Meanwhile, proponents of conspiracy allegations refer to scientific consensus as a "global warming hoax" or "global warming fraud" (Solomon 2008; Spenser 2010).

The climate change debate reached a fever pitch in 2009 when the climate research unit at Britain's University of East Anglia had its emails hacked, and claims were made, but never substantiated, that climate researchers manipulated and concealed data (supposedly of a global cooling trend) and suppressed critics by failing to publish their claims in scientific journals. The controversy blew over, but "Climategate" (as it became known) had the effect of tarnishing climate research yet did not disprove the central climate change argument. Certain supporters of the so-called mainstream consensus responded, alleging that some conspiracy proponents were part of a well-funded misinformation campaign aimed at stirring up controversy, undercutting scientific consensus, and downplaying global warming's projected effects. Greenpeace, for instance, initiated the Exxon Secrets Project to show that the energy industry explicitly funds climate change denial.

Authoritative IPCC reports contain numerous dire predictions about the impact of climate change on Africa. IPCC (2007) warns about heavier precipitation, storms, and droughts that could wipe billions off economies and destroy lives. Some of the most important predictions advise that global warming could reduce crop yields in some African countries by half by 2020; water stresses will be aggravated; the spread of diseases will increase due to more heavy precipitation events in areas with poor water supplies and overtaxed sanitation infrastructure; and water volume is decreasing dramatically in Lake Chad (shrinking from 8,494 square miles [22,000 $km^2$] to a meager 115 square miles [300 $km^2$]), affecting fisheries, farming, freshwater supplies, and the livelihoods of 20 million people in the basin. Less potable water from the lake is resulting in an increasing number of cases of diarrhea, cholera, and typhoid fever throughout the basin. The rate of decline of snow on Mount Kilimanjaro is so high that local inhabitants could be deprived of their main supply of freshwater within two decades (see next section).

Some campaigners and members of the media contend that climate change is an intractable force, a "fifth horseman of the apocalypse" that could lead to an age of ruin. Climate change has also attracted major interest and leadership from civic organizations. One such prominent civil society organization (affiliated with 300 nonprofit organizations that span the globe) is Tck Tck Tck. It has orchestrated very successful media campaigns to raise awareness of global climate change. Ahead of the UN's Copenhagen climate talks (COP15) in 2010, they organized a petition of 17 million signatures in a worldwide call to action. Preceding the Durban COP17 talks, they organized a convoy from Dar es Salam (Tanzania) to Durban (South Africa) (a 4,350-mile [7,000 km] trek across 10 countries) and carried 229 African farmers, pastoralists, and members of women's and youth groups to raise awareness of global warming in Africa.

It is not easy for the general public to come to grips with climate change. First, global environment change is extremely complex, involving interactions between two profoundly different worlds: science and human affairs. Second, the politics of climate change makes it difficult to forge connections between different political arenas: multilateral–civil society, national–local, and so forth. Third, there is concern that climate change is becoming a Global North-driven agenda and yet another vehicle for external interference with Africa. Fourth, there is a possibility (expressed in the United Kingdom's 2006 Stern Report) that climate change might be happening faster than thought and that climate-induced damages in Africa may be underestimated. Fifth, there is a real danger that the general public will become "climate fatigued" in the long process of discovery (as data collection is improved and expanded and scenarios are refined, and as the best minds in physical science, social sciences, government, and civil society come to grips with climate change).

## ENVIRONMENTAL CHANGES ON MOUNT KILIMANJARO

Mount Kilimanjaro, Tanzania's iconic volcanic mountain, carries an 11,700-year-old glacier even though the mountain is situated near the Equator. Kibo summit, soaring almost four miles into the sky, is the highest point on the African continent and often referred to as "the roof of Africa." Hemingway famously wrote about the mountain in 1934 in his novel *The Snows of Kilimanjaro* (later turned into a film), capturing the world's imagination (Fig. 11.1). The mountain took center stage in 2006 as a riveting example of how the natural environment was being altered in compressed time. Al Gore's movie, *An Inconvenient Truth*, detailed an impressive series of then-and-now images that documented the widespread retreat of many glaciers over the past century, most dramatically illustrating that Mount Kilimanjaro icecaps had almost disappeared (the icecap was once 164 feet [50 m] deep). Kilimanjaro became a poster child for demonstrating climate change on the African continent.

Scientists confirmed that during the 20th century the extent of Kilimanjaro's ice fields decreased by 80% (NASA imagery confirmed a 26% contraction between 2000 and 2007 alone). This physical transformation also has human consequences: the glacier is an important source of water for local agriculture as well as a magnet for tourists (the mountain is a major foreign currency earner, attracting about 20,000 tourists [and their money] annually). Gore claimed that Kilimanjaro's glacier would be gone "in a decade," and global experts emphasized that if current climatic conditions persist the remaining ice field would likely disappear sometime between 2015 and 2020 (IPCC 2007:440). The shrinking snows were shown widely to warn humankind about impending rapid environmental changes. However, this icon of global warming became embroiled in a heated scientific controversy that attracted global attention and illustrates the complexities in understanding the science of environmental change. Despite their dire predictions, the glacier is still very much in evidence atop Kilimanjaro today.

**FIGURE 11.1** Snow-capped Kilimanjaro. *Source:* © Charles Bowman/Robert Harding World Imagery/Corbis. Corbis 42-30873449.

Scientists with more extensive Kilimanjaro glacier expertise than IPCC specialists are uneasy about the volcano's poster-child status. They concur that global warming is indeed responsible for nearly every other glacier around the globe melting away; however, Kilimanjaro just happens to be the exception. Glacier–climate interactions in the tropics exhibit very different patterns and peculiar characteristics compared with mid- and high-latitude glaciers. Indeed, there are many competing ideas about why Kilimanjaro's snow is melting so rapidly. Some posit that declining snow there is more a result of declining moisture since the 1880s than increasing temperature (Kaser et al. 2004). Scientists note that temperatures on Kilimanjaro's summit are always below freezing, so a singular warming explanation of glacier loss seems unlikely. Instead, it is argued that a combination of factors, including changes in weather elements (humidity, temperatures, and precipitation), fewer clouds, and more sunlight hours, creates a layer of warmer air at the glacier's surface, which is considered to govern the fluctuations of tropical glaciers. Scientists emphasize ice loss is attributed to sublimation (ice turning directly into water vapor at below-freezing temperatures without proceeding through the liquid-water stage), which occurs when temperatures are cold and the air is extremely dry, as is the case at the top of Kilimanjaro (where the glacier is subjected to a moisture-sapping "freezer burn"). This process is constant, but when there is less precipitation to rebuild the glaciers, the result is a net loss of ice. This diverging scientific opinion hypothesizes that glacial retreat on Kilimanjaro may well be a response to the drying of the atmosphere and lack of precipitation, as part of a natural process of alternating wet and dry periods, rather than a consequence of global warming per se (Verschuren et al. 2009).

Many questions about Kilimanjaro remain. Glacial melting occurred in previous centuries, so it is hard to pinpoint what triggered the present ablation. Localized processes may be linked to temperature variations in other world regions (e.g., the Indian Ocean, where a large-scale connection between sea-surface temperatures and East African rainfall has been observed [Kaser et al. 2004]). Recent field research has shown that Kilimanjaro's glacial contraction is not accelerating; absolute loss has decreased more in the past few years. Climate scientists have long maintained—and empirical evidence confirms—that warming does not result only in higher temperatures: it leads to changes in weather patterns, including more precipitation in some areas and more droughts in others, and this is what may be occurring on Kilimanjaro. Additional drivers of change on the mountain include fire, land-cover transformation, and human modification. For example, deforestation on its slopes sucks moisture out of the upslope winds and helps dry out the snow.

The Kilimanjaro debate is far from resolved and may become even more complicated on the basis on future climatic models that predict global warming and increasing rainfall in eastern Africa, which could paradoxically be what saves Kilimanjaro's snowy crown. In the meantime, farmers near Kilimanjaro slopes have found water bountiful and have begun to produce water-hungry crops (e.g., tulips) for export to Europe. However, this horticulture concentration cannot be sustained if the glacier/groundwater waters dry up or if climate change brings significant increases in precipitation in the region.

## CLIMATE IMPACTS ON AFRICA

Climate modeling suggests that no major area of Africa will be a winner as the natural environment changes. In contrast, other regions of the world—for example, parts of Russia and China—will experience more favorable farming conditions based on 20- to 30-year climatic forecasts. According to leading economists, climate change will have a unique and catastrophic impact on Africa (Collier, Conway, and Venables 2008).

To begin with, Africa is warming faster than the global average (in all seasons), with drier subtropical regions warming more than moister tropics, a trend likely to continue. Median temperatures will have risen between 3 and 4 degrees Celsius from 1900 to 2100, roughly 1.5 times the global mean (Christensen et al. 2007:867). Second, Africa is an enormous landmass that lies squarely astride the Equator, stretching from 35N to 35S. These climatic effects are heterogeneous and vary by region; there is no single Africa-wide climate effect. Two thirds of Africa's land surface is classified as arid or semiarid, and the region is home to myriad fragile land and coastal ecosystems. Moving south or north from the Equator, there is a sharp decline in average annual rainfall, accompanied by an increase in variability. Africa is subdivided into numerous climatic regions, producing a geographical mosaic of climate change that is highly complicated, both within subregions and even within large countries. Third, agriculture accounts for the dominant share of economic activity in Africa (approximately 60%–70% of employment and up to 50% of gross domestic product [GDP] in some countries), making the region's economy particularly sensitive to climate. Furthermore, current African agriculture activity may be nearing the limits of plant tolerance, and further change could push farmers to the brink. Fourth, Africa's economies are unequipped to adapt to climate change. Technical progress has been slower than in other world regions, and most national economies concentrate on a narrow range of export commodities. Households have coped with temporary shocks, but current livelihood strategies are not readily able to adapt to new circumstances that require adoption of new technologies (mobile phones are an exception).

Large parts of Africa have always been exposed to a high degree of climate variability on an annual or inter-annual basis, and further climate change is projected to accentuate variability, resulting in greater climatic extremes. Annual precipitation variability can be 40% in the Sahel and almost 25% in most parts of Africa (Fig. 11.2). In the 20th century, African temperatures as a whole warmed by approximately 0.7 degrees

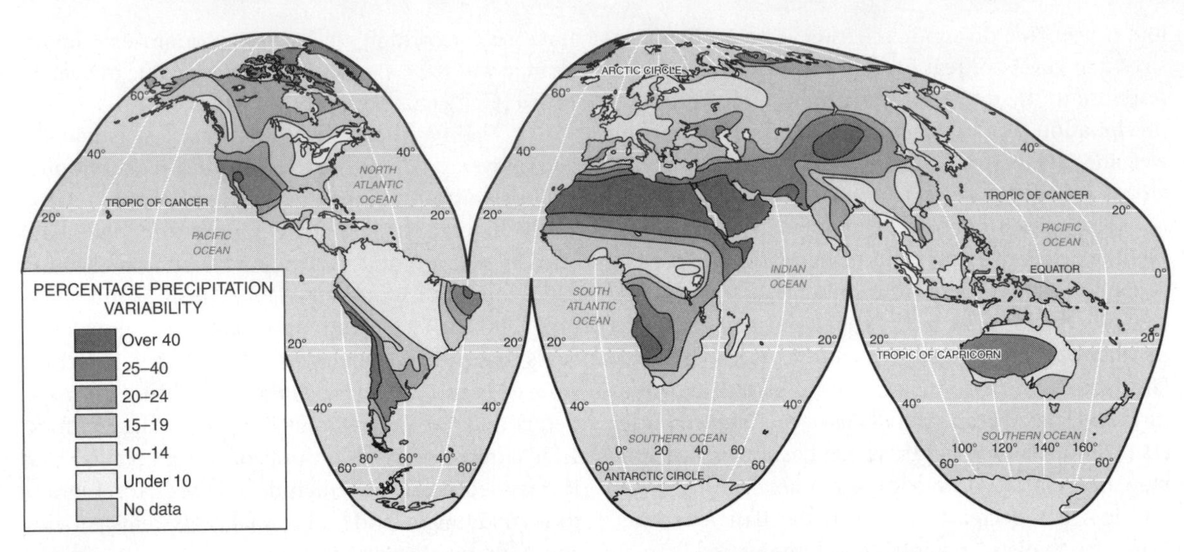


**FIGURE 11.2** Annual Precipitation Variability in Africa. *Source:* Map 1 from De Blij et al., 2013:190.

Celsius, but climate change is far from uniform. For instance, in South Africa and Ethiopia, minimum temperatures increased slightly faster than maximum or mean temperatures, and the number of warm spells in Southern and Western Africa has increased while the number of extremely cold days has decreased. Needless to say, observed temperature changes cannot be applied beyond their monitoring areas. Climate station coverage in severely limited throughout Africa, and the paucity of observational data is a constant and critical challenge. Even in South Africa, which has the region's most geographically extensive network of stations, there are gaps in coverage (Rouault, Sen Roy, and Balling 2013). Africa has eight times fewer weather stations than the minimum recommended level, with vast areas in Central Africa completely unmonitored and sparse coverage in the Horn of Africa (Fig. 11.3).

Moreover, knowledge about climate change in Africa rests on a lower scientific base than elsewhere. Until recently, many writings on climate change in Africa were highly speculative. In African contexts, social sciences research on climate change is emerging but is so far well behind the quality of natural science research. For example, the IPCC reports (particularly the Africa chapter of the 2007 report) have been severely criticized because of the absence of peer-reviewed sources, relying too heavily on "gray" materials (e.g., reports from government agencies or scientific research groups, working papers from research groups or committees, white papers). The highly respected InterAcademy Council (IAC), an umbrella organization of national academies of sciences, reported that 84% of the sources for the physical sciences sections of the IPCC 2007 report were peer-reviewed, but only 59% of the social sciences content (IAC 2010:16). This means that 41% of the latter content was based on less rigorously reviewed materials. The IAC notes that IPCC authors report a high confidence in statements for which there is little evidence: for example, the widely quoted statement that agricultural yields in Africa might decline by up to 50% by 2020. There have been other criticisms about how the IPCC appoints its regional expert panels. The government–research nexus is always politicized, and politicians do not always nominate the best researchers in the author selection process. Moreover, the sometimes-questionable selection of experts can be compounded as the regional chapters do not make use of experts outside the region (IAC 2010:18). The bottom line is that the IPCC may be the body undertaking the most

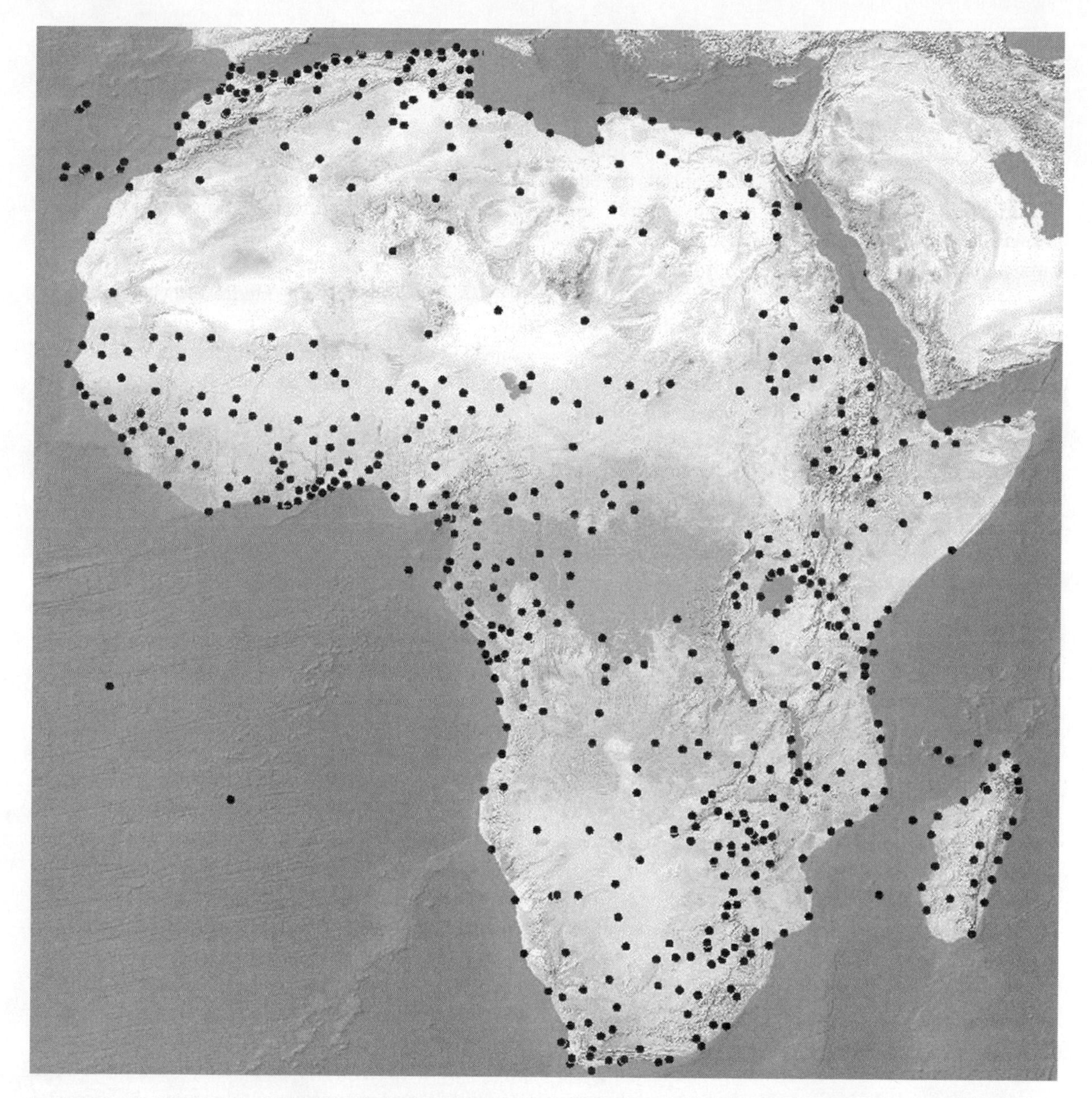

**FIGURE 11.3** Distribution of Hourly Weather Stations in Africa. *Source:* Generated from National Climate Data Center 2013.

comprehensive reporting on the region, but its current climate models and the data leave much to be desired in terms of forecasting accuracy and geographical precision. The Fifth Assessment Report is scheduled for release in 2014, but the advance summary for policymakers released in 2013 is unequivocal about human induced change on rising global temperature in the range of 1.5 to 4.5 degrees Celsius and argues that the global 2-degree-Celsius threshold will be breached within 30 years and that sea-level rises could reach 3.3 feet (1 m) sometime between 2100 and 2300 (IPCC 2013).

Climate science on Africa is patchy, with a large amount of extrapolation based on modeling. Some aspects of climate are known and relatively well understood, but there is still uncertainty about the key climatic processes, and much remains simply unknown. For example, it is not well understood how three of the global drivers of African climate (the intertropical convergence zone, the El Niño–Southern oscillation, and the West African monsoon) interact and, and, in turn, are affected by climate change. There also are major gaps in the scientific understanding of Africa's regional climate systems (e.g., the Congo Basin). To make matters worse, African researchers have so far contributed minimally to the literature on global environmental change, so most established research was conducted outside the region.

There is, however, increasing agreement on some general region-wide scenarios. In Eastern Africa, the Horn, and parts of Central Africa, average rainfall is expected to increase (by 15% or more). Southern Africa will become hotter and drier, with precipitation falling by 10% to 20% or more. Warming (in the range of 0.2–0.5 degrees Celsius per decade) will be greatest over the interior and semiarid margins of the Sahara and most of central Southern Africa. The extent of arid and semiarid land in Africa will increase within the range of 148 million to 173 million acres (60 million–70 million hectares). Droughts and floods will occur with greater frequency and intensity: flooding is anticipated in the Nile Delta and along parts of the eastern and western African coastlines.

There is still considerable uncertainty about the broad regional impacts of climate change. There is no consensus on what its effects will be in the pivotal Sahel region—drier or wetter? How climate will affect the drainage of Africa's great river basins is unknown; for example, conflicting models of the Nile Valley show greater and lesser flows. Most analyses emphasize a long list of adverse impacts, but there are a few that could be potentially favorable. For instance, a carbon fertilization effect on plant growth produced by elevated levels of carbon dioxide is capable of producing a substantial increase in crop yields for responsive crops such as maize and sorghum. Therefore, carbon fertilization, combined with rainfall increases, could yield higher agricultural productivity. Also, other crops such as cassava prosper in hotter climates.

There appears to be increased confusion over information about climate change in Africa. Datasets are incomplete, climate models rely on varying databases, and data intervals and assumptions are not always well explained. As for findings and conclusions, most of them are opaquely written. Media and civic organizations often communicate technical details in overly simplified and non-nuanced ways, so it is difficult to gauge what the general public understands.

## THE HUMAN IMPACTS OF CLIMATE CHANGE

Model projections of the physical effects of climate change in Africa are highly uncertain, particularly at the national and subnational scales where political processes operate. Within states, forecasting global environmental change becomes highly complex because the consequences of such change have the potential to combine and multiply. Security experts are now advancing the concept of "threat multipliers," whereby negative impacts create stressors that threaten to undermine both national and international security. The Pentagon, for example, predicts that flows of refugees from climatically challenged areas will be a major 21st-century security threat (Schwartz and Randall 2004).

Climate change also will exacerbate many existing problems. For example, African governments are already dealing with weak institutions, rapid population growth, widespread problems of water supply and quality, and the prevalence of malaria and diarrheal diseases. With heavy reliance on rain-fed agriculture, which accounts for a large share of most countries' economic production, pastoralists are already grappling with the effects of warmer temperatures, decreasing rainfall, and more frequent droughts (See Box 11.1 on whether the Darfur crisis was the first African climate change war?). They may need to be far more flexible in the future and increase their mobility in the search for sufficient water and grazing lands for livestock. Competition among pastoralists and cultivators for scarce resources is likely to intensify and spark resource wars.

## BOX 11.1 IS THE CONFLICT IN DARFUR THE WORLD'S FIRST CLIMATE CHANGE WAR?

The 2003–07 conflict in Darfur is etched in people's minds as an atrocious conflict during which at least 300,000 people were killed and more than 2 million displaced. John Aston, the UK Special Representative for Climate Change, labeled Darfur the "first modern climate change conflict." A climate conflict explanation has gained traction with the UN and the global media. In April 2007, the UN Security Council held its first-ever debate on climate change as a global security issue, noting Darfur as a conflict driven by resource shortages in the context of climate change. Jeffrey Sachs (2008:248–249) summed up the environmental basis of the conflict: "The only reliable growth in Darfur was its population, from less than one million at the start of the 20th century to an estimated six to seven million today. But as the population has soared, the carrying capacity of the land declined because of long-term diminished rainfall. . . . The striking pattern is the decline of rainfall starting at the end of the 1960s, a pattern that is evident throughout the African Sahel. The results have been predictably disastrous. Competition over land and water has become lethal." Repeated periods of drought were heightened by the intensification of human misuse of land and environmental degradation; increasing population pressures resulted in greater stress on land cultivation; and overgrazing and forest clearing worsened the predicament. Some forests around cities such as Nyala and El Geneina have disappeared entirely. Pastoralists in the region were designated as the first group unable to sustain their way of life in the current climate situation. The ruling elite in Khartoum, the capital of Sudan, also touted the climate/conflict explanation of the events in Darfur, but this account requires closer examination.

Darfur (Arabic: "Land of the Fur") has historically been called Billād al-Sūdān (Arabic: "Land of the Blacks") and corresponds to the westernmost portion of present-day Sudan. Sudan is a complex mosaic of hundreds of ethnic groups, and the vast majority of population embraces Islam and speaks Arabic (in addition to local dialects). Religious diversity is suppressed. The ethnic geography of the Darfur region is marked by a major fault line that corresponds to livelihood patterns: Arabs dominate pastoralism, and African ethnic groups (Fur, the Zaghawa, and Massalit) dominate in subsistence farming. Geographically, the Fur and other agriculturalists are concentrated in the southern portion of Darfur and the Marrah highlands, an agriculturally rich area suitable for cereal, rice, and fruit cultivation. Northern Darfur, by contrast, is arid and dominated by Arabs whose livelihoods revolve around herding (camels, goats, and sheep). Ethnic tensions have long simmered between the nomadic Arab herders and the sedentary Fur farmers, but large-scale violence was not unleashed until the 1980s.

Darfurians had adapted to environmental change for centuries, but changes since the late 1960s occurred at a faster pace and on a wider scale. Proponents of the climate/conflict thesis emphasize that adaptation to current and impending climate change is the key conflict trigger. Decreased rainfall and a burgeoning population mean that pastoralists were migrating to better-watered areas, pitting them against the farmers to gain access to newly scarce resources. A case in point is wealthier Arab camel nomads who ranged farther south not because they faced starvation but because they were attempting to take advantage of the southward retreat of the tsetse fly belt and the opening up of new grasslands. The exceptional drought and famine of 1984–85 were responsible for killing a large number of cattle, left many farmers and herders destitute, and shifted the migration contours again. Young pastoral men found themselves in the demeaning position of having to find work as hired herders and wage laborers, creating, for some, an economic incentive for banditry. Violence intensified during the 1985–86 dry season, when the Baggara Arab tribes initiated large-scale raids into southern Sudan to seize cattle to make quick profits. As it turns out, the government's military intelligence leadership had orchestrated these raids. In response, rebels from the agricultural community ratcheted up tensions by attacking government installations in 2003 to protest the Sudanese government's disregard of the western region and its non-Arab population. Heavy-handedly, the Khartoum government responded by creating a new Arab militia force—the Janjaweed. This group originated as a coalition of Chadian militia and their Sudanese hosts (mainly failed nomads), with arms supplied by Libya, and they began attacking sedentary groups in Darfur. Within a year, tens of thousands of people (primarily Fur and other agriculturalists) were killed, hundreds of thousands fled westward to refugee camps in neighboring Chad, and many others remained, internally displaced. More than 400 villages were destroyed in the war years (see Amnesty International [2011] before-and-after maps of villages: http://www.amnestyusa.org/sites/default/files/aiusadarfursatelliteevidence.pdf). Darfur's violent conflict worsened its ecological crisis.

The calamitous events in Darfur have been interpreted as an omen of the climate–security connection: a bleak future of people fighting for dwindling resources around the world in the context of radical environmental change. The security community predicts that in the climate change era violence will erupt both between countries and within countries, particularly in resource-poor regions of the world (for example, many parts of Africa).

There is an opposing and entirely different narrative on the causes of conflict in Darfur. This is far more than an academic debate because the explanation affects the kinds of development options that are being implemented at present and, in turn, shapes the development trajectory that will determine Sudan's future. The opposing argument emphasizes the internal political situation and reduces climate to no more than a contributing factor. An emphasis on national and local power struggles shows how the climate/conflict narrative masks the real

(Continued)

**BOX 11.1 (*Continued*)**

political-economic dynamic that underpinned the crisis. According to de Waal (2007:7), "in all cases, significant violent conflict erupted because of political factors, particularly the propensity of the Sudanese government to respond to local problems by supporting militia groups as proxies to suppress any signs of resistance. Drought, famine and the social disruptions they brought made it easier for the government to pursue this strategy." Depleted natural resources and livelihood transformations cannot on their own account for armed conflict. The fighting in Darfur was racially motivated, pitting Arab pastoralists against African farmers, but land envy more than tribal and ethnic hatred determined the events. In 2004, President George Bush described the events unfolding in Darfur as genocide, and civil society organizations such Save Darfur, an international pressure group, tried to keep the focus on the genocide, which occurred in waves of intensity followed by calm. Some of the most barbarous acts included torture, the murder of pregnant women, the rape of young girls, the abduction of individuals during attacks, and kidnapping of persons for sexual slavery (see Amnesty International 2004). Ethnic strife alone is not a satisfactory explanation of the conflict. Most villages are multiethnic, and, despite ethnic differences, there is a history of peaceful coexistence. Far more critical was the failure of the government on many levels: failure to adapt to climate change and failure to stem the conflict (instead, the government instigated violence). An outcome of the climate/conflict narrative is that it overlooks the winners in the Darfur crisis—the Khartoum elite and their international investment partners.

Any single-cause explanation of Darfur's conflict is misguided and represents an oversimplification of reality. Indeed, the opposing narratives suggest very different solutions: (1) climate adaptation and (2) regime change. In the case of Darfur, the former has held sway but does not explain the level and intensity of violence. This perspective has also led to solutions emphasizing economic development and climate adaptation while leaving internal power dynamics underexamined and intact. The solution implemented supports Darfur with much outside development funding that is deployed through Khartoum patronage networks (Verhoeven 2011). Khartoum's elite is now tapping into the climate change and global food crisis narratives to go on an offensive to recalibrate Sudan's political economy in order to increase patronage, foreign partnerships, and financial resources for the party elite to command. However, this depoliticized development solution fails to address the underlying problems. On balance, it appears that climate change is an insufficient explanation of the conflict, but it does exacerbate current tensions and inequalities. The critical factor leading to violence appears to be the degree of political and economic marginalization. As such, biophysical events (e.g., droughts) are not straightforward, and the way they play out can only be understood within their particular local, national, and international contexts.

Africa's leading vulnerabilities to climate change are outlined in Table 11.2.

There is no way that we can assign precise values to the impacts of climate change, and experts in different disciplines focus on particular dimensions of climate impacts. UNEP (2005) undertook a preliminary effort to map those regions of Africa that are most vulnerable to specific impacts of climate change (Fig. 11.4). It is a useful starting point for broader dimensions of the regional impacts of climate change, but it tells us little about the impacts within regions and/or how states might cope with those challenges. General regional trends are not uniformly distributed within the regions or even within countries. Prioritizing limited resources and targeting climate hotspots (areas facing particularly severe impact from climate change and most vulnerable to its deleterious effects) would help to more precisely identify the most vulnerable places.

Mapping vulnerability and its connections to human security is a frontier in climate research. There is no agreed definition of human vulnerability to climate change or how best to measure it. Approaches have proliferated and extensive suites of vulnerability maps have been produced. Human vulnerability centers on the exposure of individuals, communities, and societies and their susceptibility to losses, all of which are highly context-dependent. Development and poverty researchers operationalize vulnerability as an aggregate measure of human welfare and integrate environmental, social, economic, and political exposure with a range of harmful situations (Bohle, Downing, and Watts 1994). Considerable challenges remain as to how to develop robust yet practical maps that incorporate climate change and diverse attributes of risk and vulnerability.

Busby, Raleigh, and Salehyan (2014) map vulnerability in Africa based on four criteria: physical exposure, household and community resources, governance and political violence, and population density (See Figure 11.5). Based on these indicators, areas with the greatest vulnerability are concentrated in the Horn of Africa (Somalia, Ethiopia, Djibouti, Eritrea) and the

**TABLE 11.2 IMPACTS OF AFRICA'S CHANGING CLIMATE**

- Biodiversity and forests: 20% to 30% of plant and animal species are at risk of extinction. Certain bird and reptile species face outright extinction: in the Congo Basin, the world's second largest rainforest, 80% of its plant species are found nowhere else. The livelihoods of millions of Africans who depend on trees and wood products (rope from baobab bark; fuel; foodstuffs based on leaves and fruit; medicinal ingredients; tools [e.g., mortars and pestles used in food preparation such as millet]) are at risk.
- Food security: Rain-fed agriculture employs 60% of the population, and a high percentage of the population relies on livestock. A decrease in inland fisheries (e.g., Lakes Victoria and Tanganyika) threatens millions of livelihoods. A severe drop in crop yields from their current levels and a disappearance of wheat production by the 2080s threaten segments of the population. The nutrient content of food may diminish due to increased levels of pests and diseases, leading to malnutrition in poor populations. Changes in food access affect health and susceptibility to disease.
- Health: There is an increased likelihood of climate-sensitive disease outbreaks (e.g., malaria, cholera, and meningitis) and malnutrition; a southward and eastward expansion of the malaria transmission zone; and malaria incursion into the previously malaria-free highland zones of Ethiopia, Kenya, Rwanda, and Burundi. The distribution of the tsetse fly (which can cause trypanosomiasis or sleeping sickness) may extend from Central Africa's heart (the Democratic Republic of Congo [DRC]) into Western and Eastern Africa. Diarrheal diseases may increase. Animal health is also impaired by heat stress. The pollen season may be lengthened.
- Water: Higher temperatures increase evaporation from rivers, lakes, and ponds; increase soil drying and intensify evapotranspiration; and enhance crop stress. Moderate to extreme decreases in water flow via runoff are projected, with an extreme increase projected for Eastern Africa. 75 million to 250 million Africans will be exposed to water stress. Inadequate water supply infrastructure (e.g., the small number of dams [Tanzania has only two]) puts many at risk. The mismatch between water availability and need increases. Water-sharing agreements will be even more important in major river basins (e.g., the Niger Basin, which spans ten countries).
- Cities: Fragile urban areas are undergoing rapid expansion. A range of challenges will be triggered by sea-level rise, heat waves, urban heat island effects (particular parts of cities experiencing elevated temperatures), flooding, wildfires, pollution, and widespread poverty (66% of residents live in slumlike conditions). Urban governments lack the resources, infrastructure, and information to be able to cope. Lack of effective planning exposes urban residents to a range of hazards, such as the expected frequency of storms, floods, epidemics, constraints on water supply, and higher food prices. Urban pollution would be amplified.
- Coastal zones: Rising sea levels and more frequent storm surges and flooding threaten coastal and riverine habitats in Western and Eastern Africa, especially mangrove-rich shorelines. 25% of Africans live within 6.2 miles (10 km) of a coast. The Accra–Niger Delta corridor, with a projected population of 50 million in 2020, faces serious flood problems.
- Conflict or cooperation? Water and climate change wars? Armed conflict would further complicate climate-change risk management. However, it appears that intercommunal conflict tends to be resolved without erupting into warfare and that violence is not the norm.
- Migration: States and urban governments have minimal capacities to manage climate-induced migration. Movements of large numbers of climate refugees can enhance tensions, increase competition and distrust, and aggravate existing stresses (e.g., economic degradation, disease, and conflict), which can exacerbate conflict or spark new outbreaks in receiving areas.

Great Lakes region (Burundi, DRC, Kenya, Rwanda Tanzania, Uganda), both regions with a long history of civil strife, political violence, health epidemics, famine, and economic instability. Pockets of coastal vulnerability occur in Western Africa, Mozambique, and Madagascar. Somalia and Ethiopia stand out with the largest zones of composite vulnerabilities. Somalia has been an anarchic state since civil war in the early 1990s, and the prolonged period of violent conflict, endemic poverty, lack of viable government, poor land-use practices, and increasing climate variability (including a very intense drought in 2011) have exacerbated existing problems and aggravate potential climate change effects. With an estimated population of 94 million in 2013, Ethiopia is a very populous state, and much of its population is concentrated in the drought-prone western highlands of the country, where a lack of irrigation infrastructure means that people depend on rain-fed agriculture; the government has exhibited both weak responses to disaster and low adaptability to changing conditions. In general, areas with political conflict exhibit very high vulnerabilities to climate change. For example, DRC and the Central African Republic are less exposed to natural disasters than other parts of Africa, but these countries score very low on governance and political violence and are less likely to be able to adapt to climate-related crisis under changing conditions of the future.

More specialized vulnerability maps show how food insecurity heightens under different climate

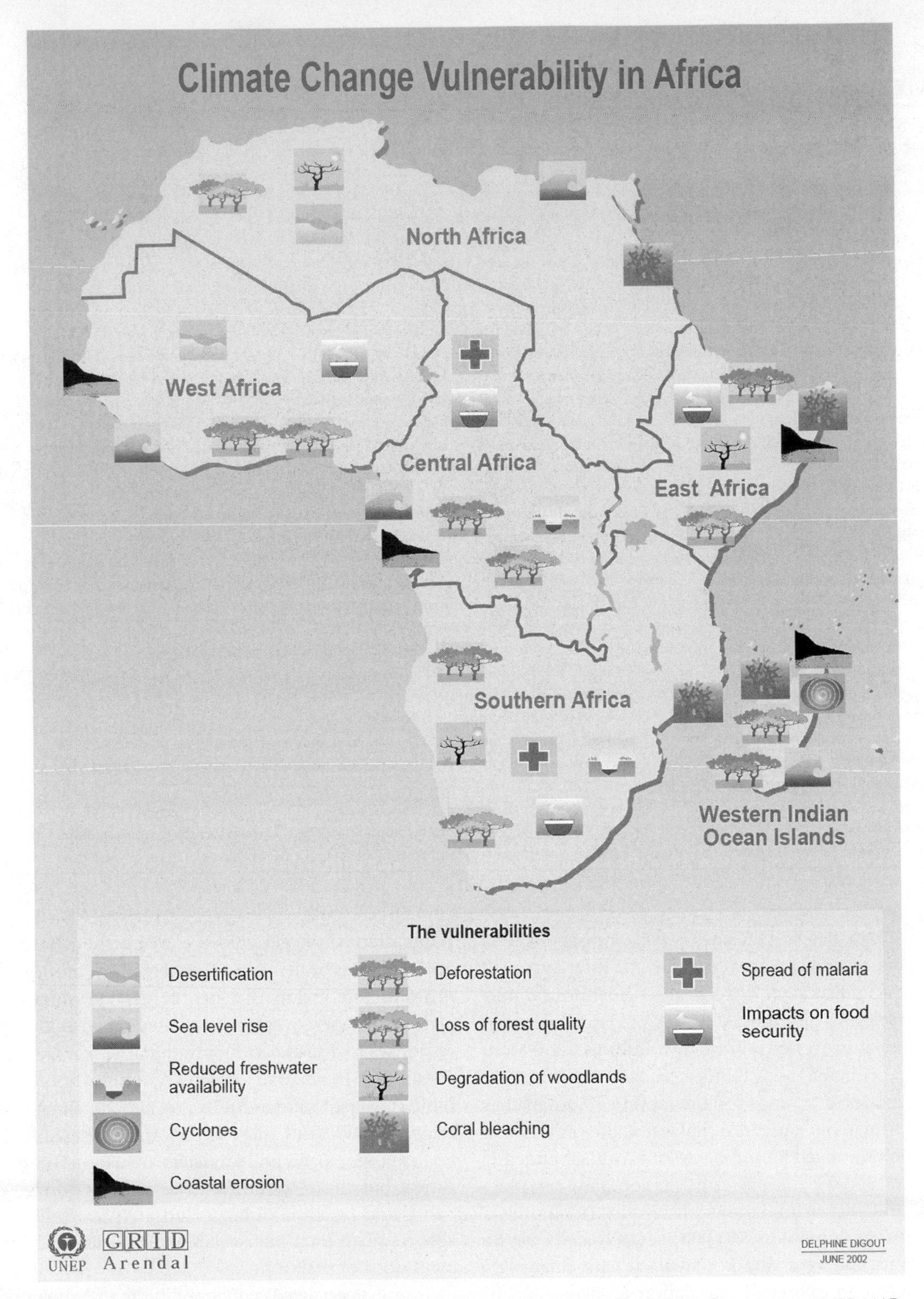


**FIGURE 11.4** United Nations Environmental Programme, Climate Change Vulnerability, Africa. *Source:* UNEP MAP 2005. Reprinted courtesy of GRID-Arendal Maps and Graphics Library.

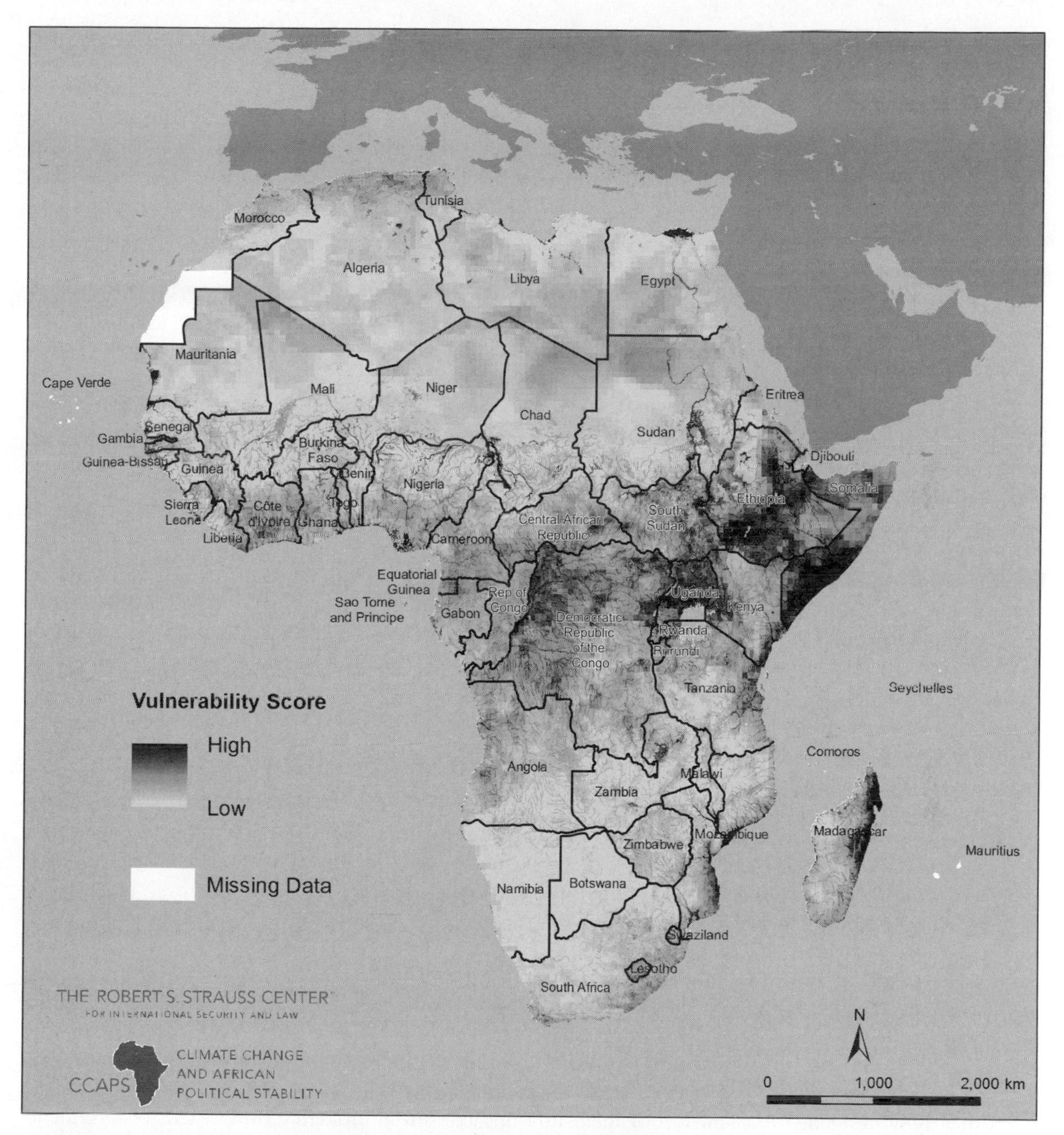


**FIGURE 11.5** Climate Vulnerability in Africa Composite Map. *Source:* Reprinted courtesy of Busby, J., C. Raleigh, and I. Salehyan. 2014. "The Political Geography of Climate Vulnerability, Conflict, and Aid in Africa." In *Peace and Conflict*, eds. D. Backer, P. Huth and J. Wilkenfield. Boulder: Paradigm Publishers. Forthcoming.

scenarios (Ericksen et al. 2011). Mapping food insecurity indicators and relating them to projected climate change in 2050 reveals a number of hotspots. The most vulnerable areas include parts of West Africa, the Nile Delta, and Southern Africa (in addition to South Africa). The distribution of food insecurity hotspots changes according to different temperature scenarios; the maximum present-day temperature during the primary growing season is 30 degrees Celsius but this temperatures will rise by 2050 producing significant crop stresses. Most climate models predict enhanced food insecurity in Southern Africa and for the countries of the Central African Republic, Ethiopia, and Cameroon.

Vulnerability mapping is useful for grounding discussions of climate change and its human spatial consequences. Although such mapping is based on soft data and numerous assumptions, it does provide a starting point for policymakers to think about the climate change conundrum at the local level, and it is likely that vulnerability mapping will become more sophisticated over time.

Climate change has the potential to alter Africa as we know it today. It may redraw the maps of water availability, food security, disease incidence, and coastal boundaries. It could increase involuntary migration, escalate tensions, and trigger conflicts—or it could foster more cooperation between nations.

Climate change has not featured prominently in development programs in Africa (e.g., Millennium Development Goals [MDGs] and National Poverty Reduction Strategies were developed around the turn of the century without reference to climate change). Nevertheless, a vast case-study adaptation literature has emerged in each of the climate-impact areas listed above, and important lessons have been learned. There is a wide spectrum of adaptation strategies, from indigenous modifications to urban flooding in Accra (Ghana) slums (furniture makers have designed wardrobes that allow residents to sit on top of them in times of flood) to an internationally funded farmers' helpline in Kenya ("M-Kilimo"), where farmers can receive expert agricultural advice, climate and weather information (in Swahili and other local languages), and advice on climate-related adaptations for land preparation, pest management, and the like. (See UNEP 2009 for a listing of other types of adaptation programs in Africa.) The major emphases in climate adaptation programs have been agriculture and disaster risk management, water resource use management, and coastal resources; in contrast, urban adaptations have barely been examined (McGray et al. 2007). In general, the links between climate change adaptation and policy were quite tenuous before this decade. Few projects set out to have a climate change focus from the initial planning stages. Rather, adaptation concerns are more serendipitous or are presumed to be addressed with existing policy, as opposed to implementing climate adaptation strategies in a more focused manner.

There is an adaptation deficit throughout Africa, meaning the inability to deal with climate change (i.e., weak institutional capacity, minimal linkages between public and private sectors, and governance structures not organized to deal with climate change). One of the core issues is inadequate physical infrastructure (e.g., lack of all-weather roads, unreliable or nonexistent electricity, the poor quality of buildings), and it is impossible to "climate-proof" infrastructure that is not there. Not surprisingly, African policymakers have prioritized urgent development challenges such as endemic poverty, lack of capital, and responses to natural disasters. Concerns about climate change that will manifest their full effects in 30 to 100 years are of low priority and beyond the electoral calendar. African policymakers have yet to begin to integrate climate change into their development agendas.

## AFRICAN CITIES AND CLIMATE CHANGE: AN URGENT AGENDA

For years, the focus of proposed strategies for dealing with climate change has been the nation-state (e.g., IPCC approaches provide advice to national governments). Nevertheless, countries have been mostly unsuccessful in brokering comprehensive agreements and in taking action. City-based climate change activism has emerged as a new approach, and progress has been quite rapid.

Several international city-based climate networks have come into existence. The C40 Cities Climate Leadership Group, launched in 2005 (with 40 original members but now expanded to 56), is the foremost reporting group on large cities and climate. Johannesburg is on the elite steering committee, and Addis Ababa and Lagos also participate in the C40 group. The World Mayors Council on Climate Change, a 60-member group, was also launched in 2005 and includes several African mayors, from cities in South Africa, Cameroon, Nigeria, Kenya, and Uganda. In addition, international urban research networks with African links have become more prominent in contributing knowledge and data on a range of cities and climate change issues. For instance, the Urban Climate Change Research Network, led by the Earth Institute at Columbia University, focuses on integrating climate risk into city development strategies and plans. This group engages leading Africa-based researchers in smaller African cities, for example Kampala, Dar es Salaam, Maputo, Durban, and Nairobi.

Climate change is already affecting cities and their residents, and greater impacts are projected as climate extremes and variability increase. Africa's urban poor are more likely to live in high-risk zones and are less able to move in the event of a disaster. A sea-level rise of 39.4 inches (1 m) could displace 3.6 million people in Lagos alone (Rosenzweig et al. 2011) and change the spatial distribution and density of both formal and informal urban settlements. Two critical urban climate change issues are (1) how cities contribute to climate change and (2) how cities are affected by climate change.

It is estimated that cities contribute in the range of 31% to 41% of total GHG (UN-HABITAT 2011:51). African cities contribute less to global GHG than other cities, based on available data. The latest C40 African city data show that Johannesburg releases 34 metric tons of emissions, followed by Lagos at 27 metric tons and Addis Ababa at 3 metric tons (World Bank 2010:68). Comparatively, New York City releases 196 metric tons, Tokyo 174, and Beijing 110. Lagos is among the C40's most inefficient cities in producing GHG as a share of GDP, just behind the world's least efficient large cities (Beijing, Shanghai, and Moscow). The three most inefficient cities share a similar global manufacturing concentration, which is absent in Africa. Instead, most emissions in urban Africa result from the transportation and energy sectors. Furthermore, by necessity, the majority of urban populations are poor and engage in informal economic activities that contribute fewer GHG. The large carbon footprint in Lagos, the most populous and leading economic center in the oil-producing state of Nigeria, is largely due to vehicular traffic, which contributes the lion's share of emissions. Government petroleum subsidies and the spatial organization of Lagos, with extensive urban sprawl, add to its high emissions profile.

Beyond the C40 cities, the city of Cape Town has conducted the most comprehensive African profile of local emissions (UN-HABITAT 2011). The Cape Town study determined that electricity use accounted for the largest share of emissions (69%), mainly because most electricity was coal-fired and inefficient. Transport accounted for the second highest GHG amount (petroleum 17%, diesel 9%) (UN-HABITAT 2011:50). Emissions in Cape Town were distributed very unevenly across the city. Areas of the urban poor accounted for very low levels of GHG emissions. Indeed, informal urban livelihoods that center on recycling waste may actually generate negative emissions (using recycled materials for building dwellings [shacks], and arts, crafts, and curios, and for extracting components and materials from discarded computers, phones, and other electronic devices). In comparison, affluent Capetonians consume a disproportionate share of resources (e.g., fuel for heating, cooling, transportation, and other items with embedded carbon, such as imported food). The Cape Town study illustrates that urban environments display highly differentiated industrial, residential, and income patterns, with each producing complex carbon footprint mosaics.

Three categories of impacts of urban climate change can be distinguished. The first involves physical risks, such as sea-level rises and frequency of extreme weather events (cyclones, heavy precipitation, flooding, landslides, and heat waves and droughts). For example, Johannesburg is likely to become significantly hotter and more humid in the future, with average maximum temperatures rising by 2.3 degrees Celsius in the near future and projected increases in extreme storm events and the length of the rainy season (City of Johannesburg 2011).

Second, cities may face difficulties maintaining basic services. African cities are already severely stretched in terms of supplying basic services such as water and sanitation to urban residents: many residential areas are not served by municipalities and must rely on informal delivery operators. Climate change will most likely have negative effects on water supply, sanitation systems, energy provision, and transportation systems, collectively triggering great disruptions in urban livelihoods (many of which are already focused on mere survival). In some cases, massive migration induced by climate change could take place; climate forecasting, however, is not yet able to effectively predict when and how cities could empty and where migrants would go—to rural areas, to bigger or smaller urban centers, and/or across international borders.

A third challenge is how to deal with existing and ongoing urbanization in conjunction with future climate-related issues. For example, a key concern is how to plan for new infrastructure, taking climate change adaptation and mitigation into account while retrofitting existing infrastructure to make it more climate-resilient.

Cities must play a more central role in global responses to climate change. At present there are limited pathways for cities to engage directly in global climate change policy and to receive funding for adaptation and mitigation activities. UN-HABITAT is piloting a climate change initiative in four cities, two of which are in Africa: Kampala and Maputo. Maputo is a coastal city affected by frequent flooding and thus at risk from sea-level rise; Kampala lies in the interior but is also affected by flooding and the degradation of precarious hill slopes. A key goal of the UN-HABITAT initiative is to share knowledge on adaptation strategies with cities, but a broader and larger concerted effort is needed. Urban governments still have a limited presence in global climate talks (e.g., COP17). Clearly, the linkages among climate change, cities, and development require much rethinking about the future and how to connect with more diverse groups in Africa. UN-HABITAT (2011:183) is now boldly calling for new approaches to ensure that responses to climate change are "catalysts of socially inclusive, economically productive and environmental friendly development." In the past, urban planning could be based on a fairly constrained set of predictable futures, but not anymore: climate change has driven humanity into uncharted territory.

## GLOBAL CLIMATE STALEMATE AND FUNDING INTRANSIGENCE: AFRICAN VOICES

It is unjust that Africa has contributed the least to world carbon emissions yet is the most vulnerable to the effects of climate change. Africa is responsible for 4.67% of global emissions, and if South Africa is excluded (the country accounts for 42% of Africa's emissions), the region's contribution to GHG is only 3.2%. Western European carbon dioxide emissions per person are 4.5 tons compared with Africa's (excluding South Africa) 0.6 per person. Africa is also unique in that more than 60% of its emissions result from deforestation and land degradation.

African emissions are out of sync with the global climate change consequences that Africans are already facing. For example, Namibia contributed 0.05% of global emissions, but temperatures there rose at three times the global average during the 20th century. In a fair world where all people have equal rights, Africa would have considerable rights to emit GHG; although these have not been exercised, they could have a tradable value (this has been called the cap-and-trade system). Meanwhile, the largest polluters, the European Union (26%), the United States (23.3%), and China (15.3%), continue to produce the lion's share of global emissions and to promote coping and resilience for Africans.

For the most part, the Global North favors a market ideology of emissions quotas, but these may not be agreed upon before 2015 and, in the best-case scenario, would not become effective until 2020. Despite the warnings of scientists, the policy dimension of global climate binding agreements is a procrastinator's paradise. Climate justice activists contend that stalling on emissions agreements is detrimental for Africa. Their slogan at the COP17 meetings was "Don't Kill Africa." They argue that corporate interests are drowning out the voices of ordinary people in the ears of global leaders as the world sleepwalks toward temperatures that average several degrees higher than today's. In the meantime, the Global North colonializes atmospheric space in the same manner that they colonized land space some time ago (Bassey 2012).

Obviously, the climate change challenge is not strictly a technical one: serious ethical issues are involved. Industrial activities are an important engine of the global economy, but they cause serious environmental problems; many of these are most evident in poor countries, which are the least equipped to deal with them. A central dilemma of climate change is that the rich and the poor have different perspectives. The wealthy emphasize mitigation and adaptation strategies, which transmits the message to the children of Africa that their inheritance will be environmental disasters and offers advice on how to cope and adapt. One could not imagine the wealthy Global North sending a signal of gloom and doom to its own children. It is in Africans' interest to reverse the argument on global climate change, from being a threat to being an opportunity to address and resolve a host of urgent problems to allow for a better and brighter future.

From an African perspective, the critical question is whether the Global North, after reaping long and sustained industrialization benefits without emissions controls, owes anything to African countries, which will experience the harshest impacts of climate change. An answer to this question raises serious financial obligations. It is often quoted that US$60 billion to $65 billion would be necessary annually to fund mitigation and adaptation in developing countries, and some suggest US$100 billion if climate-proofing the MDGs becomes the priority (Frankhauser and Schmidt-Traub 2010). Whatever the total, the costs are going to be extraordinarily high. The Durban COP17 talks produced an agreement to set up a Green Climate Fund; however, monies have yet to be allocated to the fund. Africans are now calling for direct transfers; nongovernmental organizations (NGOs), however, argue that the transfers should be directed to communities rather than to national governments.

African leaders are vocal about the obligations of the Global North to Africa and call for more funding for adaptation and mitigation. Funding priorities differ: African leaders prefer to see a large amount allocated to adaptation, whereas the Global North prefers mitigation and wants to extract something in return for any financial contributions. According to Oxfam (2010), less than 10% of climate funds are spent on helping people in vulnerable countries adapt to the impacts of climate change. Most of the money goes toward climate research and to climate-proofing existing development projects. At present, some 20 climate funds operate without any central authority to coordinate the distribution of their monies. Missing is global climate leadership that can marshal worldwide climate financing, reduce duplicate efforts and ensure an adequate balance among adaptation, mitigation, and vulnerable places.

African leaders already dealing with the consequences of climate change are acutely aware that future climate change threatens to roll back recent development gains. Yoweri Museveni, the President of Uganda, labels climate change as "an act of aggression by the developed world against the developing world" and demands compensation for Africa for global warming-inflicted damage (quoted in Brown, Hammill, and Mcleman 2007:1142). Kaire Mbuende, the Namibian representative to the UN, has called developed countries' GHG emissions "low intensity biological or chemical warfare." African leaders routinely make strong arguments to alert the Global North to forget about "making poverty history" as climate change will "make poverty permanent." However, global interests are known to attempt to assign a lower priority to African unity than climate summits. For example, the media report (Bassey 2010) that Kenya's prime minister had a key climate speech written by a Japanese economic advisor on secondment to the prime minister's office, and Wikileaks cables suggest that promises of aid were tied to Ethiopian support of the Copenhagen Accord.

Leading Nigerian environmental activist Nnimmo Bassey (Right Livelihood Laureate—"the Alternative Nobel Prize") argues that failure to reach an international agreement on carbon emissions and delays in decisions on emissions until 2020 are akin to "a death sentence for Africa." Bassey describes the UN climate change conference in Durban in December 2011 as "a big platform of hypocrisy, a lack of seriousness, a lack of recognition that Africa is so heavily impacted."

Indigenous declarations on behalf of groups within countries (e.g., pastoralists and slum-dwellers) are now more frequent as civilian society organizes to speak out on climate issues (See Box 11.2). Activists in Kenya and South Africa have played a leading role in articulating positions centering on climate justice. The indigenous vision of nature centers on a spiritual and cultural connection: Mother Earth is to be respected and revered

## BOX 11.2 INDIGENOUS KNOWLEDGE IN CLIMATE CHANGE: MITIGATION AND ADAPTATION STRATEGIES IN THE AFRICAN SAHEL

The Sahel is a transitional semiarid region of Africa that lies across major parts of ten countries (Senegal, the Gambia, Mauritania, Mali, Burkina Faso, Chad, Nigeria, Cameroon, Eritrea, and Sudan). The word "Sahel" is derived from Arabic and means "shore or border," describing the appearance of vegetation that borders the southernmost fringes of the Sahara. The term was added to the geography lexicon in the 1970s following widespread famine in the area (250,000 people and 3.5 million cattle perished), caused by rapid and unforeseen desiccation in this area known as the Sahel. Much earlier, however, Stebbing (1935) had written a pioneering article to warn colonial governments about the "encroaching Sahara" as one of the principal environmental challenges that West African colonials will encounter.

The Sahel extends from the Atlantic Ocean in the west to the Red Sea in the east, and the belt averages 200 to 700 miles wide (320 to 1,120 km) (roughly twice the size of Texas). A zone of often-extreme conditions, the Sahel has two hot seasons—from approximately February to April and September to October—punctuated by a short rainy season between May and August. It is a transitional area of semiarid grasslands, pocked with shrubbery and occasional trees; vegetation density increases markedly in a southerly direction. Perhaps most critically, the Sahel experiences major weather variations and fluctuations with a highly irregular rainfall: annual rainfall varies from around 7.8 inches (200 mm) in the north to 23.6 inches (600 mm) in the south. Since the 1960s, rainfall has been gradually declining, but whether this trend indicates that the Sahel is undergoing desertification (human-induced degradation of land in dry areas) is a subject of scientific debate.

Despite its harsh climate, the Sahel is home to nomadic pastoralists, notably the Tuareg and the Fulani. There are several prominent Sahelian cities (e.g., Bamako, Ouagadougou, and Niamey), and urban residents have long depended on rain-fed agriculture, irrigation from year-round water sources (e.g., Lake Chad and Niger River), and long-distance trade. Rapid population growth (about 3.1%) is under way in the region, and urbanization is increasing at very high rates (about 7%) (Nyong, Adesina, and Elasha 2007). Growth rates in the context of recent experiences with severe droughts raise concerns about the sustainability of the Sahel to support its populations, especially since most inhabitants rely on land-based livelihoods. Food insecurity, malnutrition, and chronic poverty are endemic. For example, the 2012 drought put more than 16 million people directly at risk (FAO 2012).

Up to the 1980s, scientists and researchers conventionally placed the blame for climatic and environmental change in the Sahel on indigenous land-use practices (overgrazing, overcultivation, deforestation, and mismanagement of irrigation), even though traditional methods had been employed successfully for millennia. Desertification received much attention, conjuring up images of the Sahara advancing southward, smothering villages and engulfing farmlands and pastures. Encroachment was recorded as so advanced that large swathes of Mali and Mauritania were forecasted to become uninhabitable (peri-urban residents now take on a daily chore of shoveling sand to keep the desert at bay) (De Blij et al. 2013). However, evidence from field observations in the area have suggested that the desertification theory was disseminated even in a situation of scientific uncertainty; it has proven hard to dislodge even when new evidence undermined that interpretation. It is not so much that desertification did not exist (there are well-defined areas that have suffered from land degradation), but that it is not prevalent over the entire area. The fringe of the Sahel is more akin to an ebbing and flowing tide, shifting south with dry spells and moving north in wetter periods.

In the 1990s, a counter-narrative emerged on the basis of first-hand observations of dryland farmers and herders who showed adaptation strategies in various regional contexts with regard to dynamic and unpredictable environments. Case studies in various subareas of the Sahel emphasized the role of indigenous knowledge bases passed on by word of mouth from one generation to the next.

Indigenous knowledge systems (which integrate intricate subsystems of gathering, predicting, interpreting, and decision making) are applied to weather forecasting. Farmers have been shown to conserve soil carbon by using no-tilling practices in cultivation, mulching, and other soil management techniques. For example, natural mulches moderate soil temperatures and extremes, suppress diseases and harmful pests, and conserve soil moisture. Adaptation strategies used by pastoralists in severe droughts include the use of emergency fodder for livestock, culling weak livestock for food, and shifting toward multi-stocking herds (when affordable), whereby herd composition can be rebalanced from cattle to sheep and goats because of their smaller feed requirements. Nomadic mobility reduces pressure on low-carrying-capacity grazing areas through cyclical movement from dry northern areas to wetter southern areas. Farmers recognize and respond to changes in climate by maintaining flexible strategies with short- and long-cycle crop varieties. Some Malians collect wild foods and consume the less palatable ones during droughts (Baro and Batterbury 2005). The baobab tree, for instance, is a source of food (leaves are used for vegetable sauce and the fruit pulp is used to make porridge and to flavor drinks), clothing, and medicine, and the root bark is used to making fishing nets, mats, and sacks. Coping with the 1997 drought in Burkina Faso, farmers implemented a range of food-saving strategies and borrowed and mortgaged against the following year's crops. In subsequent years, they readjusted their farming strategies and planted more varieties of drought-resistant crops (Mertz et al. 2009).

Some farmers, however, misinterpreted short-term weather changes and established land-use practices that accentuated environmental damage. Some farmers desperate to earn extra income and feed growing populations expanded cash cropping and cattle herding into marginal lands during wetter years, plowing furiously and introducing too many grazing animals. Such careless practices, deviating from indigenous customs, led to swift soil erosion and overly intensive agricultural methods that were unsupportable in dry years (De Blij et al. 2013).

There is a growing consensus that agricultural adaptation practices are only part of livelihood diversification strategies. Within the region, migration is a common diversification strategy during the dry seasons, but recently there has been an increase in population flows out of the Sahel, and not just to neighboring countries. For example, Sahelians are employed as mine workers in South Africa and as oilfield workers in Gabon. Remittances are critical for many families; one research study reported that Sahelian migrants in France remitted up to 15% of their incomes (twice the rate of migrants from other developing countries) (Scheffran, Marmer, and Sow 2011:123). Rural livelihoods are negotiated within the context of climatic and economic uncertainty, conflict, and poor governance in many parts of the region. Petty trading and handicrafts are among the nonfarm diversification and coping strategies. In Sudan and Ethiopia, farmers have diversified livelihoods with income from gum and resins. In Sudan, gum is obtained from *Acacia senegal,* a tree that grows throughout the Sahel, and gum arabic is a cash crop sold mainly to European and North American markets as an ingredient for soft drinks and medicines. Overall, livelihood research has shown a mixing and matching of strategies that explore both local and more distant opportunities. Whatever the successes of short-term adaptation, there can be no doubt that, over the longer term, reducing vulnerability to drought will depend further on developing the region's national and urban economies, which remain among the poorest and least diversified in the world.

rather than dominated and controlled. Climate justice advocates recognize a planet in crisis and lay the blame on the global economic system that drives unsustainable industrialization, food raising, forestry, and agricultural practices. Declarations made by indigenous groups focus on local attachment rather than on detached disregard for local activities and decisions (O'Lear 2010). At the Durban climate talks, activists argued (invoking the spirit of the Occupy movement, as Occupy COP17) that the voices of 99% of the world's population were not being heard, and that UN climate talks were taken over by governments corrupted by corporate influence.

## CONCLUSIONS

Climate change is a complex and multidimensional global challenge. Scientific studies on Africa are patchy, and current regional climate models need considerable refinement. A general acceptance that global climate and regional models are the best way to understand climate change is mismatched to the needs of communities and individuals and their support structures (e.g., civil society), all acting locally (O'Lear 2010). Current climate models prioritize state-level data and decision making even though it is apparent that climate change is not confined to national territories. A glaring weakness of state-centered analyses and state responses is that they underplay complex relationships and spatial flows across borders: peoples, technology, pollution, ideas, financial resources, etc., all move with ease, and responses to climate change may trigger significant international movement. Far more creative thinking is needed. Tackling climate change requires that people both act together and act differently. The C40 climate leadership group is one small step in the right direction, but there is a pressing need for city networks to be expanded.

For Africans to deal effectively with climate change, they should be not merely the recipients of advice but also participants in discussions on how best to deal with the challenge. Africans suffer from an excess of advice and a paucity of opportunities for real engagement in global issues. Africa's low level of climate science expertise needs to be remedied. Few African scientists publish regularly in international journals. Expertise on Africa (which includes a large number of expatriates) is located outside the region, with African experts on the human dimensions of climate change scattered across the world. The leading cutting-edge climate research centers are all now located in the Global North. Severely lacking are climate research centers on the African continent, beacons that could attract the brightest African minds, boost self-confidence, and help Africans find their own voices on climate change issues. The first African-based center for climate research, the Climate Systems Analysis Group at the University of Cape Town, only came into existence in 2009. A more concerted effort to tackle climate understanding in Africa will require several additional and even competing centers of Africa-based knowledge: the Urban Climate Change Research Network may be a model worth replicating. The advancement of knowledge is going to require significant initial investments to collect and analyze climate data from all of Africa.

According to IPCC predictions, Africa will be the region most severely affected by climate change. The stakes are particularly high in Africa because the majority of people earn their livelihood from the land, and rapid urbanization means that urban infrastructures are poorly equipped to deal with climate change. African poverty contexts mean that the types of adaptation programs suitable in the Global North are unlikely to be transferable to the region. Financial costs of adaptation to climate change in Africa will be substantial, and, of course, the amount of funding required for climate-proofing will depend on the magnitude and timing of climate changes and their subsequent impacts. Tens of billions of dollars in annual funding are needed to help Africans prepare for and respond to the unavoidable impacts of climate change.

Africa's projected economic and population growth rates indicate the region's contribution to global warming will change. If adequate energy can be supplied to urban Africa, African cities are going to become more important in terms of contributing GHGs. A continent containing close to 2 billion people (in 2050), more than half urban, will consume far more energy than it did at the end of the 20th century. Precisely because Africa today is the poorest and least industrialized world region, it may surpass the others in terms of the expansion of its energy needs over the next 50 years. Therefore, it is also in Africa that the battle against global warming will be waged.

The rapid and spectacular rise of our species, *Homo sapiens*, occurred in an earlier area of climate change. Human ingenuity has always been a hallmark of progress, from the invention of farming to the building of cities to the electronic marvels of today. Humans have been capable of fantastic feats in the past, so there is no reason why they should not be able to adapt to climate change and improve climate resilience everywhere—but the process is only just beginning.

## REFERENCES

Amnesty International. 2004. "Sudan: Darfur. Rape as Weapon of War: Sexual Violence and its Consequences." Report, Available at: http://www.amnesty.org/en/library/info/AFR54/076/2004 (accessed April 2, 2012).

Baro, M., and S. Batterbury. 2005. "Land-Based Livelihoods." In *Toward A New Map of Africa*, eds. B. Wisner, C. Toulmin, and R. Chitiga, pp. 51–65. Sterling: Earthscan.

Bassey, N. 2010. "Cancun Climate Summit: How Africa's Voice Has Been Hijacked." *The Guardian*, December 10. Available at: http://www.guardian.co.uk/environment/2010/dec/10/cancun-climate-change-summit-africa) (accessed April 3, 2012).

Bassey, N. 2012. *To Cook a Continent: Destructive Extraction and Climate Crisis in Africa*. Oxford: Pambazuka Press.

Bohle, H., T. Downing, and M. Watts. 1994. "Climate Change and Social Vulnerability: The Sociology and Geography of Food Insecurity." *Global Environmental Change* 4:37–48.

Boko, M., I. Niang, A. Nyong, C. Vogel, A. Githeko, M. Medany, B. Osman-Elasha, R. Tabo, and P. Yanda. 2007. "Africa." In *Climate Change 2007: Impacts, Adaptation and Vulnerability. Contribution of Working Group 11 to the Fourth Assessment Report of the Intergovernmental Panel on Climate Change*, eds. M. Parry, O. Canziani, J. Palutikof, P. van der Linden, and C. Hanson, pp. 433–467. Cambridge: Cambridge University Press.

Brown, O., A. Hammill, and R. Mcleman. 2007. "Climate Change as the 'New' Security Threat: Implication for Africa." *International Affairs* 83:1141–1154.

Busby, J., C. Raleigh, and I. Salehyan. 2014. "The Political Geography of Climate Vulnerability, Conflict, and Aid in Africa." In *Peace and Conflict*, eds. D. Backer, P. Huth and J. Wilkenfield. Boulder: Paradigm Publishers. Forthcoming.

Christensen, J., B. Hewitson, A. Busuioc, A. Chen, X. Gao, I. Held, R. Jones, R. Kolli, W. Kwon,

R. Laprise, V. Magaña Rueda, L. Mearns, C. G. Menéndez, J. Räisänen, A. Rinke, A. Sarr, and P. Whetton 2007. "Regional Climate Projections." In *Climate Change 2007: The Physical Science Basis. Contribution of Working Group 1 to the Fourth Assessment Report of the Intergovernmental Panel on Climate Change*, eds. S. Solomon, D. Qin, M. Manning, Z. Chen, M. Marquis, K. Averyt, M. Tignor, and H. Miller, pp. 849–940. New York: Cambridge University Press.

City of Johannesburg. 2011. "Adaptation and Vulnerability." Presentation at the C40 Summit, Sao Paulo, Brazil, 31 May–2 June 2011. Available at: http://c40saopaulosummit.com/cidadesc40/pdf/11_Johannesburg_Adaptation%20%20Vulnerability.pdf (accessed September 30, 2013).

Collier, P., G. Conway, and T. Venables. 2008. "Climate Change and Africa." *Oxford Review of Economic Policy* 24:337–353.

De Blij, H, P. Muller, J. Burt, and J. Mason 2013. *Physical Geography*. New York: Oxford University Press.

De Waal, A. 2007. "Is Climate Change the Culprit for Darfur?" Available at http://africanarguments.org/2007/06/25/is-climate-change-the-culprit-for-darfur/ (accessed March 14, 2012).

Ericksen, P, R. Thorton, A. Notenbaert, L. Cramer, P. Jones, and M. Herrero. 2011. "Mapping Hotspots of Climate Change and Food Insecurity in the Global Tropics." CCAFS Report no. 5. Available at http://ccafs.cgiar.org/resources/climate_hotspots (accessed March 14, 2012).

FAO. 2012. "Drought, High Food Prices and Chronic Poverty." Available at http://www.fao.org/crisis/sahel/en/ (accessed March 20, 2012).

Frankhauser, S., and G. Schmidt-Traub. 2010. "From Adaptation to Climate-Resilient Development: The Costs of Climate-Proofing the Millennium Development Goals in Africa." Available at http://www.cccep.ac.uk/Publications/Policy/docs/PPFankhauseretal_costs-climate-proofing.pdf (accessed March 16, 2012).

The Guardian 2011. "Top 50 Twitter climate accounts to follow." Available at http://www.guardian.co.uk/environment/blog/2010/may/11/top-50-twitter-climate-accounts (accessed December 6, 2013).

InterAcademy Council. 2010. *Climate Change Assessments: Review of the Process and Procedures of the IPCC Third Assessment Report*. Amsterdam: InterAcademy Council.

Intergovernmental Panel on Climate Change (IPCC). 2013. "Working Group 1 Contribution to the IPCC Fifth Assessment Report. Climate Change 2013: The Physical Science Basis. Summary for Policymakers." Available at http://www.climatechange2013.org/images/uploads/WGIAR5_WGI-12Doc2b_FinalDraft_All.pdf (accessed September 30, 2013).

Kaser, G., D. Hardy, T. Molg, R. Bradley, and T. Hyera. 2004. "Modern Glacier Retreat on Kilimanjaro as Evidence of Climate Change: Observations and Facts." *International Journal of Climatology* 24:329–339.

McGray, H., R. Bradley, and A. Hammill, with E. Schipper and J. Parry. 2007. *Weathering the Storm: Options for Framing Adaptation and Development*. Washington, D.C.: World Resources Institute.

Mertz, O, C. Mbow, A. Reenberg, and A. Diouf. 2009. "Farmers' Perception of Climate Change and Agricultural Adaptation Strategies in Rural Sahel." *Environmental Management* 43(5):804–816.

Nyong, A., F. Adesina, and B. Elasha. 2007. "The Value of Indigenous Knowledge in Climate Change Mitigation and Adaptation Strategies in the African Sahel." *Mitigation and Adaptation Strategies for Global Change* 12:787–797.

O'Lear, S., 2010. *Environmental Politics: Scale and Power*. New York: Cambridge University Press.

Oxfam. 2010. "Righting Two Wrongs. Making a New Global Climate Fund Work for Poor People." Report available at http://www.oxfam.org/sites/www.oxfam.org/files/righting-two-wrongs-global-climate-fund-061010.pdf

Parnell, S., and R. Walawege. 2011. "Sub-Saharan African Urbanization and Global Environmental Change." *Global Environmental Change* 21:12–20.

Rosenzweig, S., W. Solecki, S. Hammer, and S. Mehrotra. 2011. *Climate Change and Cities: First Assessment Report of the Urban Climate Change Research Network*. New York. Cambridge University Press.

Rouault, M., S. Sen Roy, and R. Balling 2013. "The Diurnal Cycle of Rainfall in South Africa in the Austral Summer." *International Journal of Climatology* 33:770–777.

Sachs, J. 2008. *Commonwealth: Economics for a Crowded Planet*. London: Allen Lane.

Scheffran, J., E. Marmer, and P. Sow. 2011. "Migration as a Contribution to Resilience and Innovation in Climate Adaptation: Social Networks and Co-development in Northwest Africa." *Applied Geography* 33:119–127.

Schwartz, P., and D. Randall. 2004. "An Abrupt Climate Change Scenario and Its Implications for United States National Security." Report to the Pentagon. Available at http://www.gbn.com/articles/pdfs/Abrupt%20Climate%20Change%20February%202004.pdf (accessed March 26, 2012).

Solomon, L. 2008. The Deniers. *The World Renowned Scientists Who Stood Up Against Global Warming, Hysteria, Political Persecution, and Fraud and Those Who Are Too Fearful to Do*. Minneapolis: Richard Vigilante Books.

Spencer, R. 2010. *Climate Confusion: How Global Warming Hysteria Leads to Bad Science, Pandering Politicians and Misguided Politicians that Hurt the Poor*. New York: Encounter Books.

Stebbing, E. 1935. "The Encroaching Sahara: The Threat to the West African Colonies." *Geographical Journal* 85(6):508–524.

Toulmin, C. 2009. *Climate Change in Africa*. New York: Zed Press.

United Nations Environmental Programme (UNEP). 2005. "Climate Change and Vulnerability in Africa." Available at: http://www.grida.no/graphicslib/detail/climate-change-vulnerability-in-africa_7239 (accessed April 2, 2012).

UNEP. 2009. *A Preliminary Stocktaking: Organization and Projects Focused on Climate Change Adaptation in Africa*. Available at http://www.unep.org/roa/amcen/docs/AMCEN_Events/climate-change/UNEPAfricaStocktaking.pdf (accessed March 20, 2012).

UN- HABITAT. 2011. *Cities and Climate Change: Global Report on Human Settlements 2011*. Available at http://www.unhabitat.org/pmss/listItemDetails.aspx?publicationID=3085

Verhoeven, H. 2011. "Climate Change, Conflict and Development in Sudan: Global Neo-Malthusian Narratives and Local Power." *Development and Change* 42:679–707.

Verschuren, D., J. Sinninghe Damsté, J. Moernaut, I. Kristen, M. Blaauw, M. Fagot, G. Haug, and CHALLACEA project members. 2009. "Half-Precessional Dynamics of Monsoon Rainfall near the East African Equator." *Nature* 462: 637–641.

World Bank. 2010. *Cities and Climate Change: An Urgent Agenda*. Washington, D.C.: World Bank.

Yanda, P., and C. Mubaya. 2011. *Managing a Changing Climate in Africa: Local Level Vulnerabilities and Adaptation Experiences*. Dar es Salaam: Mkuki na Nyota.

## WEBSITES

C-40 Cities Climate Leadership Group http://live.c40cities.org (accessed December 6, 2013). Network of mega-cities to combat and mitigate the effects of climate change.

Climate Skeptic http://www.climate-skeptic.com (accessed December 6, 2013).

Climate Systems Analysis Group (CSAG) http://www.csag.uct.ac.za (accessed December 6, 2013). Climate science analysis group from University of Cape Town.

Greenpeace http://www.greenpeace.org/usa/en/campaigns/global-warming-and-energy/exxon-secrets/faq/ (accessed December 6, 2013).

Intergovernmental Panel on Climate Change (IPPC) http://www.ipcc.ch (accessed December 6, 2013).

KILN, "*The Carbon Map*" http://www.carbonmap.org (accessed December 6, 2013).

M-Kilimo http://www.m-kilimo.com (accessed December 6, 2013). Kenyan farmers helpline that provides agricultural information and support.

Save Darfur http://www.savedarfur.org (accessed December 8, 2013). NGO coalition to raise awareness and advocate for the people of Darfur, Sudan.

Tck Tck Tck http://tcktcktck.org (accessed December 6, 2013). Network of NGOs that call for action on climate change.

United Nations Environment Program (UNEP) 2005. "Climate change vulnerability, Africa." Available at http://www.grida.no/publicafrications/vg/africa (accessed March 11, 2014).

Urban Climate Change Research Network http://uccrn.org (accessed December 6, 2013). Consortium of individuals and institutions dedicated to the analysis of climate change mitigation.

World Mayors Council on Climate Change http://www.worldmayorscouncil.org (accessed December 6, 2013).

CHAPTER 12

# CHINA AND AFRICA

## INTRODUCTION

In September 2000, when Chinese President Jiang Zemin attended the United Nations (UN) Millennial Summit in New York (which led to the Millennium Development Goals [MDGs]), no doubt he kept in mind that Beijing would hold its own meeting a month later to launch the new Forum on China–Africa Cooperation (Brautigam 2009). Such a parallel mode of operating outside of Western international frameworks characterizes the China–Africa dynamic.

The Forum on China–Africa Cooperation has set the stage for institutionalizing Sino-African relations. Over the years, its activities have expanded to include commitments to increase trade, investment, and aid; strengthen development cooperation; facilitate debt forgiveness; build a broad range of infrastructure (roads, schools, hospitals); and cooperate in natural and human resource development (from mineral extraction to education). The rising importance of the region to China was further codified in the first-ever "White Paper on China's Africa Policy" in 2006, and regular updated white papers have been issued since (the most recent in 2013). Since 2000, there appears to be a deliberately constructed convergence between Africa's development needs and Chinese economic interests (Alden 2012).

China's position in Africa has surged; it has become Africa's largest trading partner (moving from modest trade relations 20 years ago), the leading investor, the major aid donor (surpassing the World Bank), a significant migrant-sending region, and a holder of significant interests in oil leases, mining, and timber concessions. Its two-way trade with Africa reached US$200 billion in 2012 and is overwhelmingly based on the extraction of oil and strategic minerals, which are exchanged for manufactured and consumer goods. Questions are being asked about whether China is displacing the United States and European powers as a major influence in the region (Carmody and Owusu 2007). At the very least, the "Washington consensus" is now counterbalanced by an alternative approach known as the "Beijing consensus," the essence of which is a more malleable infrastructural investment-with-aid engagement (Cheru and Obi 2010)—although it must be acknowledged that China does not require African states to emulate its values.

There is widespread belief among African policymakers and some scholars that increased global economic interest in the region can provide strategic options and a policy space for Africans that they had been compelled to surrender in the 1980s with the introduction of punitive structural adjustment programs. China's engagement with African countries has helped reignite the region's economies, and it has made some headway toward restoring Africa's agency within the international system (Alden 2012). South–South engagement offers a different enabling environment, options for maneuverability, and the possibility for Africans to chart a different development course and, many hope, their own.

China's engagement with Africa is a provocative topic. There have been waves of misinformation, hype, anecdotal reporting, and hasty and speculative conclusions about what the Chinese are doing in Africa. Hyperbolic media describe the relationship as "a love affair," "a fatal attraction," "China's African safari," "a Chinese takeaway," "new colonialism," "a new scramble," "a partnership," a "mixed blessing," and so forth. There is clearly a continuity of historical tropes for understanding China's relations with Africa, and China's actions are rarely framed within a 21st-century context of a different time and place, and a world where Africans have agency. Added to this, many hold a cultural bias and expect China to operate and behave in a similar way to Western powers (despite their different language and cultural traditions and lack of familiarity with Africans and vice versa). Clearly, East Asia and the United States rank as higher investment priorities for China, but Sino-African relations are becoming increasingly important, making Africa a pivotal theater for China's emerging 21st-century role. Some analysts claim it is ushering in Africa's second liberation from Western dictates.

At the same time, evidence is accumulating that African leaders are pulling back from specific Sino-African arrangements; reconfiguring mining and exploration territories and renegotiating agreements with Chinese corporations; enforcing laws that restrict non-national artisanal mining activity; and expelling Chinese residents engaged in illegal activities. An intensified crackdown on illegal non-national mining activities in Ghana has resulted in the repatriation of many Chinese nationals, the exact number of which is not public (somewhere in the range of 166 to 4,592 persons) (Hirsch 2013). African governments (e.g., Niger, Chad, and Gabon) are directly challenging Chinese oil companies over environmental destruction. For instance, the Gabonese government withdrew a Chinese permit for a significant oil field and handed the territory over to a newly created national company (Nossiter 2013). This evidence indicates that African governments are no longer the passive partners they were 10 years ago.

Relations between China and Africa bring excitement and anticipation as well as unease and suspicion about Chinese aid and state-sponsored projects. China has many things going for it: Beijing can engage African governments without the colonial hangover, and its stellar economic performance over the last 25 years and significant advances on many development indicators provide a different set of ideas for informing African leaders. China has become the world's largest holder of capital, with over US$2.4 trillion in foreign reserves, so it seems logical that it would deploy some of these funds in capital-starved and resource-rich regions like Africa (Alden 2012). Chinese interests are propelled by Africa's resource bounty (oil, natural gas, timber, coltan, copper, land, etc.) and by its own scarcity of resources (oil, food, land, strategic minerals). China's return to Africa (see next section) also coincides with the improved economic performance of many African states and the "Africa rising" narrative. Indeed, China's investments have contributed to African economic growth, and a growing Chinese media presence has helped shape a new narrative of Africa—"a land of opportunity."

China's growing footprint is transforming Africa in dramatic ways. The willingness of the Chinese government to provide an entire package of inducements, alongside a range of leasing or supporting agreements designed to meet elite-defined needs (ranging from sports stadiums to large-scale infrastructure projects), has proved to be crucial to securing all sorts of deals, many of which pertain to resource access. The Chinese interest in Africa also happened at a time when Africans were pondering their failure to develop over the past half-century and to keep pace with the Asian success stories. The search for an alternative transformative development model has led many Africans to inspect closely and learn from the lessons of Asian giants. Chinese companies are investing heavily in the neglected African infrastructure (building major transport, communication networks, dams) and filling critical infrastructure gaps with less bureaucracy and less funding and in less time. As a result, the West no longer has a monopoly over contemporary and future development in the region.

Even though China's footprint in Africa is extensive, data are sparse and notoriously unreliable. As a result, there is lots of conjecture about the development impacts of increased Chinese trade and investments as well as about Chinese migration to Africa. The China–Africa arena is very large and dynamic, and major new initiatives and pullbacks occur almost monthly. Some excellent sources include *The China in Africa Podcast* (http://china.buzzsprout.com) and Deborah Brautigam's blog *China in Africa: The Real Story*. This chapter reviews the history of China in Africa and concentrates on examining high-profile Chinese infrastructure projects (e.g., roads, stadiums, large residential developments), and it assesses Chinese entrepreneurs' forays into African informal economies and Africans' attempt to better position themselves in their own informal economies by establishing a toehold in an African enclave in Guangzhou, China, that has become known as "Chocolate City." Specific dyads such as the China–Zambia one are examined in detail to show how this relationship between China and Africa is ever-evolving.

## CHINA RETURNS TO AFRICA

China's engagement with Africa has a longer history and different origins than that of the Europeans. The Chinese recount how during the Ming Dynasty its mighty fleet (63 vessels and 28,000 men, each ship larger than Christopher Columbus') sailed to East Africa several times between 1418 and 1433 (Brautigam 2009). However, the Chinese neither colonized Africa nor took "one inch of land, not a slave, but a giraffe for the emperor to admire" and the "giant ships also took back African herbs and local medicinal compounds, perhaps to combat a series of epidemics raging in China at that time" (quoted in Brautigam 2009:23).

In a modern epilogue to this nonintrusive narrative, a Chinese delegation traveled to an ancient trading town off the coast of Kenya, Lamu, in 2002 and confirmed the claims of a local girl, Mwamaka Sharifu Lali ("Chinese Girl"), that she was descended from 15th-century Chinese sailors shipwrecked in the archipelago (Brautigam 2009:23). Subsequently, the Chinese Embassy in Nairobi awarded Mwamaka a trip to her ancestral homeland and provided a scholarship for her to study medicine (Brautigam 2009). This gesture signals the changing face of Chinese engagement in Africa.

Scholarly literature on China's emerging relationship with Africa can be divided into five prominent themes: (1) China's image of the new face of globalization; (2) its role in African development vis-à-vis its own development trajectory; (3) its quest for allies and friends; (4) its counterbalance to the West; and (5) its emergence as a responsible power, taking its rightful place in the international community.

There are several reasons for China's deepening of links to Africa. First, after the Tiananmen Square incident of June 4, 1989, China's attitudes toward the African region changed from benign neglect to renewed interest. Former Chinese Minister of Foreign Affairs Qian Qichen stated in his memoirs that "it was . . . our African friends who stood by us and extended a helping hand in the difficult times following the political turmoil in Beijing, when Western countries imposed sanctions on China" (quoted in Taylor 2012:26).

Second, the incredible expansion of the Chinese economy in the 1990s and 2000s compelled Chinese firms and corporations as well as ordinary Chinese entrepreneurs to embark on a concerted effort to discover markets and commercial opportunities overseas. In 1999. the Chinese government initiated a "Go Out" policy to encourage Chinese companies to invest overseas, and it became a cornerstone of China's own development strategy after 2001. This policy encourages Chinese firms to do business abroad while making Chinese firms more competitive by acquiring strategic assets, securing access to natural resources, and establishing new markets for Chinese exports. Initially, large state-owned construction companies led the way, but recently private firms in all sectors have been participating. For example, telecommunication giants (Huawei and ZTE) are building telecom infrastructure and networks throughout Africa, enabling wide network coverage even to rural locations.

Third, China solicited the support of African states at the UN, acquiring a numerical advantage that prevented hostile votes against China vis-à-vis its human rights record and ensuring that the Republic of Taiwan remains an unrecognized international outcast (Taylor 2012). Beijing maintains a One-China Policy that requires aid recipients to break official relations with Taiwan (the Republic of China) and to engage exclusively with the People's Republic of China (China) as the sole and legitimate representative of all of China, and, further, to accept Beijing as the legitimate representative of Taiwan or at least acknowledge its stance on the issue. China wages its diplomatic battle against Taiwan on African soil with ample enticements, and it has led to many defections. Only four African countries recognize Taiwan (Burkina Faso, Swaziland, Gambia, and São Tomé and Príncipe). African support for China in the International Olympic Committee helped Beijing secure the hosting of the 2008 Olympics.

Fourth, China wants to counterbalance the United States as an unchallenged hegemon. Beijing is keenly aware that China needs support from African and other developing world countries to enhance its global status and to avoid U.S. coercion and outside interference in internal Chinese affairs.

Fifth, China has political ambitions to be taken seriously as a responsible power, and Africa plays a key role in achieving this. China's capital infrastructural spending in Africa and its media infrastructure (print, Internet, and TV) are moving in tandem. Indeed, Chinese media coverage of the region is expanding while Western coverage is decreasing. China's news agency, Xinhua, established an African presence in Nairobi in 1986 and now operates in 28 African countries, and its services have expanded into the free mobile phone and African news market. Chinese news media in Africa enlarged in 2012: China Central Television launched CCTV Africa, and the newspaper *China Daily* launched *Africa Weekly*. There is now an impressive array of Chinese media presenting diverse Chinese viewpoints on Africa. They offer an alternative to, and come into tension with, traditional Western media and content. There is considerable debate about the Chinese media and whether it is a vehicle for pro-China agendas or whether it operates through a different cultural lens emphasizing positive reporting, with some censorship (e.g., Dalai Lama, Falun Gong, Tiananmen Square). Chinese media deliberately portray Africa in a more positive light, with narratives of the continent as a land of opportunity, contrasting sharply with the despair-and-tragedy narrative. In many ways, the Chinese news media narrative may be seeking to rebalance the negative Africa narrative that has prevailed for so long.

China has disbursed aid to almost every country in Africa, except Swaziland because of its position on Taiwan. Chinese economic cooperation emphasizing infrastructure with complete project financing is at odds with traditional donors, who have moved away from these "traditional" instruments toward more esoteric development goals such as good governance, transparency, and lack of corruption. China maintains a non-interference policy with aid recipients, which means that Beijing pays less attention to the negative effects of aid spending and that China is willing to support both rogue regimes (e.g., Zimbabwe and Sudan in the past) and authoritarian governments (e.g., Angola); human rights violations within African states are never questioned.

Beijing's contribution to Africa is a matter of interpretation. A key problem stems from China's lack of transparency about its aid disbursements. Official aid in China is regarded as a state secret, which heightens Western and (some) African concerns. Foreign aid is sensitive due to the ongoing diplomatic dispute with Taiwan, and it is culturally insensitive to call attention to overseas assistance given that China is also a developing country with its own poor. Brautigam (2009) calculates that China committed US$2.5 billion in 2009 to Africa (which includes debt relief), a contribution considerably lower than reported in the Western media. Nevertheless, China surpassed the World Bank as a lender to Africa in 2006 and overtook the United States as Africa's major trading partner in 2009. Africa now supplies 25% of China's oil.

Although Beijing has issued an official Africa policy, there are actually many "Chinas"—as, indeed, there are many "Africas" (a major argument of this book). Disaggregating them is essential when assessing what the Chinese are doing in Africa. The notion that China or the Chinese are "colonizing" Africa—an allegation raised by some Western and African commentators—is misleading. Such a claim is based on the assumption that there is an overarching grand strategy on the part of Beijing. It is reasonable to acknowledge that there are certain aspirations focused on quite specific facets of Sino-African ties (Taylor 2011). The most obvious example is state-owned enterprises' (SOEs') investment in African resource industries clearly connected with China's own present and future energy demands, but even here there are rivalries and competition between different energy companies. Beyond the energy sector, rivalries among Chinese provinces, municipalities, companies, and individuals play out on a daily basis, revealing as a myth the idea of a monolithic China relentlessly pursuing a single agenda (Taylor 2012).

Sino-African relations are more accurately understood as processes of globalization and the reintegration of China and Africa into the global economy, projects that initially enjoyed the enthusiastic support of Western capitalism. Paradoxically, the West now criticizes the Chinese for expanding into Africa and for using market principles.

The idea of China as a model for prosperity has captured the imagination of many ordinary Africans, although others worry about being overrun by the Chinese industrial juggernaut and swamped by dominant Chinese traders in African informal markets (Brautigam 2009). It is prudent for Africa's policymakers and scholars to approach China as an alternative development model realistically and with caution. The stages of growth/modernization model emphasized in the 1960s based on the U.S./Western experience was an abysmal failure. The intensification of ties with China may overstate the ability of Beijing to shepherd an alternate development model. Indeed, China's extraordinary growth was achieved with a capable strong state, nothing comparable to which exists in Africa. Indeed, some scholars (Taylor 2012:30) think that "China's model in Africa is not having a model," and it has widespread appeal because the ideology of neo-liberalism is widely resented, but a Chinese model per se does not exist.

## CHINESE MIGRATION TO AFRICA

Early Chinese migration to Africa is associated with the Dutch East India Company and its business interests in the Cape (South Africa). Small numbers of company slaves, contract workers, and convicts were sent to the area in the later part of the 17th century. A larger stream of 63,000 contract workers arrived in the late 19th century during the gold and diamond mining boom (Park 2009). Most contract miners were repatriated to China, but those who stayed were gradually integrated into South Africa's mixed-race population.

Contemporary China-to-Africa migrations have their roots in China's geopolitical strategy for Africa during the Cold War, when Beijing supported independence struggles. In the colonial aftermath, China promoted solidarity with newly independent African countries, and between 1960 and the 1980s, Beijing sent over 150,000 technicians and workers to engage in agricultural and infrastructural development.

The current phase of Chinese migrations is linked to Beijing's economic reforms, the liberalization of emigration legislation after 1985 (Park 2009), and the "Go Out" policy. Simultaneously, domestic Chinese perceptions of overseas Chinese switched from viewing them as traitors to treating them "as new vanguards of Third World anti-colonialism" (Mohan 2012:13). Chinese aid disbursements are linked to the temporary migration of Chinese project workers. China has sent approximately 20,000 health professionals, 10,000 agro-technicians, hundreds of teachers, and tens of thousands of workers for large-scale projects (e.g., the Lagos-to-Kano railway [11,000 workers] and the Sudan pipeline) (Park 2009). This migration activity is economically rather than ideologically driven.

Several heterogeneous streams of Chinese migrants in Africa coexist. Importantly, Chinese workers do not enter the local wage labor markets but work for Chinese businesses or establish their own businesses. Six types of Chinese workers can be discerned: (1) SOE personnel managing projects, contracts, and investments; (2) owners of private sector formal firms; (3) temporary labor migrants involved in public building and large infrastructural development projects undertaken by SOEs; (4) entrepreneurs, consisting of import/export agents, wholesalers, small business owners, and large numbers of small-time traders; (5) transit individuals who enter the region (but explore relocation opportunities in Europe and North America) because of less stringent regulations; some are legal entrants but many are not; and (6) agricultural workers (Spring and Jiao 2008).

Mung (2008) calculates that Chinese firms bring in 80,000 contract workers, but others (e.g., Park 2009) believe that the real number is more than double that. Most Chinese contract workers stay for the duration of their assignment (typically one to three years) before returning home. Labor migrants can be further subdivided by profession: most are semiskilled workers, and a smaller number are managerial and professional staff. A proportion of higher-skilled professionals stay on in host countries when their contracts end, joining the swell of migrants opening small businesses (Mung 2008).

The 700 to 800 large Chinese companies operating in Africa hire local labor but rely heavily on Chinese migrant labor for public works and oil, timber, and mining operations. Of this group, about 150 are SOEs that report directly to the central Chinese government. The remainder fall into a gray area, as subsidiaries of SOEs, companies owned by provincial and municipal governments, and partially privatized companies with the state participating as an influential shareholder. For example, the China National Offshore Oil Corporation and the State Grid Corporation of China are clearly state-owned enterprises under the SOE rubric, whereas the situations of the computer maker Lenovo and the appliance giant Haier are less clear-cut: although the state is the dominant shareholder, they operate more like private companies. In contrast, the West generally has a black-and-white definition of companies as government owned or privately owned.

Since 2005, thousands of Chinese firms entered the region as private firms (Gu 2011). More than 80% of Chinese firms with African operations are private small and medium-sized enterprises. These firms are more motivated to develop a new market or a manufacturing base rather than concentrate on resource extraction and infrastructural development (the focus of SOEs). Many private sector firms have no parent firm on the mainland, and others are only marginal players in the Chinese market. These firms are motivated by profit and compete intensely with each other. They are driven to tap African market opportunities and/or to establish positions to gain preferential market access to the European Union and United States (Gu 2011). They cannot be seen as conforming to a grand Chinese strategy.

The use of Chinese contract laborers is a hot-button issue in the China–Africa dynamic, given high unemployment rates among Africans, the low salaries paid by many Chinese companies, accusations of unfair labor practices, poor track records in corporate social responsibility, disregard for health and safety standards in African workplaces, and self-segregation of Chinese workers in employer-built compounds. Part of this tension is fueled by cultural differences. Another part is driven by surges in economic nationalism, which is a response, in part, to the erosion of light manufacturing jobs (especially textiles) in the African region in the face of global competition. This competition is saliently expressed by the presence of immigrant entrepreneurs who benefit from international networks that enable cheaper imports with more variety but that decimate local industries. Besides trade, the tension is manifested in fierce competition between Chinese and local professionals, seen, for example, in medical services between Western-trained versus Chinese-trained medical professionals/clinics (Spring and Jiao 2008).

Xenophobia can be exploited by individuals seeking to gain (politically and in other ways) from an anti-Chinese undercurrent. There have been flash points in several locations in the Democratic Republic of Congo (DRC), Angola, South Africa, and Ghana—and in Zambia, in particular. Most commonly, the tension is expressed by union opposition. In 2012, South African transport unions thwarted a large Chinese investment for a Johannesburg manufacturing plant to produce minibus taxis for South African/Africa markets, despite backing from the South African government, which courted the investment as part of a national reindustrialization strategy. Opinion polls about the Chinese in Africa taken among Africans reveal complex perceptions, with both positive and negative responses (with more positive overall perceptions). Pew opinion polls show that most Africans (except South Africans) perceive Chinese influence as a positive development and more positive than U.S. influence (Sautman and Hairong 2009).

It is difficult, if not impossible, to obtain an accurate global picture of Chinese migrants in Africa (Ma 2008) due to the scarcity of data, misidentification, poor baseline data, illegal entry, and the temporary status of many Chinese in Africa. Neither African nor Chinese governments produce reliable data on the number of Chinese immigrants in the region. Scholarly estimates of the number of Chinese in Africa range from 580,000 to 800,000 (Park 2009). The media and Africans, in general, often speak about a larger Chinese invasion.

Many Africans cannot distinguish people from different East Asian countries and routinely categorize Japanese, Koreans, and others all as Chinese, perhaps resulting in inflated numbers. The numbers of Chinese workers even in very high-profile projects like the construction of the new African Union headquarters building in Addis Ababa (which was a gift) produced widely varying reports on the ratios of Chinese to Ethiopian workers, ranging from unspecified mixes of Chinese managers and Ethiopian workers; for example, from all Chinese labor (Al-Jazeera) to equal participation (Chinese press) to one quarter Chinese of the total workforce (academic field research) (Brautigam 2012). What happens in lower-profile, remote projects is unknown. Differences in national labor laws, work permits and their enforcement, the availability and cost of domestic skilled labor, and the structuring of individual investment deals also need to be taken into consideration.

The largest concentrations of Chinese immigrants are in South Africa (200,000–400,000), Nigeria (100,000), Zambia (80,000), and Mauritius (40,000) (Park 2009). Communities of over 10,000 Chinese are found in Sudan, Angola, Tanzania, and Zimbabwe (Mohan 2012; Park 2009). Still, even in South Africa, the Chinese account for less than 1% of the entire population, and immigrant Chinese communities are concentrated in the Johannesburg and Pretoria areas. Cyrildene, in eastern Johannesburg, is a recent Chinatown, having made the transition since 2000 from a predominantly Jewish neighborhood. Older Chinatowns are found in Port Luis (Mauritius) and Antananarivo (Madagascar).

## CHINESE ENTREPRENEURS IN AFRICA'S INFORMAL ECONOMIES

Chinese SOEs capture international media and scholarly attention, but there is an equivalent drive by small companies and individual entrepreneurs into the region that may be just as consequential. Chinese traders may number several hundreds of thousands across Africa, but neither accurate nor reliable data exist (McNamee 2012). Most saliently, their swelling presence is expressed in the proliferation of Chinese retail shops in urban and rural communities. Most of these migrants were not entrepreneurs in China but became entrepreneurs in the process of moving to Africa. Chinese traders have led the way in introducing high volumes of low-cost goods to African markets, often for the first time, illustrating how Africa's poor communities are being integrated into international economic circuits anchored to China. Traders and high penetration rates of Chinese goods into the informal economies represent a different Chinese presence in Africa. To date, the Chinese government has not defined its relationship with Chinese traders on the ground in Africa, and they rarely speak for themselves or are given a voice in media coverage, so they remain an enigma.

Imports of low-cost Chinese goods have led to the establishment of "China shops" all over Africa. Imports contribute to China's burgeoning trade with Africa, and revenues generated from them and traders' permits benefit national and local governments. China shops sell almost identical entry-level merchandise throughout the region: for example, China-made textiles and clothing, imitation leather goods (shoes, belts, handbags), sportswear, luggage, baby accessories (strollers, playpens, feeding chairs), small household appliances, electronic goods, and so forth. Shops are typically single-room establishments in city centers to begin with, close to Chinese-owned warehouse facilities, and are often operated by a single family or several families. Chinese entrepreneurs also establish restaurants, herbal shops, laundries, and medicine clinics or outlets. Chinese businesses are known for having extended trading hours, more points of sale, and smaller margins.

Chinese entrepreneurs enter the African marketplace in two main ways. First, they undertake site reconnaissance, traveling to the region to gauge the market, return to the mainland to produce high-demand goods, and subsequently relocate and set up shops retailing made-in-China goods. This cycle is repeated over and over throughout markets in the region and is predicated on access to cheap supplies and labor in mainland China and flexible product lines. Second, Chinese entrepreneurs obtain information about business opportunities from overseas families or members of their kinship networks, maneuvering on the basis of chain migration. Much of the merchandise is sourced directly from manufacturers in China, reducing the number of intermediaries, the cost, and ultimately the selling price of the final good.

Some alarmists claim that "China Inc." is pursuing a bottom-of-the-pyramid strategy to corner all parts of the African marketplace. However, most contend that there is no coordinated strategy and that Chinese entrepreneurs are doing what they have done for centuries—developing small businesses wherever opportunities exist and profiting from them. Not all Chinese entrepreneurs in Africa are successful; some return home with less money than they had taken with them. Many, however, report making three times what they would earn in China (McNamee 2012).

Chinese wholesalers have become dominant players in supplying low-cost merchandise, establishing entrenched positions in import trade, much like the Lebanese, Syrians, and others accomplished in earlier times. Chinese suppliers have been able to reach deep into Africa's informal economies: made-in-China goods are now ubiquitous, hawked by Chinese and Africans traders alike. Chinese supply chains rarely connect with local firms or local supply chains. Bringing in new goods to the market has enabled Chinese wholesalers to stimulate new booms in a range of items (e.g., mobile phones, bicycles, radios, watches, motorcycles) previously out of the reach of the majority of the poor. For instance, Chinese plastic sandals within a few years have conquered the entire Africa market, altering the daily life of ordinary women and children and relegating a shoeless citizenry to the historical past.

The Chinese commercial presence is changing the face of streets and towns throughout Africa. The Chinese footprint is well established in South Africa, where Chinese wholesalers in Johannesburg operate shopping hubs that serve suppliers and consumers across South Africa and the southern African region. Of course, this has been facilitated by an earlier Chinese immigrant presence in the country, and opportunities in the lucrative South African market have attracted well-financed Chinese entrepreneurs (McNamee 2012). Johannesburg is home to several large Chinese wholesale complexes (e.g., China City, China Mart, Hong Kong City, African Trade Center, Burma Orient City) that are major suppliers to both formal and informal enterprises. In Nairobi, the Luthuli Avenue Wholesalers (Chinese) Market occupies a five-story building and sells 200,000 low-cost Chinese (mainly counterfeit) mobile handsets per month (Hu 2012). In Accra in 2006, the Chinese Wholesale Market opened within Makola Market and in a short time cornered the market on low-value (and low-quality) merchandise. At the same time, some bonds have developed between Chinese and Accra traders based on similarities, resourcefulness, and social networking, contrasting with more established immigrant traders, such as the Lebanese, who have remained separate (Liu 2010). In Dakar, Senegal, Chinese traders organized the regional supply, marketing, and transport of goods, eliminating the need for many customers to travel to the city to shop (Marfaing and Thiel 2011). The remarkable success of Chinese entrepreneurs in Africa has encouraged more entrepreneurial migrants to relocate for business ventures and jobs (which were plentiful until recently). Chinese firms prefer Chinese workers and hire locals only sparingly, typically for front-of-the-shop duties. Job opportunities for Chinese workers in Africa stand in stark contrast to the predicament of most Africans, who find it difficult to secure any work. Chinese dominance in the consumer retail trade has driven many Africans out of business.

Some African companies have reacted by launching marketing campaigns to persuade consumers to buy African-made products. For example, Holmes Bros, a Durban-based South African clothing company, puts a tag on all of its T-shirts saying, "Holmes Bros say Cheers China. Thanks for choosing a locally inspired and made product, there's a whole factory of South Africans who're stoked [excited] you did. Bye-bye China" (Fig. 12.1).

Chinese penetration into the informal economies of Africa is a source of considerable resentment among African traders, shopkeepers, unions, and activists. While customers are largely tolerant, many complain that Chinese merchants do not respect consumer or worker rights. Resentment has ignited xenophobic attacks against Chinese traders on several occasions (e.g., in South Africa, Angola, and Zambia in particular). In other places (e.g., Nairobi), the poor quality of Chinese mobile (counterfeit) handsets has led to anti-Chinese demonstrations. Chinese traders have become the whipping boys for Africa's politicians, merchants,

O

Holmes Bros say

CHEERS

CHINA

Thanks for choosing a locally inspired and made product, there's a whole factory of South Africans who're stoked you did. Bye-bye China ... by not buying something from China, you're preventing massive carbon emissions that they produce shipping their products all over the world! Home is where the heart is ... so keep it real and keep buying Holmes. Schweet

**FIGURE 12.1** Anti-Chinese Backlash: Holmes Bros "Cheers China" Campaign, South Africa.

and unions. These incidents have not gone unnoticed in China. Chinese companies and the Chinese media have complained that Chinese fly-by-night operators selling counterfeit goods (e.g., US$10 mobile phones) in African markets are souring Africans' image of China (Hu 2012). In South Africa, the xenophobic atmosphere prompted many Chinese retailers to display and sell "proudly made in South Africa" merchandise.

Cutthroat competition among Chinese migrants and African firms is driving down prices in the marketplace, and profit rates are in decline. Chinese traders can face uncooperative environments on the ground, encountering xenophobic and/or corrupt local officials, and they compete directly with other Chinese and African firms in a difficult environment. Chinese traders in Africa are forging their own pathways, and there is no way for Beijing to control the multitudes of Chinese migrants (McNamee 2012). Surveys of Chinese traders in Africa reveal that many feel abandoned by Beijing and Chinese Embassy officials in their host countries (McNamee 2012). Indeed, competition and social fragmentation among Chinese migrants run deep in Africa, further dispelling a myth of a monolithic China. Growing tension is observed among Chinese companies aiming to produce goods with brand recognition and those specializing in counterfeit global or Chinese brands. Some Chinese firms are directly counterfeiting other Chinese companies' merchandise.

Most analysts assume that the Chinese trader phenomenon in Africa will continue on a cyclical path. When the current crop leave Africa, they will be replaced by new immigrant entrepreneurs who will endure similar hardships to earn incomes that would have been very difficult to secure back home. At the same time, the lower circuit in Africa's informal economies is becoming more crowded with Pakistani, Bangladeshi, Vietnamese, and Korean traders, as well as Somalis. Somalis have consolidated their positions in many markets (e.g., townships in South Africa and inner-city Johannesburg) outside their own country. It remains to be determined whether the networks and structures that have expedited the Chinese flow to Africa will remain as potent and profitable. Some African countries have undercurrents of growing popular support for indigenization strategies. For example, Ugandan local shopkeepers unsuccessfully petitioned the government to impose stiff regulation on Chinese traders, but in Malawi, a 2012 law bars Chinese traders from operating outside of the four main cities.

The Chinese have been made significant inroads into the economic life of ordinary Africans in an extraordinary short period of time. Indeed, the rapidity with which Chinese actors adapt to changing circumstances in African locations—no doubt a reflection of how the Chinese themselves are able to adapt to fast-paced change in China—continually challenge assumptions about their standing in Africa, and it means that the narrative of the Chinese in Africa is always in need of updating (Alden 2012).

## THE AFRICAN DIASPORA IN CHINA: AFRICAN ENTREPRENEURS IN THE "DRAGON'S DEN"

Chinese entrepreneurial successes have spilled over and enticed large numbers of Africans to get involved in China–Africa trade. A counterflow of African migrants and itinerant traders (mainly West Africans, especially Nigerians) centering on the fast-growing city of Guangzhou, in southern China, has emerged since the late 1990s. African migrants arrive despite difficulties in obtaining travel documents and/or short-term visas. By contrast, Chinese entrepreneurs enjoy unobstructed travel between Africa and China (Haugen 2011).

Steady Africa-to-Guangzhou migration has led to the emergence of "Africatown" (locally, districts are labeled "Chocolate City," "Little Africa," and "Guangzhou's Harlem") (Fig. 12.2). Guangzhou's African immigrant population numbers between 20,000 and 50,000 permanent residents, most of them bachelors (Haugen 2011). Tens of thousands (60,000 in 2007) of other Africans are sojourners arriving for extended stays to purchase goods, wait on order fulfillments (often six weeks or more), and complete shipment logistics (check goods to avoid being cheated and complete customs paperwork to their advantage, so as to reduce import duties) (Haugen 2011). Some African traders use networks back home to smuggle goods, especially counterfeit merchandise, into their countries (Yang 2012). The presence of enclaves of African traders in Chinese cities indicates that some Africans are playing an increasingly important role in the exchange of goods between China and Africa (Haugen 2011). Significantly, North Americans, Argentinians, Filipinos, and others also travel to Guangzhou and other Chinese cities to engage in trade relations, but only Africans have created immigrant enclaves. The experience of Africans in China is monitored by Roberto Castillo in his blog *Africans in China*.

**FIGURE 12.2** Africans Seeking their Dream in Guangzhou, China ("Chocolate City"). *Source:* © Imaginechina/Corbis. Corbis 42-50360355.

Besides entrepreneurial motivations and learning by replicating the successes of Chinese entrepreneurs, Africans migrate to China because they mistrust Chinese intermediaries and want to avoid the communication difficulties involved in arm's-length trade. African entrepreneurs who travel to China can gain an edge in trade back home because they are informed buyers and understand local African fashion tastes and styles, even though they are at a well-recognized bargaining disadvantage when dealing with Chinese suppliers compared to the Chinese traders, who drive hard bargains with suppliers. Once in China, African entrepreneurs can obtain product samples and use their African networks to obtain consumer feedback before committing to bulk purchases (Haugen 2011). Staying in China allows African traders to establish direct relations with Chinese factory showroom people and wholesalers. Importantly, Africans in China are buyers and intermediaries but never producers. Some African entrepreneurs operate as the middle people skilled in dealing with clients arriving from Africa and/or Africans switching sourcing points (e.g., from Bangkok to Guangzhou).

There is some evidence that African traders based in China are shifting their focus to higher-quality goods, targeting the African middle-class market, and moving more toward complementary trade (Haugen 2011). It is reported that African traders are more easily able to hide the origins of goods imported from China (because of customs contacts), and creative labeling allows them to pass goods off as European at higher prices (Haugen 2011). Made-in-China goods have a reputation for low quality, so African consumers are unwilling to pay higher prices for "Fong Kong," whereas European and North American merchandise enjoys a much better reputation. However, newer Chinese brands such as Huawei and ZTE are building stellar reputations with African consumers, so reputation is also subject to change.

Researchers have documented African entrepreneurs traveling to China to diversify their portfolios. A direct China link provides an informed opportunity to diversify manufacturing and sales back home. They detail it as happening in the following way: a Nigerian producer manufactures shirts in Port Harcourt, but because he cannot compete with Chinese prices and quality, he buys one run in China and produces a lower-quality imitation of this run in his own factory back home (Lyons, Brown, and Zhigang 2012).

Business and life are hard for African entrepreneurs in China. Many struggle and live in humble single rooms with cramped conditions and operate within a very different cultural context. There is a striking similarity with the way many Chinese traders operate in Africa. African entrepreneurs have to deal with currency fluctuations vis-à-vis their national currency and the dollar, and the dollar and the Chinese yuan. Thus, a Nigerian entrepreneur has to convert nairas to dollars and dollars to yuans, and currency fluctuations can be a boom or boon depending on exchange rates (Neuwirth 2012). Despite all the odds, some Africans are making it and reaping considerable profits from this dimension of China–Africa trade.

## CHINA'S INFRASTRUCTURE INVESTMENTS TO AFRICA: BUILDING BRIDGES?

Chinese infrastructure investments capture most of the news headlines about China's engagement in Africa. Despite the importance of Chinese finance for African infrastructure projects, relatively little is known about their precise value and about surrounding infrastructural commitments. China has invested in more than 50 infrastructure projects with over a half a million dollars. Thirty-five countries are negotiating with China on infrastructural finance deals (Foster et al. 2009). Some of the highest-value projects include an US$8.5 billion railway modernization project in Nigeria; a US$1.5 billion transport infrastructure upgrade (includes 754 roads) in Angola, and the building of the US$500 million Entebbe-to-Kampala highway in Uganda. Several recent announcements have been made for regional projects, such as building the West African Highway (across nine countries along the coastlines of a region stretching from Senegal to Nigeria) and rehabilitating the Zambia-to-Tanzania (Tazara) railway. China's infrastructural investments rose steadily from US$500 million in 2001 to US$7 billion in 2006 (Foster et al. 2009), and Beijing has pledged US$20 billion for investments in infrastructure and farming for 2013–15.

Importantly, in several cases, Chinese infrastructural finance is packaged with natural resource development, making use of the mechanism referred to as the "Angolan mode." With such resources-for-infrastructure swaps, African governments do not have to wait for their infrastructure until they have the money; instead, building starts immediately on the basis of a natural resource guarantee. China has underwritten a diverse range of infrastructural projects (e.g., dams, government buildings, stadiums, roads, railway lines, housing, industrial estates), gaining the goodwill of African governments, and has supplied some of the infrastructure that allows the resources to be exported (See Box 12.1 for a case study of China's stadium spending and Box 12.2 for China's lead in developing an industrial estate or special economic zone just outside Lagos (Nigeria)). This has helped fuel a commodity boom and windfall to African governments and has provided needed supplies to Chinese industries. Some Chinese infrastructural projects have been caught up in corruption scandals, and questions about the quality of construction often surface, but for the most part, Chinese funding is delivering badly needed infrastructure in a timely manner.

Infrastructure is a key priority in Africa, and improving the base is critical to achieving the MDGs. The region suffers from critical infrastructure deficits compared to its world peers. The absence of infrastructure (from roads to railways, ports, airports, dams, electricity, sanitation, and informational systems) deters foreign investment, hinders economic development, and shaves 2% off countries' per capita gross domestic product (GDP) growth rates per year (Foster et al. 2009). Infrastructure deficits make it difficult to achieve economic diversification and they impede national and regional integration efforts. As a consequence, Africans pay a premium for power, water, phone, mobile phone, and Internet services compared to other

## BOX 12.1 CHINA'S STADIUM DIPLOMACY

Turnkey stadium projects funded and built by Chinese aid dollars are now common throughout Africa (and Latin America and Asia). China has been building sports arenas in Africa (referred to as "Friendship Stadiums") since the 1970s. The Chinese built the Stade des Martyrs de Pentecost (formerly Kamanyola Stadium) in Kinshasa, DRC, in 1993. It was regarded as "the African cathedral of football" until South Africans built state-of-the-art facilities for the 2010 Fédération Internationale de Football Association [FIFA] World Cup. In terms of total costs, stadiums are a minor component of Chinese infrastructural investments, but they are immensely symbolic and popular.

China had built 58 stadiums in 31 African countries by 2012. Its stadium diplomacy intensified after 2000 and stadium support is now enlarging. Beijing built the stadiums for the most recent African Cup of Nations (e.g., Mali, 2002; Ghana, 2008; Angola, 2010; Equatorial Guinea/Gabon, 2012), which is the regional international soccer competition held every two years. China is also behind stadiums for the All-African Games (Abuja, 2003, and Mozambique, 2011), a multisports international competition held every four years. In addition, Beijing funds the upgrading of old stadiums (e.g., US$12 million to upgrade Moi International Sports Stadium in Nairobi, Kenya). Typically, nominations to host pan-African sports spectacles are put forward by bidders who have already obtained a commitment from the Chinese to underwrite the costs of new stadiums. The state-owned Shanghai Construction Group and Beijing Construction are the big players in building these stadiums. Five stadiums were completed in 2012; among them were two stadiums in Equatorial Guinea and Gabon to enable them to co-host the 2012 African Cup of Nations, and Ndola Stadium was completed in the regional capital of the copper belt in Zambia. More stadiums are slated for construction (e.g., a 40,000-seat stadium in Lilongwe, Malawi).

The construction of stadiums is included under a public building rubric, one of six key areas prioritized in Chinese foreign aid budgets. The Chinese assume responsibility for the entire process, from initial planning to completion. They take responsibility for feasibility studies, design and construction, proof of reliability of equipment and construction materials, and provision of technical and construction experts to organize, lead, and execute the project.

New stadiums benefit both recipient and donor. The recipient state and its people obtain a new stadium, which would otherwise be unaffordable or possible only through a reallocation of national spending priorities. In most cases, China gifts the stadiums or provides interest-free assistance so that the recipient does not have long-term interest repayment obligations. The recipient gets immediate access to capital and technical expertise, which is valuable to governments with an eye on the electoral calendar. Moreover, the physical structure holds symbolic significance, serving as a permanent reminder of China's goodwill and as a source of national pride and achievement that allows for hosting international spectacles and provides opportunities for place promotion. A stadium with a capacity of 42,000 was built by the Chinese in Maputo (Mozambique) (Fig. 12.3). Outside this stadium, a sign proclaims, in Chinese and Portuguese, that "the friendship between China and Mozambique will last forever like heaven and earth" (quoted in Will 2012:37).

Despite these benefits, China's stadium diplomacy has been heavily criticized. Chinese-built stadiums serve as catalysts for advancing Beijing's One-China policy. Although the Chinese are admired for their engineering expertise and efficient construction practices to deliver stadiums on time (or ahead of time), many observers note that local workers are involved sparingly and only in unskilled capacities. For example, in an Angolan stadium project, the Chinese brought in 700 overseers, engineers, and laborers and hired only 250 Angolan laborers (Will 2012:37). Beijing justified its disproportionate reliance on imported professional staff and skilled labor on the basis that few skilled workers were available in Angola, a country that only recently emerged from civil conflict. (Other SOEs also operate with few local linkages to domestic suppliers and without the training of domestic workers.) Stadiums appear to be components of secret swap agreements that repay the investor with natural resources. Besides the architectural feat and display of a shiny new stadium, it is not clear that stadiums actually benefit sports in respective countries. Sport teams have not been able to change their operating models to make the large venues viable. National and club teams cannot afford to take on maintenance of these stadiums, and many become expensive white elephant projects, falling into decay and disrepair after staging showcase international events. Questions have also been raised about construction standards. For example, FIFA pointed to serious flaws and weaknesses in DRC's national stadium and threatened to close it until substantial improvements were made.

**FIGURE 12.3** National Stadium, Maputo, Mozambique. *Source:* © Liu Dalong/Xinhua Press/Corbis. Corbis 42-27129159.

China's stadium diplomacy is a vehicle for soft power and a bargaining chip for access to African markets and resources. China conveniently responds to African elites' desire for showcase tangible deliverables and to the oversupply of infrastructure firms/labor within its own country by offering them opportunities to enter new markets. Many Chinese companies are successful in staying on after projects end, when they can compete for work in infrastructural projects (Alden 2012).

China deliberately assumes a low profile during sporting events hosted on African soil; its benefits are realized in other ways. Indirect benefits to Chinese companies were seen, for instance, when the Angolan government contracted Zoomlion (a Chinese waste management subsidiary company headquartered in Ghana) for stadium waste collection during the 2010 African Cup of Nations. China's stadium policy delivers tangible projects that have high visibility and initially are viewed positively. However, over the longer term, if stadiums fall into decay, China may be viewed as carting off African raw materials and neglecting to train domestic laborers. Bad feelings about broken gifts may well arise.

## BOX 12.2 A PATHWAY TO ECONOMIC DEVELOPMENT: CHINA'S SPECIAL ECONOMIC ZONE IN LEKKI, NIGERIA

The export of Chinese special economic zones (SEZs) to Africa was adopted as an official policy at the Forum on China–Africa Cooperation in 2006. Six experimental Chinese SEZs were, as of 2012, in the development phase: two in Nigeria (Lekki and Ogun), two in Zambia (Chambishi and Lusaka), one in Mauritius (Jinfei), and one in Ethiopia (Oriental Eastern). A framework for introducing Chinese SEZs evolved through a working relationship among the Chinese government, African governments, and Chinese enterprises. China's light manufacturing as an emerging footprint is the latest chapter in China's engagement with the region: it signals an entirely different development phase (Gu 2011).

Chinese SEZs were introduced despite the lackluster prior experiments by African governments to implement 90 national export-processing zones in 20 African countries. Export-processing zones in Mauritius and Madagascar are regarded as African successes and exceptions. In some countries (e.g., Ghana, Senegal), zones are operating partially, and in other countries (e.g., Kenya, Zambia), zones have been abandoned. Nigerian export-processing zones have not delivered to date. Poor infrastructure, limited political support, planning shortfalls, lack of a critical mass of investors, and deficiencies in the skill levels of local labor explain many of the major problems encountered in the early African export-processing zone experiments.

Chinese motivations are multidimensional. Officially, Beijing touts the mutual benefits and partnership for development, but it is clear that Chinese small and medium-sized enterprises face intense competition in their home markets, motivating their strong entrepreneurial spirit and investment in African zones (Gu 2011). Enclave development is not part of China's "Go Out" strategy because the latter was aimed at SOEs and market leaders. Instead, SEZs are pitched at small and medium-sized enterprises, offering an alternative for these manufacturers in the context of rising land and labor costs on the mainland. Moreover, clustering Chinese firms within a new industrial enclave offers benefits of mutual support and synergy.

The most ambitious SEZ in Africa is the Lekki Free Trade Zone, located 37 miles (60 km) east of Lagos. Lekki, with the Atlantic Ocean to the south and the Lekki Lagoon to the north, represents an area that is prime for tourism and real estate development. Lagos state government provided the land (40,772 acres [16,500 hectares]), and a Chinese consortium is building the project. The free trade zone is a key component of an ambitious multiuse development plan on the Lekki Peninsula (four times the size of Manhattan) that includes a plan for a new city (the estimated future population of metropolitan Lagos is 110,000), serviced by major infrastructure to be completed by 2015, including a new airport (Lekki-Epe International Airport), a deepwater port (Lekki-Tolarum Port), and a modern expressway directly linked to Lagos (a new light rail link is also under consideration). The Lekki development aims to be the Nigerian future, "the Dubai of West Africa." US$1.1 trillion has been reportedly invested by 48 international/domestic investors (Business & Maritime West Africa 2012), but reports on the ground indicate a much slower uptake than one would expect.

The Lekki Free Trade Zone is a joint venture between Chinese companies (60% share) and Nigerian interests (40% share) (divided between the Lagos state government and Lekki Worldwide Investments). The Chinese consortium comprises the China Railway Construction Corporation, the Nanjing Jiangning Economic and Technological Development Corporation, and Nanjing Beyond Investment.

The zone targets four sectors: (1) light industry (e.g., furniture, textiles, garments, footwear, household appliances, construction and building materials, and other consumer products); (2) motor vehicle assembly; (3) warehousing and logistics for a range of industries, including petroleum; and (4) real estate development. The aim is to revitalize the Nigerian manufacturing sector and to diversify the economic base to help reduce the country's dependence on imports (90% of consumer goods are imported). Lagos officials believe that they can benefit from Chinese experience and expertise in zone development and management as well as from Chinese investments in manufacturing and services, which offer the potential to link domestic manufacturers to China's global supply chains.

*(Continued)*

**BOX 12.2 (*Continued*)**

Generous incentives are offered to attract firms (100% tax holiday from all Nigerian duties, permission to sell 100% of goods in the Nigerian market, 100% repatriation of profits and royalties, duty-free imports, raw materials, and components for goods to be re-exported, and, controversially, a prohibition on strikes and lockouts for 10 years) (World Bank 2011). Developers tout the geographical advantages of Lekki, which can provide excellent access to serve as a gateway to the growing West African middle-class market.

Still under development, it is premature to assess the this project. Developers acknowledge struggling to attract Chinese private investors as well as others. It remains to be determined whether Chinese expertise can be harnessed to jump-start local investment and boost manufacturing capacity. Development experts are skeptical about Chinese zones' ability to contribute to an African industrial revival. Many fear that Lekki may be primarily used for the reshipment of Chinese products to be relabeled as "African products" in order to take advantage of trade concessions and gain access to U.S. and European markets. There is evidence that Chinese private firms prefer to locate outside China's zones to remain free from Chinese authorities' oversight. Moreover, many Chinese and international firms are reluctant to bear the cost of the zone (rents and utilities) should Lekki remain underpopulated.

It may take more than trophy projects to solve African manufacturing problems. To begin with, African political and economic contexts are very different than China's, and the conditions that dictated the success of SEZs in China are not present in Africa. These include long-terms coordination between development authorities and provincial and national governments, integration of industrial policy within national development, a well-developed manufacturing base, and good transportation infrastructure. Nevertheless, the prospect of China shifting some of its manufacturing base to Africa is not as far-fetched as it sounds: deindustrialization and reindustrialization have been consistent features of global industrial geographies for some time now.

consumers in the developing world. Substantial investments (over US$100 billion per year) are needed to build new infrastructure, refurbish dilapidated assets, and maintain all existing and new installations.

Sectors that have been heavily targeted include power (mainly hydropower), transport (especially railways), and telecommunications. According to the U.S. intelligence analysis company Stratfor, since 2010 railroads and roads account for 19% of total Chinese infrastructural investments, followed by hydroelectric dams at 9% and civic construction projects at 6% of total foreign development investment (Fig. 12.4). Ghana, Nigeria, and Zambia have over US$10 billion of projects under current development, and the DRC, Angola, and South African, all resource-rich countries, are also prioritized. Despite high levels of Chinese FDI anti-Chinese feelings intensify from time to time in numerous African countries, especially Zambia (See Box 12.3 for a case study of China-Zambia relations).

Chinese investments in African infrastructure have produced a complementarity among (1) African economic needs, (2) China's large globally competitive construction industry, which was encouraged to "go out" given competition in China's markets, and (3) the emergence of what some media have called China's own "ghost projects," where the construction market appears to be saturated as projects have been built far ahead of demand. China's infrastructural investment in Africa has greatly facilitated the resource boom in Africa, and some of this infrastructure will be important to kick-starting manufacturing industry in the region. Mutual benefit seems to drive these infrastructure deals, where new infrastructure will generate business that will be important to both Africans and the Chinese. Most importantly, China has built infrastructure at a time when Western countries have shifted their emphasis away from large-scale development projects.

Nova Cidade de Kilamba is a new town being built 18 miles (30 km) outside Luanda, the capital of Angola (See Figure 12.5). The site is close to a Chinese international airport under construction that, upon completion, will be largest international airport in Africa. Built by the SOE China International Trust and International Trust and Investment Corporation (CITIC), this US$3.5 billion project involves 12,000 Chinese workers (CITIC Construction 2012). The development spans 12,355 acres (5,000 hectares) and is the first of Angola's new "satellite cities." Consisting of 750 eight-story apartment buildings (2,000 will eventually be built), a dozen schools, and 100 retail units in 2012 (Redvers 2012), the project will be the largest new residential project on the continent when complete.

Most facts about the project are murky. Reports vary on how many people will be accommodated in Kilamba. (The BBC estimates 500,000, whereas CCTV

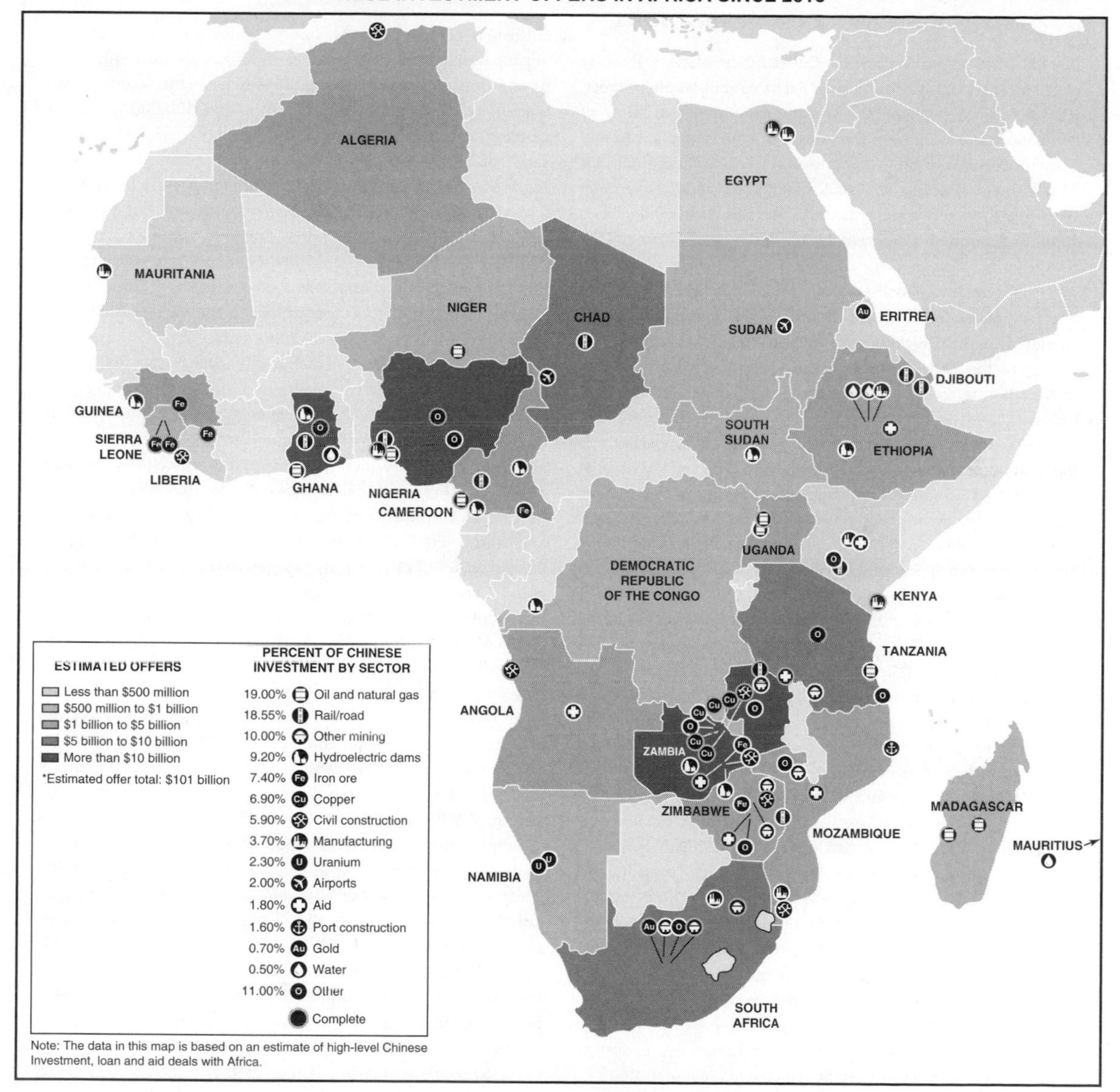


**FIGURE 12.4** Chinese Sector Investments in Africa, 2010–2012. *Source:* Adapted from Stratfor 2012.

puts the number at 210,000 people.) CITIC and the Angolan government claim that the project is a social housing effort (to provide housing for the poor), but it seems otherwise: apartments are listed at between US$125,000 and US$200,000, out of the reach of most Angolans (who earn on average US$2 per day). The slow uptake in purchases (due to costs, lack of mortgages, and delays in titling) has given rise to media claims of "Africa's first ghost city" (Medina 2012).

The project is mired in controversies on both sides. Initially promoted as a government social housing project, it was handed over to the state-run oil company

## BOX 12.3 CHINA AND ZAMBIA RELATIONS

Zambia is a high-profile player in China–Africa relations. It is one of the most urbanized countries in Africa yet one of the poorest. GDP per capita is US$1,500, and approximately two thirds of Zambians live below the poverty line. Zambia is landlocked and is highly dependent on resource exports, copper accounting for 74% of them (Gadzala 2012:43). Mining has traditionally been the engine of economic growth. Following independence in 1964, the country made great strides in developing the copper belt from an area of bush to a dynamic industrial region. By 1969, Zambia was classified as a middle-income country, with one of highest GDPs in the region. At the time, Zambia was labeled an African success story in industrial development.

Despite this early progress, development halted in the early 1970s with the collapse of copper prices. Zambia was forced to borrow heavily, leading to serious indebtedness. Between 1974 and 1994, per capita income declined by 50%, and Zambia became the 25th poorest country in the world (2012). The country became heavily reliant on foreign aid conditioned on the privatization of state-owned enterprises. Complying with the new policy environment, the Zambian government implemented liberal investment acts in 1995 and 2006, allowing the repatriation of all profits and royalties and permitting investors to bring in labor. These policy inducements opened up the economy to outside investors at the time when China's "Go Out" strategy was in full swing.

Beijing has an early track record of engaging people in this part of Africa. Ties began when China provided financial assistance and support to the liberation struggle against Britain, the colonial ruler of the territory known at that time as Northern Rhodesia. China deepened its ties by providing assistance to newly independent Zambia. A benchmark in Sino-Zambian collaboration was the 1965 bilateral agreement to build the Tanzania-to-Zambia railway (Tazara), linking the port of Dar es Salaam (Tanzania) to the town of Kapiri Mposhi (Zambia). Project financing was shared in equal parts, and work began in 1968 and was completed by 1976. Construction of this 1,160-mile (1,860 km) railway system involved 50,000 Chinese laborers. The railway line established a direct connection between the landlocked Zambian copper belt and a port, thereby avoiding transportation across territories controlled by white governments in Rhodesia and South Africa. Tazara facilitated the export of copper, providing an important mineral to China, and helped Zambians obtain good port access, which became very important following the closure of traditional routes when sanctions were imposed on apartheid South Africa and Southern Rhodesia (Carmody 2011). More recently, Beijing forgave some of the debt associated with Tazara's construction, no doubt motivated by China's (the world's largest consumer of copper) strategic need to maintain access to Zambian copper and by its considerable on-the-ground mining investments. The latter appear to be intensifying since the 2008–09 global economic downturn, when Chinese capitalists moved to acquire many foreign operations.

China became a major investor in textiles in Zambia, establishing the Mulungushi textile factory in Kabwe in the 1980s. In its heyday, the plant was the largest textile mill in the country, manufacturing 17 million meters of fabric and 100,000 pieces of clothing annually, employing 2,000 workers, and linking thousands of cotton growers; in addition, the factory won international acclaim for the quality of its cloth. Initially the factory was a turnkey operation for the Zambian government, but it was run as a joint venture after 1997, when it ran into operational difficulties. By 2007, the factory could no longer compete with cheaper textiles coming out of Asia, and it closed its doors (2007–14). In 2014 a Tanzanian private investor leased the Mulungushi textile factory and plans to resume full production.

Chinese investments in Zambia intensified after 2000, and more than 200 Chinese companies operated in the country as of 2009 (Carmody 2011:161). Chinese investors target mining, but they are also pouring money into construction (roads, railways, airports, schools, hospitals), tourism, and agriculture. Chinese investments in agricultural land have accelerated in recent years; at least 23 Chinese farms are operational in Zambia (Spieldoch and Murphy 2009:42). An investment to grow jatropha on 2 million acres (809,271 hectares) has generated significant controversy (Carmody 2011:162). Older investments, however, have been deemed successful; for example, China State Farm and Agribusiness Corporation's investment in the China–Zambia "friendship farm," a 1,648 acres (667-hectare) farm located outside Lusaka that produces wheat, soybeans, and corn for the local market. China's SEZs in Chambishi (concentrating on copper/cobalt processing and electronics) and Lusaka (concentrating on garments, appliances, tobacco, and electronics) are mixed success stories, with slower-than-expected uptake of investors (Brautigam and Xiao Yang 2011).

Zambian presidents (e.g., President Mwanawasa in 2007) have routinely described China as an "all-weather friend." Despite China's official declarations of "mutual friendship and cooperation," investments totaling US$21 billion in 2012, and the purported creation of 50,000 jobs (Embassy of the People's Republic of China in the Republic of Zambia 2012), anti-Chinese tensions have been simmering for years, and Sino-Zambian relations have trouble spots. For example, in 2005, an explosion at a Chinese explosives company serving the Chambishi mines killed 51 workers, making it one of the Zambia's worst-ever accidents. In 2011, two Chinese managers were charged with 13 counts of attempted murder after firing into a crowd of miners during a pay dispute riot. Even though the charges were dropped, rumors about the mine owners settling out of court with the families of the victims have fanned anti-Chinese sentiments (Taylor 2012). In 2012, relations threatened to plummet again when Zambian miners killed a Chinese manager and shot another during a wage protest at a Chinese privately owned coal mine in Sinazongwe, 200 miles (325 km) south of Lusaka. Reports confirm that miners in Chinese-owned operations are paid between one-third and one-sixth less than workers at non-Chinese operations (but still above national minimum wage levels) (Human Rights Watch 2011); wage differentials serve as a basis of ongoing labor tensions. Human Rights Watch (2011:1) produced a scathing report of conditions in Chinese mining operations throughout the country, emphasizing poor health and safety standards and labor conditions: "poor ventilation that can lead to serious lung diseases, hours of work in excess of Zambia law and the threat of being fired should workers refuse to work in unsafe places. Injuries and negative health consequences are not uncommon, although many incidents are not reported to the government, in violation of Zambian and international labor law. The troubling situation stems largely from the attitudes of Chinese-owned and run companies in Zambia, which have tended to treat safety and health measures as trivial." Furthermore, popular claims that the Chinese create the least-skilled jobs persist; for example, in building the Ndola "China-Zambia Friendship Stadium," "the bricklayers for the project were Chinese while Zambians were confined to mixing cement" (Carmody 2011:172). Routinely, workers employed by Chinese firms work 78 hours per week—some 30 hours in excess of the national law—and working 365 days per year is not atypical.

Anti-Chinese feelings have become a consistent theme is recent national Zambian elections. Michael Sata, a challenger in several presidential elections, took an anti-China stance in campaigning, feeding on popular sentiments about the underpayment of Zambian workers and the unfair competition that forces Zambian firms out of business. Sata's campaigning was especially popular in the copper belt and among the urban poor. He claimed that "Zambia was becoming a province—no, a district—of China" and noted that "the Chinese are not here as investors but invaders" (quoted in Sautman and Hairong 2009:756, 751). Sata toned down his rhetoric in successive campaigns (especially the 2011 campaign), and after several attempts, he won the presidency in 2011. On taking office, however, he moderated his anti-China platform and adopted a pragmatic approach to the broad relationship, moving more toward constructive engagement after two years in office. Nevertheless, the Sata government has raised minimum wages and is pushing for increasing the government's share in mining ownership. Chinese companies are keenly aware that using Chinese labor is controversial but justify their use of mainland China workers based on their productivity levels, higher skills, and cultural affinity.

Perhaps the greatest problem in Sino-Zambian relations stems from the confidential nature of the bilateral agreements between the two countries; the details are not available for public scrutiny, even though both countries are, on paper, committed to public transparency and accountability. Analysts consider these bilateral agreements to be tilted in China's favor, and project identification and selection are based more on ad hoc consultations rather than on any coherent development strategy. It is worrisome that many or most Zambian authorities lack the capacity and skills for national plan formulation, coordination, and monitoring.

There are winners and losers on the Zambian side of Sino-Zambian relations. The Zambian government reaps increased investments, improved infrastructure, and more trade taxes, but, in the process, it forgoes income tax revenues and undermines domestic companies and jobs that cannot compete with global competition. The key losers are local producers, who in the short term benefit from less expensive Chinese inputs but in the long term cannot compete with Chinese producers, who dominate the local market with cheaper imports. Local labor, trade unions, domestic firms, and their workers lose out because they cannot compete. Local contractors also lose out on the basis of the conditions written into the bilateral deals and, in turn, are disadvantaged by their lack of access to inexpensive capital. Academics and some civil society organizations take a more middle-ground perspective, realizing the balance of positive and negative factors involved in China's engagement (Mwanawina 2008).

Although Zambia is an interesting case study and is often used as a telling example in Western media's negative portrayal of China–Africa relations, it cannot be taken to represent all of Africa. Indeed, Kenyans, Sudanese, and Nigerians, to mention a few, have highly positive and nuanced views of Chinese activities in their respective countries (Sautman and Hairong 2009). Sautman and Hairong (2009) contend that negative attitudes toward the Chinese have more to do with opposition parties playing the China card, Western media picking up on this and reinforcing it, and then the African media reaffirming the bias in the Western media. Indeed, studies show that the British media are generally positive about Western actions in Africa and disparaging about Chinese activities (Mawdsley 2007). Critical reflection shows that the media have a geographically uneven and partial focus by which positive elements are downplayed (debt cancellation, investment, commodity price impacts, support for a greater international voice) and problems are highlighted (in line with Western media coverage of the general region) (Mawdsley 2007).

Sonangol, which in turn contracted its real estate arm, Delta Imobiliária, to sell the units (see http://deltaimobiliaria.co.ao/detalhe_emp.aspx?empid=1). Opposition newspapers and political party members believe high-ranking government officials have usurped the project and claim that the government is paying for the project with oil credits. Attempting to deflect criticism, the government decided to speed up occupancy of the units by publishing a list of 550 priority-access individuals (academics, doctors, journalists, and entertainers). However, none on the list qualified for social housing, and inclusion on the list of a deceased newspaper editor raised alarm bells (*Mail and Guardian* 2012). It is further alleged that Manuel Vicente, the vice

**FIGURE 12.5** Kilamba Kiaxi Housing Development.
*Source:* © SIPHIWE SIBEKO/Reuters/Corbis. Corbis 42-49899646.

president and potential successor to President Dos Santos, is a major shareholder in Delta (*Mail and Guardian* 2012).

Chinese players in the project have also faced controversy (Murray et al. 2011) regarding the role of a mysterious Hong Kong-based company, the China International Fund, which offers financing for infrastructure projects for oil and mineral concessions. The China International Fund is allegedly linked to the Chinese government and is criticized heavily in the Chinese media for not delivering on its infrastructural promises.

Kilamba is a vanity project that the government shows off to many visiting dignitaries, such as the Secretary-General of the UN, Ban Ki-moon. Many critics charge that high-rise apartment living is unsuited for larger Angolan families and cultural contexts. The project appears to have been rushed to serve a political agenda, and local architects and communities were not consulted. Bad press about the project on social networking sites raises a question about whether this type of project improves Angola's or China's reputation.

## CONCLUSIONS

The pace and scale of Chinese undertakings in Africa are breathtaking. For the first time in the region's history, countries' development trajectories are being driven by another developing country and its citizens rather than by developed countries (even though the West has continuing influence). There are no historical precedents in the region. We cannot compare China's engagements in Africa to those of France, Britain, or the United States: Chinese activities in the region are simply different and exceptional. It is not productive to label China's relations with Africa as either positive or negative; the reality is very complicated, with both challenges and opportunities (Table 12.1). As Chinese players become further entrenched in Africa affairs, they, too, will undergo the learning processes that other external powers have experienced. Many underlying issues and tensions in the China–Africa dynamic remain to be examined and managed.

**TABLE 12.1 WEIGHING UP CHINESE ENGAGEMENTS**

| *Opportunities* |
|---|
| • Investment |
| • Infrastructural projects (e.g., establishment of key intra- and interstate trade routes where none previously existed and of an information superhighway) |
| • Special economic zones |
| • Business acumen (work ethic, entrepreneurial spirit, global supply networks, business planning) |
| • Expertise and global supply models and access to European and U.S. markets and trade concessions |
| • Increase in consumer purchasing power (availability of a range of affordable Chinese goods) |
| • Creation of jobs (limited) |
| • Counterbalance to the idea that the West is best or knows best |
| • Possibility for South–South cooperation |
| *Challenges* |
| • Poor labor relations |
| • Poor environmental and health and safety standards |
| • Displacement of African workers and offloading of formal workers into the informal economy, thereby increasing unemployment |
| • Lack of skills transfer |
| • Competition with local industries and limited local supplier linkages |
| • Use of second-hand industrial equipment, resulting in lower-quality production |
| • Enclave developments |
| • Lack of transparency and bilateral deals with the elite (civil society is excluded) |
| • Uncertainty about who is in charge and what the role of Africans is |
| • Reproduction of dependency relations |

The official language of China–Africa engagement has been updated over time and differs greatly from the language that was associated with European powers' colonial incursion into Africa. The 2013 White Paper by Beijing's China-Africa Trade Cooperation includes language about expanding foreign investment and trade cooperation to support sustainable development in Africa, increasing development assistance to benefit more African people, supporting the African integration process, helping Africa enhance its capacity for overall development, and strengthening people-to-people friendships to lay a solid foundation of public support for enhancing common development and promoting African peace and stability (The Peoples' Republic of China 2013).

Engagements by China and other BRIC (Brazil, Russia, India and China) nations with Africa offer an unprecedented opportunity to deploy all means available to achieve development. Chinese ties can provide a welcome alternative to entrenched patterns of trade for investment in the region. While Sino-African relationships come without the historical baggage of the West, they bring other challenges; for example, cultural differences, shorter histories of face-to-face interactions, unrealistic expectations, and international pressures to conform to Western development policy advice. Official Chinese government engagement has operated via African government elites, and this is a contentious and potentially explosive issue that has so far been deflected. There will be mounting pressures from diverse constituencies, especially civil society, to engage with the China–Africa dynamic.

Change and adaptability are going to be the hallmarks of African development in the 21st century. African governments and civil society must learn quickly from experiences on the ground and ensure that dynamism is built into development partnerships. A shift is well under way: the Chinese state is being superseded by the Chinese private sector in Africa. This dynamism is very different from relationships of the past, which tended to be static (e.g., colonial and Development Assistance Committee relationships). In many ways, China's own development experience shows that it is possible to adapt and to tackle national development in the context of global integration, but we know that Africa is a different region and that what worked in China may not work, directly applied, to Africa.

China's extending footprint raises many important issues. Who in Africa is engaging with whom in or from China, and where, is a critical consideration. The prospect of Chinese engagement can be a tantalizing opportunity or a terrifying threat (Naidu and Mbazima 2008:759). For some of Africa's citizens, it decreases livelihood security. For some political elites, Chinese investment levels the playing field and enhances Africa's international leverage. For others, China's involvement is yet another example of an outside country appropriating Africa's resources while consolidating its own power. Ultimately, as Sino-African relations deepen, the complex interactions must be assessed in terms of how they contribute to poverty reduction and achievement of the MDGs and how they affect Africans' lives and agency at home as well as in the international arena.

Some aspects of the Africa–China trade and investment dynamic appear to be in an early process of restructuring, which may mean that Chinese investments in African mineral resources will not increase at the accelerated rate that they did from the 2000s onwards. As China's economy slows, its priorities are shifting from resource capital spending to domestic consumer spending. Already this switch is decreasing many commodity prices (e.g., copper and iron ore). Africa's resource economies are likely to be doubly affected by reductions in Chinese resource investments and declines in commodity receipts (except oil). Slowdowns in the Brazilian and Indian economies could add additional pressures and undermine Africa's growth trajectory. However, Beijing's emphasis on consumer spending could result in different adjustments, such as increasing Chinese demand for African oil and natural gas and switching from an emphasis on resource investment toward African light manufacturing and textile production.

Although China has an African policy and individual African country strategies, African states do not have the equivalent. African states must develop a China strategy to manage the challenges and opportunities. To date, Africans have been acting toward

China's expansion in a reactive and ad hoc manner rather than within national development frameworks. There is mounting evidence that many aspects of the China–Africa dynamic are dependent relationships in the making: African agency is much less evident. Given this tilt, it may be productive for African states to move beyond dealing with China bilaterally. Perhaps there would be some room for maneuverability and agency if African states engaged China on a regional basis as well as bilaterally (within a national development framework). However, the prospect for civil society engagement with China appears unlikely at this point.

## REFERENCES


Alden, C. 2012. "China and Africa: a Distant Mirror of Latin America." *Colombia Internacional* 75(1):19–47.

Brautigam, D. 2009. *The Dragon's Gift: The Real Story of China in Africa*. New York: Oxford University Press.

Brautigam, D. 2012. "Africa's New AU building: How Many Chinese Workers?" Available at http://www.chinaafricarealstory.com/2012/01/africas-new-au-building-how-many.html (accessed December 5, 2012).

Brautigam, D., and T. Xiao Yang. 2011. "African Shenzhen: China's Special Economic Zones in Africa." *Journal of Modern African Studies* 49(1):27–54.

Business & Maritime West Africa. 2012. "Lekki Free Trade Zone (LFTZ): A Pathway to Economic Development in Nigeria." Available at http://businessandmaritimewestafrica.com/investment/lekki-free-trade-zone-lftz-a-pathway-to-economic-development-in-nigeria (accessed December 15, 2012).

Carmody, P. 2011. *The New Scramble for Africa*. Malden: Policy Press.

Carmody, P., and F. Owusu. 2007. "Competing Hegemons? Chinese versus American Geo-economic Strategies in Africa." *Political Geography* 26:504–524.

Cheru, F., and C. Obi, eds. 2010. *The Rise of China and India in Africa*. New York: Zed Books.

CITIC Construction. 2012. "Angola Social Housing Kilamba Kiaxi Phase." Available at http://www.cici.citic.com/iwcm/null/null/ns:LHQ6LGY6LGM6MmM5NDgyOTYyMDEzNmIwMzAxMjAxMzdlYzA-1YTAwMDEscDosYTosbTo=/show.vsml (accessed December 25, 2012).

Embassy of the People's Republic of China in the Republic of Zambia. 2012. H.E. Chinese Ambassador Zhou Yuxiao lecture for Zambian University students, April 3, 2012. Available at http://zm.chineseembassy.org/eng/sgzxdthxx/t920669.htm (accessed December 5, 2012).

Foster, V., W. Butterfield, C. Chen, and N. Puhak. 2009. *Building Bridges: China's Growing Role as Infrastructure Financier for Sub-Saharan Africa*. Washington, D.C.: World Bank.

Gadzala, A. 2012. "From Formal- to Informal-Sector Employment: Examining the Chinese Presence in Zambia." *Review of African Political Economy* 37 (123): 41–59.

Gu, J. 2011. "The Last Golden Land? Chinese Private Companies Go to Africa." Institute of Development Studies (IDS) Working Paper no. 365. Available at http://www.ids.ac.uk/files/dmfile/Wp365.pdf (accessed December 5, 2012).

Haugen, H. 2011. "Chinese Exports to Africa: Competition, Complementarity and Cooperation between Micro-Level Actors." *Forum for Development Studies* 38(2):157–176.

Hirsch, A. 2013. "Ghana Deports Thousands in Crackdown on Illegal Chinese Goldminers." *The Guardian* July 15. Available at http://www.theguardian.com/world/2013/jul/15/ghana-deports-chinese-goldminers (accessed September 27, 2013).

Hu, S. 2012 "Mobile Phone Souring Africa's Image of China." *Caixin Online,* December 13, 2012. Available at http://english.caixin.com/2012-12-13/100472037.html (accessed December 18, 2012).

Human Rights Watch. 2011. *You'll Be Fired If You Refuse: Labor Abuse in Zambia's Chinese State-owned Copper Mines.* Available at http://www.hrw.org/sites/default/files/reports/zambia1111ForWebUpload.pdf (accessed December 5, 2012).

Liu, J. 2010. "Contact and Identity: The Experience of 'China Goods' in a Ghanaian Marketplace." *Journal of Community & Applied Social Psychology* 20: 184–201.

Lyons, M., A. Brown, and L. Zhigang. 2012. "In the Dragon's Den: African Traders in Guangzhou."

*Journal of Ethnic and Migration Studies* 38(5): 869–888.

Ma, M. 2008. "Chinese Migration and China's Foreign Policy." *Journal of Chinese Overseas* 4(1):91–109.

*Mail and Guardian*. 2012. "Angola's Trophy City: A Ghost Town." Available at http://mg.co.za/article/2012-11-23-00-angolan-trophy-city-a-ghost-town (accessed December 5, 2012).

Marfaing, L., and A. Thiel. 2011. "Chinese Commodity Importers in Ghana and Senegal: Demystifying Chinese Business Strength in Urban West Africa." German Institute of Global and Area Studies (GIGA) Working Paper no. 180, Hamburg, Germany.

Mawdsley, E. 2007. "Fu Manchu versus Dr. Livingstone in the Dark Continents? Representing China, Africa and the West in British Broadsheet Newspapers." *Political Geography* 27(5):509–529.

McNamee, T. 2012. "Africa in Their Words: A Study of Chinese Traders in South Africa, Lesotho, Botswana, Zambia and Angola." Brenthurst Discussion Paper 2012/03. Available at http://www.thebrenthurst-foundation.org/files/brenthurst_commisioned_reports/Brenthurst-paper-201203-Africa-in-their-Words-A-Study-of-Chinese-Traders.pdf (accessed December 13, 2012).

Medina, S. 2012. "A Ghost City in Angola, Built by the Chinese." *Atlantic Cities*, July 17, 2012. Available at http://www.theatlanticcities.com/design/2012/07/ghost-city-angola-built-chinese/2608/ (accessed December 17, 2012).

Mohan, G. 2012. "China in Africa: Impacts and Prospects for Accountable Development." Effective States and Inclusive Development (ESID) Working Paper No. 12, School of Environment and Development, University of Manchester, UK.

Mung, M. 2008. "Chinese migration and China's foreign policy in Africa." *Journal of Overseas Chinese* 4:91–109.

Murray, L., B. Morrissey, H. Ojha, and P. Martin-Menard. 2011. "African Safari: CIF's Grab for Oil and Minerals." *Caixin Online*, October 17, 2011. Available at English.caixin.com/2011-10-17/100314766.html (accessed December 15, 2012).

Mwanawina, I. 2008. "China-Africa Economic Relations: The Case of Zambia." Africa Economic Research Consortium Working Paper. Available at www.aercafrica.org/documents/china_africa_relations/Zambia.pdf (accessed December 5, 2012).

Naidu, S., and D. Mbazima. 2008. "China-Africa Relations: A New Impulse in a Changing Continental Landscape." *Futures* 40(8):748–761.

Neuwirth, R. 2012. *Stealth of Nations: The Rise of the Informal Economy*. New York: Anchor.

Nossiter, A. 2013. "China Finds Resistance to Oil Deals." *New York Times*, September 17. Available at http://www.nytimes.com/2013/09/18/world/africa/china-finds-resistance-to-oil-deals-in-africa.html?pagewanted=all&_r=0 (accessed September 26, 2013).

Park, Y. 2009. "Chinese Migration in Africa." South African Institute of International Affairs (SAIIA) Occasional Paper no. 24. Johannesburg.

People's Republic of China. 2013. China-Africa Economic Trade Cooperation 2013. Available at http://news.xinhuanet.com/english/china/2013-08/29/c_132673093.htm (accessed September 30, 2013).

Redvers, L. 2012. "Angola's Chinese-Built Ghost Town." *BBC News Africa*, July 2, 2012. Available at http://www.bbc.co.uk/news/world-africa-18646243 (accessed December 15, 2012).

Sautman, B., and Y. Hairong. 2009. "African Perspectives on China-Africa Links." *The China Quarterly* 199(9):728–759.

Spieldoch, A., and S. Murphy. 2009 "Agricultural Land Acquisitions: Implications for Food Security and Poverty Alleviation." In *Land Grab? The Race for the World's Farmland*, eds. M. Kugelman and S. Levenstein, pp. 29–56. Washington, D.C.: Woodrow Wilson International Center for Scholars.

Spring, A., and Y. Jiao. 2008. "China in Africa: Views of Chinese Entrepreneurship." International Academy of African Business and Development. Proceedings of the 9th Annual Conference, University of Florida, pp. 55–64.

Stratfor 2012. "Chinese investment offers in Africa." Available at http://www.stratfor.com/image/chinese-investment-offers-africa

Taylor, I. 2011. *The Forum in China-Africa Cooperation (FOCAC)*. New York: Routledge.

Taylor, I. 2012. "The Ongoing Development of Chinese Ties with Sub-Saharan Africa." In *Africa South of the Sahara*, pp. 25–31. London: Europa Publications.

Will, R. 2012. "China's Stadium Diplomacy." *World Policy Journal* 29:36–43.

World Bank. 2011. *Chinese Investments in Special Economic Zones in Africa: Progress, Challenges and Lessons Learned*. Washington, D.C.: World Bank.

Yang, Y. 2012. "African Traders in Guangzhou. Routes, Reasons, Profits, Dreams." In *Globalization from Below: The World's Other Economy*, eds. G. Mathews, G. Ribeiro, and C. Vega, pp. 154–170. New York: Routledge.

**WEBSITES**

Lekki Free Trade Zone. http://www.youtube.com/watch?v=AL8cReBAfoA (accessed December 8, 2013).

The Guardian. "The Price of Gold." http://www.theguardian.com/global-development/video/2013/apr/23/price-gold-chinese-mining-ghana-video (accessed December 6, 2013). Video on China's involvement in artisanal mining in Ghana.

Roberto Castillo. *Africans in China*. http://africansinchina.net (accessed December 6, 2013). Blog on Africans in China.

CHAPTER 13

# AFRICAN FUTURES

## INTRODUCTION

2018 will be a benchmark year for Africa: the contemporary independence era (1951–2018) will have lasted as long as the colonial interlude (1884–1951). Despite the passing of time, the broad subordinate and asymmetrical relationship that tied Africa to the great powers of the 20th century, including the United States and the former Soviet Union as well as the Europeans (in terms of economic specializations, trade, foreign investment, brain drain, and external oversight), has remained intact until recently.

There are encouraging signs that some elements are changing. The rise of the Brazil, Russia, India and China (BRIC) and the waning of the Western development model as the only game in town mean that African leaders currently enjoy more options in international relations. BRIC membership even includes one African country—South Africa—that has been elevated within the elite group of Southern powers (even though Pretoria is the much lesser power within the grouping). BRICs are promoting South–South cooperation to shape the 21st century.

In the 2000s, African states brought the decolonization episode to a close (Severino and Ray 2011). Casting off the Western development model and oversight from the International Monetary Fund and the World Bank, African leaders took control of their own economic and international policies. This allows African leaders to form economic and political alliances of their own choosing. Such enhanced agency has enabled Africa states to become full-fledged players in international relations rather than remaining objects of other nations' foreign policies (Severino and Ray 2011). This new era is the dawn of a second independence for Africa.

Long perceived as marginal to international relations, Africa is becoming a strategic region. There is increasing competition among an enlarged group of powers to engage Africa at a scale not witnessed since the original "scramble for Africa." Important decisions are still made in London, Paris, Berlin, and Washington, D.C.—capitals of 20th-century great powers—but equally consequential decisions are now made in BRIC capitals—Brasília, Moscow, New Delhi, Beijing, and Pretoria. China, in particular, is pursuing a spectacular expansion and rise.

These days, there is not a middle-ranking power (rapidly developing powers with growing international influence) that is not deepening its ties with Africa. Brazil's embassy footprint in the region in 2013 (operating diplomatic embassies in 37 countries) is as large as the United Kingdom's (with missions in 39 African countries). The Gulf states are increasing their investments in Africa, particularly in terms of land and other natural resources, and trade between the two regions is growing. Canada and the Nordic countries are maintaining their ties with strong humanitarian support. Moscow is reengaging with Africa (after turning away at the end of the Cold War) and has written off US$11.3 billion of sovereign debts accumulated between 2005 and 2012. Even nontraditional partners are firming up ties. Tehran declared 2008 as the year of expanding relations between Iran and Africa, and in the same year Ankara hosted the Turkey–Africa Cooperation Summit (Taylor 2010). Largely unnoticed, in 2011, Malaysia became the third largest investor in the region: almost 25% of Malaysian foreign development investment flows to Africa (Miles 2013).

For their part, African leaders are extending their horizons. Some African oil producers are cementing partnerships with the United Arab Emirates to learn from Abu Dhabi's vast experience in oil production. Kenya's Vision 2030 strategy document borrows heavily from the experiences of Malaysia and Singapore. Mozambique is implementing a staff-training partnership with Brazil to learn from its experience. African countries are sending numerous delegations to participate in Chinese training programs (more than 129 programs have been offered in railway construction, planning and development, coastal region economic development, "commercial Chinese" language instruction, etc.) (Brautigam 2012).

The African tilt toward democratization has also broadened opportunities for society's input into various policy arenas. The first generation of African presidents pursued foreign policies strongly tied to the interests of former colonial rulers. The current generation of leaders is more willing to accommodate demands of their own populations in determining the best interests of the state. More swayed by popular sentiment, governments these days, in particular instances, have canceled Chinese building contracts, advocated for the indigenization of industries, and established curbs on foreign ownership of business operations.

Over the past decade, African leaders have become more prominent actors in high-level international politics, for example in climate negotiations and a range of South–South coalitions (China-India–Brazil–South Africa known as the Basic Group and South America–Africa Cooperation Forum). More prominence is also accompanied by a rising assertiveness of Africa-based institutions, for instance the African Union (AU) since its 2002 launch. AU's vision is for an "Africa integrated, prosperous and peaceful, an Africa driven by its own citizens, a dynamic force in the global arena" (African Union Commission 2004:7). Registering successes in peacekeeping missions in Burundi (2003), Darfur (2004), Somalia (2007), and Mali (2012) and creating the African Standby Force (a 15,000-person force) are noteworthy achievements. Importantly, the AU spoke with one voice at the 2009 United Nations (UN) Climate Change Conference (Copenhagen Summit). The most important statement of intent is AU's adoption of the New Partnership for Africa's Development, a blueprint for economic development that confirms African leaders' faith in foreign development investment and global partnerships. The AU, however, stills needs considerable development and resources to become a regional force.

Africa countries are enjoying a greater ability to influence bilateral relations. The postconditionality era has ushered in greater divergences in African states' relations with various donors. For instance, Uganda, Rwanda, and Ethiopia are taking ownership of aid agendas and priorities. Kigali has demonstrated considerable successes in steering Development Assistance Committee donors toward its aid priorities and in substituting Chinese aid for Western aid following Development Assistance Committee (the twenty-nine wealthy Organization for Economic Cooperation and Development donors) aid cuts in 2012 after accusations of Kigali's involvement in the Democratic Republic of Congo (DRC).

Africa is a sought-after diplomatic partner. In a world where numbers matter, the votes of 54 African states count in such areas as trade, climate, and international security negotiations. Moreover, as democracy makes further headway throughout the world and as the principle of democratic representation becomes more entrenched, Africa's burgeoning population (by 2050 Africans will outnumber U.S. and European populations by two to one) and number of national states will make it harder for a handful of 20th-century great powers to justify their monopoly. Democratization of world governance will have to be reapportioned to include more historically underrepresented regions such as Africa.

According to Ian Taylor (2010:6), "Contra to the notion that Africa is a passive bystander in global processes, African elites have generally proven themselves excellent arch-manipulators of the international system." Nevertheless, the collective presence of Africa on the international stage is still a work in progress. Indeed, historical legacies still weigh heavily. African states remain minor powers shackled by a high level of poverty and underdevelopment, degraded urban and rural environments, inadequate institutional capacity, corrupt political systems, and subservient roles in setting global agendas. Despite increased space and maneuverability produced by significant shifts such as the rise of the BRICs and of Gulf states' investments in the region, African leaders operate in a tight space bequeathed by the state system inherited at independence, their colonial incorporation into the global economy, and the structure of weighted voting allocations at the World Bank and the International Monetary Fund (approximately 6% in 2011) (Brown 2012). While the media flip-flops between simplified notions of extremes of "hemmed in" or "seizing the 21st century" (e.g., *The Economist* magazine's flip from "hopeless" [2000] to "rising" [2013a]), it is more accurate to portray African leaders as operating in a space somewhere between these extremes.

## WASHINGTON'S ENGAGEMENT WITH AFRICA

The United States is Africa's largest foreign aid provider. Its support of the region increased from US$1.4 billion in 2001 to US$9.2 billion in 2010, a level equivalent to more than one quarter of all U.S. international economic assistance (U.S. GAO 2013:31) (Fig. 13.1). The leading category of development assistance is health (Washington is the largest contributor to the Global Fund, which targets combating HIV/AIDS, tuberculosis, and malaria) followed by humanitarian aid. Support for infrastructure is only a minor part of U.S. foreign aid. The U.S. strategy is driven not only by African development concerns: Washington is also motivated to secure access to Africa's natural resources (Carmody 2011) and to combat pandemics that could pose global security threats (Taylor 2010).

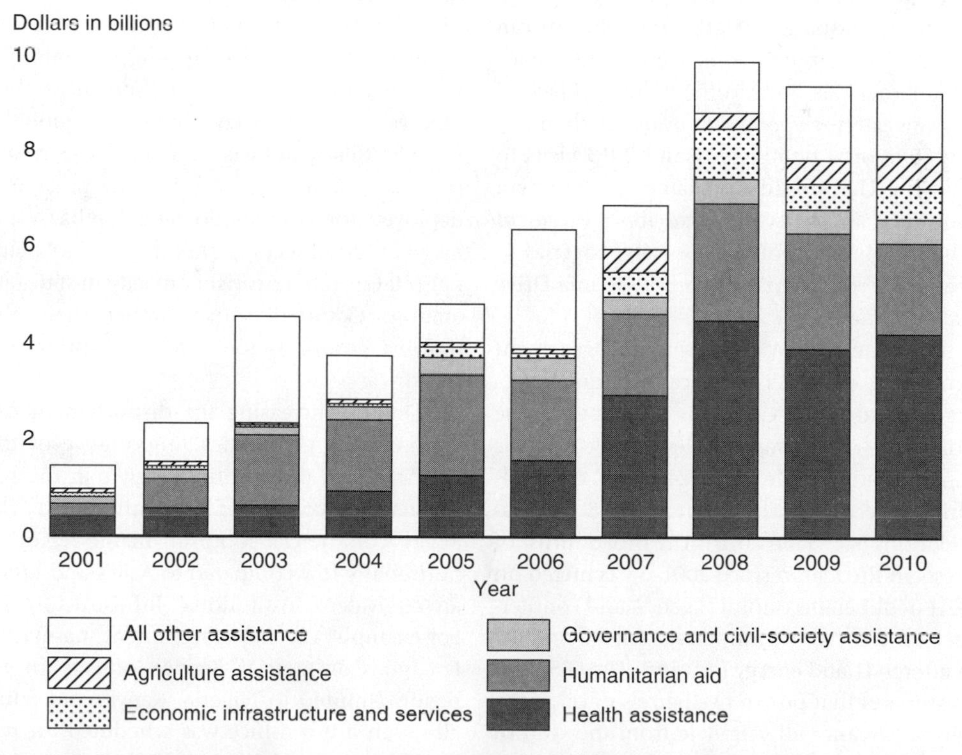


**FIGURE 13.1** U.S. Government Development Assistance to Africa, 2001–2010. *Source:* U.S. GAO, p. 32, Fig. 10.

By comparison, the Chinese government aided Africa to the tune of US$6 billion between 2001 and 2009 (Brautigam 2011), but Beijing is more deeply engaged in infrastructure-for-mineral swaps and in financing African infrastructural projects (roads, buildings, stadiums, dams) via interest-free loans and gifts. When available national data are compared (e.g., Angola, Ghana, Kenya), Chinese government loans and related financing exceed the level of U.S. aid (U.S. GAO 2013). Moreover, Chinese loans are more attractive to host governments: the borrower's costs are lower, and Beijing allows for greater flexibility in using such local resources as labor and materials. China's more intense involvement is focusing global attention on Africa and thwarting U.S. regional interest.

Despite a surge of involvement by middle powers (i.e., states with considerable international power but lesser than superpowers) in the region, the United States still exerts substantial hard-power advantages on the basis of its longstanding relationships, generous development assistance, economic prowess, and global status. It also maintains a soft-power advantage from its African diaspora ties and from the African heritage of its commander-in-chief. President Barack Obama (his father was born in the Kenyan village of Kogelo) is an extremely popular leader with many Africans, underscored by his 2009 and 2013 visits to the region. The U.S. president's name is prominent throughout rural and urban Africa, from President Barack Obama High School (Benue State, Nigeria) to Hotel Obama (Accra, Ghana) to Barack Obama Drive (Dar es Salaam, Tanzania).

Still, the context of Washington's reengagement with Africa is occurring after two decades of disengagement after the end of the Cold War. At that time, the absence of nuclear weapons, low levels of economic activity, and a minimal role in international trade relegated Africa to secondary importance in U.S. geostrategic calculations (de Walle 2010). The broad thrust of U.S. interests in the region since 2001–13 centered on advancing a neoliberal economic agenda and countering threats to U.S. security interests (most notably terrorism after 9/11 and energy security). The U.S. government estimates that one in five barrels of oil entering the global stream will originate from the Gulf of Guinea, West Africa (Carmody 2011:48), so Africa now fits into the United States' newer energy security nexus, whereby increasing oil flows from the region allow for reduced dependence on Middle Eastern oil.

Post-9/11, Washington became preoccupied with terrorism and especially the influence of Al-Qaeda in the region. A good example of the recent upgrading of Africa's strategic place in U.S. geopolitical calculations is the establishment of Washington's Africa Command (AFRICOM). The Pentagon originally planned to base AFRICOM within the region (Liberia and Botswana were considered), but African opposition resulted in the command being relocated to Stuttgart, Germany. However, two field operational sites—Ascension Island and Djibouti—serve as forward region supply bases. The United States operates two drone bases in the region, one in Djibouti and the other in Mali, and Washington may be running drone missions out of other African states but the details are classified. However, using drones for surveillance in Africa's conflict zones is becoming more common. For example, the UN is deploying drones in the DRC to monitor militias and cross-border movements. For the most part, the United States is engaging in focused counterterrorism. This involves relying on units within the U.S. First Infantry Division (known as the Big Red One, based in Fort Riley, Kansas), a regionally aligned brigade that conducts counterterrorism and rapid-response missions in Africa. This brigade has been deployed for commando raids, embassy protection, and UN peacekeeping training and to help develop national counterterrorism capacity in state–state cooperation. Counterterrorism rather than eliminating terrorist groups is the core U.S. anti-terror strategy for Africa.

Despite increasing the disbursement of aid and elevating the geopolitical importance of the region, the American diplomatic presence in the region has, surprisingly, contracted (de Walle 2010). The United States' commercial footprint in the region is disproportionally low compared to Asia's and Europe's, and an equivalent institutional infrastructure is lacking. For example, the U.S. government maintains only six Foreign Commercial Service officers in the entire region (limited to Nigeria, Kenya, and South Africa, although a new office was scheduled for reestablishment in Ghana in August 2013) (Coons 2013:13).

Commercial Service officers play a key role in providing information on potential African opportunities for U.S. businesses (e.g., evaluating potential in-country partners, facilitating U.S. business contacts with local firms, and providing local financing options).

Washington is raising the importance of U.S. commercial engagement in its Africa strategy, calling for increased trade and investment. This is motivated by a perception that Washington has lost ground in Africa, having already surrendered economic opportunities (China became Africa's largest trading partner in 2009, overtaking the United States), and is in danger of ceding political leadership to its global competitors. According to Chris Coons, Chair of the Senate Foreign Relations Subcommittee on African Affairs, the United States "has been asleep at the wheel" in its Africa policy and has failed to implement "a more aggressive approach to economic engagement in sub-Saharan Africa." As a consequence, "Chinese competitors are securing long-term contracts that could lock American companies and interests out of fast-growing African markets for decades to come" (Coons 2013:2, 13). Some (e.g., de Walle 2010) claim that American foreign policy in Africa "lacks ambition" and that Washington is ineffective at developing policies toward regions regarded as secondary in importance.

Quite surprisingly, President Obama distanced himself from Africa during his first term. To date, a distinctive set of Obama Africa policies is barely discernible, even though the Senate Foreign Relations Subcommittee on African Affairs has articulated a new policy with economic engagement as the centerpiece. The prospect of continued and sustained economic growth in the region, abundant supplies of minerals and other natural resources, and the emergence of an African middle class is behind a U.S. reprioritization of its Africa goals, but to date a bold and attentive Obama policy for Africa is missing. The Obama administration is planning, in 2015, to reauthorize and strengthen the African Growth and Opportunity Act, a set of measures to extend trade preferences, but unfortunately the economic and fiscal crises of the second Obama term are inhibiting a more expansive U.S.–Africa engagement. Reenergizing U.S.–Africa policy will require much more contact with African leaders (and not just the youth) and regular visits by senior U.S. officials (as Chinese leaders have been doing for years). Howard French (2013:3) observes that the Obama administration "must put an end to the belittling, small ball ritual whereby African leaders are invited to Washington in groups of three or four (as if an African country by definition does not merit a one-on-one discussion), offered a quick photo opportunity, a few homilies about democracy and governance and sent on their way."

## AFRICA'S TERRORISM THREATS

Africa has a recent history of terrorist strikes on Western targets that precede the 9/11 attacks in the United States in 2001. For example, in East Africa, the bombings of U.S. embassies in Nairobi and Dar es Salaam in 1998 killed 224 people—including 12 Americans—and injured 5,000. Intermittent attacks have occurred since on diverse domestic as well as Western targets. An attack on soccer fans watching the 2010 World Cup final on TV in Kampala killed 74 people, but most have been smaller, lower-risk attacks on churches, nightclubs, bars, and police and army officials. Terrorism has also been accompanied by various efforts to generate funds for extremist agendas, such as the kidnapping and abduction of Westerners (e.g., Kenya, Mali, Nigeria and Somalia), payment extortions from aid organizations, bank robberies, and human trafficking. Remittances, donations, and contraband smuggling raise the bulk of terrorism funds.

The attack on the Westgate shopping mall in Nairobi on September 21, 2013, by Al-Shabaab insurgents ("Shabaab" means youth in Arabic) killed 67 shoppers and sent shockwaves about African terrorism around the world (Fig. 13.2). A week later, a radical Islamic group murdered 40 college students while they slept in their dorm in Gujba, northern Nigeria, showing another vicious side of African extremism. In the same week Al-Qaeda in the Islamic Maghreb (AQIM) detonated a car bomb in Timbuktu, killing two people, an incident that followed their series of kidnappings of Westerners in Mali that began in 2003. Figure 13.3 shows the most active terrorism corridors in Africa.

Some consider Africa to be a new frontline for international terrorism. Contributing factors to this viewpoint are fears about new attacks on soft targets, an African momentum for jihad, a growing international dimension to African terror organizations, cooperation among terrorist groups, and the possibility of a coordinated and catastrophic event.

Others warn us not to jump to conclusions about the geographical shift of terrorism to Africa. One countering viewpoint emphasizes that African extremist groups affiliate with global terror groups only when they are on the wane. Accordingly, the Westgate bombing and other terrorist incidents should be interpreted as desperate efforts to elevate weakened groups. According to another viewpoint, political rebellions and criminal networks are intertwined in controlling routes for contraband trade and other activities, and although alliances may be brokered at times with jihadist extremists, the main undercurrent remains unresolved national political issues rather than an African spiral toward a jihad.

**FIGURE 13.2** Terrorism, Westgate Mall, Nairobi, 2013. *Source:* © GORAN TOMASEVIC/Reuters/Corbis. Corbis image 42-51574593.

News stories tend to reflect emotions and fear more than informed research. Naming African extremist groups and giving them a coherent identity makes them scarier. The reality is that many of Africa's current extremist groups are not uniform entities but factionalized organizations that are constantly evolving in response to internal power struggles and global responses. Hard data about these groups are rare: contact with these organizations is difficult and their agendas change and are kept secret. Hansen's (2013) fascinating study of Al-Shabaab reveals that it is constituted

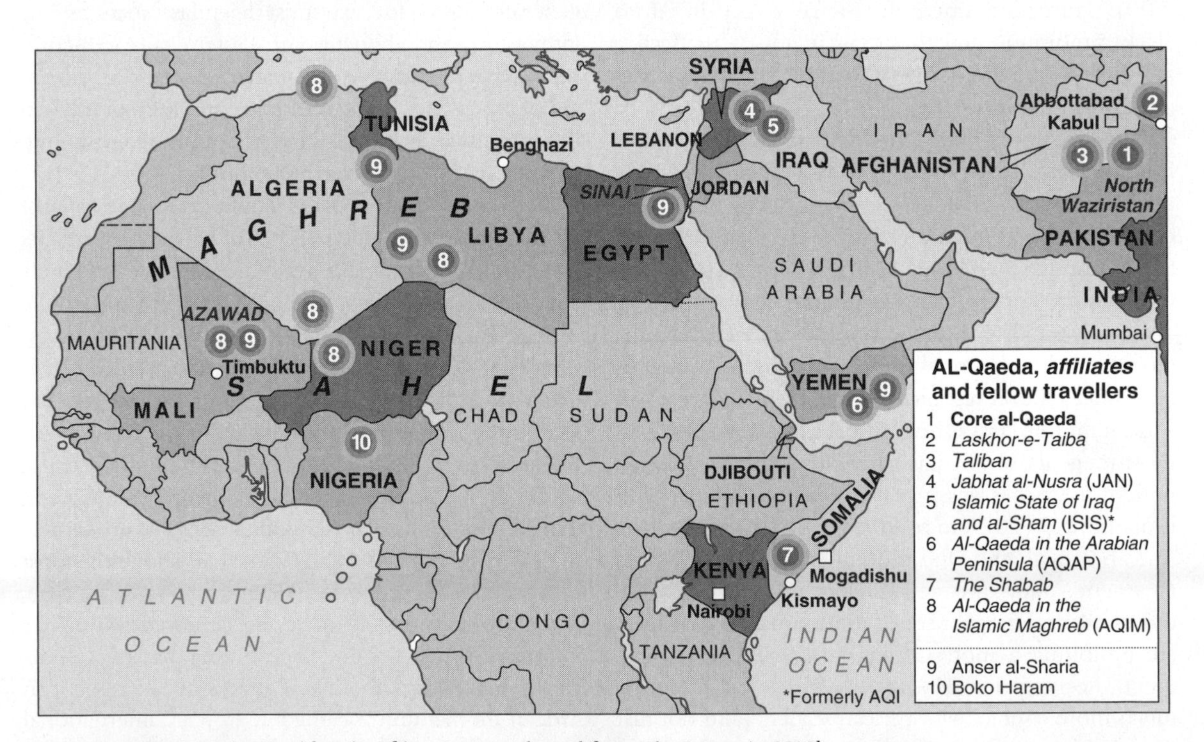


**FIGURE 13.3** Terrorist Corridors in Africa. *Source:* Adapted from *The Economist* 2013b.

by loose ties, political/security alliances of clans, sub-clans, sub-sub-clans, pseudo warlords, and global jihadist networks. Extreme localism, nationalism, pan-Islamism, and global networks are intermingled, making it hard to decipher the goals of the network. As a result, the subject of terrorism in Africa is always going to be a highly speculative one, driven by particular partial readings of African complexity and heavily influenced by emotion and perception of risk.

Several hotspots of terrorism activity are evident in sub-Saharan Africa. First, there is an East African corridor where Al-Shabaab operates. This organization has been a U.S.-designated foreign terrorist organization since 2008. Al-Shabaab emerged during a period in 2006 when Ethiopia invaded Somalia. The organization was attractive to recruits, and in a power vacuum it consolidated an extensive geographical base. From 2007 to 2011, the organization controlled much of south and central Somalia, including parts of Mogadishu and the port city of Kismayo. Controlling Kismayo was very profitable for all kinds of trade ventures. For example, the illegal charcoal trade, which resulted from burning the acacia tree (prohibited in Somalia) and shipping it to the Arabian Peninsula, brought in US$25 million per year (Gettleman and Kulish 2013). Al-Shabaab has become involved in the illegal ivory trade, which has grown in recent years to become one of the organization's main funding sources. Repeated Kenyan and Ethiopian troop invasions, combined with a 2010 intervention by the UN-backed AU force, eventually weakened Al-Shabaab's geographical control of urban centers but not its rural grip. Subsequently, under the leadership of Ahmed Abdi Godane, the organization shifted its focus from territory to terrorism. The organization is very successful in recruiting and operating internationally, even within the United States (40 Americans are known to have joined Al-Shabaab as fighters, and several U.S. citizens have been convicted of raising funds for the organization), and it has had some successes in exporting jihadists (e.g., Kenya and Nigeria) (Hansen 2013). Since 2010, the organization has received strong backing from Al-Qaeda as well as support from Eritrea.

AQIM operates in a second corridor, largely in the Sahel, with a logistical base centered in northern Mali and a zone of activity that extends to Mauritania, Niger, Burkina Faso, and Algeria. This amorphous group thrives in areas weakly controlled by governments. Vast distances, dispersed populations, and deep poverty enable AQIM to engage in cross-border crime and insurgency. This group emerged out of the Algerian civil war in the 1990s. AQIM was placed on the list of U.S.-designated foreign terrorist organizations in 2002 and it announced its global jihad in 2007. In 2012 AQIM allied with Tuareg rebels in northern Mali and exported its power center to the vast ungoverned north of Mali. The Tuareg, a nomadic group that represent 10% of the population of Mali, have been engaged in a political campaign for self-determination since the end of the French colonial presence in the region, A French-led military intervention in 2013 was needed to thwart the southward advance of the joint AQIM–Tuareg group; at its peak they controlled 60% of the national territory. More hardline AQIM extremists broke away from the Tuareg in January 2013. AQIM aspires to rid the region of Western influence and to install fundamentalist regimes that will function under Islamic religious (Sharia law), whereas the Tuareg are more moderate Islamists with political ambitions focused on states with a Tuareg presence (e.g., Mali, Niger, Algeria, and Libya). AQIM, in contrast, maintains transnational ties and has global ambitions. Some AQIM volunteers were active in the Iraq war and participated in suicide bombings against the U.S. armed forces. AQIM also participates in international narcotic smuggling, providing a secure waystation between suppliers in South America and the European market.

Groups under the banner of the "Congregation of the People of Tradition for Proselytism and Jihad" operate in a third corridor. The Nigerian government labels these groups "Boko Haram," a label that has stuck. Boko Haram is a Hausa translation of "Western education is sinful." To be fair, there is neither one group (but several) nor a single platform but rather a political and religious Islamist agenda to prevent people from being influenced by Western values, corruption, and easy money. Loose groupings of radical Islamists operate within a zone centering on northeast Nigeria. The group emerged in 2002 and set up a separatist community on the basis of Islamic fundamentalism in the village of Kanama on the Nigeria–Niger border. Labeled "the Nigerian Taliban," the following year the group began to attract notoriety. A 2011 suicide bomber attack on a UN compound in Abuja killed 23 people and launched Boko Haram onto the international stage as an outfit with the technical and doctrinal capacity to upscale its activities. At that time many considered the UN bombing to be an outlier event that detracted from the group's central mission to advocate for an Islamist state as a response to the moral and financial corruption of Nigeria's establishment, widespread poverty, and the poor economic situation in northern Nigeria. However, an escalation of attacks on government institutions, churches and secular schools since has resulted in the killing of over 4,000 people (2011–2014) and student abductions are on the rise. A good proportion of recent deaths have been students so that government has been forced to close five high schools in March 2014 (Council on Foreign Relations 2014). Mainly supported through bank robberies this group has neither attracted support nor financing outside of northeastern Nigeria. During its first decade the group mainly focused on internal northern Nigerian targets, and this kept them off the list of U.S.-designated terror groups. However, in 2013 Washington designated the group a foreign terrorist organization, signaling a shift in U.S. intelligence thinking about the evolution of terrorist activities in Nigeria and Africa in general. Analysts differ, however, over whether Boko Haram maintains a jihadist agenda or not (Walker 2013).

There are widely diverging perspectives about the emergence of Africa as a base for terrorism. Beyond these three zones, terrorist activity has been recorded in other African states. For example, Al-Qaeda cells in Ethiopia have been dismantled, and Tanzania and South Africa have ongoing counterterrorism efforts focusing on Al-Shabaab activity. Military and security experts have been warning for years that the region will assume a more central role in international terrorism (Chillers 2003), and this is reflected in the perspective that Africa has emerged as a terrorism front. The specter of terrorism in the region is enhanced by an abundance of physical safe havens, porous borders, and endemic poverty, all against a background of protracted ethnic and other conflicts and undertrained and underequipped national military forces. Specific spaces in slums, rural deserts, maritime zones, and so forth operating outside of government control and purview are potential hotbeds for radicalization (Davis 2010a). Obi's (2007) assessment that the Sahel could serve as a trans-African highway and recruitment ground for terrorists to move back and forth between Africa and other regions to attack Western interests and/or to destabilize African states is being realized.

Examples of pariah states that serve as hotbeds and refuges for terrorists include Somalia and Sudan. One U.S. official referred to Sudan (1991–96) as a "Holiday Inn for terrorists" (quoted in Menkhaus 2009:97). A number of international terrorists, including Osama bin Laden, used Sudan as a safe haven and operations base until the notoriety of pariah status became too costly for the Khartoum government in the mid-1990s. At that time, Islamist leader Hassan al Turabi was sidelined and "foreign guests" were requested to move on. Given the complexity of political and development challenges in the region, infiltration of terrorists in search of sanctuary is always a strong possibility.

It is important to emphasize that poverty does not breed terrorists. Many Africans have long experienced conditions of prolonged and extraordinary poverty and have still managed to survive. Nevertheless, frustration is growing over a lack of employment options and opportunities (especially as perceived by an enlarged youthful cohort) as macroeconomic conditions improve and rising expectations fail to materialize. As a consequence, alienation, marginalization, and (possible) radicalization are new challenges for Africans and the international community to manage.

A second perspective points to the radicalization of Africans both within and beyond Africa (Menkhaus 2009). One Minneapolis-based member of the Somali diaspora, Shirwa Ahmed, served as a suicide bomber as part of a series of terrorist attacks in Somalia in 2008. He was among a dozen or so Somali-American young men who suddenly went missing in Minneapolis but resurfaced in Somalia after being allegedly indoctrinated and trained by the Somali jihadist group Al-Shabaab. According to the U.S. Federal Bureau of Investigation (FBI), Ahmed was the first U.S. citizen to commit an act of terror for an Al-Qaeda–affiliated group. There is considerable speculation about how this group of young men in the United States become radicalized and mobilized. Other American jihadists have fought and died for Al-Shabaab. No doubt Wahhabist and jihadist websites exposed Somalis and others to this worldview, and such exposure appears to have been important in radicalizing the Minneapolis group, in combination with other drivers of radicalization (e.g., after-school youth group leaders or clerics in local mosques). The Minneapolis misfits may have been a precursor to the rise of homegrown terrorists taking on extremist jihadist worldviews. The Boston Marathon bombings of 2013 indicate that individuals with extremist tendencies (or weakly connected to international extremist groups) may act independently and carry out heinous acts.

A far greater terrorist threat emanates from non-African operatives (or non-African diasporas) exploiting weak state security, high levels of corruption, a booming illicit economy, and myriad soft (Western) terrorism targets. The region provides a permissive operating environment for terrorists from other regions of the world, and terrorist networks are engaging in illicit economic activities to generate funds. Smuggling of precious metals is a lucrative endeavor that attracts various clandestine networks, including terrorists. Al-Qaeda has been active in the smuggling of conflict diamonds in West Africa. Hezbollah, an organization often perceived as supporting or engaging in terrorist acts, has participated in diamond smuggling and other illicit economic activities in the DRC via a small network of Lebanese members of the diaspora. Non-African terrorist networks also engage in "legitimate" African commerce to provide cover, to build local contacts, and to use local partnerships to launder money.

A third perspective contends that the prospect of Africa emerging as a critical arena in global terrorism is overblown (Davis 2010a). This appears to be the case especially in regard to jihadist extremism. Africans, of whatever faith, exhibit very low rates of suicide and are culturally averse to suicide attacks as a terrorist tactic. The dominant African version of Islam (Sufism) is more resistance to radical interpretation, and this may, in part, explain why African Muslims in the diaspora are less predisposed toward radicalization. It is significant that Africa's large numbers of Muslim emigrants have produced few recruits (leaders or foot soldiers) for Al-Qaeda and other international terrorism networks. (Aberrations include the London killing of an army reservist in 2013, a botched London subway bombing in 2005, and the 2008 recruitment of U.S.-based Somali diaspora members to the Al-Shabaab jihadist group in Somalia, a few of whom participated in the Westgate 2013 bombing.) The global picture is that Africans in the diaspora demonstrate "exceptionalism" in that its poor members suffer from the same conditions conventionally cited to explain the rise of radicalism and terrorism (segregation, isolation, marginalization, victimization) among some members of the Muslim diaspora, but Africans are far more likely to shun radical activism and to participate only in mainstream political and civil society. Indeed, guest workers' exposure to more radical Islam in the Gulf states has not radicalized them once they return home. The African Muslim diaspora in the United States enjoys greater levels of social and economic mobility, whether arriving as a privileged elite or not: a positive immigrant experience yields very different results in the United States as opposed to the United Kingdom (where radical activism is more frequent).

In general, it seems that Africans' obligations to send remittances to their homeland requires them to concentrate on working and earning income and avoiding any activities that could jeopardize their ability to send and receive remittances. Thus, any activity that undermines the flow of remittances faces a strong, knee-jerk reaction in their host and home communities. Generally, African diaspora involvement in Al-Qaeda is nowhere near the level generated by diasporans originating from North Africa, the Middle East, and South Asia.

However, the African *exceptionalism thesis* does not explain the existence of longstanding subnational terrorism. Spectacular attacks have captured the global imagination, and Africa (1990–2002) recorded the second highest number of casualties from internal terrorism of any region after Asia (Davis 2010a:135). Ethnic-based insurgencies, classified as terrorist acts in their own right, reveal that Africa's most lethal terrorists are not Islamic jihadists but pathological insurgents. Examples abound, including the activities of the Revolutionary United Front (Sierra Leone), the Lord's Resistance Army (Uganda), various warlord conflicts in DRC, numerous ethnic manifestations of terrorism such as clan/sub-clan violence in Somalia, and the

Interahamwe, which perpetrated the Rwandan genocide. Notably, Africans have been far more willing to support and fund terrorist activities linked to local and national movements than international/global agendas.

In African policy contexts, issues that result in more deaths, such as HIV/AIDS, ethnic conflict, and endemic poverty, eclipse terrorism. Nevertheless, a concern about the destabilizing role of terrorism has risen on national and regional agendas. It is widely acknowledged that terrorism exacerbates the difficulties of economic and sustainable development. Attacks or even threats of attacks discourage trade, investment, and tourism, deplete national income, and displace African communities. Not surprisingly, this vast region has provided ample opportunities for attacking soft Western targets—hotel and mall bombings (Kenya), tourist abductions (Mali), oil worker kidnappings (Nigeria, Sudan), and the like—but the landscape of terrorism in Africa is constantly changing.

## POST-MILLENNIUM DEVELOPMENT GOALS: WHICH ALTERNATIVES?

Most policymakers, academics, and practitioners believe it essential to develop a global consensus around a framework to update and/or replace the Millennium Development Goals (MDGs) in 2015. There is considerable discussion about whether the MDG framework should continue along the same lines, be expanded, or be restricted. There is a consensus that the post-2015 framework needs to include more African ownership by diverse African stakeholders, but widely diverging views exist on how to realize this.

A compelling case is made to allow African states to set their own goals and to have external powers align their development support and financing with these local agendas. Changing general MDG targets to geographically sensitive targets in time and specific places to reflect the diversity of conditions in Africa is also suggested. Broadening the engagement of stakeholders (civil society, trade unions, environmentalists, businesses, etc.) in the post-MDG process seems essential to move beyond the reliance on expert insiders, particularly beyond the donor communities in the Global North. Some argue that more of the indigenous knowledge base should be integrated into economies, agricultural systems, science, and governance. Others maintain that more steadfast support can be marshaled by setting "zero targets" (e.g., zero hunger and zero food waste) as opposed to goals that aim for 50% reductions (the standard target of various MDGs in 2000).

One perspective argues for moving toward an "MDG-minus" framework (Nayyar 2012). Duplication and overlapping targets confuse the development process, so the case is made to reduce the number of indicators and to eliminate subtargets where there is a paucity of data. However, the opposing perspective—an MDG-plus framework—has greater momentum (Nayyar 2012). MDG-plus proponents have come up with various lists of what should be incorporated into the post-MDG agenda. Table 13.1 lists the most important recommendations. Obviously, 18 goals are too many, so hard decisions will have to be made about priorities.

At the same time, there are durable strengths that Africans can draw on (Table 13.2), and African agency and ownership must be part and parcel of a new global framework.

**TABLE 13.1 MDGS POST-2015**

1. Sustainable cities (including sustainable transport, carbon-neutral buildings, air quality, no toxic waste, zero waste)
2. Global warming stabilization and renewable energy
3. Protection of biodiversity and habitats (forests, mineral-rich environs, rivers, deltas, oceans)
4. Civil society empowerment and participation
5. Capacity building at all levels
6. Education
7. Food and land security (hunger elimination, nutrition, sustainable agriculture, organic farming, land reform)
8. Green economy (sustainable production, consumption, and livelihoods)
9. Health (disease control, reproductive health, public health systems, mental health, preventive medicine, and integration of Western and traditional systems)
10. Open migration (borders, remittances, reduce environmental refugees and human trafficking)
11. Poverty elimination and inequality reductions
12. Social justice (human rights, income equality, gender equity, democracy, transparency, accountability, social security)
13. Water access and community rights
14. Decent work, worker protection, job security
15. Global tax reform and elimination of tax havens
16. Information access to technology
17. Industrial and small-enterprise development
18. Sustainable agriculture and sustainable farmer livelihoods

**TABLE 13.2 DURABLE STRENGTHS THAT AFRICA CAN DRAW ON**

- Complex and resilient agro-pastoral systems
- Rich indigenous knowledge
- Water, land, forests, fisheries, animal and mineral wealth
- Cultural, linguistic, and heritage assets
- Creative and talented populations and a connected diaspora
- Long history of resistance in the face of adversity (e.g., slave trading, colonialism and the struggle for independence)
- Strong collective structures at family and village levels
- Vigorous entrepreneurship in the informal economy
- Effective and vocal civil society
- Vocal and active women and women's groups
- Solidarity among African governments and South–South partnerships
- New initiatives in African regional economic cooperation
- New impetus for African peacemaking and peacekeeping roles
- Growing democratization

*Source:* Excerpted from Samatar et al. 2005:331.

## NEW URBAN FUTURES

Thanks to its resource boom, emerging middle class, and rapid urbanization, Africa is being marketed as a new real estate frontier (Watson 2014). Property developers; architecture, planning, and engineering firms; and investment funds have turned to Africa after the property crises in the Global North. Many developers are partnering with international architecture and engineering firms and local property development firms with financial institutions to contribute to the future of cities in Africa. Many urban projects in Africa are inspired by cities like Dubai, Singapore, and Shanghai. Until recently, global cities of the North (London, Paris, New York) were the places to aspire to, but contemporary ambitions now look East.

Numerous large property development companies have launched prototype mixed-use development projects that draw on the expertise of urban designers from South Africa, Brazil, Europe, and the United States. Some projects have obtained jump-start financing from Chinese, Russian, and/or South African investors, who join with local speculators and developers. A few projects even involve Chinese construction companies and others are entirely African entrepreneurial efforts. Figure 13.4 presents the highest-profile projects in terms of area, investment, and international publicity.

Large-scale property developers are embarking on far-reaching projects designed to reshape the configuration of several African cities. Africa has not witnessed such a surge of new planning since the construction of colonial cities in the late 19th and early 20th centuries. These new visions are expressed in two types of urban projects with some overlap: (1) satellite cities, (2) new central business districts, and there is also some effort to retrofit existing development through new green initiatives. Most projects make liberal use of "eco-city" characteristics (e.g., wind and solar power, energy-efficient building, green space).

Many new developments involve building satellite cities for tens of thousands of workers outside the boundaries of existing metropolitan areas. There is a West African corridor of projects stretching from King City to Eko Atlantic as well as an East African cluster that pivots around Tatu City and Konza Techno City (on the outskirts of Nairobi). There is competition between West and East Africa to build Africa's new leading-edge technological city. Hope City in Ghana and Kenya's Konza Techno City are major rivals in this regard: Hope City is touted as "Africa's Silicon Valley" and Konza is promoted as a "Silicon Savannah" and both seek to replicate Bengaluru's rise to global prominence in the information technology field. Satellite cities tend to be located approximately 18 to 37 miles (30–60 km) from major capital cities, although La Cité du Fleuve (DRC) and Malabo II (Equatorial Guinea) are island developments closer to existing urban centers. Many satellite cities are envisaged in the context of polycentric development that will relieve congestion around the city center while offering new world-class infrastructure and unlimited possibilities. Broadly coherent in structure but distinct in design, these elite self-contained developments will be separate from the "mother city."

East Africa is the continent's most active region in planning satellite cities. This may be explained, in part, by the fact that many of the fastest-growing cities in the world are located here (e.g., Kigali, Kampala, and Dar es Salaam) (Society for International Development 2010). Most notably, there are plans to build several satellite cities around Dar es Salaam and Nairobi. These developments would offer greenfield sites for mixed-use developments complete with ultramodern

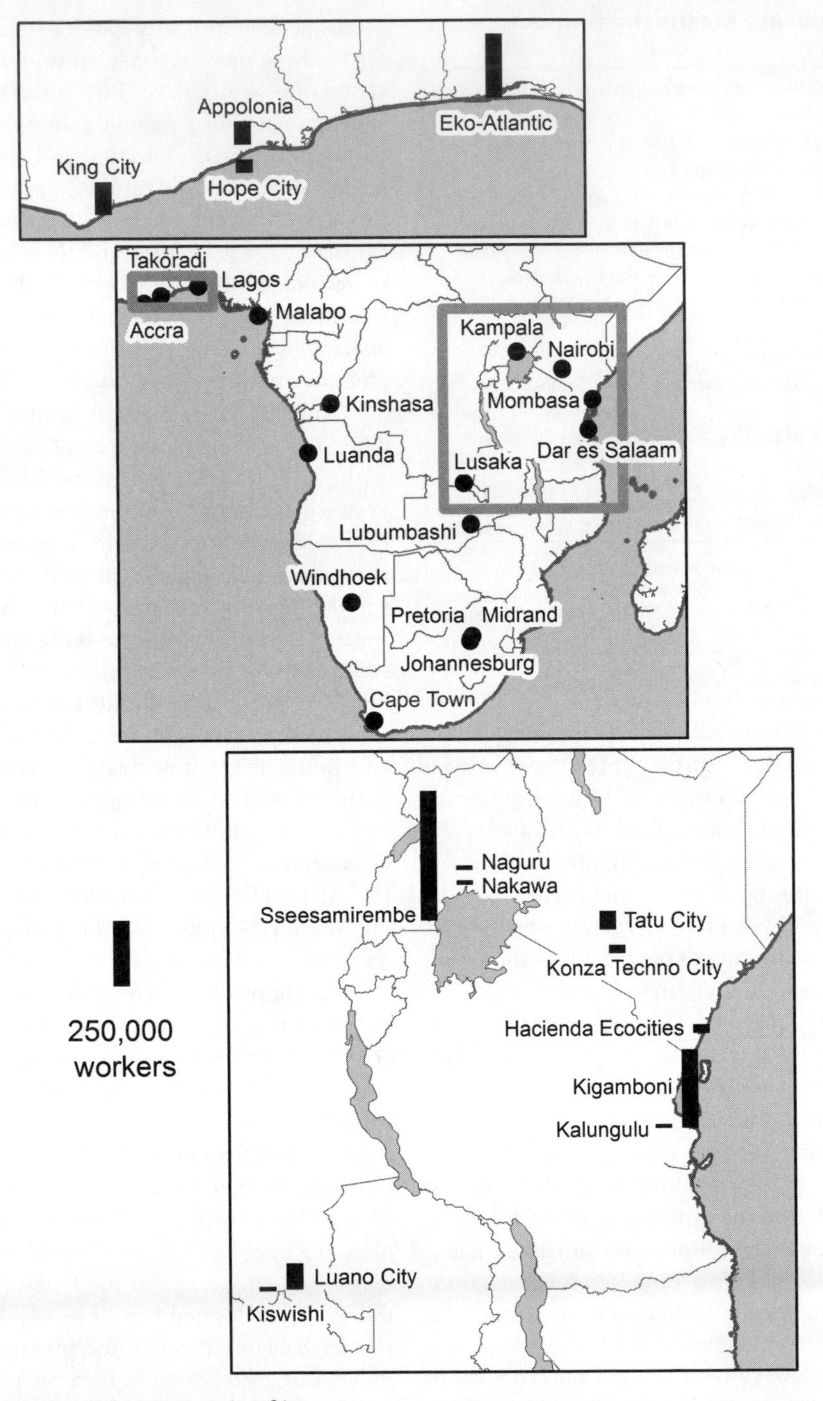


**FIGURE 13.4** New Mega Urban Projects in Africa.

services, high-quality infrastructure, safe and secure environments, and a lack of land disputes. The use of private investors reduces overall costs for governments and, on paper at least, allows governments to concentrate their resources on providing services.

Anticipating the future of urban development in Africa, Johannesburg-based Rendeavour Africa—a subsidiary of the Russian investment bank Onexim, one of the foremost urban land developers in Africa—has quietly purchased land exclusively for building satellite cities. Four coffee plantations have been acquired close to Nairobi as well as agricultural lands on the urban fringes of Accra, Lubumbashi, and Lusaka. Other property developers have negotiated substantial land purchases around capital cities such as Dar es Salaam, Harare, Kampala, Kigali, and Luanda. Real estate developers are positioning these new satellite cities to maximize their value in anticipation of the property booms in areas experiencing rapid population and high economic growth rates. Local elites, successful expatriates, and affluent members of the African diaspora are particularly being targeted as likely investors.

Proposals to construct entirely new city centers are increasingly popular. Most projects aim to replace the current central business district with a 21st-century globally-connected one. For example, Malabo II (Equatorial Guinea) will be an area anchored by government facilities and global corporations. In the DRC, La Cité du Fleuve ("River City") is an exclusive development to be situated on two artificially islands in the Congo River that will serve as the "mirror city" of Kinshasa, just offshore from Kinshasa's central business district. The planners are attempting to build "an oasis of tranquility" linked to Kinshasa by causeways.

Whereas most projects focus on new developments, efforts to retrofit existing development are also evident (e.g., Nakawa [Uganda] and Menlyn Maine [South Africa]). Not surprisingly, the scale and scope of urbanization in Africa have led many visionaries to point out that retrofitting is inadequate, essentially a Band-Aid for a broken leg. The pace and intensity of urbanization require African responses that diverge from European and North American experiences. Accordingly, many African proposals represent a movement away from the urban upgrading and regeneration, which predominated in earlier decades (based on Western planning experiences). The African shift is a reaction to retrofitting that was never brought up to scale and represents an invigorated effort to bring positive attention to African urban projects.

So-called eco-cities are very much in vogue in Africa. A precise global or African definition of an eco-city does not exist. The UN Environmental Programme (2012:VI) distinguishes eco-cities from conventional cities on the basis of "their superior urban environmental quality and livability to be achieved *inter alia* through compact, mixed-used developments, low-energy transportation, renewable energy generation and a reduced overall ecological footprint." Virtually all of Africa's eco-city projects use boilerplate language about the high quality of buildings and infrastructure, connectivity, and livability compared to existing urban settlements.

Implementing green urban projects has become "an (almost) ubiquitous global phenomenon" (Joss, Cowley, and Tomozeiu 2013:1), although the intensity of greening efforts varies from project to project. Joss, Tomozeiu, and Cowley (2011) report 174 eco-city projects in 2011 globally (developments at least the size of a neighborhood, encompassing several green policy areas including transport, energy, housing, and the environment). A substantial proportion of eco-city initiatives are currently at the planning stage or under construction, with implementation typically anticipated to be 10 to 15 years (but longer in practice).

Many eco-city projects are under development in Africa. The highest profile project is Eko Atlantic (Lagos), a project that has received international acclaim and was awarded a Clinton Global Initiative Commitment Certificate (See Box 13.1 for a detailed description of Eko Atlantic). Propelled by the growing desire of the Nigerian government and private-sector project designers to put such new projects on the global map; Eko Atlantic is promoted as a green development creating a world-class city in the heart of Africa, and, at the same time, it seeks to update and transform Lagos' notoriously bad international reputation as a poster-child of crime and urban dysfunction.

Africa's new urban projects have been roundly criticized as "fantasy projects" (Murray 2011; Watson 2014). Some projects have already displaced people. Filip

## BOX 13.1 EKO ATLANTIC (NIGERIA)

Popularly called Eko, the original name of the sleepy coastal settlement on the Atlantic coast of Nigeria, this area was transformed into the megacity of Lagos. Estimated to house 15 million residents in 2013, Lagos is projected to contain 25 million people by 2025, which will make it one of the biggest metropolitan areas in the world (Africa Research Bulletin 2013). Despite losing its status as federal capital in 1993 (when Abuja became Nigeria's capital), Lagos has continued to grow (60,000 settlers arrive every day), even as national urban development projects are being concentrated elsewhere. For the most part, it has evolved independent of urban planning through a process of "amorphous urbanism" (Gandy 2005). As a result, Lagos has a burgeoning population marked by high rates of poverty; residents are crowded into a limited space threatened by sea-level rise, ocean surges, and extreme weather events. Despite these massive challenges, the city remains the economic pulse of the second-largest economy in Africa (many believe it will soon overtake South Africa's as the largest one).

Eko Atlantic (*Eko* is the Yoruba word for Lagos) is the most ambitious urban development project in all of Africa (Fig. 13.5). Promoters expect the mini-city to refute negative images by showcasing a different side of Nigeria (Eko Atlantic 2013). Eko Atlantic is promoted nationally and internationally as a model public–private partnership between the Lagos state government and South EnergyX Nigeria (a subsidiary of Chagoury Group, a sprawling industrial and financial conglomerate operated by a highly influential family of Lebanese origin) to protect Lagos from storm flooding and rising sea levels. It is managed by a 78-year lease, based on funding from local financial institutions (e.g., First Bank and Guaranty First Bank) as well as international banks (e.g., BNP Paribas Fortis and KBC Bank). Participants include various experienced international architectural design and engineering firms (e.g., Dar Al Handasah Nazih Taleb & Partners, MZ Architects [Lebanon], and Dredging International [Belgium]).

This model city project aims to develop a new space for businesses and residents and to boost employment opportunities. When completed, the development will accommodate 3,000 new buildings zoned in 10 separate districts over 6 square miles (10 km$^2$) on land reclaimed from the Gulf of Guinea. The mini-city is expected to emerge as the gateway to Africa, the financial hub of Nigeria and the rest of West Africa. Promoters seek to develop an "ocean-front city that will be one of the wonders of the 21st century" (Eko Atlantic 2013). Presidents of Nigeria and Lebanon, as well as former U.S. President Bill Clinton, have strongly endorsed Eko Atlantic.

Inaugurated in 2006, most work to date has been preparatory: dredging, constructing a sea wall 4.3 miles (7 km) long around the reclaimed land (colloquially termed "the Great Wall of Lagos"), and preparing foundations at the site for high-rise, high-density urban development. The next stage involves building power-generating and sewage management facilities, light rail, and a network of roads designed to encourage free-flowing traffic within its borders. When complete, Eko Atlantic will accommodate 250,000 residents and 190,000 commuters, reducing pressure on the already-crumbling infrastructure of Lagos. This Dubai-like project is controversial. It will be privately administered throughout and provide its own security. There has been no effort to incorporate green (energy-efficient) building design into the project, which was conceived as a standalone development, neither part of the Mabogunje Lagos Master Plan (2006) nor the adjacent Lekki Master Plan.

**FIGURE 13.5** Eko Atlantic, Nigeria. *Source:* © Sunday Alamba/AP/Corbis.

The project has not always followed the letter of the law. An environmental impact assessment (EIA) was not undertaken before dredging commenced in 2008. An EIA took place subsequently, and approval was granted only in 2012, in violation of the 2004 EIA Act. Moreover, the EIA acknowledges that the project may shift the erosion of barrier islands to the east of the project, with adverse effects radiating outward up to 6 miles (10 km). Local residents blame the project for causing storm surges and flooding at nearby beaches (16 people were killed in a 2012 storm surge on Kuramo Beach). It is discouraging that stakeholder consultation was limited to communities on the eastern side of the project and that the western-side communities were not included. Moreover, there is no provision to hold the developers responsible for any negative environmental impacts. Left to fend for themselves, poor communities in adjacent areas have neither the resources to elevate their homes nor the ability to evacuate in times of disaster. There will also be repercussions on land: neighboring communities will experience greater traffic congestion created by the four-lane access road linking the mainland to Victoria Island.

Indeed, the lessons of dredging in Dubai reclamations may be misapplied to Lagos. The rate of sand replenishment along the West African Atlantic coast is entirely different than it is in the Persian Gulf (where there is no lack of sand, and alternative supplies are plentiful in nearby desert environments). The Gulf of

Guinea's rougher waters behave differently near the West African coast, and erosion is more challenging. Furthermore, the current sand supply offshore from Lagos is not being replenished naturally: sand drifts eastward along the West African coast and port construction to the west (e.g., Lomé and Cotonou) reduce the quantity of sand available for deposition.

Beyond the employment generated by the project's construction, it is unclear how many jobs will be created: the model city appears to absorb and centralize existing financial and service employment within the new activity center. The needs of support staff (e.g., security guards, housekeepers, clerical workers) are not part of the master plan, so access to affordable housing, transportation, and food and basic services will be virtually impossible.

Some of the project's backers have been implicated in past improprieties. For example, Gilbert Chagoury (Chair of Chagoury Group) was a financial confidant to former Nigerian dictator Sani Abacha and was found guilty of money laundering in Switzerland in 2002 in a complex financial web that siphoned funds from Nigeria's treasury. Chagoury eventually agreed to return US$66 million transferred out of Nigeria in return for immunity from prosecution (Africa Research Bulletin 2013). The Chagoury Group is also a major contributor to the Clinton Global Initiative (Africa Research Bulletin 2013).

There are questions about whether Eko Atlantic will become yet another symbol of the gap between poor Nigerians (most of whom live on US$2 per day) and the corrupt elite. The project may be emblematic of Africa's new urban frontier with global cachet, stable electricity, clean water, private security, and world-class buildings and amenities, but this kind of enclave development will not improve the livelihoods of the vast majority of Nigerians.

De Boeck (2011:272) notes that certain government and municipal authorities are participating in heavy-handed "politics of erasure," tearing down the informal city as well as unruly business and residential structures, tidying urban spaces by banning containers/kiosks that typically accommodate informal enterprises, and concentrating on grandiose new developments. Huchzermeyer (2011) emphasizes that privileging showcase private sector projects reinforces the urgency of the perceived need to free African cities of "slums" and/or to abandon the city housing the majority of Africans. Collectively, these projects may be producing a landscape of elite enclaves and splintering urbanism, thus initiating resettlement processes that widen the spatial exclusion of the poor. Satellite cities could incorporate affordable housing, but planners opt not to focus on this. Thus, low-paid workers will not be able to afford to live in satellite developments and will need to commute long distances, thereby adding a cross-commuting component that indirectly enlarges the overall carbon footprint.

Martin Murray (2011:1) labels these "fantasy-projects" as "city doubles" because these projects represent diametrical opposites of existing African cities. Constructing entirely new urban landscapes enables city builders to bypass the problems associated with everyday urbanism in Africa: inadequate infrastructure and inferior service delivery, overcrowding, poor aesthetics, lack of zoning and code regulations, traffic gridlock, and burgeoning informal economic activity. Colonial planners developed master plans for a different place and time, marked by colonial functions, smaller populations, and nonintegrated spaces. Colonial ideas about the ordering and layering of social and economic space bear little resemblance to contemporary urban realities, but continuities and legacies of the colonial spatial framework prompt modern visionaries to start anew. However, the new cycle of spatial fragmentation based on walls, patrols by private security firms, and skyrocketing prices are segregating people by income and represent the contemporary equivalent of the colonial regulations that mainly segregated people by race.

The eco-city rubric comprises considerable diversity in green initiatives, forms, and functions, but the eco-city label is too loosely applied. There is no uniformity of minimum green standards. Many project proposals are given a "green sheen" by designers who want to create autonomous sustainable cities for elites, who, in turn, desire to secede from unsustainable cities and to live in safe, low-carbon cocoons with the latest green gadgets and buildings. Such bold urban visions run the risk of providing only a thin green veneer on developments, privileging the developers' profit motivations and short-term goals. Some projects fail to gain traction beyond their launch (e.g., Sseesamirembe Eco-City, or Lake Victoria Free Trade Zone) as adequate foreign and domestic investment failed to materialize (See Box 13.2 for a discussion of the stalled Sseesamirembe Eco-City project).

The strong technological focus of eco-city projects (given the central concern to effect a rapid transition to

## BOX 13.2 SSEESAMIREMBE, UGANDA: SUSTAINABLE UTOPIAS GOING AWRY IN A GLOBAL WORLD

Sseesamirembe Eco-City, or Lake Victoria Free Trade Zone, was initiated in 2006 by the Ugandan government with local partners and Chinese backing of US$1.5 billion, making it one of the largest Chinese investment projects in Africa. The project was located in the Rakai District in southwest Uganda on the border with Tanzania (Fig. 13.6).

The proposed sustainable project was to be implemented in a large area (200 square miles [323 km$^2$]), and the development plan integrated multiuse urban developments, an airport, a port on Lake Victoria, sustainable agriculture and forests, green belts, and nature reserves. Planned as a low-carbon zone with efficient infrastructures (pedestrian-friendly, energy-efficient urban design and solar power), administrative support, and high-tech services, the development was to serve as a hub for attracting new businesses to the region. Promoters claimed it would be a "new Hong Kong of Africa" and "another Dubai" (Sseesamirembe 2013).

The project was suspended in December 2009 following a presidential commission of inquiry's investigation into rumors about the healing practices of the project's promoters: members of the Sserulanda Foundation religious group with a following in the Rakai District (officially registered as a religious group in Uganda in 1987) and the United States (headquartered in Fairfax, Virginia). The group is described both as a local "sect" based on cult leadership and rituals (e.g., the preservation of bodies in special rooms, akin to the preservation of mummies in Egyptian pyramids) and as a "new-age religion" linking African and Asian traditions that subscribe to vegetarian diets and living by ecological principles.

Its spiritual leader, Joseph Mugonza, later renamed Bambi Baaba, reportedly had a vision in early childhood of a modern city linking ecology and urban development. During the civil war in Uganda, the area served as a refuge, and notable Ugandans such as current Uganda President Museveni allegedly took refuge there. On this basis, it is claimed that some members of the current Ugandan government (2011–present) cemented contacts with Sserulanda religious leaders and that behind-the-scenes brokering resulted in the unveiling of a highly ambitious business venture, which gained traction and publicity but fell apart under closer scrutiny.

The project took off because the Sserulanda Foundation assembled and obtained title to a large tract of land. Leaders of the religious movement claimed to control 46 square miles of land (75 km$^2$) of land by 1995, but their actual acquisition was more in the range of 3 to 31 square miles (5–50 km$^2$) (Médard and Golaz 2011). During the government inquiry, a witness accused the foundation of robbing its followers and new converts and of organizing the confiscation of their land by hired legal experts (quoted in Médard and Golaz 2011:14).

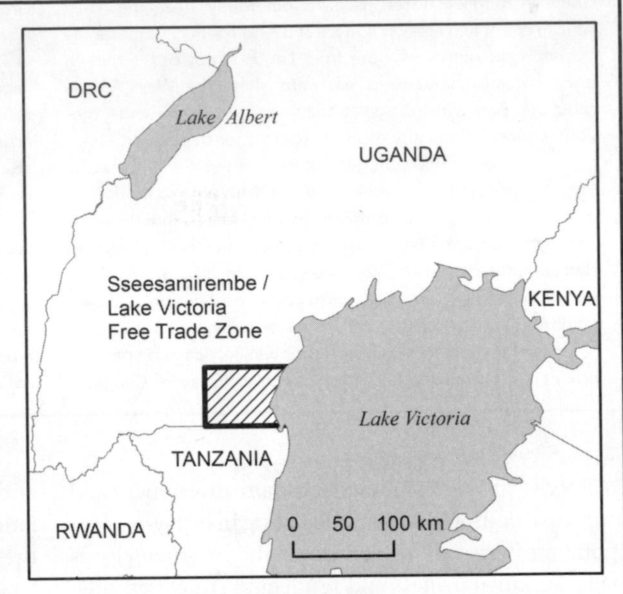


**FIGURE 13.6** Sseesamirembe, Uganda.

In 2000, the Sserulanda Foundation established the Sserulanda Development Association (SDA) and the religious leadership took on an economic development role, no doubt with others pulling many strings. Subsequently, the SDA asked Uganda's government for permission to operate the Lake Victoria Free Trade Zone as an autonomous territory. The Sseesamirembe city charter proposed that the Ugandan government secede to the SDA the authority to administer the 200 square miles (322 km$^2$) under a separate governmental structure, with powers to enact separate laws, ordinances, and procedures and to maintain its own security force.

In 2011, local residents, farmers, and ranchers demanded access to and use of their former lands and denounced the mega-project as a Chinese land grab. The designation of Chinese financial front man Liu Jianjun raised alarms. This larger-than-life character was fond of dressing in African garb for interviews and prone to boasting about his African credentials, principally his leadership and self-reported successes in Baoding villages (the supposed sites of 10,000 transplanted Chinese farmers, but Sino-African experts dispute their existence) and throughout Africa as well as his honorary title of African chief (albeit it from an unnamed African country). Liu's grand plans to spearhead China's new export to Africa—farmers—heightened concerns and galvanized opposition to the mega-project, resistance that has taken on a distinctively anti-Chinese tone.

Despite a farcical element to Mr. Liu's persona, he appears to have become the private face of Beijing's interest in African agriculture land. To many people's surprise, Liu held a press conference in Beijing (2008) to announce the Lake Victoria Free Trade

Zone, an event that included representation and speeches from high-ranking officials from both the People's Republic of China and the Ugandan government. Liu and his Baoding villages seemed to have official support.

The Sseesamirembe story reflects a phenomenon that is occurring with greater frequency in 21st-century Africa: the mixing of virtual and real worlds, the entanglement of players from different worlds and with different visions of sustainable development. There is nothing extraordinary about a religious movement wanting to construct an ideal city, but getting the project off the ground is another matter. Project promotion (or exaggeration and vagueness concerning details such as the amount of land under control) is a convenient ploy to attract interest and foreign investors seeking to profit from selling sustainable utopias. Neither is there anything extraordinary about shadowy foreign investors getting involved in new mega-projects. Entanglements of religious and "development" utopias—the association of business investors, government officials, religious leaders, followers, and hustlers—create a milieu where each actor tries to use the others by feeding into different utopian desires with essentially virtual elements. Sseesamirembe, as conceived in the original master plan, may never see the light of day. However, its short history from unveiling to implementation (though the project was suspended in 2009, media reports indicated it restarted in 2011) is a harbinger of the immense challenges in putting into action green mega-urbanism. Ultimately, suspended or failed green experiments undermine the global tilt toward sustainable development.

low-carbon urban infrastructures and economies) translates into high-risk endeavors. At present, urban sustainability is being primarily advanced by technological systems, innovation, and corporate investment. Outside of academic circles, the sociocultural and political dimensions of the livability of eco-cities are overlooked.

If the projects mapped in Figure 13.4 are implemented, major metropolitan areas will be fundamentally restructured. New satellite-cities will reshape the urban hierarchies established in earlier times. There is a real danger of abandoning the traditional African city by promoting detached projects geared explicitly toward serving elites and satisfying the desires of the leisure classes. Despite the claims of promoters, whose projects still need to raise investments, investor shortfalls may be compensated through buy-ins from governments desperate to realize 21st-century urban visions in the high-stakes global arena. Fear of failure and negative international publicity may result in municipal and regional infrastructure budgets being diverted away from service provision for the poor. Moreover, satellite developments will add to urban sprawl by doing little to reduce urban Africans' dependence on automobile transportation, and will exacerbate the difficulties of everyday Africans who travel on foot. Watson (2014) contends that the scale envisioned in these new plans might be sufficient to mobilize a broad-based coalition of slum dwellers, unemployed youth, and business and civic organizations in order to effectively counter these interventions and advocate for a more democratic and representative agenda of urban priorities.

## URBAN SUSTAINABILITY

The World Commission on Environment and Development (1987:43) defined sustainable development as that "which meets the needs of the present without compromising the ability of future generations to meet their own needs." Its Brundtland Report (after Gro Harlem Brundtland, who led the commission) addressed the interconnections among population growth, resource depletion, pollution, and excessive consumption and emphasized that current patterns of development were unsustainable. The report attempted to reconcile the ecological limits to growth articulated by the Northern green movement and the need to eliminate poverty as voiced in the Global South (Swilling and Annecke 2012). Absorbing "sustainable development" into global development discourse, the term reconciled the tension and opposing interests of North and South. This highly influential report provided the strategic foundation for subsequent high-level global forums that produced agreements to implement sustainable development (e.g., Earth Summits and UN Climate Change Conferences).

The sustainable development rubric has been accompanied by a burgeoning literature on sustainability, with African perspectives represented (Burns and Weaver 2008). Within this discourse, there is a consensus that in African contexts sustainability must incorporate improvements in housing, living, and working conditions of the poor to complement environmental initiatives (Satterthwaite 2001). Sustainable development as it is framed within Western thought processes and based on Global North experience cannot be directly applied to African contexts. Initiatives that are not well adapted to local contexts will be neither resilient nor sustainable. Development experience is littered with examples of failed projects that underestimated the realities of existing cultural, political, environmental, and economic environments. For example, technological fixes introduced without adequate maintenance and supplies of spare parts will break down and lead to abandonment and stalled development. However, as discussed in Chapter 6, sustainable prototypes such as ishacks and solar-powered toilets are being developed by local and international collaborations to make development contributions.

There are four major critiques of the Brundtland conception of sustainable development (McMichael 2009; Pieterse 2011). First, it assumes that all conflicts and contradictions can be resolved through rational, democratic deliberation. Pieterse (2011:310) argues that "the prevailing discourses provide a foil for biased and shallow policy decisions that generally reinforce the status quo in unsustainable and unjust situations." A second problem is that it reinforces assumptions of mainstream development economics and the emphasis on economic growth. Third, "urban sustainability" has been hijacked as a real estate development project with urban-ecological security provided to enclave consumers at a premium price and not available to many. Fourth, the concept is popular precisely because it lacks definition and measurability: malleability allows for a wide variety of interpretations and manipulations to jump on the green bandwagon.

The Global North's conception of sustainable development may not be directly transferable to Africa. A key question is whether African traditions can contribute to an indigenous or more authentically African-rooted conception of sustainable development. Aspects of African historical traditions and social values (e.g., group mutuality, spirituality, self-help, communal bonding, and social responsibility) should be recovered and appropriated for an African approach to sustainable development. An African sense of being human has at its core values of *ubuntu*: compassion, hospitality, and generosity (Dei 1993). Such a value system emphasizes human interactions with nature and the environment as opposed to controlling them.

Asomani-Boateng (2011) calls for recovering the positive characteristics of early African cities (e.g., Kumasi, Great Zimbabwe, Kano) and for adapting aspects of indigenous urban form and architecture to help shape thinking about Africa-imbued sustainable development. Development strategies based on indigenous urban forms, embodying the culture, aspirations, experiences, and values of local people, are consistent with the central tenets of sustainable development and can be adapted to the 21st century. Unfortunately, for too long anything indigenous has carried a negative connotation: a perspective sorely in needs of reexamination.

The indigenous urban forms and architecture in early African cities were sustainable, incorporating dense, mixed, and multiuse development, principles championed by today's green proponents but not widely attributed to their precolonial African antecedents. Several key elements—for example, urban agriculture, indigenous architectural models such as traditional compounds, active grassroots participation, green urban boundary buffers—especially pertinent to an African sustainable development agenda. This involves transplanting elements of the rural village into an urban environment. Agriculture was a pervasive activity in indigenous cities, and "the greening of these cities with edible plants for both the aesthetics and nutritional needs of the city residents remains one of the most sustainable innovations by indigenous urban builders" (Asomani-Boateng 2011:258). In a total rupture with the past, colonial and contemporary municipalities have provided no policy role for urban agriculture. Instead, they undermined urban farming, clinging to the view it is a transitory, interim, and traditional rural activity with no place in modern cities and economies. In reality, urban agriculture can provide fresh, locally grown food, clean air, and jobs for urban dwellers.

Courtyard architecture in traditional compound housing, with rooms organized around an open-sky courtyard, maximized urban space and facilitated mixed and multiple uses of spaces in a compact form. This housing was affordable, and the economics of compound construction remain intact (one room in a compound costs approximately 33% of a comparable room in a standard villa/bungalow) (Asomani-Boateng 2011). Lower construction costs are due to the incorporation of local building materials and basic and efficient building techniques, perfected by informal artisans. Compounds promote higher urban densities due to the large number of rooms contained in a single courtyard. Some compound housing in the past (e.g., Benin City) incorporated roof technologies that allowed for the collection of surface water, which was used for drinking, cooking, and planting. Of course, compound architecture would need to be adopted to incorporate a vertical element to accommodate higher densities. Project developers of the Hope City Project in Accra, aiming to build the tallest building in Africa (75 stories) in a US$10 billion project, describe it as a 21st-century realization of the African compound in a tall vertical structure.

## EVIDENCE OF AN UNSUSTAINABLE WORLD

Human activity is breaking through dangerous thresholds by its destructive and wasteful practices. Biophysical limits are being breached on several counts, and more affluent populations continue to expand rather than reduce their ecological footprints. Overconsumption by a minority elite ultimately translates into less being available for the majority by degrading and using up natural resources. The elite's unsustainable usage of the world's primary resources and ecosystems is undermining the ability of all countries to achieve sustainable development.

Global society has been operating under an assumption of infinite resources. Economic growth, profit, and expansion drive the global capitalist system. Cities and industries depend heavily on access to resources, and planning runs on assumptions of limitless supplies. The current trajectory (based on present-day widely used technologies) is driving environmental degradation that is bound to result in decline, a reversal in standards of living, and eventual collapse. A world threatened by famines and floods, scarcity of fresh water and energy, and displacement of populations is not the brave new world typically imagined. Ultimately, the vitality and survival of people and their homelands will be compromised. These threats are real: some African countries, especially in the Sahel, are already experiencing the staggering impacts of human-induced environmental change.

Swilling and Annecke (2012:27–28) summarize seven global documents in the policy domain that confirm the looming threats to society. These threats challenge the wisdom of continuing along a business-as-usual capitalist development path.

First, the UN's Millennium Ecosystem Assessment (2005), compiled by 1,360 scientists from 95 countries, confirmed for the first time that 60% of the ecosystems upon which humans depend for survival are becoming degraded. For example, marine ecosystems are used unsustainably and are becoming degraded because of overfishing, pollution, warming, acidification, and other factors driven by human-induced change (UN 2005). To meet demands for food, fresh water, timber, fiber, and fuel, humans have altered the ecosystem more extensively over the past 50 years than in previous millennia (UN 2005). Degradation is forecast to worsen over the next 50 years.

Second, global warming has been scientifically established. The Intergovernmental Panel on Climate Change (IPCC) confirms that the warming trend results from human activity, especially the release of greenhouse gases into the atmosphere, largely attributable to the burning of fossil fuels, deforestation, and agricultural production. The IPCC (2013) warns that the average temperature will rise by between 0.3 and 4 degrees Celsius (0.5–8.6 degrees Fahrenheit) by the late 21st century, triggering major ecological and socio-economic changes. Extensive destructive consequences will occur in Africa: 75 to 250 million Africans will suffer the consequences of increased water stress by 2020; output from rain-fed agriculture will decline by half by 2020; and sea-level rise will damage most low-lying African coastal cities by 2099.

Third, oil discoveries as well as oil production have peaked in some countries, but new discoveries in the United States and Mexico and several Africa countries (e.g., Kenya and Uganda in 2012) are impeding the transition toward alternative energy. There are now 50 countries in Africa with active oil and gas explorations. Such activity has forced a revision of the peak oil thesis (popular 2000–11). The sustained and increasing demand from industries and consumers (e.g., fuel for transport and production of cement for buildings, various other oil derivatives, such as polymers in manufacturing plastics, and fertilizers and herbicides for growing food on commercial farms) may perpetuate our reliance on oil. For example, oil is used for 60% of the global economy's energy needs.

Fourth, inequality is rampant: 20% of the global population (largely residing in the Global North) account for 86% of total private consumption expenditure (UN Development Programme 1998). Oxfam (2013) reported that "the richest 1 percent has increased its income by 60 percent in the last 20 years, with the financial crisis accelerating rather than slowing the process," while the income of the top 0.01% increased even more. Sharp divergences in consumption patterns illustrate the degree of inequality. For example, the wealthiest 20% of the global population consume 45% of the total meat and fish, 58% of the energy, 84% of the paper, and 84% of the motor vehicles; the poorest 20% consume relative shares of less than 5%. Conspicuous consumption has been taking precedence over meeting basic needs on a global scale.

Fifth, the majority (62%) of urban Africans reside in slums (UN-HABITAT 2012). UN-HABITAT's landmark publication *The Challenge of Slums* reveals that most people are living in appalling conditions in self- or community-built dwellings without many opportunities to improve their living conditions. Governments, international organizations, and nongovernmental organizations are failing to provide decent housing and decent jobs. Population and urbanization projections predict that the number of people living in African cities will increase, particularly the portion of new residents who will reside in slums, trapping the latter in deep poverty with few escape routes (Cazzavillan, Donadellu and Persha 2013).

Sixth, food insecurity is a longstanding problem that will worsen in the context of various resource impairments. Almost one quarter of global land is degraded to some degree, and declining soil fertility is a major driver of food price increases (Watson, Wakhungu, and Herren 2008). In addition, rising oil prices affect industrial agriculture, which depends heavily on fertilizer and pesticide inputs derived from the oil supplied by global chemical conglomerates. Agricultural experts propose diverging solutions that rely on an African green revolution or indigenous agricultural practices.

Finally, the global economy used 7 billion tons of primary resources (biomass, metals, industrial and construction minerals) in 1900; consumption increased to 60 billion tons by 2005, with the rate of increase accelerating rapidly after 1980 (quoted in Swilling and Annecke 2012:44). The UN Environmental Programme's International Resource Panel (2013) notes that extracted materials of the Earth's resources should be around 6 tons per capita as opposed to the current 8 tons. An expanding population will only add to the pressures. The International Resource Panel is developing quantitative assessments of absolute resource limits, but in the meantime, prices will in all likelihood increase due to this resource depletion.

Global society thresholds may be breached unless there is a shift to low-carbon energy; use of water, energy, and other resources becomes more efficient; and sustainable technologies are adopted in building, transport, and agriculture. Unfortunately, there is scant evidence that policymakers are heeding or registering the full implications of these warnings and their multiple impacts. Economic growth, human settlement expansion, and exploitation of natural resources continue almost unabated. Reversing the current trends will require significant changes in policy, institutions, and practices related to global development.

## POST-MDGs: THE SUSTAINABLE DEVELOPMENT AGENDA

The Millennium Declaration (2000) put forth a central focus with key goals. Although the MDG targets will not be met when the implementation timetable expires in 2015, this global policy framework demonstrated the power of universal goals for building momentum to coordinate global, national, and local action. The process of MDG implementation is revealing the need to improve the quality of data to track progress and to support capacity-building efforts.

There is a groundswell to declare a sustainable development agenda for 2015–30. The Rio+20 Summit in 2012 concluded by outlining a sustainable development framework with four dimensions: economic development (including the end of extreme poverty), social inclusion, environmental sustainability, and good governance. At the same time, UN Secretary-General Ban Ki-moon launched the UN Sustainable Development Solutions Network to mobilize global scientific and expert knowledge to turn sustainable development aspirations into practical actions that could be applied to all countries while respecting national policies, priorities, and differences as well as different national, regional, and local capacities. The special session of the UN General Assembly in September 2013 concluded by agreeing on a two-year road map of negotiations about sustainable development goals. The year 2014 will be used to identify sustainable development priorities, and these will be operationalized; in 2015 global sustainable development goals will be launched, and these goals will be carried forward to 2030.

A greater effort is under way to mobilize a broader constituency, drawing on national and local governments, academia, science, civil society, and business. Business engagement is regarded as key to achieving sustainable development because it is directly responsible for two thirds of natural resource use. It remains to be determined whether business goals can be realigned with sustainable development; that is, to move beyond profit as the sole or main motivation, which more realistically will happen as the changing political climate compels businesses to realign their operations to sustain their profits. Mobilizing diverse stakeholders (e.g., government at all levels, civil society, international organizations, businesses, local communities) is the only way to build collective partnerships to realize sustainable development.

The case for moving to a sustainable development pathway is predicated on scientific evidence that the world's current trajectory is dangerous and that remaining on that track is unsustainable. Many key ecosystems essential for human and societal well-being are under threat or have already been destroyed, and an unprecedented mass extinction of species is well under way. Climate change is no longer a future threat but a stark present reality for Africa and elsewhere. For example, the measurement of carbon dioxide concentrations in the atmosphere at the Mauna Loa Observatory (Hawaii) were at the highest recorded levels since measurements began in 1958, with a reading of 400 parts per million in May 2013. Our scientific understandings confirm that humans are having a greater impact on Earth (along critical dimensions of climate, ecosystems, land use, and food and primary commodity scarcity) than any other agent, and social scientific research confirms that inequality, social exclusion, and gaps in educational attainment are widening.

Key sustainable development challenges include (1) how to harness available technologies to decouple economic growth and living standards from environmental resource use and pollution; (2) how to decouple agricultural productivity from unsustainable land-use conversion and wasteful uses of water, energy, and fertilizers; and (3) how to create greener cities and more environmentally friendly infrastructure.

Africa's current development does not follow a sustainable development pathway. Even though the region is experiencing high economic growth rates, many Africans are falling deeper into poverty, which may feed into resentment that could be potentially explosive. There are numerous distressed subregions (the Sahel, the Horn of Africa, and the Great Lakes), where underdevelopment, conflict, natural resources depletion, and disease challenges are compounded by landlocked status (but not in the Horn) and insufficient domestic resources to end extreme poverty. Africa's vulnerable regions need special international support to break vicious cycles of underdevelopment, poverty, insecurity, environmental degradation, and rapid population growth.

Property developers in Africa are looking to Asia to provide fast-track urban transformation, with projects based on the experiences of Shanghai, Singapore, and Dubai. Unfortunately, lessons from China's urban transformations have not been heeded: Chinese cities have been built in ways that waste energy and create traffic congestion. The Chinese approach to urban layout, with "superblocks" (0.3 square miles [0.5 km$^2$]) interspersed with large boulevards and the separation of residential, business, and shopping districts, complicates travel patterns and creates traffic congestion and hostile environments for pedestrian and bicycle traffic. The unheeded urban planning lesson is that smaller blocks, combined with smaller streets and mixed uses, produce more livable cities.

Africa's urban populations will continue to grow, and cities will continue to use massive amounts of resources (water, food, energy, minerals, timber, consumer and industrial commodities). New city land use and spatial infrastructures (e.g., buildings, energy, water, and transportation) last for decades, so planning decisions are vital in determining the trajectory of carbon dioxide emissions, water use, etc. Well-designed green buildings with proper insulation and ventilation and energy-efficient systems use only 25% of the energy of their non-green counterparts, and the energy savings last for decades. African investment in top-quality green buildings, backed by strong building codes and the discipline to enforce them, will be necessary.

Adequate financing is also essential. One painful lesson of the MDGs is that many goals were underfunded. Moreover, sustainable development goals need to be bolstered by significant improvements in data collection and processing (e.g., incorporating existing tools such as geographic information systems, remote sensing, and social networking) to provide place-based, real-time information that is gender and culturally sensitive to analyze progress and to promote fairness and inclusion. Even these combined efforts may fall short. Global regimes of trade, finance, taxation, business accounting, and intellectual property may need to be reformed to be in harmony with sustainable development.

Implementing sustainable development policies and practices across the globe will require a significant transformation of energy, industrial, urban, and agricultural systems and a movement away from resource- and energy-intensive growth. Such a transition will represent one of the greatest technical, organizational, and societal challenges that humanity has ever faced. These challenges are monumental and require a recalibration from our current thinking and behavior toward nature toward a creative process for imagining and operating within more livable parameters and spaces. Finding ways to validate and use qualitative measures is essential, but quantitative measures of progress must also be considered. The emergence of new indicators—for example, the "extended Human Development Index" (adding environmental sustainability, gender inequality, and human security to the development toolkit) and the "Happiness Index," proposed by Nobel Prize winners Amartya Sen and Joseph Stiglitz (Stiglitz, Sen, and Fitoussi 2009)—are movements in the right direction.

Swilling and Annecke (2012) argue for a "double movement": a turn toward sustainable development as well as a "just" transition. The latter embraces everyone's needs, in particular those of the poor, and connects their basic needs to the global challenge of innovating, investing, and intervening to use resources more sustainably. Mahatma Gandhi's comment that there is enough in the world to provide what we need but not enough to satisfy human greed is certainly relevant to our times.

## RETHINKING URBANISM

Cities contain a dense concentration of overlapping actors engaged in imagining, mediating, contesting, and deploying sustainable development architecture (Swilling and Annecke 2012). Nevertheless, the burgeoning literature on environmental sustainability focuses heavily on climate change, deforestation, and desertification and has only begun to grapple with urban environments and their role in broader processes. At the same time, the urban studies literature is a latecomer in acknowledging the physical-environmental foundations of urbanization processes (Heynen, Kaika, and Swyngedouw 2006).

Throughout the 19th and 20th centuries, various urban movements (e.g., Garden City, New Town, Techno City, New Urbanism) attempted to reinvent the city. Green urbanism is the urban movement of the 21st century. It encompasses a wide array of approaches: eco-cities (discussed earlier), smart cities (emphasizing information technology networks and efficiencies in utility and service provision), carbon-neutral/low-carbon cities (focusing on energy, transportation, infrastructure, and building efficiencies), as well as broader approaches such as ecoregions (prioritizing city sustainability within the context of natural flows, such as in river basins) and narrower approaches such as solar cities (where solar energy replaces fossil fuels).

Green urbanism has its origins in the mid-1970s, when it emerged in the context of the rising environmental movement. Throughout the 1980s and early 1990s, it remained mainly an innovative concept with few practical examples. The UN's 1992 Earth Summit in Rio de Janeiro and the resulting sustainable development program (Agenda 21) formed the background for the first wave of green urban initiatives. Curitiba (Brazil), Waitakere (New Zealand), and Schwabach (Germany) are examples of first-generation green urban initiatives. Since 2005, the green city phenomenon has become mainstream, and numerous green initiatives are under way in Africa.

Green urbanism aims to curb the environmental impacts of cities (greenhouse gas emissions, water pollution, wastes), to reduce dependence on increasingly costly and insecure long-distance inputs (fossil fuels, minerals, food, and construction materials), and to roll out ecological modernization (efficiency, smart grids). Green urbanism spans a range of initiatives from soft mitigation efforts such as retrofitting buildings for energy efficiency, expanding mass transit and freight rail, and constructing smart grids to manage electrical systems to medium mitigation initiatives (e.g., alternative energy investments in wind, solar, biofuels) to the complete greening of entire cities.

There are major challenges in greening African cities. African governments are challenged to provide more jobs and prefer to focus on jobs now rather than the long-term employment that may be created from green urbanism (Simon 2013). It is complicated to move green urbanism from a technical piecemeal fix, emphasizing greening elite residential enclaves and business parks, to focus on the most deprived high-density areas (lacking amenities and green open space).

Pieterse (2009) calls for the conceptual inversion to understand the city from the bottom-up, or through the eyes of the majority of residents who appropriate the city for their own ends. Slum urbanism can be better understood as the quiet encroachment of the masses to build their own neighborhoods piece by piece, needs to be incorporated into green urbanism. Slum dwellers are active manipulators of their environments for good as well as bad. Their lack of access to various infrastructures means they must devise ways to access water, energy, food, transport, and outlets for their sewage and solid waste. As such, slum dwellers participate in appropriating cities' ecological resources (e.g., rivers for conveying water and waste, soils for planting food as well as platforms for erecting dwellings and commercial facilities,, biomass for fuels and building materials, forests for charcoal, and fossil fuels for accessing mobility and supplying power for refrigeration, lighting and mobile phone recharging) (Swilling and Annecke 2012). The urban modernity of the poor is, therefore, another costly affair that works against sustainability, but on balance, slum dwellers live more within the carrying capacity of the planet. To their credit, slum dwellers recycle, reuse, and repurpose materials to construct shacks; they engage in urban agriculture by capturing nutrients from sewage and reuse it as fertilizer. The global nongovernmental organization Shack/Slum Dwellers International (detailed in Chapter 2) and others (reviewed in various issues of *Environment and Urbanization*) are major players in the quiet greening of the activities of the poor (e.g., setting up waste management systems that collect, sort, and recycle materials) that goes largely unnoticed outside of nongovernmental organization and certain scholarly circles. Any future sustainable development framework needs to harness the positive green aspects of slum dwellers' community initiatives.

Moving toward sustainable urbanism will require fundamental changes in how society operates. Table 13.3 highlights sustainable development aspirations.

**TABLE 13.3 SUSTAINABLE DEVELOPMENT ASPIRATIONS**

- No travel by privately owned automobiles
- Penalties for long-distance travel by automobile and air travel
- Compulsory living in high-density buildings that generate more energy than is consumed
- Local food production and intensification of urban agriculture
- Enforced waste separation at source
- Ban on pollution and transshipment of hazardous materials (which could result in bans on certain kinds of plastics and compounds, or in incentives to move to greener technology)
- Price hikes for energy-intense food items (e.g., meat) and for consumer goods that incorporate rare metals (e.g., mobile phones)
- Massive investments in "green-collar" jobs in clean-tech industries and technologies
- Regulatory/market interventions to restrict carbon-intensive imports and to enforce zero waste
- Encouragement of green slum urbanism
- Massive investments in the restoration of ecosystems services (e.g., aquifers, rivers, coastal areas, soils, forests, wetlands, and other biodiverse sites)

*Source:* Swilling and Annecke (2012:129).

## LIVABLE URBANISM

Pieterse (2011) advocates an alternative conceptualization of sustainability that focuses on sustainable lives and livelihoods rather than just sustaining development and specific techno-green fixes. A sustainable livelihood approach is sensitive to processes of social and ecological reproduction within diverse contexts. These processes are nonlinear and context-specific and can be attained through multiple pathways; thus, there is not a single overarching model or pathway to sustainable development. People and locales cannot fully escape their contemporary conditions, but this is all the more reason to work on locally specific agendas and solutions.

Allen (2002:16–17) argues that we should focus on five interrelated dimensions of sustainability and for thinking of the interrelationship among the dimensions as the most fruitful way to tackle sustainable development:

- *Economic sustainability:* the ability of a local economy to sustain itself without causing irreversible damages to the natural resource base on which it depends and without increasing the city's ecological footprint
- *Social sustainability:* a set of actions and policies aimed at improving the quality of life but also at fair access and the distribution of rights over the use and appropriation of the natural and built environment
- *Ecological sustainability:* the impact of urban production and consumption on the integrity and health of the city region and global carrying capacity
- *Physical sustainability:* the capacity and aptitude of urban and rural environments and techno-structures to support human life and productive activities
- *Political sustainability:* the quality of governance systems guiding the relationships and actions of different actors among the other four dimensions; involves the democratization and participation of civil society in all areas of decision-making

## CONCLUSIONS

Urbanization in developing countries represents the single greatest transformation of this century. It is projected that developing countries will triple their built-up urban area—from 124,000 square miles (200,000 km$^2$) in 2000 to 372,000 square miles (600,000 km$^2$) in 2030 (Suzuki et al. 2009:1). An added 248,000 square miles (400,000 km$^2$) will be constructed in just 30 years. Humankind is building a whole new urban world at warp speed. Africa is ground zero in this urban revolution, and the region faces severe resource constraints—fiscal, administrative, technical, and natural—not to mention underdevelopment, increased inequalities, rapidly growing populations, and a perception that extremist groups are becoming more active and embedded in the region.

A paradigmatic shift toward a greener frontier is occurring. Sustainability in its various forms is being mainstreamed in policy and practice. Perhaps the greatest achievement of agreeing on sustainable development goals will be the creation of a new norm. Several features will be salient; for example, centrality of carbon and climate change discourses, increased transfer of knowledge and cooperation, and a preponderance of green-smart technological solutions. Critics from the region (e.g., Pieterse 2011; Swilling and Annecke 2012) believe the shift to sustainability does not go far enough. Davis (2010b:29) believes it is little more than "green washing" and poses a profound question: "Who will build the ark" to save humanity from the impending tragedy? Certainly, the focus on economic growth as the basis of development has not been displaced. Instead, the emphasis is on incorporating more environmentally sensitive, technological, and design solutions to slow unsustainable processes rather than on making a fundamental shift to a completely different way of thinking, operating, and living.

Multiple urban projects are under way in the region. Ambitious to varying degrees, some have a high chance of realization (e.g., Eko Atlantic) and a few (e.g., Sseesamirembe) are struggling to gain traction. But most are not sufficiently emboldened to make more than an isolated sustainable development impact within their project borders. Each project proposes one solution in a limited geographical area, as opposed to solutions to the extensive challenges that face contemporary African metropolitan areas. Although isolated projects are important, there is little evidence that they can be scaled up and integrated, even though comprehensive sustainable urban development plans are urgently needed.

Various city vision plans (e.g., Nairobi 2030 and Johannesburg 2040) seek to address urban development in more holistic ways, but critics contend that existing plans reproduce top-down thinking and are not inclusive of all stakeholders. Regardless of their specific limitations, the track records of implementation for the plans of urban visionaries have been particularly weak since the colonial interlude. Current heavily promoted "showcase" urban projects are unsustainable in the extreme. By ignoring slums, poverty, and the appalling deficiencies in overall urban infrastructure, the sum of these projects will hardly make a dent in developing more sustainable urban environments.

Cities provide a plethora of opportunities to solve such pressing global issues as poverty and education and to initiate sustainable development. Cities are environments where services can be more efficiently delivered because of their density. Access can be provided to better-quality education and health services, and jobs can more easily be created within metropolitan concentrations than in rural areas. Despite the many challenges of African urbanization, cities function as crucibles of creativity, innovation, and ingenuity, and creative problem solvers will always be motivated to devise new solutions to various challenges. There is much logic to the sustainable development forward-planning premise that savings accruing over the medium term from building sustainable infrastructures will eventually free up resources for poverty eradication.

The sustainable development turn offers a unique opportunity for cross-collaboration among nontraditional partners and for adopting a more holistic approach to the African development experience, emphasizing human well-being, social and economic inclusion and resilience in the face of climate change and greater environmental sustainability. With the United Nations adopting Sustainable development Goals (SDGs) that will guide international development policy from 2015–2030, a unique window of opportunity exists to better align urban progress with the sustainable development agenda. Even though the sustainable development rubric is far from being a single unified philosophy (large differences persist among incremental reformers seeking to modify the status quo and transformationalists who push for more radical change), sustainable development offers a framework for connecting diverse projects and goals in a bigger intellectual project for rethinking Africa's future.

The sustainable development discourse evokes strong references to partnerships, and such a track is going to require steadfast commitment from all parties. To be true partnerships, they must involve mutual responsibilities for genuine, people-centered sustainable and socially just development. It is, however, difficult to envisage how many existing international partnerships related to Africa can be reconfigured as real partnerships in practice as opposed to in rhetoric, given the heavy hand of history and the prevailing asymmetrical relations. Accordingly, for sustainable development to actually take root, it is of the utmost importance to scrutinize all partnerships to ensure that the rhetoric of partnership is not just a more subtle form of external power imposition.

## REFERENCES

Africa Research Bulletin. 2013. "Nigeria: Africa's First Smart City." *Africa Research Bulletin: Economic, Financial and Technical Series* 50:19883A–19884B.

African Union Commission. 2004. *2004–2007 Strategic Framework of the African Union Commission.* Addis Ababa: African Union Commission.

Allen, A. 2002. "Urban Sustainability under Threat: The Restructuring of the Fishing Industry in Mar del Plate, Argentina." In *Development and Cities,* eds. D. Westendorff and D. Eade, pp. 12–42. Geneva: UNRISD and Oxfam.

Asomani-Boateng, R. 2011. "Borrowing from the Past to Sustain the Present and the Future: Indigenous African Urban Forms, Architecture, and Sustainable Urban Development in Contemporary Africa." *Journal of Urbanism* 4(3):239–262.

Brautigam, D. 2011. *The Dragon's Gift: The Real Story of China in Africa.* New York: Oxford University Press.

Brautigam, D. 2012. "China's Training Programs." Available at http://www.chinaafricarealstory.com/2012/02/links-i-liked-chinas-training-programs.html (accessed April 17, 2013).

Brown, W. 2012. "A Question of Agency: Africa in International Politics." *Third World Quarterly* 33(10):1889–1908.

Burns, M., and A. Weaver. 2008. *Exploring Sustainability Science: A Southern African Perspective.* Stellenbosch: Sun Media.

Carmody, P. 2011. *The New Scramble for Africa.* Malden: Polity Press.

Cazzavillan, G., M. Donadellu, and L. Persha. 2013. "Economic Growth and Poverty Traps in Sub-Saharan Africa: The Role of Education and TFP Shocks." *Research in Economics* 76(3):226–242.

Chillers, J. 2003."Terrorism in Africa" *African Security Review* 12(4):91–103.

Coons, C. 2013. *Embracing Africa's Economic Potential: Recommendations for Strengthening Trade Relationships between the United States and Sub-Saharan Africa.* Availableathttp://www.coons.senate.gov/embracing-africas-economic-potential (accessed April 21, 2013).

Council on Foreign Relations 2014. *Nigeria Security Tracker. Mapping Violence in Nigeria.* Available at http://www.cfr.org/nigeria/nigeria-security-tracker/p29483 (accessed March 24, 2014).

Davis, J., ed. 2010a. *Terrorism in Africa. The Evolving Front in the War on Terror.* Lanham, MD: Lexington Books.

Davis, M. 2010b. "Who Will Build the Ark?" *New Left Review* 61 (Jan.–Feb.):29–46.

De Boeck, F. 2011. "Inhabiting Ocular Ground: Kinshasa's Future in the Light of Congo's Spectral Urban Politics." *Cultural Anthropology* 26(2):263–286.

Dei, G. 1993. "Sustainable Development in the African Context: Revisiting Some Theoretical and Methodological Issues." *African Development* 18(2):97–110.

de Walle, N. 2010. "US Policy toward Africa: The Bush Legacy and the Obama Administration." *African Affairs* 109(434):1–21.

*Economist, The.* 2000. "The Hopeless Continent." May 11. Available at http://www.economist.com/node/333429 (accessed April 17, 2013).

*Economist, The.* 2013a. "Africa Rising: A Hopeful Continent." March 2. Available at http://www.economist.com/news/special-report/21572377-african-lives-have-already-greatly-improved-over-past-decade-says-oliver-august (accessed April 17, 2013).

*Economist, The.* 2013b. "The state of al-Qaeda. The unquenchable fire." September 28. Available at http://www.economist.com/news/briefing/21586834-adaptable-and-resilient-al-qaeda-and-its-allies-keep-bouncing-back-unquenchable-fire?zid=312&ah=da4ed4425e74339883d473adf5773841 (accessed October 28, 2013).

Eko Atlantic. 2013."The opportunity of Eko Atlantic." Available at http://www.ekoatlantic.com/about-us/ (accessed August 5, 2013).

French, H. 2013. "The China-Africa Convergence: Can America Catch Up?" *Africa Plus.* Available at http://africaplus.wordpress.com/2013/06/16/the-china-africa-convergence-can-america-catch-up/ (accessed June 26, 2013).

Gettleman, J., and Kulish 2013. "Somali militants mixing business and terror." *New York Times,* September 30. Available at http://www.nytimes.com/2013/10/01/world/africa/officials-struggle-with-tangled-web-of-financing-for-somali-militants.html?pagewanted=1 (accessed October 3, 2013).

Gandy, M. 2005. "Learning from Lagos." *New Left Review* 33 (May–June):36–52.

Hansen, D. 2013. *Al Shabaab in Somalia: The History and Ideology of a Militant Islamist Group, 2005–2012.* New York: Oxford University Press.

Heynen, N., M. Kaika, and E. Swyngedouw. 2006. *In the Nature of Cities: Urban Political Ecology and the Politics of Urban Metabolism.* London: Routledge.

Huchzermeyer, M. 2011. *Cities With Slums: From Informal Settlement Eradication to a Right to the City in Africa.* Cape Town: University of Cape Town Press/Juta Academic.

Intergovernmental Panel on Climate Change (IPCC). 2013. "Working Group 1 Contribution to the IPCC Fifth Assessment Report. Climate Change 2013: The Physical Science Basis. Summary for Policymakers." Available at http://www.climatechange2013.org/images/uploads/WGIAR5_WGI-12Doc2b_FinalDraft_All.pdf (accessed September 30, 2013).

Joss, S., D. Tomozeiu, and R. Cowley. 2011. *Eco-City Profile: A Global Survey 2011.* London: University of Westminster Eco-Cities Initiative.

Joss, S., R. Cowley, and D. Tomozeiu. 2013. "Towards the 'Ubiquitous Eco-City': An Analysis of the Internationalization of Eco-City Policy and Practice." *Urban Research & Practice* 6(1):54–74.

McMichael, P. 2009. "Contemporary Contradictions of the Global Development Project: Geopolitics, Global Ecology and the 'Development Climate.'" *Third World Quarterly* 30(1):247–262.

Médard, C., and V. Golaz. 2011. "Utopies à Rakai: Accumulation foncière et investissements internationaux." *Transcontinentales* 10–11, 1–15.

Menkhaus, K. 2009. "African Diasporas, Diasporas in Africa and Terrorist Threats." In *The Radicalization of Diasporas and Terrorism*, eds. D. Zimmermann and W. Rosenau, 83–109. Zurich: Center for Security Studies.

Miles, T. 2013. "Malaysia, Not China, Is Asia's Top Investor in Africa." Available at http://in.reuters.com/article/2013/03/25/malaysia-africa-idINDEE92O0BV20130325 (accessed April 17, 2013).

Murray, M. 2011. "City Doubles": Re-urbanism in Africa." Paper presented at the Art of Citizenship in African Cities Conference, May 6, 2011, Columbia University, New York.

Nayyar, D. 2012. "The MDGs after 2015: Some Reflections on the Possibilities." UN System Task Team on the Post-2015 UN Development Agenda. Available at http://www.un.org/millenniumgoals/pdf/deepak_nayyar_Aug.pdf (accessed April 17, 2013).

Obi, C. 2007. *West African Security in the Context of the Global War on Terror: Some Reflections*. Leipzig: Institut für Afrikanistik.

Oxfam. 2013. "The Costs of Inequality: How Wealth and Income Extremes Hurt Us All." Oxfam Media Briefing, January 18. Available at http://www.oxfam.org/sites/www.oxfam.org/files/cost-of-inequality-oxfam-mb180113.pdf (accessed September 25, 2013).

Pieterse, E. 2008. *City Futures. Confronting the Crisis of Urban Development*. New York: Zed.

Pieterse, E. 2011. "Recasting Urban Sustainability in the South." *Development* 54(3):309–316.

Samatar, A., B. Wisner, R. Chitiga, T. Smucker, E. Wangui, and C. Toulmin. 2005. "Agenda for Action." In *Towards a New Map of Africa*, eds. B. Wisner, C. Toulmin, and R. Chitiga, pp. 329–345. Sterling: Earthscan.

Satterthwaite, D. 2001. *The Scale and Nature of Urban Poverty in Low and Middle Income Nations*. London: International Institute for Environment and Development.

Severino, J-M., and O. Ray. 2011. *Africa's Moment*. Cambridge: Polity Press.

Simon, D. 2013. "Climate and Environmental Change and the Potential for Greening African Cities." *Local Economy* 28(2):203–217.

Society for International Development. 2010. "Are Satellite Cities the (Official) Future of GHEA's Urbanisation?" Greater Horn of Eastern Africa Outlook 18. Available at http://www.sidint.net/docs/GHEA18_Satellite_Cities.pdf (accessed May 30, 2013).

Sseesamirembe. 2013. "Sseesamirembe Eco-City." Available at http://www.sseesamirembe.com (accessed May 28, 2013).

Stiglitz, J., A. Sen, and J. Fitoussi. 2009. "Report by the Commission on the Measurement if Economic Performance and Social Progress." Report commissioned by President Sarkozy, Government of France. Available at http://www.stiglitz-sen-fitoussi.fre (accessed May 19, 2013).

Suzuki, H., A. Dastur, S. Moffatt, N. Yabuki, and H. Maruyama. 2009, *Eco² Cities: Ecological Cities as Economic Cities*. Washington, D.C.: The World Bank.

Swilling, M., and E. Annecke. 2012. *Just Transitions: Explorations of Sustainability in an Unfair World*. New York: United Nations University Press.

Taylor, I. 2010. *The International Relations of Sub-Saharan Africa*. New York: Continuum.

United Nations (UN). 2005. *Millennium Ecosystems Assessment: Ecosystems and Human Well-Being. A Synthesis*. New York: Island Press.

United Nations Development Programme (UNDP). 1998. *Human Development Report 1998*. New York: UNDP.

United Nations Environment Programme (UNEP). 2012. *21 Issues for the 21st Century: Results of the UNEP Foresight Process on Emerging Environmental Issues*. Nairobi: UNEP.

United Nations Environment Programme (UNEP). 2013. *International Resource Panel*. Available at http://www.unep.org/resourcepanel (accessed May 18, 2013).

United Nations Human Settlements Programme (UN-HABITAT). 2003. *The Challenge of Slums: Global Report on Human Settlements*. London: Earthscan.

United Nations Human Settlements Programme (UN-HABITAT). 2012. "Housing and Slums Are the Key Area of the UN-Habitat's Mandate." Available at http://www.unhabitat.org/content.asp?cid=11441&catid=5&typeid=6&subMenuId=0 (accessed September 26, 2013).

United States Government Accountability Office (U.S. GAO). 2013. *Sub-Saharan Africa: Trends in U.S. and Chinese Economic Engagement.* GA0 Report 13–199. Available at http://www.gao.gov/assets/660/652041.pdf (accessed April 21, 2013).

Walker, A. 2013. "What is Boko Haram? "United States Institute of Peace. Special Report No. 38. Available at http://www.usip.org/sites/default/files/resources/SR308.pdf (accessed October 3, 2013).

Watson, R., J. Wakhungu, and H. Herren. 2008. *Agriculture at a Crossroads: International Assessment of Agriculture Knowledge, Science and Technology for Development* (IAASTD). Available at: http://www.unep.org/dewa/agassessment/reports/IAASTD/EN/Agriculture%20at%20a%20Crossroads_Synthesis%20Report%20(English).pdf (accessed May 18, 2013).

Watson, V. 2014. "African urban fantasies: dreams or nightmares?" *Environment & Urbanization* 26 (11):1–17.

World Commission on Environment and Development (WCED). 1987. *Our Common Future.* New York: Oxford University Press.

## WEBSITES

Lekki Free Trade Zone. http://www.youtube.com/watch?v=AL8cReBAfoA (accessed December 8, 2013).

La Cité du Fleuve. http://www.youtube.com/watch?v=0yW8ZIDuXls (accessed December 8, 2013).

Sseesamirembe Eco-City. http://www.sseesamirembe.com (accessed December 8, 2013).

UN Sustainable Development Solutions Network (SDSN). www.unsdn.org (accessed December 8, 2013).

# INDEX

**H**

**N**



**O**

**U**



**V**